Unifying Physics of Accelerators, Lasers and Plasma

Unifying Physics of Accelerators, Lasers and Plasma introduces the physics of accelerators, lasers and plasma in tandem with the industrial methodology of inventiveness, a technique that teaches that similar problems and solutions appear again and again in seemingly dissimilar disciplines. This unique approach builds bridges and enhances connections between the three aforementioned areas of physics that are essential for developing the next generation of accelerators. A Breakthrough by Design approach, introduced in the book as an amalgam of TRIZ inventive principles and laws of technical system evolution with the art of back-of-the-envelope estimations, via numerous examples and exercises discussed in the solution manual, will make you destined to invent.

Unifying Physics of Accelerators, Lasers and Plasma outlines a path from idea to practical implementation of scientific and technological innovation. This second edition has been updated throughout, with new content on superconducting technology, energy recovery, polarization, various topics of advanced technology, etc., making it relevant for the Electron-Ion Collider project, as well as for advanced lights sources, including Free Electron Lasers with energy recovery.

The book is suitable for students at the senior undergraduate and graduate levels, as well as for scientists and engineers interested in enhancing their abilities to work successfully on the development of the next generation of facilities, devices and scientific instruments manufactured from the synergy of accelerators, lasers and plasma.

Solutions manual is included into the book

Features

- Introduces the physics of accelerators, lasers, and plasma in tandem with the industrial methodology of inventiveness
- Outlines a path from idea to practical implementation of scientific and technological innovation
- Contains more than 380 illustrations and numerous end-of-chapter exercises

Unifying Physics of Accelerators, Lasers and Plasma

Second Edition

Andrei A. Seryi
Elena I. Seraia

CRC Press is an imprint of the
Taylor & Francis Group, an **informa** business

Designed cover image: Alexandra Seraia

Second edition published 2023
by CRC Press
6000 Broken Sound Parkway NW, Suite 300, Boca Raton, FL 33487-2742

and by CRC Press
4 Park Square, Milton Park, Abingdon, Oxon, OX14 4RN

CRC Press is an imprint of Taylor & Francis Group, LLC

First edition published by CRC Press 2016

ISBN: 978-1-032-35035-6 (hbk)
ISBN: 978-1-032-35250-3 (pbk)
ISBN: 978-1-003-32607-6 (ebk)

DOI: 10.1201/9781003326076

Typeset in Nimbus Roman
by KnowledgeWorks Global Ltd.

Publisher's note: This book has been prepared from camera-ready copy provided by the authors.

Dedication

---

to

Our

Teachers –

Inventors, Scientists

and Relativistic Engineers

Contents

List of Figures

List of Tables

Foreword to the second edition

Lasers and accelerators of particle beams are two hallmarks of highly technological societies. Both technologies were developed as tools to enable fundamental inquiries into the nature of matter and the control and modification of its properties. Driven to ever more demanding levels of performance in research, both have evolved in scale and complexity and both have spread their societal impact in the form of applications that play an intrinsic role in our everyday lives. As particle accelerators were invented decades before lasers, their societal application to commerce and medicine came first with polymerization using electron beams and with radiation therapy of malignancies, both in the 1940s. Societal applications of accelerators are now ubiquitous in medicine, industry and national security, touching as much as $500B of commerce in the 21st century.

Despite their invention in 1960, several decades after the first primitive accelerators, the influences of laser technology in research, commerce and medicine have become just as commonplace as the impact of particle beam sources and accelerators. In many ways, the two technologies and their underlying physics have become symbiotic, with relativistic particle beams driving powerful free-electron lasers and with high-power, ultra-short pulse lasers stabilizing the time structure of high-power, high-energy accelerators. Perhaps nowhere are the physics and the technologies of particle beams and coherent radiation from lasers more deeply intertwined than in the research area of laser-plasma wakefield acceleration. In the commercial sector, particle beams, lasers and plasmas are playing indispensable roles in the development and manufacture of ever more advanced semiconductor products that are now ubiquitous in our modern societies.

The detailed science of these amazing intellectual achievements and the similarity (and even equivalence) of the underlying physics of accelerators, lasers and plasma was the inspiration for courses developed by Prof. Seryi for the US Particle Accelerator School, as well as for the programs of his directorship of the John Adams Institute in England. Thanks to the encouragement of his students and his colleagues in the US and the UK and of his publisher, the first edition of this book was born.

In addition to the canonical topics of accelerator physics, the first edition introduced to its readers to a topic seldom found in standard textbooks of accelerator, laser and plasma science. From its first chapter, that edition offered a detailed examination of paths of innovation, technological invention, scientific breakthroughs, and predictions of phenomena relevant to both new research tools and their new societal applications. The second edition extends that elucidation; for example, the section about scientific estimations recalls the well-known example of Enrico Fermi estimating the yield of the Trinity nuclear test[1] by dropping shreds of paper in the

[1] "Fermi at Trinity," J.I. Katz, Nuc. Tech. vol. 207, No. Sup1, (2021)

observation bunker. Indeed, Fermi was famous for demanding that his students be able to make quick, order-of-magnitude estimates.

In the world of accelerators and lasers used in fundamental research, innovation in design is frequently demanded by requirements on hardware performance that exceed the present state of sound engineering practice. With respect to such innovation, the author introduces the reader to the TRIZ method, developed in the mid-20th century in the Soviet Union. This engineering technique systematizes the approaches to cutting through the web of apparently contradictory system requirements placed on new technologies. The second edition of this text adds a new, thought-provoking chapter (15) comprising extensive examples of forty inventive principles, all relevant to the accelerator-laser-plasma area. Furthermore, each chapter of the second edition offers further examples of invention case studies, including discussions about linking the particulars of the inventions to inventive principles.

The elaboration of the symbiosis among accelerator, laser and plasma science is provided by Chapter 6, which discusses plasma acceleration, both laser and beam driven. Chapter 6 also includes a discussion of synchrotron radiation due to the betatron motion of an electron beam traversing a plasma. Since its first discussion in 1988, this phenomenon[2] now provides a commonly used diagnostic of beams accelerated in plasma. Betatron radiation also has been used to produce very high-quality, phase-contrast images of biological specimens — a likely first societal application of plasma-driven accelerators. A second symbiosis of accelerators, lasers and plasmas is provided by the production of 13.5 nm extreme ultra-violet (EUV) radiation for semi-conductor lithography either via laser-plasma interactions or by the conversion of electron beam energy into EUV by the free-electron laser process.

Among important innovations in accelerator technology, this edition discusses beam cooling (Ch. 10), energy recovery in superconducting RF cavities (Ch. 11) and laser-manipulation of particle beams (Ch.12). Also new in the second edition is an extended chapter (13) on several advanced technologies such as new RF power sources, coherent combination of laser beams, resonant plasma excitation, nonlinear injection into storage rings and superconducting RF cavities and magnets. Additional features of the new edition are a solution guide to the exercises and — in the eBook (Kindle) edition — all illustrations are in color.

The second edition of *Unifying Physics of Accelerators, Lasers and Plasma* is an ideal introduction to dominant technologies of 21st-century scientific inquiry and societal applications. It is suitable as both a graduate textbook and for self-study by more experienced physicists.

– William A. Barletta — Cambridge, November 25, 2022

[2]"Radiation from Fine Self-focused Beams at High Energy," W.A. Barletta and A.M. Sessler, published in High Gain, High Power Free Electron Lasers, (Bonifacio, De Salvo-Souza, and Pellegrini, ed.), Elsevier Scientific Publishers (North Holland), 1989

Foreword to first edition

Accelerators were invented in the early 1930s and first developed in a significant way by E.O. Lawrence at the University of California, Berkeley. Initial ideas arose in response to the needs of particle physics expressed by Ernest Rutherford as early as 1924, but right from the outset, Lawrence used his early cyclotron accelerators for both pure research in particle (then called nuclear) physics and more practical applications. Much of the development was funded by their use in the field of medicine for isotope production and therapy. After the World War II years 1939 to 1945 came the invention of the synchrotron — a clever extension of the cyclotron principle to undercut the rising cost of cyclotrons and extend their energy range to allow the production of more massive fundamental particles. Readers who follow the description of the TRIZ inventive process in this book may wonder if this invention was an accidental example of what TRIZ calls the "Russian Doll" technique.

Soon it was realized that electron synchrotrons were a prolific and controllable source of synchrotron radiation, initially in the ultraviolet spectrum and later down to the wavelengths of X-rays. A large number of these SR sources were built and their beams used for scattering experiments, to fathom the structure of new materials for engineering and elucidate the molecular structure of protein and the other complex molecules that have come to dominate our understanding of today's life sciences. To extend the scope of these studies in order to allow single molecules to be reconstructed from their scattering patterns, a completely new concept — the free electron laser — was invented. This has already been used in several countries as a basis for the construction of research facilities that rival the large colliders of particle physics in scale and budget. Finding a new approach to the production of X-rays in this way has been another leap in human imagination, opening the door to new and more powerful applications in a field where those applications bear an almost immediate return on investment. It is the purpose of this book to explore how such leaps in imagination may be stimulated.

To be significant, such inventive processes need to identify a symbiotic relationship between two apparently quite different fields — for the free electron laser these were accelerators and lasers — and then find a common factor that can be used to improve both fields. This technique has been recently developed and refined as a replacement for simple brainstorming. It is called TRIZ and has already been identified by industry as a means toward more rapid technological progress. We discuss here how it may be used to cut the Gordian Knot[3] of rising costs and complexity, which threaten to impede the development of more powerful instruments for both pure and applied research.

[3]The Gordian Knot is a legend of Phrygian Gordium associated with Alexander the Great. It is often used as a metaphor for disentangling an intractable problem.

One does not have to look far to find a candidate for a symbiotic technique to pair with today's accelerators. A laser beam penetrating a plasma generates accelerating fields many orders of magnitude greater than today's accelerators. This book for the first time puts these two techniques side by side so that we may identify commonalities of solution and stimulate today's students of accelerator and laser/plasma physics to work together and use TRIZ to solve the problem.

The book's author heads the John Adams Institute, whose participating universities include postgraduate students in both the laser and accelerator fields. We are well placed to kindle the fire of invention amongst them, and we urge those responsible for young and inventive minds elsewhere to join this mission. Our students are already applying our ideas to the Future Circular Collider project of CERN.

– Edmund Wilson — Geneva, Switzerland
March 7, 2015

Preface to the second edition

This second edition is the result of community demand. The original book was used in various courses and graduate schools, and most recently, it was picked up by CERN's "eBooks for all!" program, which selects popular books in accelerator science and works with the publishers to convert them to Open Access. While we were pleased to see this apparent success of and demand for the book, we also saw the merits of expanding on it in a second edition.

There was both a desire and a need to supplement the materials to cover new and vital areas of accelerator science which have emerged since 2015 or which were omitted in the first edition. These are, in particular, superconducting CW linacs — which are the key components of facilities such as CEBAF or the superconducting FELs such as LCLS-II — and the Electron-Ion Collider — now an approved project — which critically relies on the technologies of energy recovery, polarized beams and spin dynamics, superconducting RF and cryogenics, and sophisticated control of beam resonances, as well as advanced methods of beam cooling. These additional topics are introduced in the second edition using the same approach of building the bridges of understanding between different areas of accelerator, laser and plasma technology via inventive principles.

Furthermore, this edition includes several new chapters, one of which contains illustrations of 40 inventive principles based on examples from accelerator physics, lasers, plasma, and even biology and other fields. The ebook (EPUB) version is published in a new optimized format compatible with flowable layout and scalable fonts, so that it can be easily enjoyed on a computer, tablet or even a phone. The majority of the pictures have been changed to color versions, making the eBook of the second edition much more attractive. Based on feedback from readers of the first edition, the second edition also includes the solutions guide directly into the body of the book.

The *Breakthrough by Design* or *Innovation by Design* approach — an amalgam of TRIZ inventive principles and laws of technical system evolution with the art of back-of-the-envelope estimations, applied to several neighboring areas of science and technology — is also introduced in this edition. We believe that by developing and applying this methodology, you will be *destined to invent*.

Each chapter includes new sections of "invention case studies" that allow us to expand our discussion of the inventions made in the areas of accelerators, lasers and plasma. In these sections, as well as throughout the book, we look at these inventions from a TRIZ viewpoint. Examples that we consider include the creation of the Mak telescope for academic astronomical studies, the generation of EUV light for the semiconductor industry, the invention of the ways to increase LEP collider energy, and so on.

Additionally, we endeavor to make predictions for the future. In this edition, we analyze the lessons of *The Year 2000*, in which predictions for technological innovations were made in 1968 for the year 2000. In the spirit of *The Year 2000*, based on the knowledge of the state-of-the-art technologies discussed in this book and armed with TRIZ laws of evolution of technical systems interpreted for modern scientific areas, we made predictions for the field of accelerator, laser and plasma science and technology for the year 2050.

We believe that this book, and the *Breakthrough by Design* approach that it introduces, can help readers realize their scientific, technological and inventive dreams and make them a reality.

Finally, we would like to express our gratitude to all our colleagues, students, readers, CRC Press editors, and to our family and friends who made this second edition possible.

– Andrei Seryi and Elena Seraia
Newport News, Virginia
November, 2022

Preface to first edition

The aim of this book is to build bridges and connections between three areas of physics that are essential for developing the next generation of accelerators: accelerators, lasers and plasma. These three fields of physics will be introduced in tandem with the industrial methodology of inventiveness, a method which teaches that similar problems and solutions appear again and again in seemingly dissimilar disciplines. This methodology of inventiveness will, ultimately, further enhance connections between the aforementioned fields, and will grant the reader a novel perspective.

The text is suitable for students of various levels between senior undergraduate and graduate physics who are interested in enhancing their ability to work successfully on the development of the next generation of facilities, devices and scientific instruments manufactured from the synergy of accelerators, laser and plasma. I would also recommend this book to anyone interested in scientific innovations.

The idea for the book *Unifying Physics of Accelerators, Lasers and Plasma* came not by accident — it naturally resulted from a search for the best method to suitably train the undergraduate and graduate students at John Adams Institute and its affiliate universities. The aim is to teach several physics disciplines in a coherent way, simultaneously ensuring that this training would develop and stimulate innovativeness.

Materials for this book developed gradually, starting from a single lecture presented at JINR Dubna in March 2014, to several lectures given to Oxford undergraduates in May 2014, and eventually to a full week-long US Particle Accelerator School course given in June 2014. The text, to a large extent, follows the materials developed for this USPAS course. Interaction with the students during this course proved to be a significant inspiration for converting these materials into book form, as well as the lectures, presentations of mini-projects by student teams and tutorial sessions that were focused on analyzing key inventions that shaped the discussed scientific areas.

The style of this book is perhaps different from a typical textbook on accelerator physics or books on lasers or plasma. Here, we tend to use qualitative discussion and prefer to avoid heavy math, using back-of-the-envelope type derivations and estimations whenever possible (even if they provide approximate formulas). We believe that this is the better way to convey physics principles to the reader and is certainly preferable when discussing such a broad area of physics.

I would like to express my gratitude to my many colleagues who directly or indirectly helped with this endeavor. Particular thanks go to Professor Riccardo Bartolini, Professor Emmanuel Tsesmelis and Professor Ted Wilson for joint work on the short-option course for Oxford undergraduates, which was the first step toward the development of the USPAS course that gradually led to development of this book. I am grateful to the students of my USPAS-2014 course for inspiration and enthusiastic discussions and also to Professor Bill Barletta who supported the idea of the

USPAS course, discussed the very first time with Bill in early April 2013.

I am grateful to the Rector of Novosibirsk State University, Mikhail Fedoruk, for creating an inspiring opportunity to present a lecture on Science and Inventiveness to a cohort of 1,000 first-year students on September 1, 2014; the preparations for this lecture significantly stimulated the thought process for writing the first chapter of this book.

I am grateful to Professor Zulfikar Najmudin for help with the materials for ion plasma acceleration and thankful to Professor Peter Norreys and Professor Bob Bingham for discussions on the topics connecting chirped pulse amplification and radar inventions.

I am in immense debt to Professor Ted Wilson, who made tremendous efforts to be the first reader of the manuscript and who gave me a lot of valuable comments.

I am grateful to my John Adams Institute colleagues — Professors Riccardo Bartolini, Stewart Boogert, Simon Hooker, Ivan Konoplev, Zulfikar Najmudin and Mike Partridge for reading the manuscript or its individual chapters and giving their valuable comments. I would like to express gratitude to all of my colleagues and friends who helped with the preparation of this book and whom I failed to thank here directly.

I am also delighted to thank Francesca McGowan, CRC Press editor, who on April 30, 2013 appeared out of the blue in my office asking for directions to a meeting — this unexpected conversation helped to crystallize my thoughts toward this book. I appreciate all the aid Francesca and her CRC Press colleagues provided during the writing process.

I would like to express my deep thanks to my family, first of all to my daughter Sasha, who designed the book cover, made several illustrations and helped with grammar.

And most importantly, my eternal gratitude to my wife Elena, who put a massive amount of effort into converting lecture materials to LaTeX, created more than 200 illustrations for this book and gave me support and motivation, without which this book would never have been finished.

– Andrei Seryi — Oxford, 2015

Authors

Andrei Seryi is the current Associate Director for Accelerators Operations and R&D at Jefferson Lab and, until 2018, was the Director of the John Adams Institute and a professor at Oxford University. He graduated from Novosibirsk State University in 1986 and received his Ph.D. from the Budker Institute of Nuclear Physics in 1994. He worked at the Stanford Linear Accelerator Center until 2010, where he led the design and first stages of implementation of the Facility for Advanced Accelerator Experimental Tests (FACET) Project and coordinated the beam delivery efforts for the linear collider. He served as a deputy spokesperson of the ATF International Collaboration for the ATF2 Project — the prototype of the final focus system with local chromaticity correction. He is a Fellow of the American Physical Society. His professional interests include applying accelerator science to discovery science, industry, healthcare and energy; organizing scientific research; project management; crisis management in scientific and technological areas; inventions and innovations; and developing novel training approaches.

Elena Seraia worked until 2018 at the Medical Department of the University of Oxford, in Radiation Oncology and Biology, which strove to find comprehensive cancer treatments. After graduating from Novosibirsk State University with a degree in cell biology and genetics, she worked at the Institute of Cytology and Genetics in Novosibirsk. In the late nineties, Elena worked at the Fermi National Accelerator Laboratory (FNAL), participating in the production of new silicon detector systems for the Tevatron collider. From 2000 to 2010, she worked in the Stanford Functional Genomics Facility of the Medical Department at Stanford University, the facility which was the first to develop the principle of contact printing of oligonucleotide DNA microarrays. Elena was a key contributor to the first edition of *Unifying Physics*, creating more than 260 illustrations and writing part of the chapter devoted to the effects of radiation on DNA and cancer therapy. She was also the translator of the 2016 Russian edition of Unifying Physics and ultimately made this second edition possible.

1 Basics of Accelerators and of the Art of Inventiveness

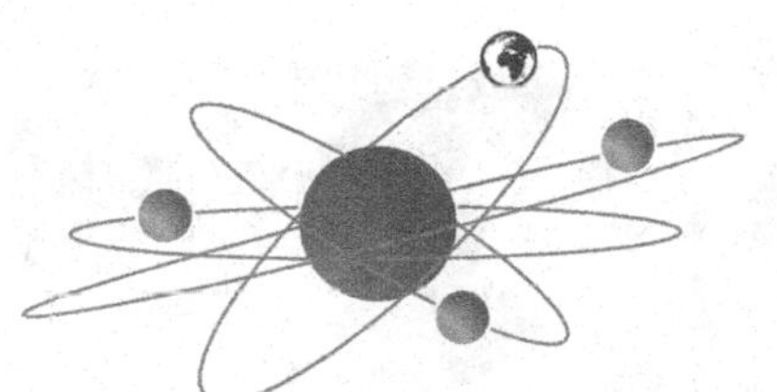

Can you imagine that electrons
Are planets circling their Suns?
Space exploration, wars, elections
And hundreds of computer tongues

Author's translation of 1920 poem of Valery Bryusov "The World of Electron"

In this chapter we will discuss the basic terms related to accelerators. We will also describe the framework and methods of the theory of inventive problem solving.

1.1 ACCELERATORS AND SOCIETY

Accelerators are essential for science and society — they are in use in high-energy physics, nuclear physics, healthcare and life science. They are important to industry, the development of new materials, energy and security, and can be applied to many other fields.

It is enough to state just three reasons why society needs accelerators: tens of millions of patients receive accelerator-based diagnoses and treatments each year in hospitals and clinics around the world; all products that are processed, treated or inspected by particle beams have a collective annual value of more than $500B; and a significant fraction of the Nobel prizes in physics are directly connected to the use of accelerators.

About 30% of Nobel prizes in Physics are due to use of accelerators.

If you are a young person thinking about your future, you might be interested to know that accelerator science can lead to attractive career prospects, because whether you are more inclined toward a theoretical approach, experiments or computer modeling, there will be an array of tasks and opportunities ready for you. Moreover, the knowledge gained and developed in accelerator science has connections and applications to surprisingly remote disciplines, from stock market predictions to planetary motion dynamics.

The recent discovery of the Higgs boson by the Large Hadron Collider — the grandest machine ever built — has once again demonstrated accelerator science's and

DOI: 10.1201/9781003326076-1

associated technologies' potential to reveal some of the most fundamental constructs of nature.

Accelerator science demonstrates a rich history of inventions, often inspired by nature itself. We are very often motivated by nature and try to compete with it, not always knowing who invented certain things first — nature or humans. For example, it is perhaps a common belief that gears were invented by humans. In fact, insects have been using gears[1] for millions of years! The mechanism allows the insect to jump in a straight line, its left and right leg synchronized by a gear-like connection — see Fig. 1.1 — which works better than a synchronization via nervous signals.

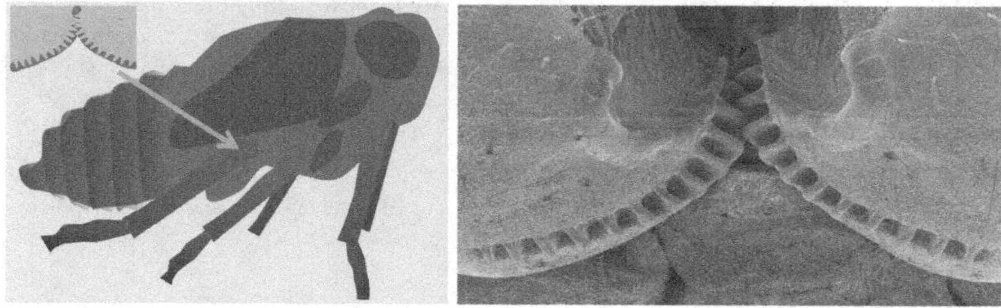

Figure 1.1 Gear-like structure in a jumping insect *nymphal planthopper* as an illustration of nature's inventiveness. Photo on the right from M. Burrows and G.P. Sutton university of Cambridge, U.K. Reproduced with permission.

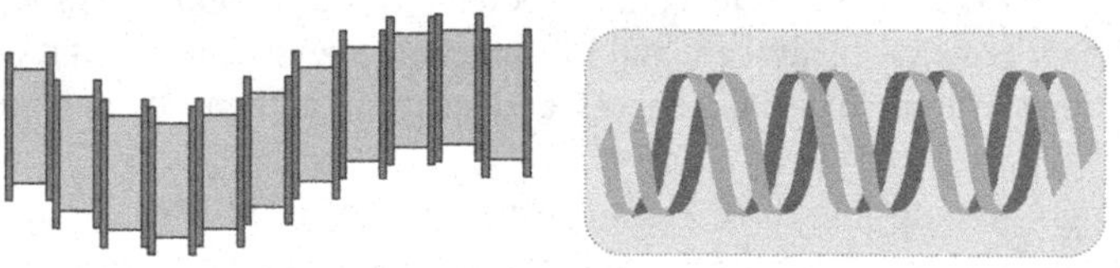

Figure 1.2 Helical solenoid channel (left) inspired by DNA dual helix shape (right).

The shapes created by nature often inspire our creativity in accelerator physics as well — e.g., the spiral-shaped Muon collider cooling channel[2] (consisting of an integrated helical solenoid and accelerating cavities interleaved with absorbers; see Fig. 1.2) was possibly inspired by the double-helical DNA. We hope that examples like this, together with a rigorous inventiveness methodology described in this book will arm the reader with a new systematic approach that will enable efficient inventiveness.

---

[1]M. Burrows and G. Sutton, *Interacting Gears Synchronize Propulsive Leg Movements in a Jumping Insect*, Science, **341**, Sep 13, 2013.

[2]Ya. Derbenev and R. Johnson, *Six-dimensional muon beam cooling using a homogeneous absorber: Concepts, beam dynamics, cooling decrements, and equilibrium emittances in a helical dipole channel*, Phys. Rev. ST Accel. Beams **8**, 041002, Apr 29, 2005.

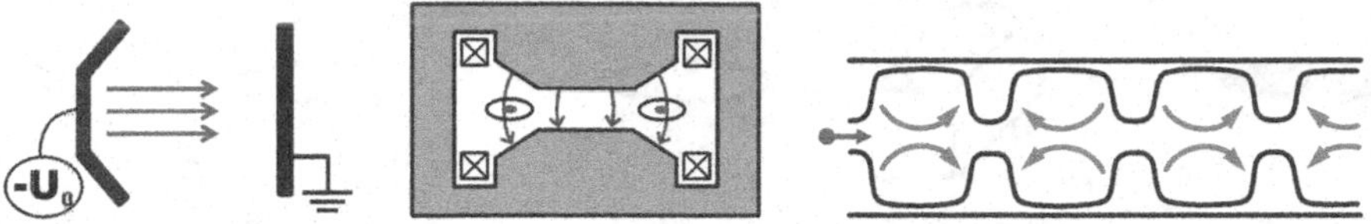

Figure 1.3 Basic principles of acceleration — electrostatic, betatron, in an EM wave in an accelerating structure.

1.2 ACCELERATION OF WHAT AND HOW

Accelerators can be either giant or tiny, but all have similarities and the same subsystems. A giant accelerator, such as SLAC's linear accelerator and a cathode ray tube TV (which is also an accelerator, albeit smaller), both have all of the main components of a modern linear accelerator or collider. This includes a source of charged particles, an acceleration area, a drift region with focusing and steering, and a target or detector (represented in the case of a TV by the phosphorous screen).

When discussing acceleration, we assume that we accelerate a *bunch* of particles — a compact cloud of, for example, electrons or positrons, protons or antiprotons, ions or any other charged particles.

The simplest accelerating mechanism is electrostatic direct acceleration — caused by DC voltage and a corresponding electric field. Another method is betatron acceleration, which is caused by a magnetic field changing in time which, according to Maxwell's equation $\oint \mathbf{E} \cdot d\ell = -d/dt \int \mathbf{B} \cdot d\mathbf{S}$, creates a curl of electric field **E** suitable for acceleration. The third method is acceleration in an electromagnetic wave; however, one should note that an EM wave in free space cannot continuously accelerate particles along the direction of its propagation, as its **E** and **B** components are transverse to the direction of the EM wave's propagation. Therefore, in order to use an EM wave for acceleration, one needs to change the structure of its fields, which can be achieved by propagating the EM wave in an appropriately shaped accelerating structure. These three methods are illustrated in Fig. 1.3.

Assuming we know how to accelerate the beam, we can ask the question of why would we want to do that? That is, how are we planning to use the accelerated beam? One can foresee at least four different uses of the beam, as illustrated in Fig. 1.4. We can direct the accelerated beam onto a target, either for scientific experiments (e.g., in nuclear physics) or for modifying or treating the target itself. We could direct two accelerated beams onto each other, as is typically done for high-energy physics experiments. Acceleration could also be used to characterize the beam or perhaps to separate it into different species or isotopes. Finally, we can use the accelerated beam to generate useful radiation.

1.2.1 USES, ACTIONS AND THE EVOLUTION OF ACCELERATORS

Having discussed in the previous section why we need to accelerate the beam and how an accelerated beam can be used, we will now take this moment to define the basic actions that can be applied to the beam: acceleration, focusing and cooling, and

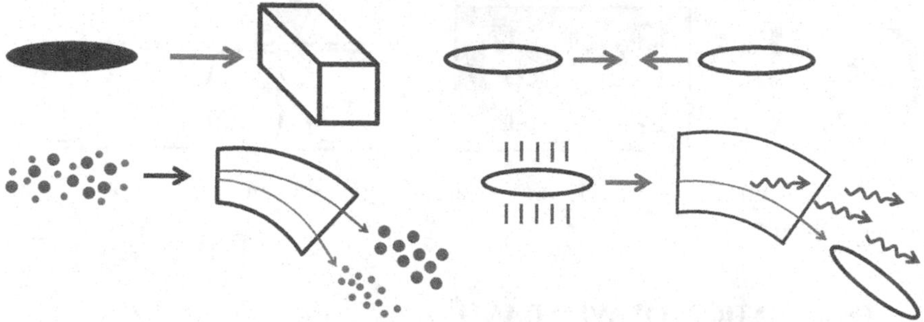

Figure 1.4 Uses of accelerated beams — sending to target, colliding with another beam, characterization of the beam or separation into species, generation of useful radiation.

the generation of radiation, as well as the corresponding parameters and characteristics of these actions.

In cases of acceleration we aim to find out the final energy of the particles and usually prefer to achieve as high a rate as possible of the energy change (usually called the accelerating gradient). If the electrostatic accelerating voltage is U_0, the final energy is $E = \gamma mc^2$ equal to $E = eU_0 + mc^2$, where γ is relativistic factor, $\gamma = 1 + eU_0/(mc^2)$, m is the rest mass of the particle, e is its charge and c is speed of light.

Whether we plan to send the beam to a target or collide it with another beam, we strive to achieve a certain flux of particles; we therefore may need to focus the beam to a small size on the target or at the interaction point with the oncoming beam. As it is with light, a sequence of focusing and defocusing lenses (in this case electromagnetic lenses) focus the beams, as is illustrated in Fig. 1.5.

Using lenses to focus the beam does not affect its so-called phase-space volume, which is usually called *emittance* ε and defined as an area of the ellipse occupied by the beam in coordinate-angle phase space (for example, x and x' as illustrated in Fig. 1.5). If the two transverse planes are independent (not coupled), then both ε_x and ε_y emittances are conserved. Emittance ε is usually defined in units of $m \cdot rad$ or $mm \cdot mrad$. If the beam is accelerated, the so-called normalized emittance $\varepsilon_n = \gamma\varepsilon$ is conserved. If the beam emittance is large (which can especially be true for positrons or antiprotons, which are created in "hot" collisions of the initial beam with the target) it can be particularly difficult to focus such a beam to a small size. Therefore, the next important action applied to the beam is *cooling*, which is intended to reduce the emittance of the beam.

In the context of the generation of radiation, we concern ourselves with reducing both the sizes of the beam in the emitting region, and its angular spread (as emitted radiation usually follows the direction of the particles). Therefore, low emittance and beam cooling may again be a necessity. The corresponding characteristic, which describes the radiation generation, is called *brightness* and is defined as the number of photons emitted per unit of time from a certain area into a certain solid angle (see Fig. 1.5). One also typically adds "into a certain bandwidth" (such as, for

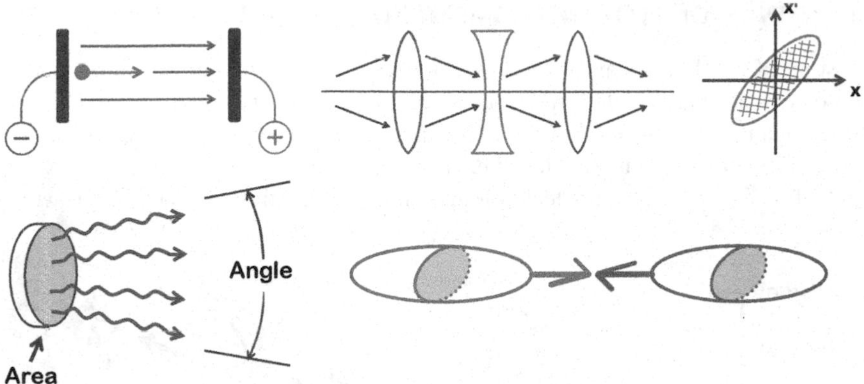

Figure 1.5 Actions on accelerated beams — acceleration, focusing, generation of radiation, colliding.

example, 1% of the central wavelength) as one is usually interested only in a certain spectral range of the emitted radiation. Brightness is therefore defined in the units of $number of photons/(s \cdot m^2 rad^2(\% \cdot bandwidth))$.

Finally, when two beams are colliding, we aim at focusing the beams into the smallest possible size at the interaction point, in order to maximize the probability of interaction. In this case, we are therefore interested in the characteristic called *luminosity* $\mathscr{L}$, which is defined in such a way that the product of luminosity and the cross section of interaction σ (which has the units of m^2) gives the number of events per unit of time. The luminosity is then defined in units of $1/(s \cdot m^2)$.

These basic actions or manipulations that can be applied to the beam help to define the evolution of accelerators.

In 1954 Enrico Fermi presented, in his lecture, a vision of an accelerator that would encircle the Earth, and would attain highest possible energies. We will use this example to discuss natural evolution of accelerators and other scientific instruments further in this book.

Scientific and technological advances in the area of accelerators have focused on an increase of energy of accelerated beams, mastering acceleration techniques (including the acceleration of different species) and on an increase of accelerating gradients. The need for smaller beam sizes facilitated the development of beam focusing and cooling methods. Demands for higher brightness stimulated mastery of the methods of radiation generation. Desires to increase luminosity of colliding beams led to improvements of a whole class of techniques, from emittance preservation to stabilization of nanometer beams. Last, one of the biggest motivations for accelerator progression was their potential to be applied in various scientific and technological areas.

1.2.2 LIVINGSTON PLOT AND COMPETITION OF TECHNOLOGIES

The history of accelerators and various accelerator technologies can be summarized in a so-called "Livingston plot" where the equivalent energy of an accelerated beam is plotted against time — see Fig. 1.6. One can clearly see that, over the course of many decades, the maximum achieved energy grew exponentially. It was the development of different accelerator technologies that enabled this exponential growth.

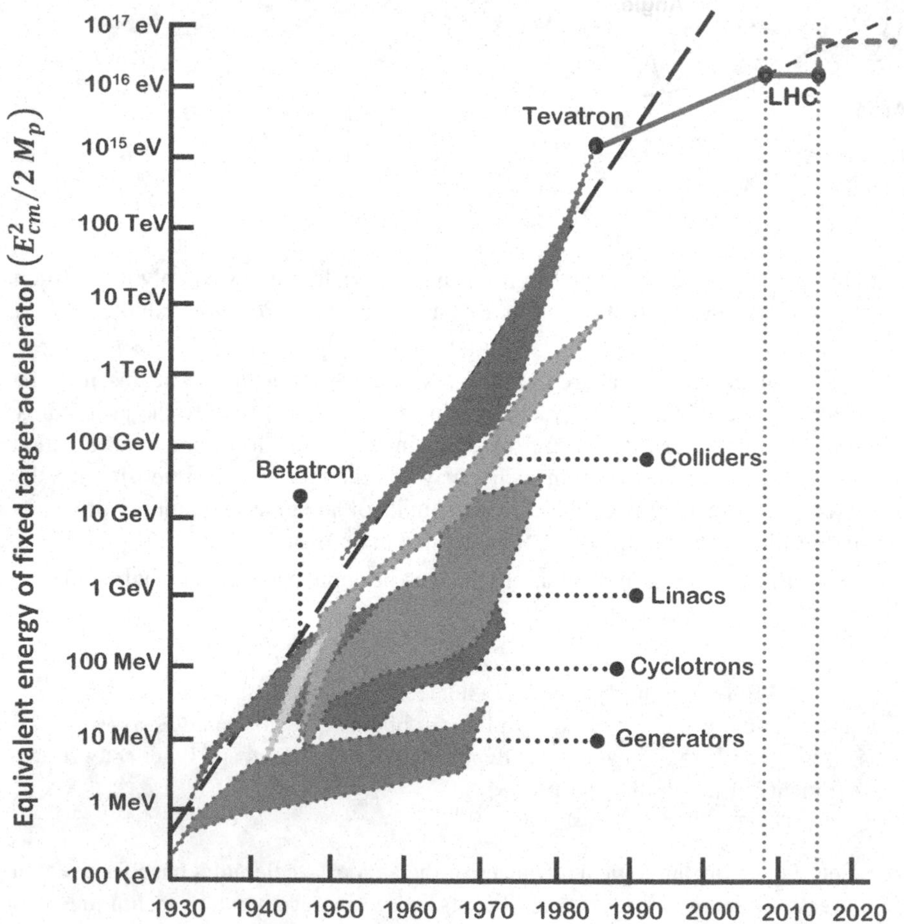

Figure 1.6 Livingston plot of evolution of accelerators.

The Livingston plot also depicts that new accelerating technologies replaced each other once the previous technology had reached its full potential. This evolution and *saturation* of particular acceleration technologies, and birth of the new technologies, is a common phenomenon in any technological or scientific field — illustrated in Fig. 1.7.

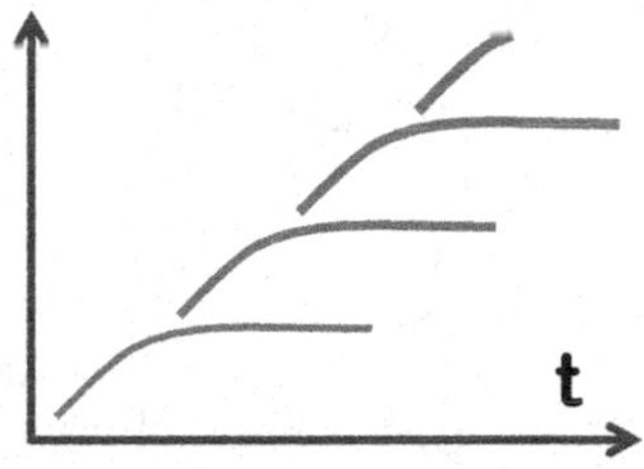

Figure 1.7 Evolution of technologics — saturation and replacement by newer technologies.

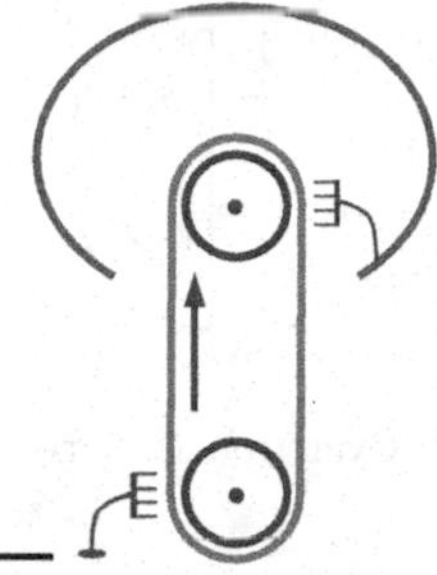

Figure 1.8 Van der Graaf accelerator.

The three most recent decades shown in Fig. 1.6, represented by Tevatron and LHC colliders, exhibit a much slower exponential energy growth over time. This may be an indication that the existing technologies of acceleration came to their maximum potential, and that further progress would demand creation of new accelerating methods — one which was more compact and more economical. There are several emerging acceleration techniques, such as laser-driven and beam-driven plasma acceleration, which can bring the Livingston plot back to the exponential path — which we will further discuss later on.

1.3 ACCELERATORS AND INVENTIONS

Accelerator science exhibits a rich history of inventions. Let us briefly skim through some of the most influential inventions. Those interested in further reading are recommended to explore the fascinating story of accelerators described in *Engines of Discovery.*[3]

Between 1900 and 1925, radioactive source experiments initiated by Rutherford created a demand for higher-energy beams.

From 1928 to 1932, Cockcroft and Walton developed electrostatic acceleration using voltage multiplication created out of diodes and oscillating voltage — ultimately reaching around 700 kV of voltage. At about the same time, Van der Graaf created a method of voltage charging wherein a rubber mechanical belt would carry charges deposited onto the belt by sharp needles via the ionization of gas molecules — which achieved voltages at around 1.2 MV (see Fig. 1.8).

Resonant acceleration development commenced in 1928 with Ising establishing the concept and Wideroe building the first linac.

In 1929, Livingston built his small prototype of the cyclotron as his PhD thesis, inspired by Lawrence, who studied Wideroe's thesis (see Fig. 1.9). This concept was then realized later in large scale by Lawrence.

In 1942, the principle of magnetic induction helped Kerst build the first betatron.

In 1944, the synchrotron (see Fig. 1.10) was invented by Oliphant, while MacMillan and Veksel independently invented the principle of RF phase stability, which made longitudinal focusing of beams possible.

---

[3]A. Sessler and E. Wilson, *Engines of Discovery, A Century of Particle Accelerators*, World Scientific, 2014.

Figure 1.9 Cyclotron accelerator.

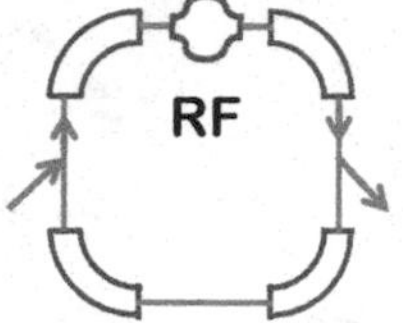

Figure 1.10 Synchrotron accelerator.

In 1946, Alvarez built the proton linac by using an RF structure with drift tubes (in progressive wave in 2π mode).

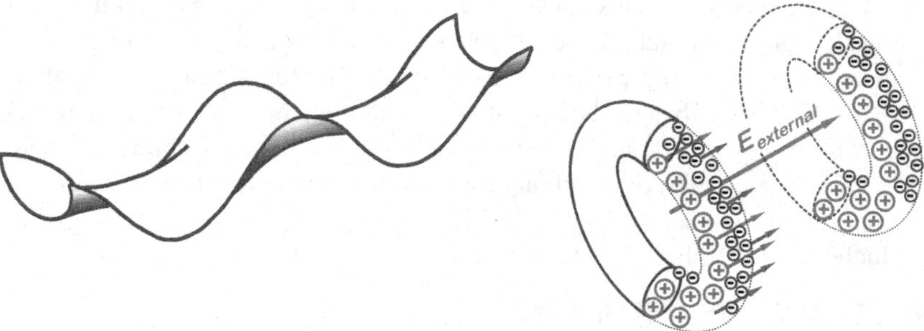

Figure 1.11 Strong focusing concept.

Figure 1.12 Collective acceleration.

In 1950, Christofilos developed strong focusing, which he later patented as the alternate gradient concept (transverse strong focusing). Fig. 1.11 illustrates the strong focusing as a gutter with its edges bent up and down in a sine-like manner. The story of the invention of strong focusing has an interesting aspect — it is usually attributed to Courant and Snyder, since the Christophilos patent was pointed out only after the Cosmotron team had announced the idea. This example is certainly relevant for anyone considering whether to publish or to patent their ideas.

In 1951, a tandem Van der Graaf accelerator was developed by Alvarez, thusly upgrading the electrostatic acceleration concept. A charge-exchange stripping foil was placed at the high voltage point and the source of negative ions was placed at ground potential (which was also much more practical for its servicing). This resulted in a voltage twice as large.

In 1953, Courant, Snyder, and Livingston built the weak focusing 3.3 GeV Cosmotron in Brookhaven, and in 1957, Veksler built a 10 GeV (which was the world record at that time) synchrophasotron in Dubna, whose magnet weighed 36,000 tons and was registered in the Guinness Book of World Records. This record in energy was overtaken by the CERN Proton Synchrotron in 1959, constructed under the leadership of Sir John Adams. The CERN PS was the first strong focusing accelerator, closely followed by the AGS at Brookhaven.

In 1956, Veksler suggested the principle of collective acceleration (see Fig. 1.12), which became the predecessor to a variety of collective methods based on plasma acceleration.

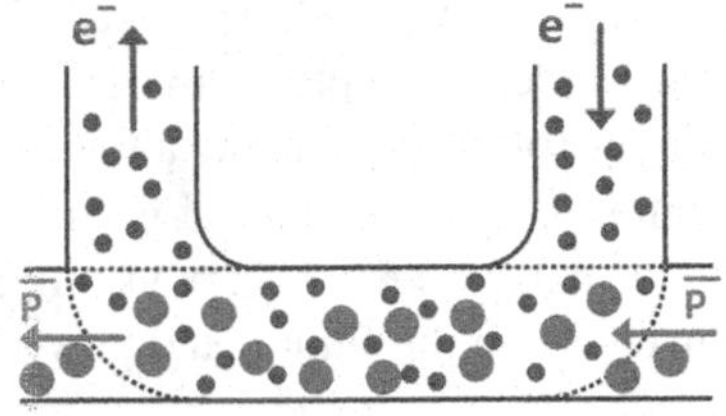

Figure 1.13 Electron cooling concept.

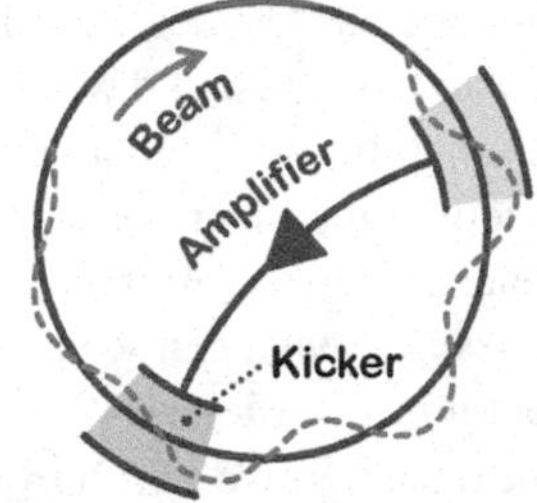

Figure 1.14 Stochastic cooling concept.

In 1956, Kerst discussed the concept of colliding beams and in 1961, the e+e- collider — the concept for a particle-antiparticle collider — was invented by Touschek, and the first colliders then built in Italy, US and Novosibirsk.

The collider concept created the need to develop the methods used to decrease beam emittances, in particular for antiparticles, and in 1967, Budker proposed electron cooling as a way to increase the proton or antiproton beam density (see Fig. 1.13). Shortly thereafter, Van der Meer proposed stochastic cooling (see Fig. 1.14) in 1968 as a way to compress the beam's phase space.

In 1970, Radio-Frequency Quadrupole (RFQ) was invented when Kapchinski and Telyakov built the first RFQ that allowed a simultaneous focusing and acceleration of the beam. This technology allowed a much more efficient and compact acceleration of ions and protons from a very low energy to an energy as high as a few MeV.

In 1971, Madey developed the principle of a Free Electron Laser (FEL), a method that allowed the production of coherent, hard X-rays of unprecedented brightness.

In 1979, Tajima and Dawson proposed acceleration of the beam in plasma waves excited by a laser (see Fig. 1.15). It was only much later that suitable lasers became available and this method started to become competitive.

From around 1980, superconducting magnets were developed in various labs around the world, which allowed for a drastic increase of beam energy in circular acceleration. Development of this technology still continues today (8 Tesla magnets are used in LHC, and 11 T magnets developed for its upgrade).

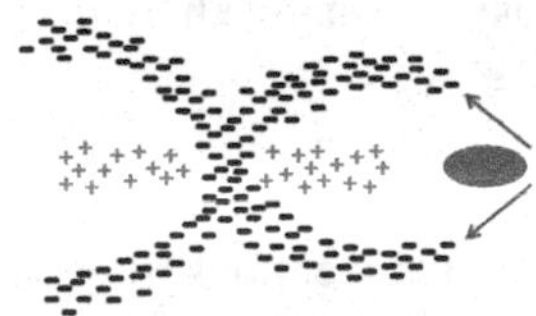

Figure 1.15 Plasma acceleration concept.

> Plasma acceleration is one of the most dynamically developing area of science and technology. Ultra-high accelerating gradients are creating a breakthrough in compact accelerators and FELs.

Likewise, the superconducting RF (radio-frequency) technology (created around 1980) is currently being used and is still being developed today in many labs and industrial companies, allowing an increase in the RF gradient and efficiency, and thus the eventual energy and power of the beams.

The most recent decades have been rich with inventions as well. The years between 1990 and the present day have seen a photon collider concept, an (experimentally verified) crab waist collision and integrable optics for rings, to name a few.

The variety of the inventions (many of which will be discussed later on in further detail) demonstrates the *past successes* of our field, but we are concerned with a different question: what can we look forward to inventing in the future, and how are we to invent it more efficiently?

In the next section we will introduce the methodical approaches to inventiveness used in industry, and will then explain how these methods can be applied to, in this case, accelerator science.

1.4 HOW TO INVENT

Methods described in this section are rarely known in science, but they are widely used in industry.

In his March 7, 2013 contribution to Forbes,[4] "What Makes Samsung Such An Innovative Company?" the author, Haydn Shaughnessy, wrote: "But it was ... that became the bedrock of innovation at Samsung. In 2003 ... led to 50 new patents for Samsung and in 2004 one project alone, a DVD pick-up innovation, saved Samsung over $100 million. ... is now an obligatory skill set if you want to advance within Samsung." The reader may speculate on what the dots signify — but read on.

What is this magic method, and what is the word intentionally omitted in the above quote? The answer will be given in just a few pages.

Formal inventive approaches rarely known in science but widely used in industry

1.4.1 HOW TO INVENT – EVOLUTION OF THE METHODS

Let's start by recalling the techniques of problem solving, starting from a well-known brute force or exhaustive search method. In this case, any potential solutions and ideas are considered and evaluated one by one. It is easy to imagine how inefficient this method may be.

An improved variation of the exhaustive search method is called *brainstorming*. Its author, Alex Osborn, introduced this method in the 1950s. Brainstorming is a psychological approach that helps to solve problems and create inventions by separating the process of idea generation from the process of critical analysis.

The method of brainstorming has its limitations — the absence of critical feedback, which is the main feature of the method, is simultaneously its handicap, as feedback is required to develop and improve an idea.

---

[4]Haydn Shaughnessy, Forbes, 2013, http://www.forbes.com/sites/haydnshaughnessy/2013/03/07/why-is-samsung-such-an-innovative-company/

The next method that emerged on the scene, the so-called *synectics*, was introduced by George Prince and William Gordon in an attempt to improve brainstorming. One of the main feature of synectics is that it is assumed to be practiced by a permanent group of experts, whose members, with time, become less sensitive to critics from among their peers and thus become more efficient in inventing. This method emphasized the importance of seeing familiarity in an unknown and vice versa, which helps to solve new and unfamiliar problems using known methods. The authors of synectics stressed the importance of approaching a problem with a fresh gaze, and also stressed the use of analogies to generate this attitude.

The analogies employed by synectics can be direct (any analogy, e.g., from nature), emphatic (attempting to look at the problem by identifying yourself with the object), symbolic (finding a short symbolic description of the problem and the object), or even metaphorical (describing the problem in terms of fairy-tales and legends). An example of such metaphorical description is given in Fig. 1.16 in application to an antisolenoid task (see Exercises for this chapter), which, however, is obviously extremely unhelpful and counterintuitive.

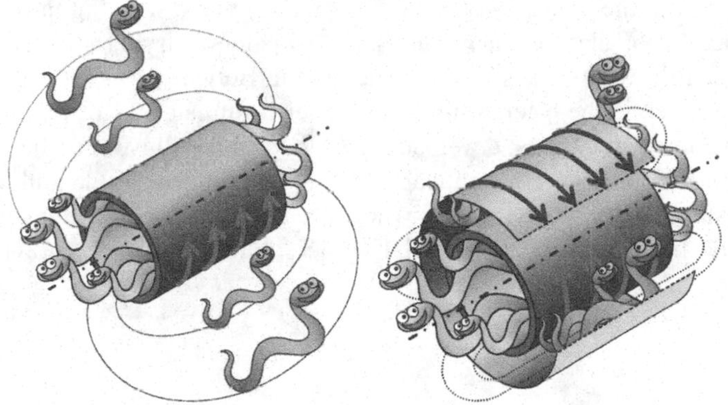

Figure 1.16 For illustration of description of problem with language of fairy tails in method of synectics. On the left — a solenoid with its field lines, and on the right — a nested solenoid where the flux is returned between two solenoids, which makes the dual solenoid force-neutral; i.e. it will not interact with other nearby solenoids.

Synectics is still a variation of the exhaustive search method and is the limit of what can be achieved, maintaining the brute force method of an exhaustive search.

There must, then, be a better inventive approach free from the irrationality of synectics. Indeed, why would one employ analogies and metaphors and other irrational factors in order to come to a formula — *the action has to happen itself* — a natural and universal inventive formula of an ideal invention? Such a formula (which means that the action needed to solve the problem should happen by itself, i.e., without introducing any additional systems) should indeed be programmed into the process of any inventive solution, aimed at an appropriately selected part of the object, and with precise identification of the physical contradiction and intended physical action. These features are the characteristics of *TRIZ*, the powerful inventive method described in the next section.

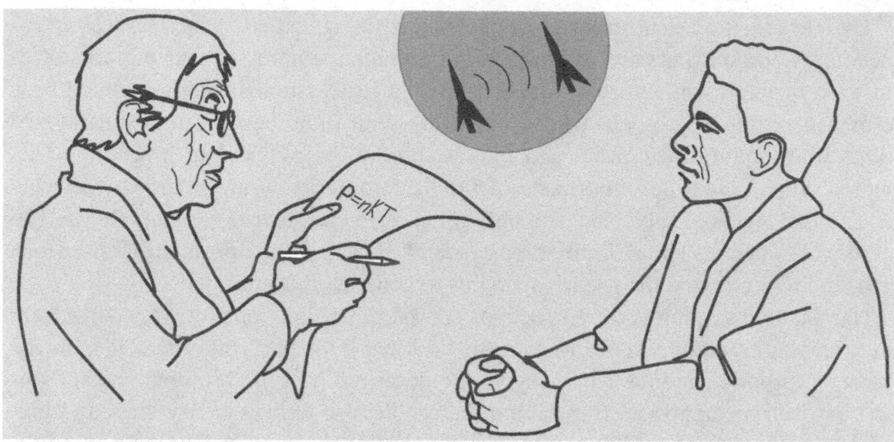

The abbreviation TRIZ (pronounced *[treez]*) can be translated as the Theory of Inventive Problem Solving. TRIZ was developed by Genrikh Altshuller in the Soviet Union in the mid-20th century. The author, while working in the patent office since 1946, analyzed many thousands of patents, trying to discover patterns to identify what makes a patent successful. Between 1956 and 1985, following his work in the patent office, Altshuller formulated TRIZ and, together with his team of supporters, developed it further. In the illustration he is imagined learning from his senior colleague while simultaneously dreaming about his sci-fi story of two space captains who were flying through the Sun on spaceships made from neutron star material while communicating via gravitational waves.

1.5 TRIZ METHOD

The creator of TRIZ, Genrich Altshuller[5] , and his team, devised four key constructs of the TRIZ methodology:

- The same problems and solutions appear again and again but in different industries.
- There is a recognizable technological evolution path for all industries.
- Innovative patents (about a quarter of the total) used science and engineering theories outside their own area or industry.
- An innovative patent uncovers and solves contradictions.

---

[5]He was also known under his pen name Altov, and famous for his short science-fiction stories, where he was also quite inventive. In one of them two space captains, named Icarus and Daedalus, attempted to fly through Sun on two spaceships made from neutron star materials, communicating via gravitational wave along the way. When engine of one failed, the other captain steered his spaceship to the first one, locked both spaceships via their gravitational attraction, and pulled both out of Sun, saving his colleague.

Together with these main clauses, the authors created a detailed methodology, which we will review in the next section before making the connection to physics.

And as the reader has likely already guessed, it was the word "TRIZ" that was skipped in the quote at the beginning of the previous section. It was TRIZ that became the bedrock of innovation at Samsung and that can do so much more if applied to other existing fields such as accelerator science.

1.5.1 TRIZ IN ACTION – EXAMPLES

Let us consider the TRIZ approach with an example that is often cited in TRIZ textbooks. Imagine that we need to polish an optical lens with an abrasive stick (see Fig. 1.17) and we need to do it quickly. However, there is a problem: polishing the lens generates heat, which degrades the optical properties of the produced lens. Existing cooling methods are ineffective, as one cannot supply adequate and uniform cooling to each abrasive particle.

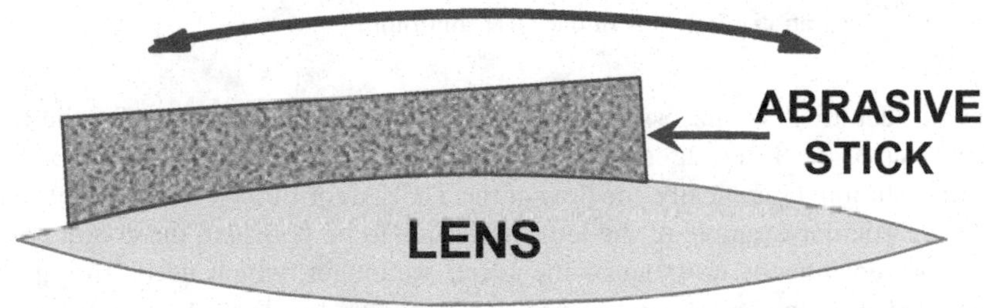

Figure 1.17 Illustration of TRIZ in action — initial specific problem.

The first step of the TRIZ approach consists of identifying the pair of contradicting parameters: the one that needs to be improved and the one that simultaneously gets worse. In the considered case of the lens polishing, these parameters are the *speed* (the one to be improved) and the *temperature* (the one that gets worse).

The second step of the TRIZ algorithm starts by identifying if anyone else has solved such a contradiction in the past. And it is here that the power of TRIZ lies — hundreds of thousands of analyzed patents allowed the TRIZ team to put together a table, *the contradiction matrix*, which helps to find relevant *inventive principles* corresponding to a particular pair of contradicting parameters.

The beauty of the TRIZ method is also illustrated by the fact that the entire field of engineering can be described by the TRIZ contradiction matrix using only 39 parameters, and that the number of TRIZ inventive principles is also remarkably small — only 40.

The second step of the TRIZ algorithm involves checking[6] the TRIZ matrix for this pair of contradicting parameters and then identifying the relevant, in this case, inventive principles.

---

[6]See, for example, http://www.triz40.com/

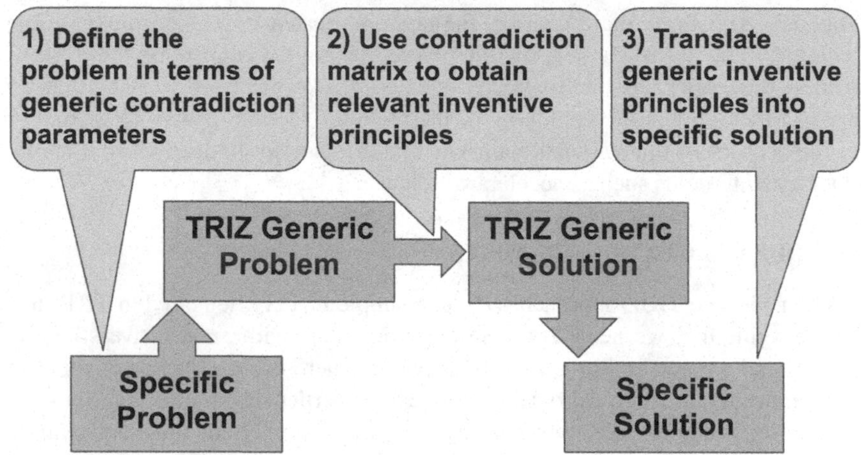

Figure 1.18 Illustration of the flow of the TRIZ algorithm.

Having identified the necessary inventive principles, we are ready for the third and final step of the TRIZ algorithm — translating a generic inventive principle into a specific solution. Graphically, the flow of the TRIZ algorithm is shown in Fig. 1.18.

In our particular example of the lens that needs to be polished, the contradiction matrix zoomed into the crossing of the *speed* parameter, which needs to improve without damaging the *temperature* parameter, is pictured in Table 1.1.

Table 1.1
TRIZ contradiction matrix for speed–temperature parameters and indexes of the corresponding inventive principles

Improving parameter	Parameter that deteriorates				
	...	9.Speed	...	17.Temperature	...
...	...	...	...	...	...
9.Speed	...	...	...	2,28,30,36	...
...	...	...	...	...	...
17.Temperature	...	...	...	...	...
...	...	...	...	...	...

The inventive principles listed (according to their numerical index) in the speed–temperature cell of the contradiction matrix are: 2-Taking out; 28-Mechanics substitution; 30-Flexible shells and thin films; 36-Phase transitions.

The next step is to select the inventive principle that is most suitable which in our example can be judged to be the latter one — 36-Phase transition (use of phenomena occurring during phase transitions, such as volume changes, loss or absorption of

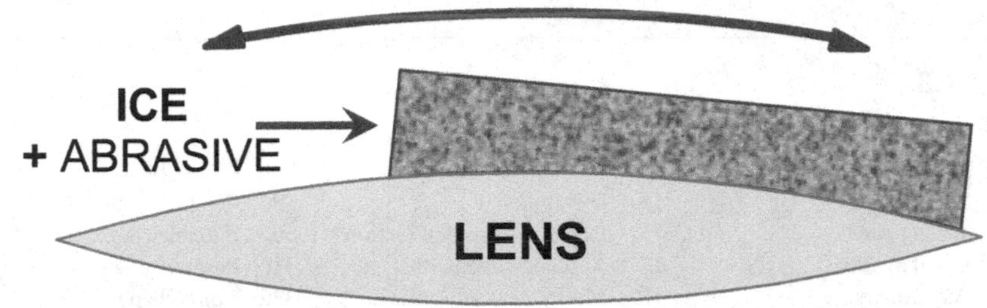

Figure 1.19 Illustration of TRIZ in action — specific solution.

heat, etc., according to the description of this TRIZ principle).

The corresponding specific solution to the lens polishing dilemma (which can be suggested according to the selected inventive principle of phase transition) is to use ice together with abrasive particles, which provides efficient uniform cooling, as illustrated in Fig. 1.19.

Before we discuss the connection between TRIZ and science, allow us to present a complete list of the standard TRIZ contradicting parameters (shown in Table 1.2) as well as the list of inventive principles (Table 1.3). A detailed description of the TRIZ principles is also available at the web page mentioned in footnote 6 of this chapter.

Table 1.2
Elements of TRIZ contradiction matrix

No.	Parameter	No.	Parameter	No.	Parameter
1.	Weight of moving obj.	14.	Strength	27.	Reliability
2.	Weight of stat. obj.	15.	Durability of mov. obj.	28.	Measurement accuracy
3.	Length of moving obj.	16.	Durability of stat. obj.	29.	Manufacturing precision
4.	Length of stat. obj.	17.	Temperature	30.	Object-affected harmful
5.	Area of moving object	18.	Illumination intensity	31.	Object-generated harmful
6.	Area of stationary obj.	19.	Energy use by mov. obj.	32.	Ease of manufacture
7.	Volume of moving obj.	20.	Energy use by stat. obj.	33.	Ease of operation
8.	Volume of stat. obj.	21.	Power	34.	Ease of repair
9.	Speed	22.	Loss of energy	35.	Adaptability or versatility
10.	Force (Intensity)	23.	Loss of substance	36.	Device complexity
11.	Stress or pressure	24.	Loss of information	37.	Difficulty of detecting
12.	Shape	25.	Loss of time	38.	Extent of automation
13.	Stability of the object	26.	Quantity of substance	39.	Productivity

1.6 TRIZ METHOD FOR SCIENCE

We have finally arrived at the section that will elucidate the meaning of the epigraph to this chapter.

Table 1.3
TRIZ inventive principles

No.	Principle	No.	Principle	No.	Principle
1.	Segmentation	15.	Dynamics	29.	Pneumatics, hydraulics
2.	Taking out	16.	Partial or excessive actions	30.	Flexible shells, thin films
3.	Local quality	17.	Another dimension	31.	Porous materials
4.	Asymmetry	18.	Mechanical vibration	32.	Color changes
5.	Merging	19.	Periodic action	33.	Homogeneity
6.	Universality	20.	Continuity of useful action	34.	Discarding, recovering
7.	Nested dolls	21.	Skipping	35.	Parameter changes
8.	Anti-weight	22.	Blessing in disguise	36.	Phase transitions
9.	Preliminary anti-action	23.	Feedback	37.	Thermal expansion
10.	Preliminary action	24.	Intermediary	38.	Strong oxidants
11.	Beforehand cushioning	25.	Self-service	39.	Inert atmosphere
12.	Equipotentiality	26.	Copying	40.	Composite materials
13.	The other way around	27.	Cheap short-lived objects		
14.	Spheroidality – Curvature	28.	Mechanics substitution		

The TRIZ inventive principle of Russian dolls (nested dolls, or *matreshka*) can be applied not only to engineering, but to many other areas, including science. A rather spectacular example is the construction of a high-energy physics detector, where many different sub-detectors are inserted into one another, like a nested doll, in order to enhance the accuracy of detecting elusive particles (see Fig. 1.20).

The reader can now see that the 1920 poem by Valery Bryusov, which describes an electron as a planet in its own world (see the figure near the epigraph in the beginning of the chapter), can also be seen as reflection of the nested doll inventive principle, this time in poetic science fiction.

TRIZ textbooks also often cite Wilson's cloud chamber (invented in 1911) and Glaser's bubble chamber (invented in 1952) as examples, in the terminology of TRIZ, of a system and anti-system. Indeed, the cloud chamber works on the principle of bubbles of liquid created in gas, whereas the bubble chamber uses bubbles of gas created in liquid (see Fig. 1.21).

It is thus reasonable to ask: would it have taken almost half a century to invent the bubble chamber had the TRIZ anti-system principle been applied?

The systematic application of TRIZ to science is indeed a valid question, and moreover, it can give us new insights. We will now discuss TRIZ in connection with accelerator science.

1.7 AS-TRIZ

The TRIZ method was originally created for engineers. However, the methodology is universal and, as the previous section demonstrated, can largely be applied to science, and in particular to accelerator science.

Figure 1.20 High-energy physics detectors, which have a layered "nested" structure, reflecting the TRIZ inventive principle of nested dolls.

This is the house that Jack built.
This is the malt
That lay in the house that Jack built.
This is the rat, that ate the malt
That lay in the house that Jack built.
This is the cat, that killed the rat, that ate the malt
That lay in the house that Jack built...

"This is the house that Jack built" of the *Mother Goose Rhymes* is an example of the nested doll inventive principle in folk poetry.

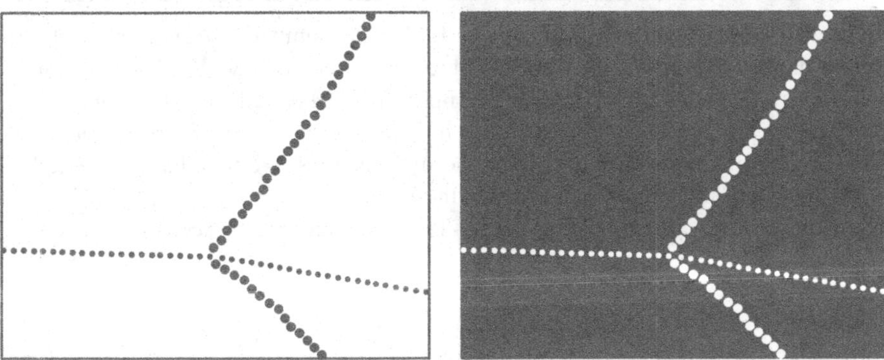

Figure 1.21 Particle interaction event observed in a cloud chamber invented by Wilson in 1911 (left), and in a bubble chamber invented by Glaser in 1952 (right).

Many of the parameters from the TRIZ contradiction matrix, as well as TRIZ inventive principles, can be directly applied to problems arising in accelerator science. Still, accelerator science is a distinct discipline, and it is only natural to add accelerator science-related parameters and inventive principles to TRIZ. We call this emerging extension *Accelerating Science TRIZ* or *AS-TRIZ*, highlighting via its name a wide applicability of the method to *various* areas of science, even those beyond the field of accelerators.

Below are just a few suggested additional parameters for the AS-TRIZ matrix of contradictions (which are based on manipulated beams and corresponding characteristics discussed in Section 1.2.1), and just a couple of inventive principles to start with — see Tables 1.4 and 1.5. We will populate this table as we move along the chapters.

Table 1.4
Emerging AS-TRIZ parameters

No.	Parameter	No.	Parameter	No.	Parameter
1.	Energy	6.	Intensity	11.	Spatial extent
2.	Rate of energy change	7.	Efficiency	12.	Sensitivity to imperfections
3.	Emittance	8.	Power	13.	Cooling rate
4.	Luminosity	9.	Integrity of materials	14.	...
5.	Brightness	10.	Time duration or length	15.	...
...	...	...	...	...	...

Table 1.5
Emerging AS-TRIZ principles

No.	Principle	No.	Principle
1.	...	2.	...
3.	Undamageable or already damaged	4.	Volume-to-surface ratio
...	...	...	...

The first inventive principle shown in Table 1.5, number 3: "Undamageable or already damaged," is illustrated by the three examples below, while 4, "Volume-to-surface ratio," naturally arises, for example, from Maxwell's equation, where an integral on a surface is connected to the integral over volume $\int \mathbf{E}d\mathbf{S} \propto \int \rho dV$, or from other similar equations, e.g., from thermodynamics. We will discuss examples corresponding to this principle later on in the book.

Let us illustrate an application of AS-TRIZ to a couple of accelerator science examples.

The first example is the beam profile monitor with its tungsten or carbon wire, shown in Fig. 1.22. In order to measure the beam's profile, the wired frame needs to cross the beam very quickly. The problem is that, as beam intensity increases, the beam and energy losses in the wire also increase, and the wire can get damaged after a single use. The physical contradiction in this case is between the parameters *intensity* (to be improved) and *integrity* (what degrades).

In order to solve the beam profile monitor problem, let us apply emerging AS-TRIZ principle 3: "Undamageable or already damaged." That is, we must replace the material that can be damaged with another media which either cannot be damaged (light) or is already "damaged" (e.g., plasma).

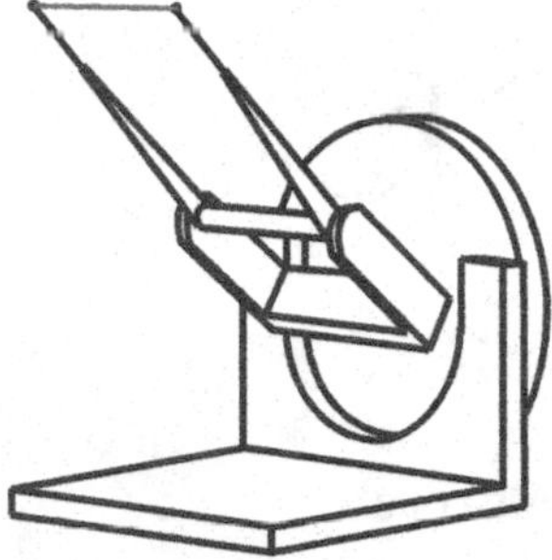

The *physical contradiction* between the parameter *intensity* and *integrity* can be resolved using the *undamageable or already damaged inventive principle*, i.e. replacing a material wire by a laser light or plasma.

Figure 1.22 Carbon wire beam profile monitor.

Selecting light as a medium that cannot be damaged naturally arrives at an already well-known solution — a beam profile monitor based on laser wire[7], shown in Fig. 1.23.

The next example is a standard glass mirror. The problem here is that, as the intensity of the laser increases, the mirror can get damaged. The contradiction is again between *intensity* and *integrity*. Let us apply the same principle 3 of AS-TRIZ, and replace the mirror with something that is already damaged — plasma. Indeed, such a solution is already known in laser and plasma science — the so-called plasma mirror. A plasma mirror is created by focusing a laser beam onto a piece of glass or gas. It will then be transmitted until the intensity reaches a certain value sufficient to produce plasma on the surface. At that moment, the reflective index of the surface changes, and the laser beam is then reflected from the plasma's free electrons.

The final example for this chapter, one that we will return to for further details later on, is the dilemma of accelerating structures. Made from metal, the acceleration structures (either normal conducting or superconducting) are susceptible to breakdowns that limit the useful accelerating gradient, typically to about 100 MeV/m.

The problem of an accelerating cavity can therefore be formulated as follows: as the rate of energy E change (accelerating gradient) increases, the surface of cavities gets damaged with occasional breakdowns. The contradiction is therefore between *rate of E change* (to be improved) and *integrity* (what degrades).

Applying again principle 3 of AS-TRIZ, and replacing accelerating structure with plasma, we come to the well-known concept of plasma acceleration, when an "accelerating structure" is temporarily produced in plasma by a driving laser pulse.

By taking note of these three examples, we already can start to fill in the AS-TRIZ contradiction matrix with inventive principles — see Table.1.6.

Many more examples related to TRIZ and AS-TRIZ will be discussed in the following chapters. However, the style of the book from this point on will change — the main text will follow the standard style of scientific textbooks, while the occasional

[7] In the laser wire the bunch interacts with optical light that acts like the physical wire of the mechanical beam profile monitor.

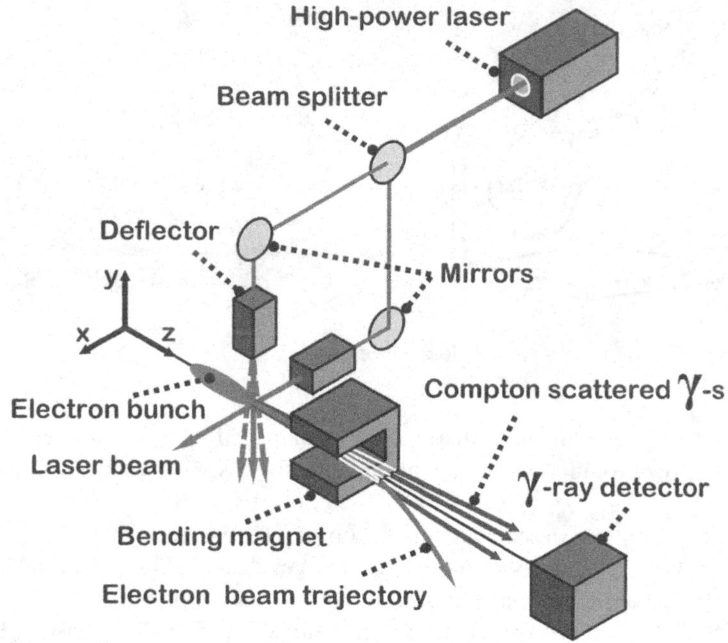

Figure 1.23 Laser wire beam profile monitor.

Improving parameter	Parameter that deteriorates			
	...	...	9. Integrity	...
...	...	...	...	...
2. Rate of E change	...	...	3, ...	...
...	...	...	...	...
6. Intensity	...	...	3, ...	...
...	...	...	...	...

Table 1.6
Emerging AS-TRIZ contradiction matrix and indexes of inventive principles

notes (primarily in the shaded boxes) will point out relevant connections to TRIZ. We will also use TRIZ in order to highlight similarities between the three discussed subjects — physics of accelerators, lasers and plasma.

These two parallel narratives, about science and about inventions, will eventually merge again in the final chapter, where we will summarize our discussions about the art of inventiveness and accelerator science.

1.8 TRIZ AND CREATIVITY

The connection of TRIZ and creativity is often discussed and a question posed if there is still room for creativity in step-by-step TRIZ process. We claim that use of TRIZ does not take away your creativity, and instead it encourages and inspires you.

Use of TRIZ does not take away your creativity! It instead strives to encourage and inspire you. It helps to find potential solutions, narrowing down the choices, but does not find the final answer for you. By arming you with a considerably deep breadth of knowledge, TRIZ can help you to be successful if you choose to apply its methods. *Looking at the world through the prism of TRIZ. Illustration by Sasha Seraia.*

TRIZ has many advantages in comparison with methods developed earlier, such as brainstorming or synectics (see Fig. 1.24), due to introduction of its inventive principles, tables of contradiction, and laws of evolution of technical systems.

Moreover, use of TRIZ will help us to build bridges between accelerators, lasers and plasma, and indeed even possibly other currently unrelated sciences, technology or engineering. Remember, one of the most important principles of TRIZ is that "the same problems and solutions appear again and again but in different disciplines."

1.9 THE ART OF SCIENTIFIC PREDICTIONS

Let's now talk about scientific revolutions — particularly about what drives them.

The two opposite views on scientific revolutions were expressed by philosopher Thomas Kuhn and by physicist-philosopher Freeman Dyson.

According to Thomas Kuhn, scientific revolutions are concept-driven and created by a paradigm shift, a completely new set of ideas.

On the contrary, Freeman Dyson claimed that scientific revolutions are tool-driven. He wrote, "The human heritage that gave us toolmaking hands and inquisitive

Figure 1.24 TRIZ vs. brainstorming. Illustration by Sasha Seraia.

brains did not die. In every human culture, the hand and the brain work together to create the style that makes a civilization. Science will continue to generate unpredictable new ideas and opportunities. And human beings will continue to respond to new ideas and opportunities with new skills and inventions. We remain toolmaking animals, and science will continue to exercise the creativity programmed into our genes."

While both of these points of view have the right to coexist, we as the authors and readers are more interested in accelerator (or plasma, laser or other) technology, and so stand closer to the second point of view. Moreover, we strive to make positive and proactive impacts on the evolution of science and technology via the creation of new tools and instruments. However, where should we focus our efforts? Can we learn from past efforts to make our impact more reliable and efficient?

Let's look at predictions made in 1968 for the year 2000, described in the book[8] *The Year 2000*. In this fascinating and encyclopedic volume, a variety of predictions were made for many fields, from demographics and politics to science and technology.

But, initially, the book asked a question — can the methodology of predictions be reliable? To illustrate a potential consequence of naive predictions, the authors presented the plot of the Gross National Product (GNP) of USA in comparison with the total R&D spending from 1945 to 1963 (which were the years of a boom in physics and R&D spending) and observed that linear approximation of the curves would result in R&D spending exceeding the GNP before the year 2000, which would be obviously unsustainable (see Fig. 1.25 to compare prediction with the actual values).

Warned by such a lesson to avoid naive extrapolations, the authors of *The Year 2000* gave a hundred predictions for the fields of science and technology, several of which we list below with our assessment as to whether they were accurate.

Example predictions made in 1968 for the year 2000:
#01 — Applications of lasers for sensing, communication, cutting, welding ✓
#31 — Some control of weather and/or climate
#35 — Human hibernation for extensive periods (months to years)
#58 — Chemical methods for improving memory and learning ✓
#67 — Commercial extraction of oil from shale ✓
#81 — Personal pagers and perhaps even two-way pocket phones ✓
#99 — Artificial moon for lighting large areas at night ~✓

As we can see, some predictions were accurate (lasers, pocket phones, memory and learning, oil extraction) and some were not (climate, human hibernation). One might wonder why a check mark is placed next to the artificial moon prediction, if it were not for the recent news[9] which claimed that "Southwestern China's city

---

[8] *The Year 2000*, K. Herman, A. Wiener (editors), ISBN 978-0025604407, 1968.

[9] *"Chengdu to launch 'artificial moon' in 2020,"* People's Daily Online, October 16, 2018, http://en.people.cn/n3/2018/1016/c90000-9508748.html

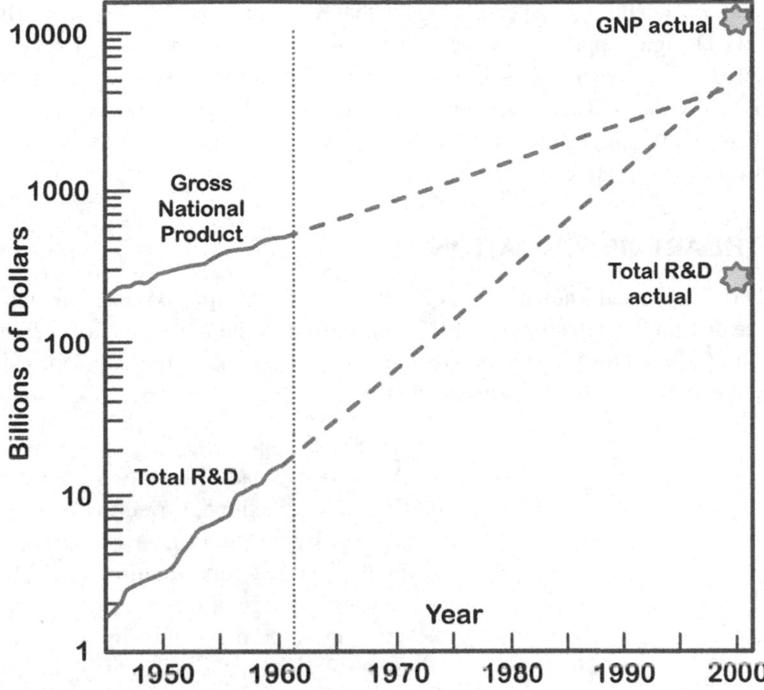

Figure 1.25 GNP and R&D expenditures vs. time illustrating failure of naive extrapolation as discussed in the book "The Year 2000" by K. Herman, A. Wiener (Editors), 1968.

of Chengdu plans to launch its illumination satellite, also known as the "artificial moon," in 2020, according to Wu Chunfeng, chairman of Chengdu Aerospace Science and Technology Microelectronics System Research Institute Co., Ltd. The illumination satellite is designed to complement the moon at night. Wu introduced that the brightness of the "artificial moon" is eight times that of the real moon, and will be bright enough to replace street lights. The satellite will be able to light an area with a diameter of 10 – 80 km, while the precise illumination range can be controlled within a few dozen meters. The testing of the illumination satellite started years ago, and now the technology has finally matured, explained Wu. Some people expressed concern that the lights reflected from space could have adverse effects on the daily routine of certain animals and astronomical observation. Kang Weimin, director of the Institute of Optics, School of Aerospace, Harbin Institute of Technology, explained that the light of the satellite is similar to a dusk-like glow, so it should not affect animals' routines."

These amazing dreams haven't yet been realized, perhaps due to recent global events; however, such specific plans certainly deserve a near-complete check mark.

We hope that the "Year 2000" examples have shown the reader that it is possible to make viable predictions and efficient research plans that are not only based on learning from the past of a particular field of science, but are also based on looking around at and across different disciplines of science and technology.

The authors would also like to suggest that we can, possibly, use the "Breakthrough By Design" approach, which we will introduce in a moment, not only for making predictions, but for pro-actively shaping the future of science and technology. This approach will be based on TRIZ, which will include understanding the general laws of evolution of science and technical systems, as well as the art of back-of-the-envelope estimations.

1.10 THE ART OF ESTIMATIONS

Enrico Fermi was well known for his ability to perform approximate calculations. One of the documented examples was his estimation of the magnitude of the Trinity test. Fig. 1.26 illustrates his ingenious approach to the back-of-the-envelope estimations when he was $\sim$10 miles from the Trinity test site.

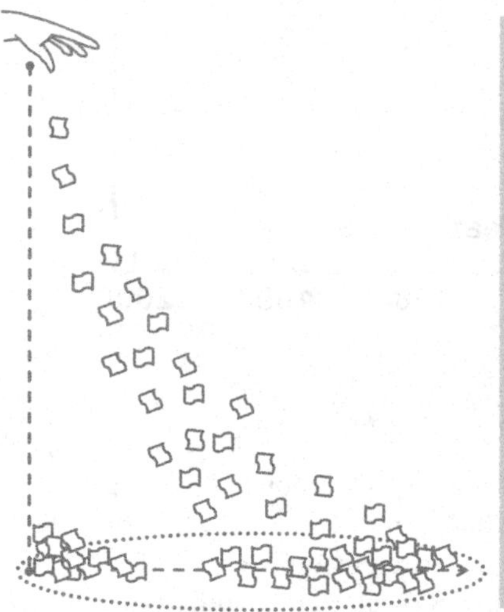

"About 40 seconds after the explosion, the air blast reached me. I tried to estimate its strength by dropping from about six feet small pieces of paper before, during, and after the passage of the blast wave. Since, at the time, there was no wind I could observe very distinctly and actually measure the displacement of the pieces of paper that were in the process of falling while the blast was passing. The shift was about 2 and 1/2 meters, which, at the time, I estimated to correspond to the blast that would be produced by ten thousand tons of TNT." — Enrico Fermi

Figure 1.26 Fermi paper experiment.

The importance of back-of-the-envelope estimations is obvious — such estimations are important because they help us to quickly understand if our idea works. But, even more importantly, such estimations allow us to improve cross-disciplinary understandings between scientists from different fields, like biology and physics. Indeed, if our derivations can be expressed in a small number of intuitively clear and simple impressions, we will be able to talk about them to colleagues from other disciplines much more efficiently, and vice versa.

One can learn how to make back-of-the-envelope estimations. To train oneself on these estimations, one can consider various questions, which do not have to be necessarily serious[10]. However, the estimates should always be based on a physical

[10] For example, can you estimate how fast you should run across the river of a certain width so that you would not sink while running through? See also the estimate task at the end of this chapter.

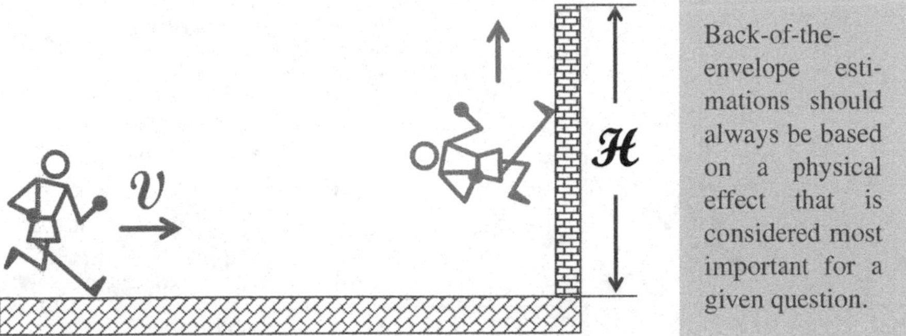

Back-of-the-envelope estimations should always be based on a physical effect that is considered most important for a given question.

Figure 1.27 For estimation of speed to run up a wall.

effect that is considered most important for a given question. Making an estimate would also allow us to figure out how to improve a system.

For example, let's estimate what speed V is needed to reach height H and get to other side of the vertical wall, as illustrated in Fig. 1.27.

The most important physical effect that we need to take into account in this case is gravity. While we run up along the vertical wall, our head is falling down. We can estimate that we can still make it to the top point of the wall, and get to the other side, if our head falls with respect to the legs by a small fraction of our height. To be more specific, let's make an estimate of V by requiring that, during the run along the wall, the head will not fall to lower than half the height of the person, as shown in Fig. 1.28. We will then find that (where g=9.8 m/s^2)

$$V = H(g/h)^{1/2}$$

or, for H=2 m and h=1.8 m, we get $V \approx 4.7$ m/s, which looks very reasonable.

Gravity is the most important effect to take into account. We can estimate that we can still make it to the other side of the wall if, during the run, the head will not fall to lower than half the height of the person.

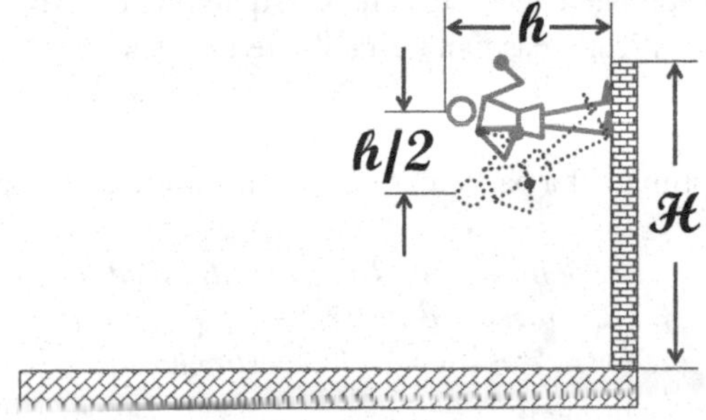

Figure 1.28 For estimations of how to run up to a wall.

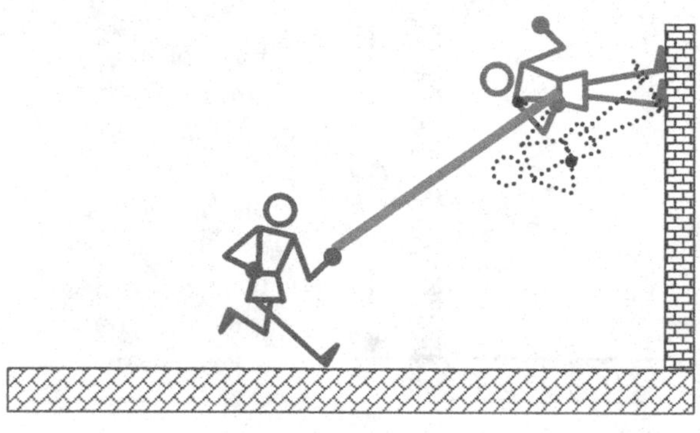

Back-of-the-envelope estimations help us understand how to solve the main challenge and what inventive principle to apply — in this case, number #9 "Preliminary anti-action."

Figure 1.29 Estimations lead to inventions.

Once we have understood the main challenge and made the estimation, we can solve the problem. To prevent the head falling while we run up the wall, we can apply an appropriate inventive principle. In this case, the principle #9 "Preliminary anti-action" will do the job, and it can be implemented with two people running in sync, with the second person holding a long stick, which will be pushing the first person's torso up, to compensate the action of gravity, as illustrated in Fig. 1.29.

As we mentioned earlier, Enrico Fermi was known for his ability for back-of-the-envelope estimations. Inspired by this example, and aiming to cultivate it, many leading centers teach the art of estimating starting from high-school levels. There are also books that can help you to master the art of estimations[11].

Together with TRIZ, the art of estimation — applied to several fields of science — can create a novel approach, which we call the *Breakthrough by Design approach* or, interchangeably, *Innovation by Design*, the subtitle of this second edition.

1.11 BREAKTHROUGH BY DESIGN APPROACH

The TRIZ conclusion that the same solutions appear again and again, but in different disciplines, shows us that widening our knowledge range will make us more inventive in technical areas. Expanding our knowledge into unfamiliar scientific and technical areas can be facilitated if exact explanations and derivations are replaced with intuitively clear back-of-the-envelope estimations. Practicing such estimations will not only help us understand other fields of science, but will also help us quickly estimate if a particular idea or invention makes practical sense.

Breakthrough by Design or Innovation by Design approach — an amalgam of TRIZ inventive principles and laws of technical system evolution with the art of back-of-the-envelope estimations, applied to several neighboring areas of science and technology.

[11] See, e.g., "Guesstimation 2.0" by Lawrence Weinstein, Princeton University Press, 2012.

EXERCISES

1.1. *Analyze the evolution of technical or scientific systems.* Discuss the evolution of any scientific or technical (or accelerator-science-related) fields, identifying successive technologies that arise, saturate and get replaced by new approaches and solutions.

1.2. *Analyze inventions or discoveries using TRIZ and AS-TRIZ.* A plasma mirror is often used when a standard metal mirror cannot withstand the power density of the laser. Analyze this technology in terms of the TRIZ and AS-TRIZ approach, identifying a contradiction and a general inventive principle that were used (or could have been used) for this invention.

1.3. *Analyze inventions or discoveries using TRIZ and AS-TRIZ.* Analyze and describe any scientific or technical invention/discovery (possibly related to accelerator science) in terms of the TRIZ and AS-TRIZ approaches, identifying a contradiction and an inventive principle that were used (or could have been used) for this invention or discovery.

1.4. *Developing AS-TRIZ parameters and inventive principles.* Based on what you already know about accelerator science, discuss and suggest the possible additional parameters for the AS-TRIZ contradiction matrix, as well as the possible additional AS-TRIZ inventive principles.

1.5. *Practice in reinventing technical systems.* Imagine that we need to provide 10 kW of electric power inside the top of a metal spherical electrode of the Van der Graaf accelerator, as shown in Fig. 1.8, to feed the electronics and the ion source. (It is not possible to bring this power via electrical wires from the low level). How do we provide this power? Remember the TRIZ principle *to use resources and energy that you already have in the system.*

1.6. *Practice in the art of back-of-the-envelope estimations.* In the book *The Three Body Problem* by Liu Cixin, the civilization of Alpha Centauri sent protons to Earth with embedded artificial intelligence, which penetrated high-energy physics experiments such as at the LHC, in order to spoil the results and stop scientific progress on Earth. Imagine that some of the AI-protons were captured, trapped, and Earth scientists decided to study their internal AI structure. They assumed that the AI was encoded into the interaction of gluons in the AI-protons and used fixed target experiments, similar to CEBAF[12], to study the AI-protons. Estimate the energy of the electron beam that Earth scientist needed to use to study the AI-protons. *(It is assumed that you can identify the most important factors in this task, define the necessary parameters and set their values, and get a numerical answer.)*

[12]CEBAF — Continuous Electron Beam Accelerator Facility, at Jefferson Lab.

2 Transverse Dynamics

Let's begin our discourse about the basics of accelerator physics with the topic of the transverse dynamics of charged particles. This will lead into a discussion of the basics of synchrotron radiation and of acceleration in the following chapters, intermediated by a dialogue on the synergies between accelerators, lasers and plasma.

2.1 MAXWELL EQUATIONS AND UNITS

We start by recalling the Maxwell equations with an emphasis on their systems of units, focusing in particular on SI and Gaussian-cgs systems. The microscopic Maxwell equations (i.e., equations in vacuum) expressed in SI units, in both differential and integral form, are

$$\nabla \cdot \mathbf{E} = \frac{\rho}{\varepsilon_0} \quad \text{or} \quad \oint_{\partial\Omega} \mathbf{E} \cdot d\mathbf{S} = \frac{1}{\varepsilon_0} \int_{\Omega} \rho dV \tag{2.1}$$

This Maxwell equation illustrates the universality of the inventive principle of changing the volume-to-surface ratio.

$$\nabla \cdot \mathbf{B} = 0 \quad \text{or} \quad \oint_{\partial\Omega} \mathbf{B} \cdot d\mathbf{S} = 0 \tag{2.2}$$

$$\nabla \times \mathbf{E} = -\frac{\partial \mathbf{B}}{\partial t} \quad \text{or} \quad \oint_{\partial\Sigma} \mathbf{E} \cdot d\ell = -\frac{d}{dt} \int_{\Sigma} \mathbf{B} \cdot d\mathbf{S} \tag{2.3}$$

$$\nabla \times \mathbf{B} = \mu_0 \left(\mathbf{J} + \varepsilon_0 \frac{\partial \mathbf{E}}{\partial t} \right) \quad \text{or}$$

$$\oint_{\partial\Sigma} \mathbf{B} \cdot d\ell = \mu_0 \int_{\Sigma} \mathbf{J} \cdot d\mathbf{S} + \mu_0 \varepsilon_0 \frac{d}{dt} \int_{\Sigma} \mathbf{E} \cdot d\mathbf{S} \tag{2.4}$$

Permittivity of free space	$\varepsilon_0 \approx 8.85 \cdot 10^{-12}\ F/m \text{ or } A^2 s^4 kg^{-1} m^{-3}$
Vacuum permeability	$\mu_0 = 1/(c^2 \varepsilon_0) \approx 1.26 \times 10^{-6}\ N \cdot A^{-2}$
Speed of light in vacuum	$c \approx 2.99 \times 10^8\ m/s$

The Lorentz force acting on a charged particle in an electric and magnetic field in SI units is expressed as

$$\mathbf{F} = q(\mathbf{E} + \mathbf{v} \times \mathbf{B}) \tag{2.5}$$

DOI: 10.1201/9781003326076-2

While SI is the standard, the Gaussian system is more natural for electromagnetism. The differential Maxwell equations expressed in Gaussian-cgs units are

$$\nabla\cdot\mathbf{E}=4\pi\rho \quad , \quad \nabla\cdot\mathbf{B}=0$$
$$\nabla\times\mathbf{E}=-\frac{1}{c}\frac{\partial\mathbf{B}}{\partial t} \quad , \quad \nabla\times\mathbf{B}=\frac{1}{c}\left(4\pi\mathbf{J}+\frac{\partial\mathbf{E}}{\partial t}\right) \tag{2.6}$$

And the Lorentz force is

$$\mathbf{F}=q\left(\mathbf{E}+\frac{\mathbf{v}}{c}\times\mathbf{B}\right) \tag{2.7}$$

The above equation highlights why the Gaussian system is so useful: the electric and magnetic fields are expressed in the same units, which stresses that these fields have a similar nature.

Deriving a formula, instead of using constants such as e or $\hbar$, express the end result via more natural quantities — $m_ec^2, r_e, \lambda_e, \alpha$, etc.

Throughout this text we will use both SI and Gaussian units. We will, however, plan to construct equations or at least the end results in such a way that allows for easy conversion between different units. As a result, we will avoid having quantities such as electric charges or Planck's constant in our equations, and instead will use only quantities of length, speed and energy.

Charge of electron in SI: $e\approx 1.6\times 10^{-19}\ C\ or\ A\cdot s$
Charge of electron in Gaussian units: $e\approx 4.8\times 10^{-10}$ *cgs units*
Classical radius of electron $r_e\approx 2.82\cdot 10^{-15} m$
Fine structure constant $\alpha\approx 1/137$

Let's write down some of the equations that will be useful for such conversions into the natural and unit-independent form.

The most useful one is for the classical radius of an electron:

$$\mathrm{SI}:\ r_e=\frac{1}{4\pi\varepsilon_0}\frac{e^2}{m_ec^2} \quad , \quad \mathrm{Gauss}:\ r_e=\frac{e^2}{m_ec^2} \tag{2.8}$$

And the one for the fine structure constant:

$$\mathrm{SI}:\ \alpha=\frac{e^2}{(4\pi\varepsilon_0)\hbar c} \quad , \quad \mathrm{Gauss}:\ \alpha=\frac{e^2}{\hbar c} \tag{2.9}$$

The latter also gives us the reduced Compton wavelength for the electron ($\lambda_e = r_e/\alpha\approx 3.86\cdot 10^{-13} m$), which is very useful whenever the Planck's constant needs to be hidden in an equation.

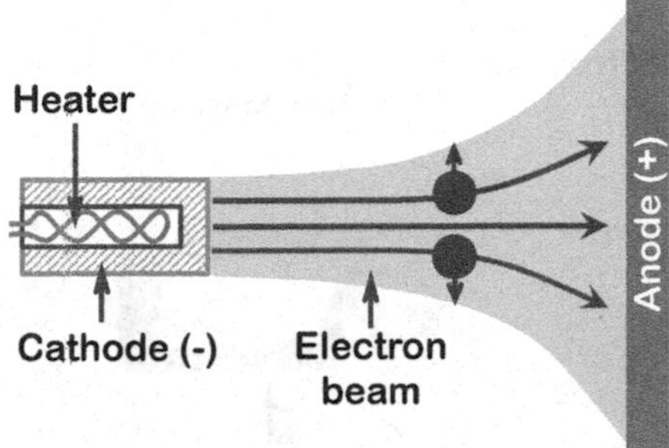

Evolution of electron guns demonstrates a remarkable path of numerous inventions. Thermionic guns, laser driven photocathodes, superconducting RF guns, inverted guns, polarized, and yet to be developed polarized SRF photoguns — we will discuss some of these inventions in this book.

Figure 2.1 Simple electron gun.

2.2 SIMPLEST ACCELERATOR

An example of a simple accelerator is a thermionic gun, in which thermionic emission from the cathode generates electrons that are accelerated across a high voltage gap to the anode. The current in thermionic guns usually behaves according to so-called "law of three-halves-power" or Child-Langmuir's law. In this case the space charge of already emitted electrons limits the current, which as the result behaves vs. applied voltage V as $I \propto V^{3/2}$.

A variety of materials can be used for the cathodes of thermionic guns. Pure metals are the most robust choice but require extremely high temperatures. On the other hand, the oxides cathodes are covered with alkaline earth metal oxides (e.g., the BaO *impregnated* tungsten cathode), and provide a higher emission at lower temperatures. However, they are more delicate as they require a better vacuum for their operation and are vulnerable to poisoning by heavier elements present in the residual gas.

In thermionic guns, a grid can be installed between the anode and cathode. Applying pulsed voltage to the grid allows for the generation of a train of pulses suitable for consequent RF acceleration.

In an electron gun, the electrons generated by thermionic emission tend to repel from each other as illustrated in Fig. 2.1, resulting in reduced beam quality. The additional focusing electrodes can help to maintain the quality of the beam. In particular, it was shown by J.R. Pierce in 1954 that in planar geometry, an electrode at the potential of the cathode inclined at 67.5° (called the Pierce angle), as shown in Fig. 2.2, will help to maintain a parallel flow of electrons in the regime when the gun current is limited by the space charge. Any intermediate accelerating electrodes or the anode would need to be placed along the equipotential lines, as shown in Fig. 2.2, to maintain the parallel flow.

The same Fig. 2.2 also illustrates the concept of the beam anode (collector) made in the shape of a Faraday cup — a useful device for the accurate measurement of electron current. Any secondary charged particles emitted from the walls are eventually absorbed and do not affect the measurements of the current.

Controlling the beam shape and beam quality in high density electron guns often requires the use of an accompanying solenoid magnetic field. Let us now consider equations for the motion of charged particles in electromagnetic fields.

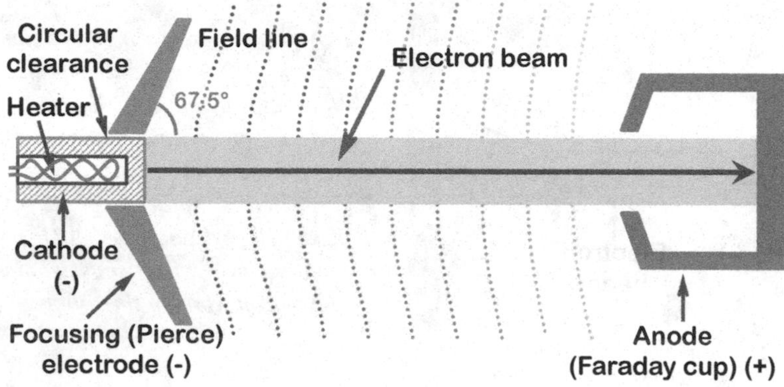

Figure 2.2 Electron gun with Pierce electrode and collector made in the form of a Faraday cup.

2.3 EQUATIONS OF MOTION

2.3.1 MOTION OF CHARGED PARTICLES IN EM FIELDS

The motion of a particle with charge q in an electric **E** and magnetic **B** fields is given by the following equations:

$$\frac{d\mathbf{p}}{dt} = q(\mathbf{E}+\mathbf{v}\times\mathbf{B}) \quad , \quad \frac{d\mathscr{E}}{dt} = \mathbf{F}\cdot\mathbf{v} \tag{2.10}$$

where the momentum **p** and energy $\mathscr{E}$ of the particle are

$$\mathbf{p} = m_0\gamma\mathbf{v} \quad , \quad \mathscr{E} = m_0\gamma c^2$$

and m_0 is the particle mass, and γ is the relativistic factor.

Let's consider the case of the uniform magnetic field when the equation of motion simplifies to

$$m_0\frac{d\gamma\mathbf{v}}{dt} = q\mathbf{v}\times\mathbf{B} \quad \text{or} \quad m_0\gamma\dot{v}_x = q\,v_y B \quad \text{and} \quad m_0\gamma\dot{v}_y = -q\,v_x B$$

which then can be rewritten as

$$\ddot{v}_x = \frac{qB}{m_0\gamma}\dot{v}_y = -\left(\frac{qB}{m_0\gamma}\right)^2 v_x$$

which has a solution of

$$v_x = v_0\cos(\omega t) \quad \text{or} \quad x = \frac{v_0}{\omega}\sin(\omega t)$$

where

$$\omega = \frac{qB}{m_0\gamma} \tag{2.11}$$

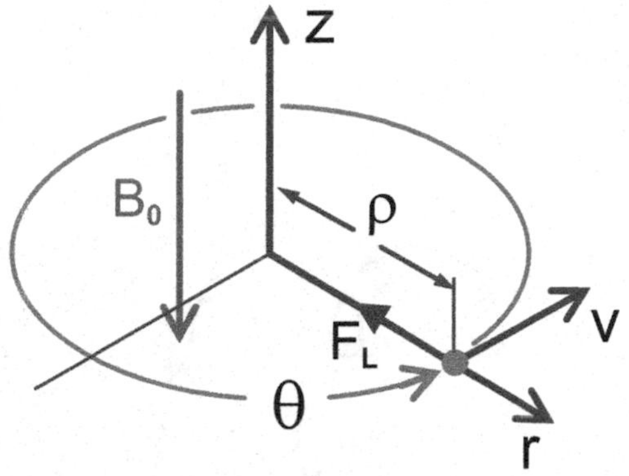

Figure 2.3 Motion of charged particles in a uniform magnetic field.

Considering parameters of an accelerator, it is convenient to use, instead of the beam energy, the beam **magnetic rigidity** *given by the following practical equation* $B\rho\,[Tesla\cdot m] \approx 3.3356\, p\,[GeV/c]$

This solution describes motion (see Fig. 2.3) with a radius of

$$\rho = \frac{v_0}{\omega} = \frac{v_0 m_0 \gamma}{qB} \tag{2.12}$$

Let's rewrite this radius (called the Larmor radius) in both systems of units:

$$\text{SI}: \ \rho = \frac{p}{qB} \qquad \text{Gaussian}: \ \rho = \frac{pc}{qB} \tag{2.13}$$

This kind of motion is observed, for example, in dipoles — magnets intended primarily for bending the trajectories of charged particles.

A quantity called *magnetic rigidity* $B\rho$ is often used to describe motion in magnetic fields. It is defined as

$$\text{SI}: \ B\rho = \frac{p}{q} \qquad \text{Gaussian}: \ B\rho = \frac{pc}{q} \tag{2.14}$$

and for a particle with the elementary charge and momentum p given in GeV/c is equal to $B\rho[Tesla\cdot m] \approx 3.3356\, p[GeV/c]$ or $B\rho[kGs\cdot cm] \approx 3335.6\, p[GeV/c]$.

2.3.2 DRIFT IN CROSSED E × B FIELDS

While we are on this topic, let's consider the case of uniform **E** and **B** fields that are perpendicular — a situation often encountered when dealing with plasma and beams.

Qualitatively, if a particle is initially at rest in these crossed fields, it is initially pulled by the electric field, and then, due to emerging velocity, the magnetic field turns it around. When the direction of the particle's motion reverses, the electric field slows it down and eventually stops the particle some distance away from its initial position. After that, the aforementioned motion begins anew. The resulting trajectory of the particle resembles Fig. 2.4. As a result, these equations of motion predict a

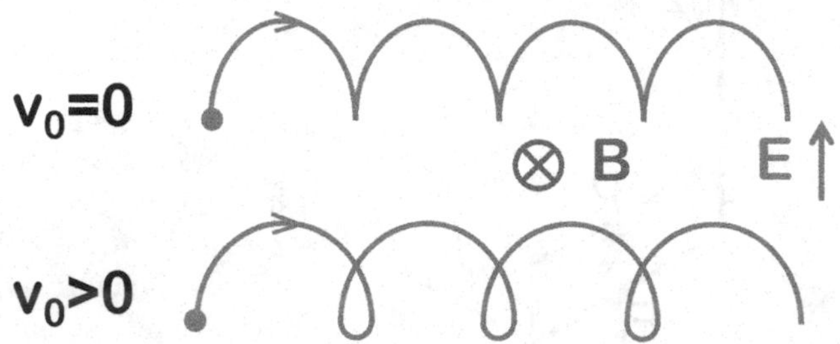

Figure 2.4 Drift in crossed E×B fields.

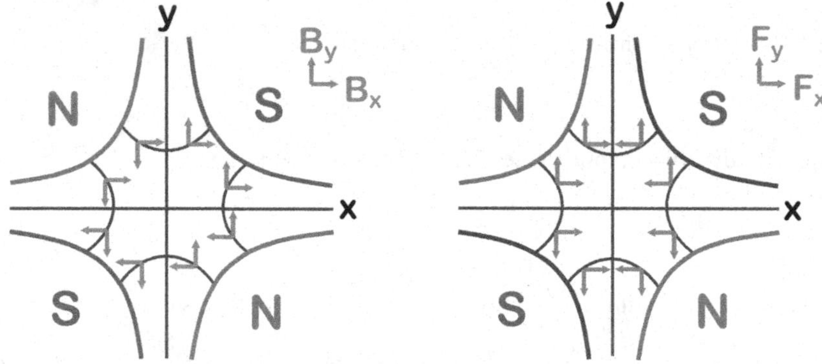

Figure 2.5 Magnetic fields and forces acting on a particle in a quadrupole.

particle drift with constant velocity, which is perpendicular to both **E** and **B** and with its value given by

$$\text{SI}: \ \mathbf{v}_d = \frac{\mathbf{E}\times\mathbf{B}}{B^2} \qquad \text{Gaussian}: \ \mathbf{v}_d = c\frac{\mathbf{E}\times\mathbf{B}}{B^2} \tag{2.15}$$

The efficiency of the Gaussian system of units continues to astound us! Not only does it give us an intuitively clear and beautiful formula, but it also immediately shows that the above equation is valid only in the assumption that the electric field is much smaller than the magnetic field $E \ll B$.

2.3.3 MOTION IN QUADRUPOLE FIELDS

Let's consider motion in a quadrupole magnet (see Fig. 2.5) where, ideally, the fields depend linearly on the distance from the center of the magnet:

$$B_x = Gz \quad \text{and} \quad B_z = Gx$$

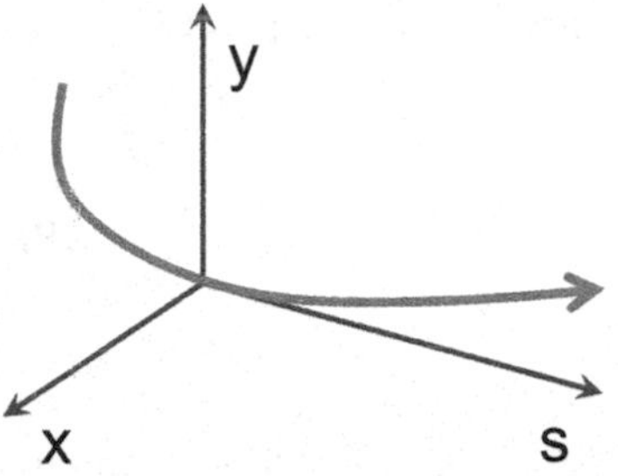

Figure 2.6 Frenet–Serret curvilinear coordinate system.

Since the particle trajectory in an accelerator is defined by bending magnets, a curvilinear coordinate system is best suited to describe the motion of the particles.

where G is the gradient of the quadrupole (and we use $z \equiv y$ in this section). We will rewrite the equation of motion as

$$m_0 \frac{d\gamma \mathbf{v}}{dt} = q\mathbf{v} \times \mathbf{B}$$

and using Cartesian coordinates:

$$\ddot{x} = -\frac{q}{m_0\gamma} G\dot{s}x \quad , \quad \ddot{z} = \frac{q}{m_0\gamma} G\dot{s}z$$

$$\text{and} \quad \ddot{s} = \frac{q}{m_0\gamma} G(\dot{x}x - \dot{z}z)$$

Changing the independent variable from time to path length, and considering only the case of small deviations from the axis, reduces these equations to

$$x'' - Kx = 0 \quad \text{and} \quad z'' + Kz = 0 \tag{2.16}$$

$$\text{where} \quad K = \frac{q}{p}\frac{\partial B_z}{\partial x} = \frac{q}{p} G \tag{2.17}$$

This brings us to the discussion of linear betatron motion.

2.3.4 LINEAR BETATRON EQUATIONS OF MOTION

The particle trajectory in an accelerator is initially defined by dipole magnets, therefore a *curvilinear coordinate system* is best used to describe the motion of the particles; see Fig. 2.6. A *reference orbit* is usually selected, corresponding to an *ideal particle*, which typically has a nominal energy and zero transverse offsets and angles.

The focusing elements, quadrupoles and higher-order elements are placed in space so that their centers correspond to the reference orbit.

The motion of a charged particle with respect to the *reference orbit* and along the *curvilinear abscissa* (s in Fig. 2.6), and influenced by the the magnetic fields of dipole magnets and quadrupole magnets, is given by the linear Hill's equations

$$\frac{d^2y}{ds^2} + K_y(s)y = 0 \tag{2.18}$$

where transverse coordinate y stands for either a horizontal or vertical axis ($y = x, z$).

Let's write these equations down for the horizontal

$$K_x(s) = \frac{1}{\rho^2(s)} - \frac{1}{B\rho}\frac{\partial B_z(s)}{\partial x} \tag{2.19}$$

and the vertical

$$K_z(s) = \frac{1}{B\rho}\frac{\partial B_z(s)}{\partial x} \tag{2.20}$$

planes. We see that the equation almost exactly resembles those derived in the previous section (see Eqs. 2.16 and 2.17) except for an additional term $1/\rho^2(s)$, which corresponds to a weak focusing of a dipole.

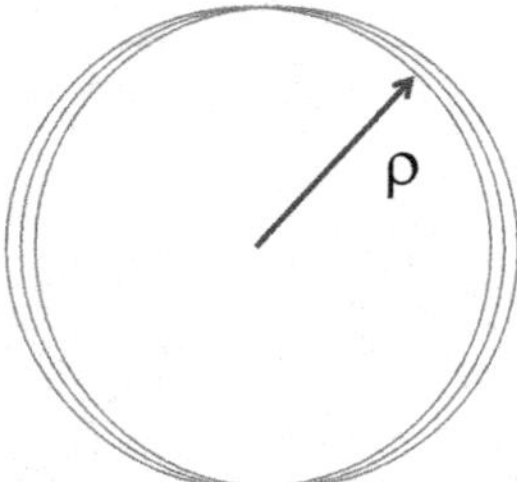

Figure 2.7 Shifted circles cross.

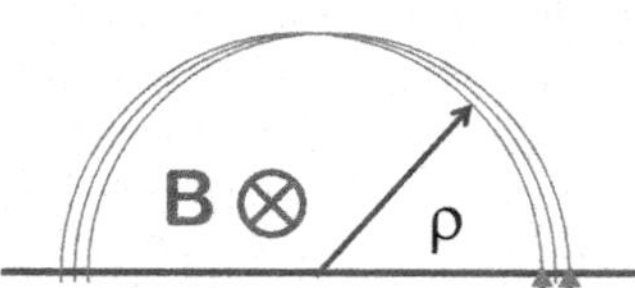

Figure 2.8 Illustration of the origin of weak focusing in dipoles.

The origin of the term corresponding to the weak focusing in a dipole can be illustrated by the following example. Consider the shifted circles in Fig. 2.7. They cross in two points (we will ignore second-order effects). Translating this example of shifted circles to a dipole magnet whose field fills a half plane as shown in Fig. 2.8, we conclude by observation that the trajectories of the particles in this dipole exhibit an equivalent "focusing" with the wavelength of motion (along the curvilinear coordinate s) given by $2\pi\rho$, corresponding to

$$x = x_0 \sin(s/\rho)$$

or to the following equation

$$\frac{d^2x}{ds^2} + \frac{x}{\rho^2} = 0 \tag{2.21}$$

which has the same term $1/\rho^2(s)$ as the Hill's equations above.

The above derivations generally assumed no periodicity; however, in a circular machine, K_x, K_z and ρ are periodic. These are linear equations and can be integrated, which will be discussed in the next section.

2.4 MATRIX FORMALISM

2.4.1 PSEUDO-HARMONIC OSCILLATIONS

Let's look for the solution of the Hill's Eq. 2.18 in following form

$$y(s) = \sqrt{\varepsilon_y \beta_y(s)} \cos\left[\phi_y(s) - \phi\right] \tag{2.22}$$

where

$$\phi_y(s) = \int_{s_0}^{s} \frac{ds'}{\beta_y(s')} \tag{2.23}$$

which describes *pseudo-harmonic oscillations*. Here the beta functions β (in x and z) are proportional to the square of the *envelope* of the oscillations.

Beta functions β_x and β_z are proportional to the square of the envelopes of the oscillations in the transverse planes x and z.

The functions $\phi(s)$ (also in x and z) describe the phase of the oscillations.

We can find the differential equation for the beta functions, by substituting the form Eq. 2.22 into the Hill's equation. We use the following equation first

$$y'(s) = \frac{\beta'(s)}{2}\sqrt{\frac{\varepsilon}{\beta(s)}}\cos(\phi(s)-\phi) - \phi'(s)\sqrt{\varepsilon\beta(s)}\sin(\phi(s)-\phi)$$

and the second derivatives

$$y''(s) = \left[\frac{\beta''(s)}{2\sqrt{\beta(s)}} - \frac{\beta'^2(s)}{4\beta^{3/2}(s)} - \sqrt{\beta(s)}\phi'^2(s)\right]\sqrt{\varepsilon}\cos(\phi(s)-\phi) -$$

$$-\left[\phi''(s)\sqrt{\beta(s)} + \frac{\beta'(s)\phi'(s)}{\sqrt{\beta(s)}}\right]\sqrt{\varepsilon}\sin(\phi(s)-\phi)$$

which we substitute to Hill's equation and proceed by equating the coefficients to zero in front of *sin* and *cos* parts. We therefore obtain

$$\frac{1}{2}\beta\beta'' - \frac{1}{4}\beta'^2 + k(s)\beta^2 = 1 \tag{2.24}$$

and

$$\phi_y'(s) = \frac{1}{\beta_y(s)} \tag{2.25}$$

which represents the differential equations for beta function and the betatron phase.

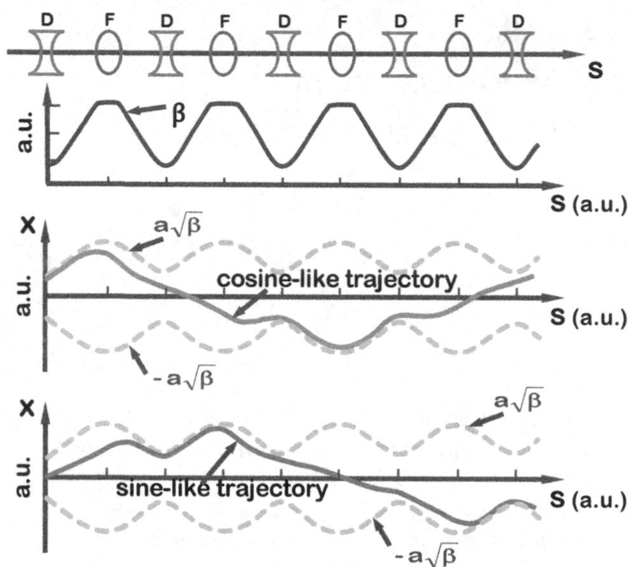

Figure 2.9 Illustration of pseudo-harmonic oscillations and cosine-like and sine-like principal trajectories.

2.4.2 PRINCIPAL TRAJECTORIES

The solutions of Hill's equation can be found in the form of *principal trajectories*. These are two particular solutions of the homogeneous Hill's equation

$$y'' + k(s)y = 0$$

which satisfy the following initial conditions

$$C(s_0) = 1; C'(s_0) = 0; \text{ cosine}-\text{like solution}$$

$$S(s_0) = 1; S'(s_0) = 1; \text{ sine}-\text{like solution}$$

The general solution can then be written as a linear combination of the principal trajectories

$$y(s) = y_0 C(s) + y_0' S(s)$$

The pseudo-harmonic oscillations, and the cosine-like and sine-like principal trajectories are illustrated in Fig. 2.9 for a FODO chain of quadrupole magnets.

Alpha function is defined as $\alpha = -\beta'/2$

Furthermore, we can derive the connection between the principal trajectories and pseudo harmonic oscillations. Let's express the amplitude and angle functions as

$$y(s) = \sqrt{\varepsilon\beta(s)}\cos(\phi(s) - \phi)$$

$$y'(s) = -\sqrt{\frac{\varepsilon}{\beta(s)}}\,[\sin(\phi(s)-\phi)+\alpha(s)\cos(\phi(s)-\phi)]$$

in terms of the principal trajectories

$$y(s) = y_0 C(s) + y_0' S(s)$$

Using simple algebraic manipulations, we find that

$$C(s) = \sqrt{\frac{\beta(s)}{\beta_0}}\,(\cos\phi(s)+\alpha_0\sin\phi(s))\ ,\quad S(s) = \sqrt{\beta(s)\beta_0}\,\sin\phi(s)$$

and correspondingly for the reverse formula

$$\phi(s) = \mathrm{arctg}\frac{S(s)}{\beta_0 S(s)-\alpha_0 C(s)}$$

$$\beta(s) = \frac{1}{\beta_0}\left\{\frac{S^2(s)+[\beta_0 S(s)-\alpha_0 C(s)]^2}{\beta_0 S(s)-\alpha_0 C(s)}\right\}$$

or, in a simpler shape

$$\beta(s) = \frac{1}{\beta_0}\left[\frac{S(s)}{\sin\phi(s)}\right]^2$$

$$\alpha(s) = \frac{-S'(s)\sqrt{\frac{\beta(s)}{\beta_0}}+\cos\phi(s)}{\sin\phi(s)}$$

Having derived the above equations, we can now see that we can describe the evolution of the particle trajectories in a transfer line or in a circular accelerator by means of matrix formalism (see Fig. 2.10). In other words, linear transformations that are enabled by the linearity of the Hill's equations express as

$$\begin{pmatrix} y(s) \\ y'(s) \end{pmatrix} = \begin{pmatrix} C(s) & S(s) \\ C'(s) & S'(s) \end{pmatrix}\begin{pmatrix} y(s_0) \\ y'(s_0) \end{pmatrix} \tag{2.26}$$

The matrix elements $C(s)$ and $S(s)$ depend only on the magnetic lattice and not on the initial conditions of the particle.
The transfer matrix is therefore given by

$$\mathbf{M}_{1\to 2} = \begin{pmatrix} C(s) & S(s) \\ C'(s) & S'(s) \end{pmatrix} \tag{2.27}$$

The described approach allows for the possibility of using the matrix formalism to describe the evolution of the coordinates of a charged particle in a magnetic lattice.

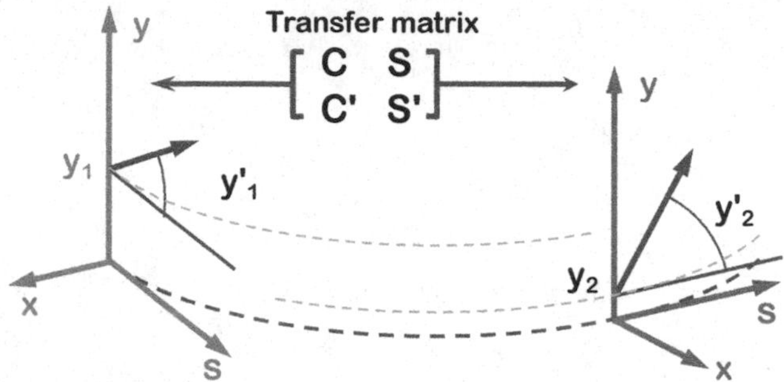

Figure 2.10 Linear matrix approach for evaluation of the evolution of the particle coordinates in a transfer line.

2.4.3 EXAMPLES OF TRANSFER MATRICES

The most common elements in accelerators are *drifts*, *bending magnets*, which steer the trajectory, *quadrupoles*, which provide transverse focusing, and the *accelerating section*, which accelerates the beam. Each of these elements can be represented by a particular transfer matrix. Let's consider here a couple of the simplest examples. The transfer matrix of a *drift space* is simply

$$\mathbf{M} = \begin{pmatrix} 1 & d \\ 0 & 1 \end{pmatrix} \tag{2.28}$$

Bending magnets, especially in large accelerators where each individual bend provides only a tiny bit of bending, have transfer matrices very close to those of drift space except when *edge focusing* (illustrated in Fig. 2.11) needs to be taken into account. In the latter, the faces of the bending magnet are not perpendicular to the reference trajectory. As clearly seen from the figure, this can provide additional defocusing on the horizontal plane[1] due to edge effects.

The matrix of a focusing quadrupole is

$$\mathbf{M} = \begin{pmatrix} \cos(\sqrt{|K|}L) & \frac{1}{\sqrt{|K|}}\sin(\sqrt{|K|}L) \\ -\sqrt{|K|}\sin(\sqrt{|K|}L) & \cos(\sqrt{|K|}L) \end{pmatrix} \tag{2.29}$$

and the matrix of a defocusing quadrupole is

$$\mathbf{M} = \begin{pmatrix} \cosh(\sqrt{K}L) & \frac{1}{\sqrt{K}}\sinh(\sqrt{K}L) \\ \sqrt{K}\sinh(\sqrt{K}L) & \cosh(\sqrt{K}L) \end{pmatrix} \tag{2.30}$$

For a thin lens, $L \to 0$ with KL staying finite, these matrices correspondingly become

$$\mathbf{M}_F = \begin{pmatrix} 1 & 0 \\ -|K|L & 1 \end{pmatrix} \quad \text{and} \quad \mathbf{M}_D = \begin{pmatrix} 1 & 0 \\ KL & 1 \end{pmatrix} \tag{2.31}$$

[1] Vertical plane edge focusing can also occur but only in the case of a finite vertical gap of the bending magnet.

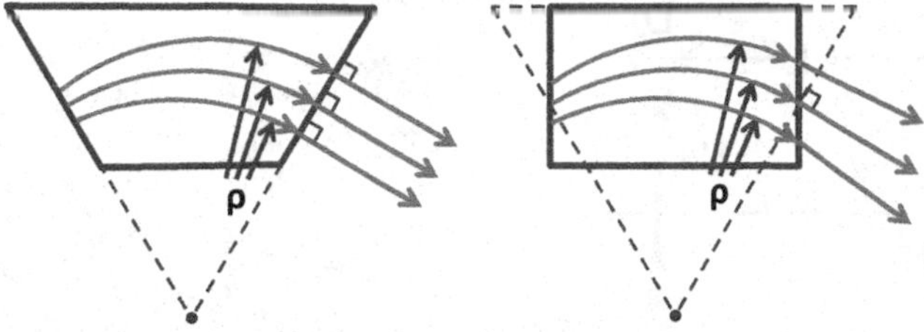

Figure 2.11 Mechanism of the edge focusing of bending magnet in a horizontal plane.

The higher-order elements such as sextupole and octupole magnets are often used in accelerators for nonlinear corrections. In linear approximation their transfer matrices are equivalent to the corresponding matrix of a drift space. They will have, however, higher-order terms and can be described by higher-order matrices (as will be discussed later in this chapter).

2.4.4 MATRIX FORMALISM FOR TRANSFER LINES

The matrix formalism is very practical for computing propagation through transfer lines, especially since the transfer matrix of each individual element of the beamline needs to be calculated only once. In a purely linear transfer matrix, the overall linear matrix of the beamline, computed as a step-by-step matrix multiplication of all individual elements, would satisfactorily describe the propagation of particles through the beamline.

As a practical example, let's consider a pair of thin quadrupoles of different polarity separated by a drift space as shown in Fig. 2.12. The overall horizontal transfer matrix of such a system is given by

$$\mathbf{M}_x^{1\to 2} = \begin{pmatrix} 1 & 0 \\ \frac{1}{f_2} & 1 \end{pmatrix} \begin{pmatrix} 1 & L \\ 0 & 1 \end{pmatrix} \begin{pmatrix} 1 & 0 \\ -\frac{1}{f_1} & 1 \end{pmatrix} = \begin{pmatrix} 1-\frac{L}{f_1} & L \\ -\frac{1}{f^*} & 1+\frac{L}{f_2} \end{pmatrix}$$

$$\text{where} \quad \frac{1}{f*} = \frac{1}{f_1} - \frac{1}{f_2} + \frac{L}{f_1 f_2} \tag{2.32}$$

The quantity f^* can be considered as an effective focal distance of a system of two lenses — such a system is usually called a focusing doublet. The overall vertical transfer matrix $\mathbf{M}_y^{1\to 2}$ is obtained by reversing the signs of f_1 and f_2. By referring to Eq. 2.32, we can see that there is a region of parameters where the sign of f^* is the same and positive for both the horizontal and vertical planes (for example, when $f_1 = f_2$), which corresponds to focusing in both planes.

2.4.5 ANALOGY WITH GEOMETRIC OPTICS

As we can see, the particle trajectories can be described using matrix formalism in a very similar way to ray propagation in an optical system. The magnetic quadrupoles

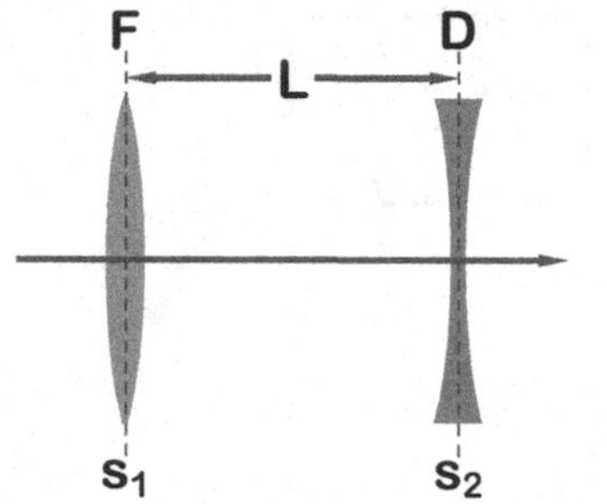

Figure 2.12 For illustration of transfer matrix of two quadrupoles separated by a drift.

A doublet consists of a focusing and a defocusing quadrupole. There is a region of parameters where the sign of its effective focal distance is the same and positive for both the horizontal and vertical directions, which corresponds to focusing in both planes.

play the role of focusing and defocusing lenses but, unlike an optical lens, a magnetic quadrupole focuses in one plane and defocuses in the other plane due to the nature of Maxwell's equations. However, as we have just shown in the previous section, a doublet of quadrupoles can focus simultaneously in both planes, as illustrated in Fig. 2.13.

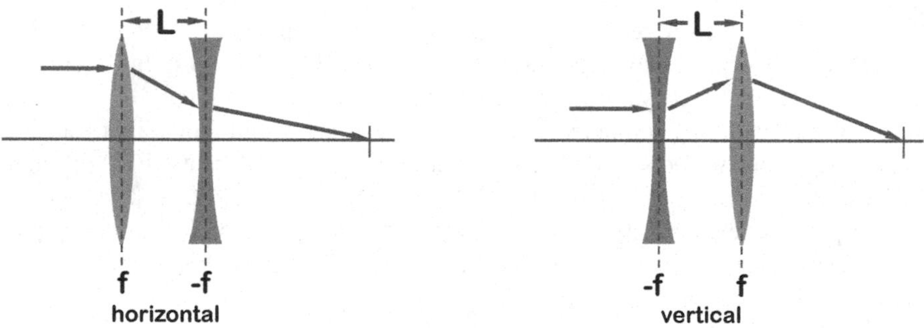

Figure 2.13 A doublet can focus simultaneously in both planes.

The similarity and difference with geometrical optics can be highlighted by the following example.

An optical telescope with two lenses can provide arbitrary demagnification.

Let's consider an optical telescope consisting of two lenses as shown in Fig. 2.14. It is intuitively clear and in fact can be proven that two lenses (in linear approximation), properly spaced and located, can provide an arbitrary *demagnification*.

In a direct analogy to geometrical optics, two focusing doublets are needed in order to create a telescope with arbitrary demagnification in the case of magnetic element optics (i.e., four quadrupoles appropriately placed and spaced).

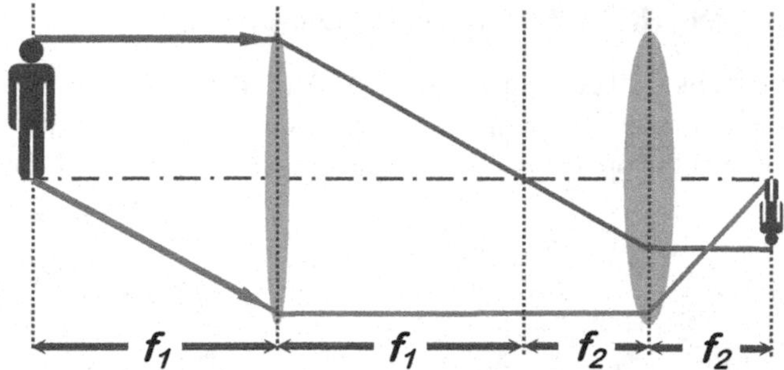

Figure 2.14 Optical telescope with two lenses.

The comparison of geometrical optics to magnetic element optics is a powerful method that often helps for back-of-the-envelope evaluations of various optical systems.

Four quadrupoles are needed to create a telescope with arbitrary demagnification for charged particle optics.

2.4.6 AN EXAMPLE OF A FODO LATTICE

Let's consider one more practical example — an alternating sequence of focusing (F) and defocusing (D) quadrupoles separated by a drift (O) — this is a so-called FODO lattice; see Fig. 2.15.

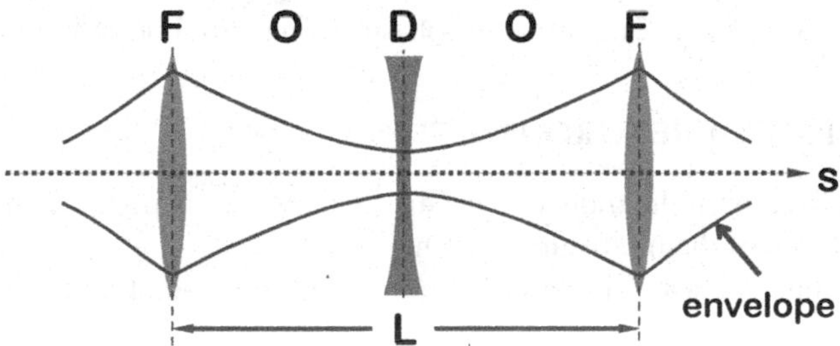

Figure 2.15 FODO lattice.

The transfer matrix of the FODO cell can be derived as

$$\mathbf{M} = \begin{pmatrix} 1 & 0 \\ -\frac{1}{f} & 1 \end{pmatrix} \begin{pmatrix} 1 & \frac{L}{2} \\ 0 & 1 \end{pmatrix} \begin{pmatrix} 1 & 0 \\ \frac{1}{f} & 1 \end{pmatrix} \begin{pmatrix} 1 & \frac{L}{2} \\ 0 & 1 \end{pmatrix} = \begin{pmatrix} 1+\frac{L}{2f} & L\left(1+\frac{L}{4f}\right) \\ -\frac{L}{2f^2} & 1-\frac{L}{2f}-\frac{L^2}{4f^2} \end{pmatrix}$$

We will use this expression in the following section to evaluate beam stability.

2.4.7 TWISS FUNCTIONS AND MATRIX FORMALISM

The optical functions — beta, alpha and gamma (defined below) are called Twiss functions. Using the formulas defined in the previous sections, the matrix elements can be expressed via the optics functions at the beginning and end of the beamline:

$$\mathbf{M}_{s_0\to s} = \begin{pmatrix} C(s) & S(s) \\ -C'(s) & S'(s) \end{pmatrix} = \tag{2.33}$$

$$= \begin{pmatrix} \sqrt{\frac{\beta(s)}{\beta_0}}(\cos\Delta\phi + \alpha_0\sin\Delta\phi) & \sqrt{\beta(s)\beta_0}\sin\Delta\phi \\ -\frac{(\alpha(s)-\alpha_0)\cos\Delta\phi+(1+\alpha(s)\alpha_0)\sin\Delta\phi}{\sqrt{\beta(s)\beta_0}} & \sqrt{\frac{\beta_0}{\beta(s)}}[\cos\Delta\phi - \alpha(s)\sin\Delta\phi] \end{pmatrix}$$

Here β_0, α_0 and the phase ϕ_0 (in $\Delta\phi = \phi - \phi_0$) correspond to the beginning of the transfer line. The expressions above are called *Twiss parameterization* of the transfer matrices.

So far, we haven't yet assumed any periodicity in the transfer line. However, if we now consider a periodic machine, then the transfer matrix over a single turn (single turn map) would reduce to

$$\mathbf{M}_{s_0\to s_0} = \begin{pmatrix} \cos\mu + \alpha_0\sin\mu & \beta_0\sin\mu \\ -\gamma_0\sin\mu & \cos\mu - \alpha_0\sin\mu \end{pmatrix} \tag{2.34}$$

where the gamma function is defined as

$$\gamma_0 = \frac{1+\alpha_0^2}{\beta_0} \tag{2.35}$$

and where we used $\mu = \Delta\phi$ to define the phase advance for one turn.

2.4.8 STABILITY OF BETATRON MOTION

Having considered periodic transfer maps in the previous section, we are now ready to discuss stability of the multi-turn motion.

Consider a circular accelerator with a transfer matrix, which for one turn equals to **M**. Let's rewrite the Twiss parameterization for **M** given by Eg.2.34 as

$$\mathbf{M} = \cos\mu\cdot\mathbf{I} + \sin\mu\cdot\mathbf{J} \tag{2.36}$$

$$\text{where } \mathbf{I} = \begin{pmatrix} 1 & 0 \\ 0 & 1 \end{pmatrix} \quad \text{and} \quad \mathbf{J} = \begin{pmatrix} \alpha_0 & \beta_0 \\ -\gamma_0 & -\alpha_0 \end{pmatrix}$$

After n turns, the particle coordinates will be given by the successive application of the one-turn transformation matrix n times, as follows:

$$\mathbf{x}_1 = \mathbf{M}\mathbf{x}_0 \quad \dots \quad \mathbf{x}_2 = \mathbf{M}^2\mathbf{x}_0 \quad \dots \quad \mathbf{x}_n = \mathbf{M}^n\mathbf{x}_0$$

The beauty of the parametrization given in Eq. 2.36 is that, as can be easily proven, $\mathbf{J}^2 = -\mathbf{I}$ and thus $\mathbf{M}^2 = \cos 2\mu \cdot \mathbf{I} + \sin 2\mu \cdot \mathbf{J}$. Similarly, one can show that

$$\mathbf{M}^n = \begin{pmatrix} \cos n\mu + \alpha_0 \sin n\mu & \beta_0 \sin n\mu \\ -\gamma_0 \sin n\mu & \cos n\mu - \alpha_0 \sin n\mu \end{pmatrix} \tag{2.37}$$

Observing the expression for this periodic transverse map, one can conclude that stability of the transverse motion necessarily requires the phase advance μ to be a real number, which ensures that the multi-turn motion represents stable oscillations. The condition of μ being real can be re-written as $|\cos\mu| < 1$ or as a more general expression involving the *trace* or *spur* (sum of its diagonal elements) of the transfer matrix

$$|\cos\mu| = \frac{1}{2}|tr\mathbf{M}| < 1 \tag{2.38}$$

The criterion defined above is a necessary condition for a transfer line to be suitable for multi-turn stable dynamics. We are now ready to apply this criterion to a practical example.

2.4.9 STABILITY OF A FODO LATTICE

Let's apply the stability criteria expressed as Eq. 2.38 to the Twiss parameterization of the matrix or the FODO cell derived in Section 2.4.6.

$$\mathbf{M} = \begin{pmatrix} 1 + \frac{L}{2f} & L\left(1 + \frac{L}{4f}\right) \\ -\frac{L}{2f^2} & 1 - \frac{L}{2f} - \frac{L^2}{4f^2} \end{pmatrix} = \begin{pmatrix} \cos\mu + \alpha\sin\mu & \beta\sin\mu \\ -\gamma\sin\mu & \cos\mu - \alpha\sin\mu \end{pmatrix}$$

The trace of this transfer matrix is given by

$$tr\mathbf{M} = 2 - \frac{L^2}{4f^2}$$

And the stability criterion thus requires

$$|\cos\mu| = \left|1 - \frac{L^2}{8f^2}\right| < 1 \quad \text{or} \quad f > \frac{L}{4} \tag{2.39}$$

The resulting criterion $f > L/4$ is intuitively very clear, and it can also be understood from the analogy with geometric optics and from considerations of the behavior of the beam envelope. Looking at Fig. 2.15, one can observe that $f = L/4$ would correspond to the situation when the size of the envelope in the defocusing quadrupole approaches zero, and even stronger quadrupoles (i.e. lower f) would make it impossible to sketch a repeatable finite envelope.

2.4.10 PROPAGATION OF OPTICS FUNCTIONS

As we have discussed above, the coordinates of the particles can be propagated via a transfer line using the matrices of the transfer line defined by the principal trajectories

$$\mathbf{M}_{1\to 2} = \begin{pmatrix} C(s) & S(s) \\ C'(s) & S'(s) \end{pmatrix} \tag{2.40}$$

Similarly, we can write down the expression that computes the propagation of the optics function along the transfer lines using a matrix based on the principal trajectories

$$\begin{pmatrix} \beta \\ \alpha \\ \gamma \end{pmatrix} = \begin{pmatrix} C^2 & -2CS & S^2 \\ -CC' & CS'+SC' & -SS' \\ C'^2 & -2C'S' & S'^2 \end{pmatrix} \begin{pmatrix} \beta_0 \\ \alpha_0 \\ \gamma_0 \end{pmatrix} \tag{2.41}$$

The initial values of the optical functions in this equation are either determined by the periodicity conditions as in the case of a circular machine, or correspond to the initial values at the entrance of the system as in the case of a transfer line.

Let's consider two cases as examples. Drift space expresses as

$$\mathbf{M} = \begin{pmatrix} 1 & s \\ 0 & 1 \end{pmatrix} \quad \rightarrow \quad \begin{pmatrix} \beta \\ \alpha \\ \gamma \end{pmatrix} = \begin{pmatrix} 1 & -2s & s^2 \\ 0 & 1 & -s \\ 0 & 0 & 1 \end{pmatrix} \begin{pmatrix} \beta_0 \\ \alpha_0 \\ \gamma_0 \end{pmatrix}$$

We can see that the β function has a parabolic behavior in correlation to the drift length.

A thin focusing quadrupole of focal length $f = 1/KL$

$$\mathbf{M} = \begin{pmatrix} 1 & 0 \\ KL & 1 \end{pmatrix} \quad \rightarrow \quad \begin{pmatrix} \beta \\ \alpha \\ \gamma \end{pmatrix} = \begin{pmatrix} 1 & 0 & 0 \\ KL & 1 & 0 \\ (KL)^2 & 2KL & 1 \end{pmatrix} \begin{pmatrix} \beta_0 \\ \alpha_0 \\ \gamma_0 \end{pmatrix}$$

In this case, it is the γ function that has a parabolic behavior in correlation to the inverse focal length of the quadrupole.

2.5 PHASE SPACE

2.5.1 PHASE SPACE ELLIPSE AND COURANT–SNYDER INVARIANT

To summarize our discussion about betatron motion, let's discuss the evolution of a phase space ellipse.

The solutions of the Hill's equation

$$\frac{d^2y}{ds^2} + K_y(s)y = 0 \tag{2.42}$$

discussed in Section 2.4.1 is reproduced below

$$y(s) = \sqrt{\varepsilon\beta(s)}\cos\left(\phi(s) - \phi\right) \tag{2.43}$$

$$y'(s) = -\sqrt{\frac{\varepsilon}{\beta(s)}}\left[\sin(\phi(s)-\phi)+\alpha(s)\cos(\phi(s)-\phi)\right]$$

This solution describes an evolution of an ellipse in phase space (y,y'). The parameters of the ellipse are described in Fig. 2.16.

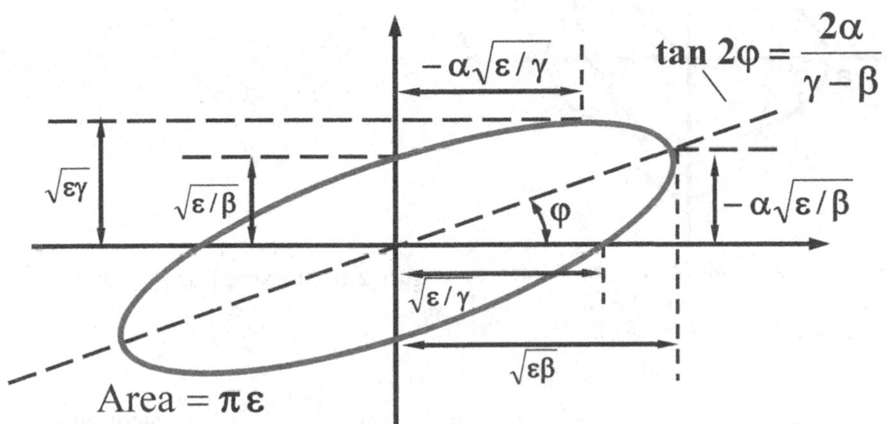

Figure 2.16 Betatron motion in phase space.

Hill's equations have a remarkable property — they have an invariant:

$$A(s) = \beta\, y'^2 + 2\alpha y y' + \gamma y^2 = const. = \varepsilon \tag{2.44}$$

This can be proven by substituting the solutions of Hill's equations Eq. 2.43 into Eq. 2.44 for $A(s)$.

The quantity $A(s)$ is called the *Courant–Snyder invariant* and is connected to the area of the ellipse in phase space with a factor of π as $Area = \pi\varepsilon$.

The Courant–Snyder invariant — and thus the area of the ellipse — stay constant independent of the optics of the beamline. As illustrated in Fig. 2.17, the ellipse rotates and its shape may change while its area remains invariant.

2.6 DISPERSION AND TUNES

2.6.1 DISPERSION

We have so far assumed that the particles of the beam have a nominal energy equal to that of the energy of the reference particle. In practice, however, there is always some energy offset or energy spread within the beam. The function that characterizes the orbit of an off-energy particle in an accelerator is called the *dispersion function*.

The primary effect of the energy offset is the difference of the trajectory in bending magnets — as illustrated in Fig. 2.18. After the propagation of an off-energy particle in a magnet, both an offset and angle of the orbit are created. In linear

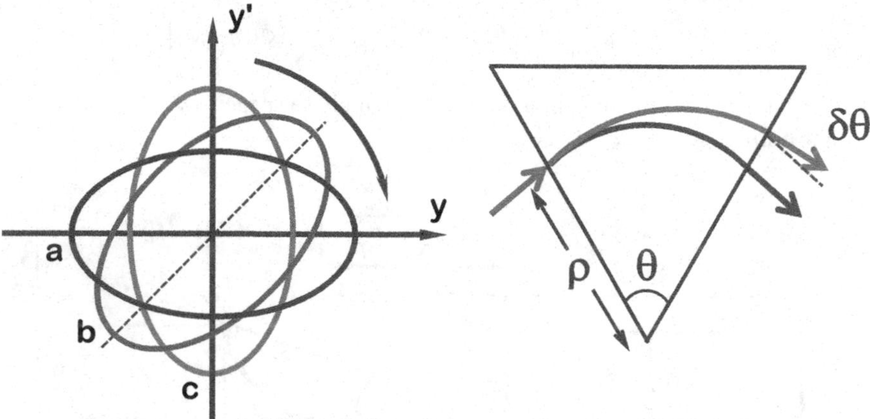

Figure 2.17 Evolution of phase-space ellipse. Locations: (a) in D and (c) in F quadrupoles, and (b) in between.

Figure 2.18 Bending magnet creates dispersion.

approximation, the radius of curvature of the trajectory for an off-energy particle expresses as

$$\frac{1}{\rho(s)} = \frac{eB}{p} = \frac{eB}{p_0\left(1+\frac{\Delta p}{p_0}\right)} \approx \frac{1}{\rho_0(s)}\left(1-\frac{\Delta p}{p_0}\right) \tag{2.45}$$

Substituting this to Eq. 2.19 gives us the Hill's equation for an off-energy particle:

$$\frac{d^2x}{ds^2} - \left(K(s) - \frac{1}{\rho^2}(s)\right)x = \frac{1}{\rho(s)}\frac{\Delta p}{p_0} \tag{2.46}$$

The answer to the above equation can be found using the following form:

$$x = x_0 + \frac{dp}{p}D \tag{2.47}$$

where D is the dispersion function and x_0 describes betatron oscillation around the dispersive orbit as shown in Fig. 2.19.

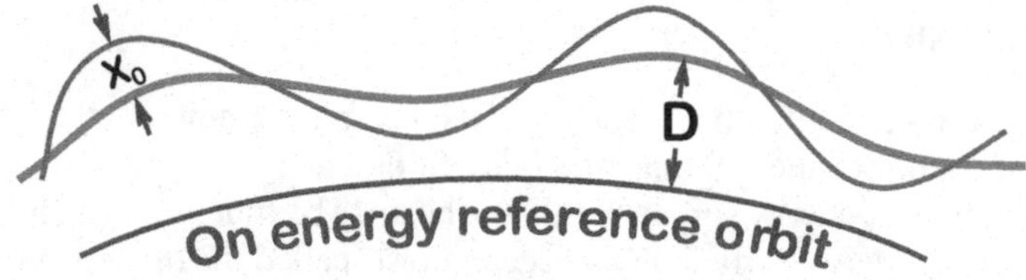

Figure 2.19 Dispersion.

Substituting this to Eq. 2.46 will yield the following equation, which governs the evolution of the dispersion function:

$$\frac{d^2D}{ds^2} - \left(K(s) - \frac{1}{\rho^2(s)}\right) D = \frac{1}{\rho(s)} \tag{2.48}$$

Dispersion function D can also be expressed in terms of the principal trajectories as

$$D(s) = S(s) \int_{s_0}^{s} \frac{C(t)}{\rho(t)} dt - C(s) \int_{s_0}^{s} \frac{S(t)}{\rho(t)} dt \tag{2.49}$$

$$D'(s) = S'(s) \int_{s_0}^{s} \frac{C(t)}{\rho(t)} dt - C'(s) \int_{s_0}^{s} \frac{S(t)}{\rho(t)} dt \tag{2.50}$$

We have assumed in this section that all bending occurs in a horizontal plane and therefore only a horizontal dispersion function is nonzero. This may not be the case in instances of vertically bending magnets or coupling, discussed below.

As dispersion affects the space location of the reference orbit for off-energy particles, it thus also affects the orbit's path length. Defining C as circumference or orbit path length in curvilinear coordinates, the deviation of the path length can be shown to be given by

$$\Delta \mathrm{C} = \frac{dp}{p} \oint \frac{D(s)}{\rho(s)} ds \tag{2.51}$$

We can also compute the so-called *momentum compaction factor* with

$$\alpha_c = \frac{d\mathrm{C}/\mathrm{C}}{dp/p} = \frac{1}{\mathrm{C}} \oint \frac{D(s)}{\rho(s)} ds \tag{2.52}$$

which will be discussed further in Chapter 5, which deals with longitudinal dynamics.

2.6.2 BETATRON TUNES AND RESONANCES

Taking into account the definition of the betatron phase in Eq. 2.23, we can write the phase advance over one turn of a circular machine as

$$\Delta\phi_C = 2\pi Q = \oint \frac{ds}{\beta(s)} \tag{2.53}$$

The quantity Q is called the *betatron tune*. The betatron tunes are essential quantities used to analyze the stability of a circular accelerator. In particular, if Q is an integer number, *resonance* conditions occur, as a tiny disturbance at some place along the orbit would repeat for many turns, accumulating into a large disruption of the particle's motion.

The general equations for resonance conditions of the betatron tunes can be written as follows:

$$m\,Q_x + n\,Q_z = k \tag{2.54}$$

where m, n and k are integer numbers and where $|m| + |n|$ is called the order of the resonance. Resonances of the lowest orders are the most dangerous for the stability of the particle motion and thus must be carefully avoided by proper machine optics design.

2.7 ABERRATIONS AND COUPLING

2.7.1 CHROMATICITY

Energy offsets of the charged particles beams create the dispersion function and also result in different focusing strengths of the magnetic elements (see Fig. 2.20).

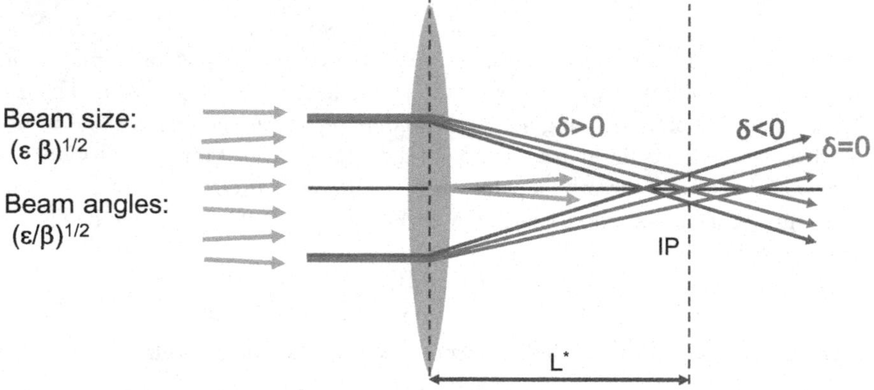

Figure 2.20 Chromaticity of a focusing quadrupole.

The quadrupole strength for off-energy particles in the first order can be approximated as follows

$$k_1 = \frac{e}{p}\frac{\partial B_y}{\partial x} = \frac{e}{p_0(1+\delta)}\frac{\partial B_y}{\partial x} = \frac{k_0}{1+\delta} \approx k_0(1-\delta) \tag{2.55}$$

Let's first consider a situation when a quadrupole is focusing almost parallel beam with emittance ε into a point (indicated as IP — Interaction Point in Fig. 2.20) located at the focal distance (defined as L^*). If the beam in front of the lens is exactly parallel, it will focus to a zero size in the focal point. Contribution to the nonzero size at the IP will consist of two components:

$$L^*(\varepsilon/\beta)^{1/2} \quad \text{and} \quad (\varepsilon\beta)^{1/2}\sigma_E$$

where the first term is given by the initial angular spread of the beam times the focal distance, and the second term is the chromatic contribution to the beam size, where σ_E is the beam energy spread. The ratio of these terms gives chromatic dilution of the beam size.

It is customary to express chromatic dilution via beta function at the IP defined as β^*. We can note that the betatron component of the beam size $L^*(\varepsilon/\beta)^{1/2}$ is equal to

$(\varepsilon\beta^*)^{1/2}$ by definition, which gives us $\beta^* = L^2/\beta$ and allows to express chromatic dilution of the beam size as

$$\frac{\Delta\sigma}{\sigma} \approx \sigma_E \frac{L^*}{\beta^*} \tag{2.56}$$

The ratio L^*/β^* is an estimate of chromaticity of a focusing beamline shown in Fig. 2.20, and it can be very large for final focusing systems in particular. We will discuss this further in the Chapter 10.

Chromatic effects of quadrupoles also affect dynamics in the circular accelerators, in particular their betatron tunes. Taking into account the equations that describe beta function evolution, we can show that the betatron tunes shift in correlation to changes in focusing strength:

$$\Delta Q = \frac{1}{4\pi} \oint \beta(s) \Delta K(s) ds \tag{2.57}$$

Using the above approximation for the off-energy quadrupole strength, we thus write the expression for *chromaticity* Q' as describing the dependence of the betatron tune on the energy offset of the particle and define it as the derivative of the betatron tunes with respect to the relative energy change:

$$Q' = \frac{dQ}{d\delta} = -\frac{1}{4\pi} \oint \beta(s) k_0(s) ds \tag{2.58}$$

Compensation of chromatic effects in focusing beamlines or in circular accelerators can be performed by sextupoles placed in dispersive regions — we will discuss this further below.

2.7.2 COUPLING

Throughout this chapter we have assumed that the motions of particles in horizontal and vertical planes are independent. This could indeed be the case if machine optics consist of bending magnets and quadrupoles that are perfectly placed in space. However, any rotational misalignments of these elements can create *coupling* of the horizontal and vertical motions.

Coupling can also be created by other magnetic elements such as solenoids (especially strong coupling can occur when a solenoid overlaps with quadrupole field, mixing different types of symmetry), or by misaligned nonlinear magnets such as sextupoles or octupoles, etc.

Some amount of coupling is unavoidable in a real machine and it usually needs to be corrected. A standard way to correct coupling (or to create it on purpose if needed) is to use skew quadrupoles — these are standard quadrupoles rotated by 45° as shown in Fig. 2.21.

2.7.3 HIGHER ORDERS

In storage rings, chromaticity is defined as a dependence of the betatron tunes on energy.

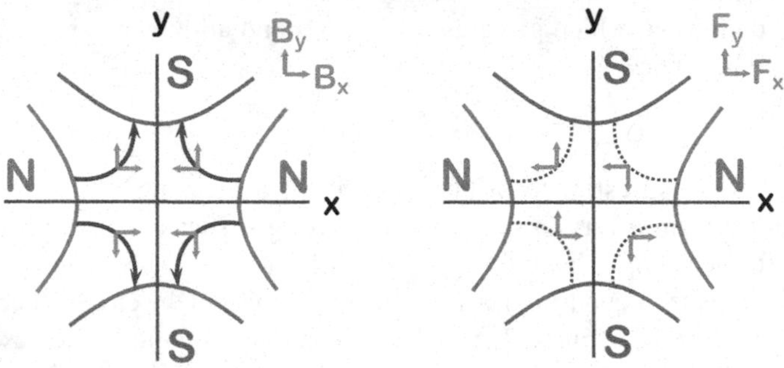

Figure 2.21 Skew quad fields and forces.

In single-path beamlines, it is more convenient to use other definitions. Let us first recall the linear matrix approach

$$x_i^{out} = R_{ij}\ x_j^{in} \tag{2.59}$$

this time noting all six components of the vector of interest, adding to the two coordinates and their angles the longitudinal offset Δl as well as the energy offset δ

$$x_i = \left(x, x', y, y', \Delta l, \delta\right)' \tag{2.60}$$

The second, third and other higher terms that can result from nonlinear elements such as a sextupole shown in Fig. 2.22 or octupole (Fig. 2.23) can be included in the matrix formalism in a similar manner:

$$x_i^{out} = R_{ij}\ x_j^{in}\ + T_{ijk}\, x_j^{in}\, x_k^{in} + U_{ijkn}\, x_j^{in}\, x_k^{in}\, x_n^{in} + \ldots \tag{2.61}$$

where T and U are the second- and third-order matrices.

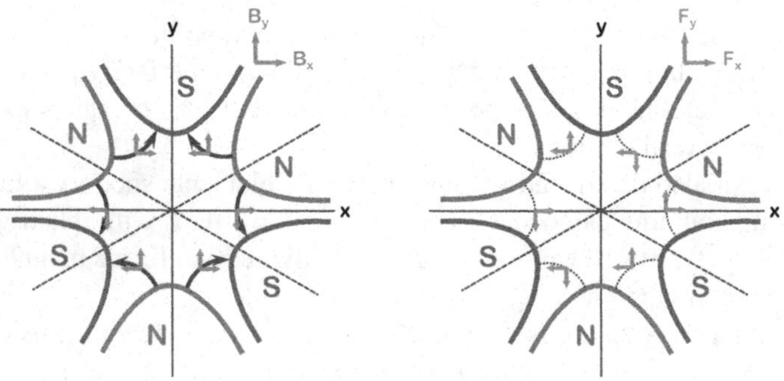

Figure 2.22 Sextupole fields (left) and forces (right).

Unlike in the storage rings, in transfer line design, one usually calls chromaticity the second-order elements T_{126} and T_{346}. All other high-order terms are just *aberrations*, purely chromatic (like T_{166}, which is second-order dispersion), or chromo-geometric (like U_{32446}).

For transport lines, such as final focus systems in particular, it is useful to define the chromaticity in terms of the so-called W functions.

Let's assume that betatron motion without energy offset is described by Twiss functions α_1 and β_1 and with the energy offset δ by functions α_2 and β_2.

Let's define chromatic function W (for each plane, x and y) as $W = (i \cdot A + B)/2$ where $i = \sqrt{-1}$, and where

$$B = \frac{\beta_2 - \beta_1}{\delta(\beta_2 \cdot \beta_1)^{1/2}} \approx \frac{\Delta\beta}{\delta\beta}$$

and

$$A = \frac{\alpha_2\beta_1 - \alpha_1\beta_2}{\delta(\beta_2 \cdot \beta_1)^{1/2}} \approx \frac{\Delta\alpha}{\delta} - \frac{\alpha}{\beta}\frac{\Delta\beta}{\delta}$$

Using familiar formulae $d\beta/ds = -2\alpha$ and $d\alpha/ds = K \cdot \beta - (1+\alpha^2)/\beta$ where $K = e/(pc) \cdot dB_y/dx$ and introducing

$$\Delta K = \frac{K(\delta) - K(0)}{\delta} \approx -K$$

we obtain the equation for W function evolution

$$\frac{dW}{ds} = \frac{2i}{\beta}W + \frac{i}{2}\beta\Delta K \tag{2.62}$$

Knowing that the betatron phase is $d\Phi/ds = 1/\beta$, we can see that if $\Delta K = 0$ then the complex vector W rotates with double betatron frequency and stays constant in amplitude. In quadrupoles and sextupoles, only the imaginary part of W will change.

From these equations on can show, in particular, that if in the final defocusing lens $\alpha = 0$, then it gives $\Delta W = L^*/(2\beta^*)$, where L^* is the distance from the final lens to the focus, and β^* is the beta function in the focus (i.e., at the IP — interaction point). One can also show that if T_{346} mentioned above is zeroed at the IP, the W function is also zeroed, illustrating the equivalency of different definition of chromaticity.

2.8 TAIL FOLDING OCTUPOLES – INVENTION CASE STUDY

When nonlinear magnetic elements such as sextupoles are used for chromatic corrections, they impact the entire beam, the core and its tails. However, very high order nonlinear elements will have very little impact on the beam core, and will affect only beam particles at the large amplitudes. This principle can allow manipulation of beam tails predominantly, without affecting the core of the beam.

An octupole is a magnet with eight poles (see Fig. 2.23), which will focus in x and y transverse planes and defocus on diagonals, with the focusing/defocusing force proportional to the cube of the offset x^3.

However, just like the quadrupole doublet (see Fig. 2.13) in a FODO, which can focus in both directions, two octupoles of different sign separated by a drift space can also provide focusing in all directions[2] for a nearly parallel beam. In this case, the focusing force provided by such octupole doublet will be proportional to the fifth power of the radial offset from the beam center r^5.

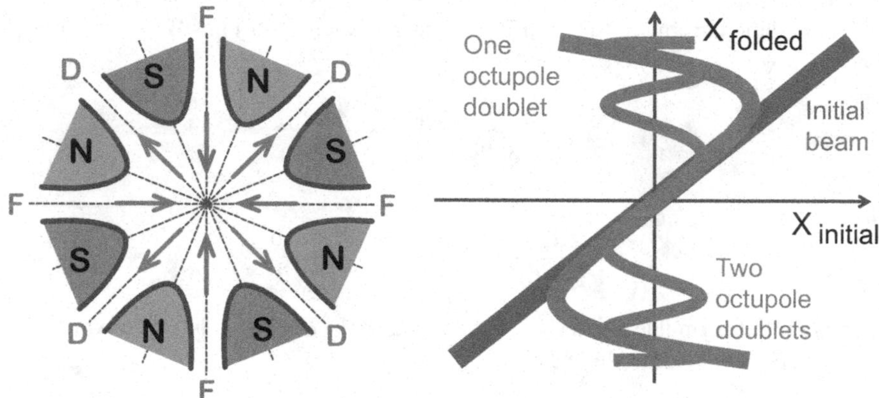

Figure 2.23 Forces in an octupole magnet. **Figure 2.24** Octupole doublet tail folding.

This feature of octupole doublets can be used for nonlinear *tail folding* in various beamlines, in particular in final focusing systems (see also Chapter 10). An illustration of such tail folding, with one or two octupole doublets, is shown in Fig. 2.24.

The octupole doublet, when one magnet cancels another and the effect comes from the next order, reminds us the inventive principle of *system and anti-system*, as well as the *nested doll* inventive principle — the combination of these two inventive principles can be observed in many inventions in technical areas.

[2]R. Brinkmann, P. Raimondi and A. Seryi, In Proc. of PAC 2001, FPAH066, (2001).

EXERCISES

2.1. *Chapter materials review.*
Define the region of parameters where a pair of thin quadrupoles will focus the beam in both planes.

2.2. *Chapter materials review.*
Prove Eq. 2.39, which defines the stability of a FODO beamline, geometrically, using the analogy of traditional geometrical optics.

2.3. *Chapter materials review.*
A parallel proton beam of E=200 MeV enters a beamline. It is necessary to focus this beam into a point at a 3 m distance from the entrance. Estimate the necessary parameters of a quadrupole system (gradients, lengths) that can perform this task.

2.4. *Mini-project.*
Consider the same proton beam as in the previous exercise, as well as the same focusing requirements. Assume that the focusing is performed by a continuous, cylindrical electron beam. Estimate the necessary electron density which can perform the focusing task. Select electron beam energy, determine the electron current and discuss and select an optimal design of the electron beam system, as well as its feasibility.

2.5. *Mini-project.*
In 1954, Enrico Fermi presented, in his lecture, a vision of an accelerator that would encircle the Earth. Design such an accelerator, assuming that it will be shaped like a polygon with N sides, that there will be N space stations launched around the Earth, located in the vertices of the polygon, and that each space station will carry one bending dipole magnet and one quadrupole magnet. Assume that, between the space stations, the accelerator's orbit will be straight, and the particle beam will propagate in the open-space vacuum (i.e., without any vacuum chambers). Determine, in particular, what will be the energy of the beam in this accelerator, for the case of an electron or proton beam.

2.6. *Analyze inventions or discoveries using TRIZ and AS-TRIZ.* Analyze and describe scientific or technical inventions described in this chapter in terms of the TRIZ and AS-TRIZ approaches, identifying a contradiction and an inventive principle that were used (could have been used) for these inventions.

2.7. *Developing AS-TRIZ parameters and inventive principles.* Based on what you already know about accelerator science, discuss and suggest the possible additional parameters for the AS-TRIZ contradiction matrix, as well as the possible additional AS-TRIZ inventive principles.

2.8. *Practice in reinventing technical systems.* Suggest a way to make an achromatic beamline consisting only of focusing and defocusing quadrupoles and drift spaces, without any bends or nonlinear elements.

2.9. *Practice in the art of back-of-the-envelope estimations.* A circular accelerator with a 600 m perimeter has a horizontal betatron tune equal to $Q_x = 5.173$. Estimate the average vertical betatron function in this accelerator. *(It is assumed that you can identify the most important effects playing roles in this task, can define the necessary parameters and set their values, and can get a numerical answer.)*

3 Synchrotron Radiation

In this chapter we will consider one of the most important phenomena that governs the behavior of accelerators — synchrotron radiation (SR).

SR can be both helpful, as it yields the creation of high brightness radiation sources, and harmful, as it can deteriorate the beam by creating additional energy spread and beam emittance growth.

Traditional derivations of SR equations are rather mathematically involved. However, in this chapter we will use simplified back-of-the-envelope style derivations, which nevertheless obtain all of the important characteristics of SR with high accuracy.

3.1 SR ON THE BACK OF AN ENVELOPE

In our simple picture, the SR is the result of the charged particle leaving part of its fields behind when it is moving on a curved trajectory. The part of the field that is left behind (or radiated) cannot catch up with the motion of the particle, as it cannot move faster than the speed of light.

Armed with this concept, let's estimate the power loss due to SR, the typical energy of the emitted photons, and other parameters of synchrotron radiation as well as the most important effects that SR inflicts on the beam.

3.1.1 SR POWER LOSS

The straightforward concept of SR described above is represented in Fig. 3.1. In this instance, the particle moving with velocity v (which is close to the speed of light) on a radius R has its field lines pointing mostly transversely, and the part of field moving further away (on the radius $R+r$) would be left behind, as it cannot move faster than c, and will be therefore radiated.

The radius r can be evaluated as

$$r = R\left(\frac{c}{v} - 1\right) \approx \frac{R}{2\gamma^2} \tag{3.1}$$

where we assumed that $\gamma \gg 1$ or $\beta \approx 1$ and thus $(1 - v/c) = (1-\beta)(1+\beta)/(1+\beta) \approx (1-\beta^2)/2 = 1/(2\gamma^2)$.

The energy in the field that is left behind (radiated) can be estimated (we use Gaussian units in this section) as a volume integral of the field squared:

$$W \approx \int E^2\, dV \tag{3.2}$$

The field E can be estimated as the field on the radius r from the particle

$$E \approx \frac{e}{r^2}$$

DOI: 10.1201/9781003326076-3

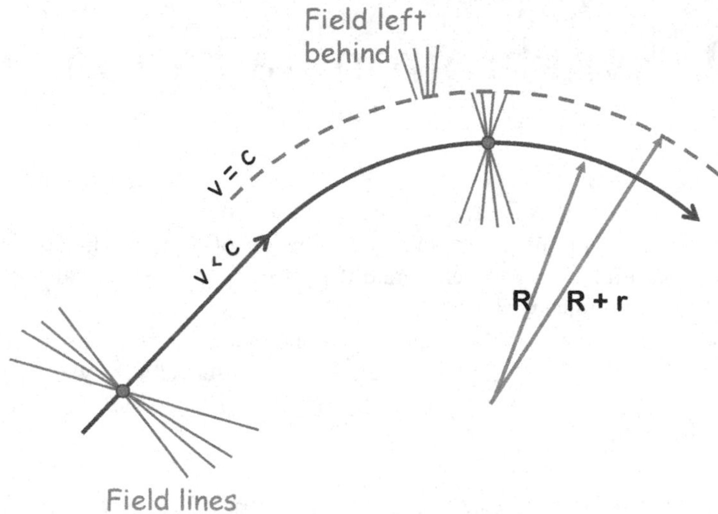

Figure 3.1 Synchrotron radiation — conceptual explanation.

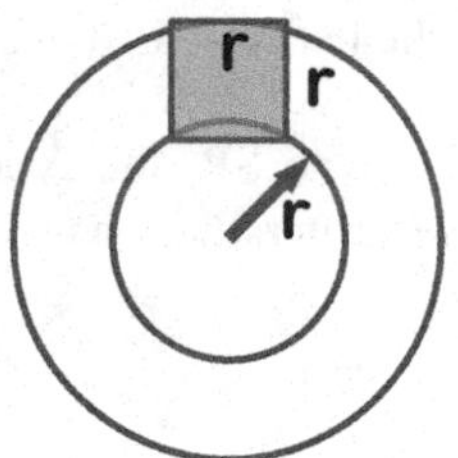

Figure 3.2 Illustrating the characteristic volume used in Eq. 3.3.

> Estimating the integral of the field over radiated volume, we ignored the dependence of the field across the considered area.

and the characteristic volume can be estimated (see Fig. 3.2) as

$$V \approx r^2 ds \tag{3.3}$$

where ds is an element of the path along the orbit, as we will eventually wish to find the power lost per unit of length.

The energy loss per unit length can thus be written as

$$\frac{dW}{ds} \approx E^2 r^2 \approx \left(\frac{e}{r^2}\right)^2 r^2 \tag{3.4}$$

and after substituting

$$r \approx \frac{R}{2\gamma^2}$$

we get an estimate of:

$$\frac{dW}{ds} \approx \frac{e^2\gamma^4}{R^2} \tag{3.5}$$

which we can compare with the exact formula:

$$\frac{dW}{ds} = \frac{2}{3}\frac{e^2\gamma^4}{R^2} \tag{3.6}$$

and conclude that our rough estimations give a very reasonable result.

3.1.2 COOLING TIME

Knowledge of the SR-caused power losses immediately allow us to estimate the SR cooling time of the beam in a storage ring.

First of all, let's — for convenience's sake — rewrite the exact formula for power lost per unit length

$$\frac{dW}{ds} = \frac{2}{3}\frac{e^2\gamma^4}{R^2}$$

in the way that does not depend on the systems of units:

$$\frac{dW}{ds} = \frac{2}{3}\frac{r_e\,\gamma^4}{R^2}mc^2 \tag{3.7}$$

The SR energy loss of a particle per turn is therefore:

$$U_0 = \frac{4\pi}{3}\frac{r_e\,\gamma^4}{R}mc^2 \tag{3.8}$$

The effect of particle cooling due to SR is based on the fact that when an electron radiates a photon, its momentum decreases. Taking into account that while the beam of particles can have a range of angles within the beam, the accelerating RF cavity would restore only the longitudinal part of momentum, whereas the transverse degrees of freedom of the particles will be cooled down as illustrated in Fig. 3.3.

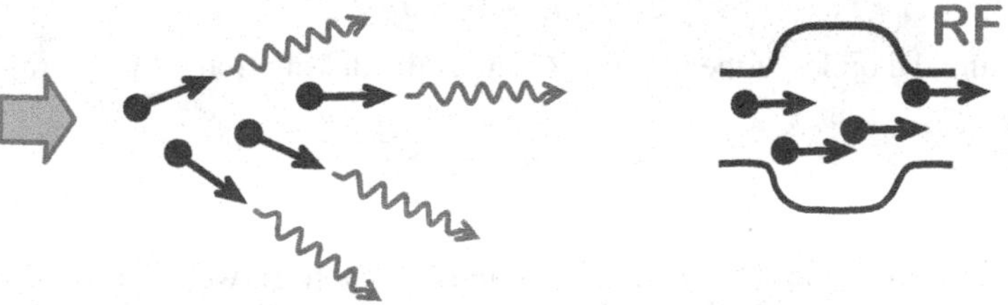

Figure 3.3 RF cavity restores only longitudinal momentum, thus other degrees of freedom are cooled due to synchrotron radiation.

Let's estimate the cooling time τ of a particle with energy E_0 cycling with revolution period T_0 in a circular machine as $\tau = E_0 T_0/U_0$, which yields

$$\tau \approx \frac{2\pi R}{c}\frac{\gamma mc^2}{U_0} \tag{3.9}$$

After substitution, the inverse cooling time can be written as

$$\tau^{-1} \approx \frac{2}{3} \frac{c r_e \gamma^3}{R^2} \tag{3.10}$$

3.1.3 COOLING TIME AND PARTITION

In the previous section we estimated the inverse cooling time as $\tau^{-1} \approx 2 c r_e \gamma^3 /(3R^2)$. Traditionally, there is a factor of 2 in the definition in the cooling time:

$$\tau = 2E_0 T_0 / U_0 \quad \Rightarrow \quad \tau^{-1} = \frac{1}{3} \frac{c r_e \gamma^3}{R^2} \tag{3.11}$$

We will use this latter definition in this section.

We can express the evolution of the beam emittance under the influence of an SR damping as

$$\varepsilon(t) = \varepsilon_0 \exp(-2t/\tau) \tag{3.12}$$

Both transverse planes, as well as the longitudinal motion in rings, are usually coupled. Thus we can expect that the damping will be distributed between these degrees of freedom in some proportion depending on details of the optics.

Distribution of cooling between the degrees of freedom is defined by the so-called partition numbers J_x, J_y and J_E, which we mention here without derivations. The cooling time of a degree of freedom is correspondingly

$$\tau_i = \frac{\tau}{J_i} \tag{3.13}$$

The total radiated power due to SR is fixed and constant, therefore

$$\sum \tau_i^{-1} = const. \tag{3.14}$$

which corresponds to the *partition theorem*

$$\sum J_i = 4 \tag{3.15}$$

for a typical accelerator

$$J_x \approx 1, \quad J_y \approx 1, \quad J_E \approx 2 \tag{3.16}$$

and adjusting the optics of the machine changes the distribution of the partition numbers.

3.1.4 SR PHOTON ENERGY

In order to estimate the typical energy[1] of the SR photons, we need to make an assumption that is based on relativistic kinematics: the radiation of relativistic particles is emitted into a cone with angular spread of $1/\gamma$.

Let's take this assumption into account when examining the radiation emitted during motion along the curved trajectory shown in Fig. 3.4 and ask a question — during what time interval Δt would the remote observer see the emitted fields?

---

[1] *For $\gamma \gg 1$ the emitted photons go into $1/\gamma$ cone.*

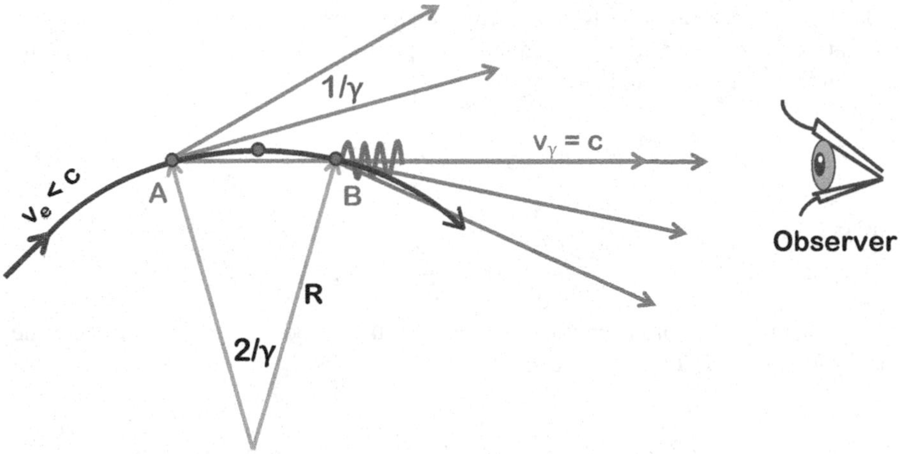

Figure 3.4 SR and remote observer.

Looking at this figure we conclude that the emitted radiation is observed during particle travel along the $2R/\gamma$ arc.

Let's now keep in mind that the radiation travels at speed c, while particles travel at v. At point B shown in Fig. 3.4 the separation between radiation and the particles is given by

$$ds \approx \frac{2R}{\gamma}\left(1 - \frac{v}{c}\right) \tag{3.17}$$

Therefore, the observer will see radiation during the following time interval:

$$\Delta t \approx \frac{ds}{c} \approx \frac{2R}{c\gamma}(1-\beta) \approx \frac{R}{c\gamma^3} \tag{3.18}$$

We can thus proceed to estimate the characteristic frequency of emitted photons as the inverse of the time duration of the flash, as seen by the observer:

$$\omega_c \approx \frac{1}{\Delta t} \approx \frac{c\gamma^3}{R} \tag{3.19}$$

Comparing the above with the exact formula that we reproduce here without derivation

$$\omega_c = \frac{3}{2}\frac{c\gamma^3}{R} \tag{3.20}$$

we can again see that our back-of-the-envelope estimations give pretty accurate results while not hiding the physics of the phenomena behind heavy math.

3.1.5 SR – NUMBER OF PHOTONS

Having estimated the characteristic energy of the SR photons, we can now estimate the number of photons emitted per unit length by a single electron.

Let's use our estimation for the rate of energy loss $dW/ds \approx e^2\gamma^4/R^2$ and the estimation of the characteristic frequency of photons $\omega_c \approx c\gamma^3/R$, and rewrite the latter in terms of the photon energy:

$$\varepsilon_c = \hbar\omega_c \approx \frac{\gamma^3 \hbar c}{R} = \frac{\gamma^3}{R}\lambda_e mc^2 \tag{3.21}$$

where

$$r_e = \frac{e^2}{mc^2} \qquad \alpha = \frac{e^2}{\hbar c} \qquad \lambda_e = \frac{r_e}{\alpha}$$

The number of photons emitted per unit length can be obtained by dividing the energy loss per unit length by the energy of the photons

$$\frac{dN}{ds} \approx \frac{1}{\varepsilon_c}\frac{dW}{ds} \approx \frac{\alpha\gamma}{R} \tag{3.22}$$

It is also practical to derive an expression for the number of photons emitted per unit of the bending angle θ

$$N \approx \alpha\gamma\theta \tag{3.23}$$

which is given by a remarkably simple and clear formula.

3.2 SR EFFECTS ON THE BEAM

The derived characteristics of SR allow evaluation of the effects of SR on the beam.

3.2.1 SR-INDUCED ENERGY SPREAD

The energy spread $\Delta E/E$ will grow due to statistical fluctuations ($\sqrt{N}$) of the number of emitted SR photons and therefore can be estimated as

$$\frac{d\left((\Delta E/E)^2\right)}{ds} \approx \varepsilon_c^2 \frac{dN}{ds}\frac{1}{(\gamma mc^2)^2} \tag{3.24}$$

which gives the following estimation

$$\frac{d\left((\Delta E/E)^2\right)}{ds} \approx \frac{r_e \lambda_e \gamma^5}{R^3} \tag{3.25}$$

Comparing this with the exact formula

$$\frac{d\left((\Delta E/E)^2\right)}{ds} = \frac{55}{24\sqrt{3}}\frac{r_e \lambda_e \gamma^5}{R^3} \tag{3.26}$$

confirms good accuracy of the estimation as the numerical factor $55/(24\sqrt{3}) \approx 1.32$.

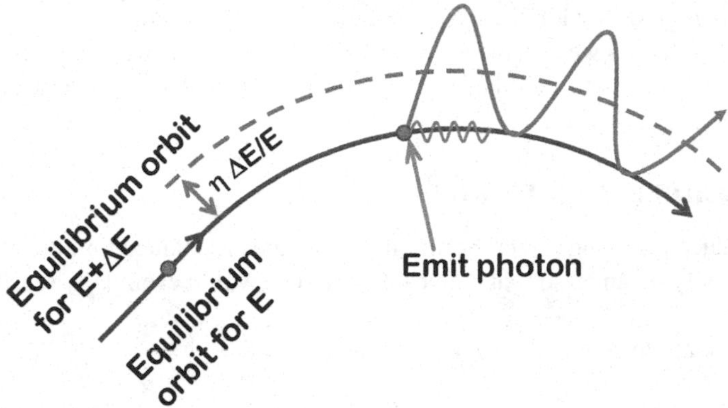

Figure 3.5 SR excites oscillation of particles and corresponding emittance growth.

3.2.2 SR-INDUCED EMITTANCE GROWTH

Let's estimate the beam emittance growth rate due to synchrotron radiation. The qualitative picture of the phenomenon is shown in Fig. 3.5.

In this diagram, the dispersion function η shows how the equilibrium orbit shifts when particle energy changes due to photon emission. Correspondingly, when a photon is emitted and the energy of the particle becomes equal to $E + \Delta E$ (where ΔE is negative), the particle starts to oscillate around a new equilibrium orbit. The amplitude of oscillation will be equal to

$$\Delta x \approx \eta \, \Delta E / E$$

Let's compare this with the betatron beam size given by

$$\sigma_x = (\varepsilon_x \, \beta_x)^{1/2}$$

and write an estimate for the emittance growth as

$$\Delta \varepsilon_x \approx \Delta x^2 / \beta$$

By expanding the equation, we obtain an estimation for the emittance growth:

$$\frac{d\varepsilon_x}{ds} \approx \frac{\eta^2}{\beta_x} \frac{d\left((\Delta E / E)^2 \right)}{ds} \approx \frac{\eta^2}{\beta_x} \frac{r_e \, \lambda_e \, \gamma^5}{R^3} \tag{3.27}$$

In the above estimation we ignored the dependence of β and η on s; however, these dependences can alter the results. The exact formula, which takes into account the derivatives of the Twiss functions, is as follows:

$$\frac{d\varepsilon_x}{ds} = \frac{\left(\eta^2 + \left(\beta_x \eta' - \beta_x' \eta / 2 \right)^2 \right)}{\beta_x} \frac{55}{24\sqrt{3}} \frac{r_e \, \lambda_e \, \gamma^5}{R^3} \tag{3.28}$$

$$\nwarrow \quad = \mathscr{H} \quad \nearrow$$

where the parenthesis with Twiss functions in front of the numerical coefficient is usually called $\mathscr{H}$. As we see, the back-of-the-envelope estimation correctly captures the most important features of the phenomenon and also produces a useful and simple expression.

3.2.3 EQUILIBRIUM EMITTANCE

The SR-induced cooling of the beam emittance and SR-induced emittance growth would naturally balance, so that the beam emittance would eventually reach equilibrium value.

Let's look at the estimated rate of emittance growth:

$$\frac{d\varepsilon_x}{ds} \approx \frac{\eta^2}{\beta_x} \frac{d\left(\left(\Delta E/E\right)^2\right)}{ds} \approx \frac{\eta^2}{\beta_x} \frac{r_e\, \lambda_e\, \gamma^5}{R^3}$$

and the SR cooling rate

$$\frac{d\varepsilon}{ds} = -\frac{2}{c\tau}\varepsilon \qquad \text{with} \qquad \tau^{-1} = \frac{1}{3}\frac{c r_e \gamma^3}{R^2}$$

and equate them to obtain an expression for the horizontal equilibrium emittance:

$$\varepsilon_{x0} \approx \frac{c\tau}{2}\frac{\eta^2}{\beta_x}\frac{r_e\, \lambda_e\, \gamma^5}{R^3} \tag{3.29}$$

or, after substitution:

$$\varepsilon_{x0} \approx \frac{3}{2}\frac{\eta^2}{\beta_x}\frac{\lambda_e\, \gamma^2}{R} \tag{3.30}$$

In the equations above, we ignored dependence of R on longitudinal coordinate s. In order to obtain more accurate formulas from these equations, one needs to use the values $\langle 1/R^2 \rangle$ and $\langle 1/R^3 \rangle$, which are averaged over the orbit period.

In the vertical plane, SR's contribution to emittance is only due to $1/\gamma$ angles of emitted photons, but usually the impact on highly relativistic beams is negligibly small.

The vertical equilibrium emittance is therefore usually defined not by SR directly, but by the coupling coefficient k (which is $\ll 1$) of x-y planes:

$$\varepsilon_{y0} \approx k\, \varepsilon_{x0}$$

In the above, we ignored partition numbers, but they can be taken into account in accurate calculations. The equilibrium energy spread of the beam can also be calculated in a similar manner.

3.3 SR FEATURES

Performance of SR-based light sources depends on spatial and spectral characteristics of synchrotron radiation. Below we will introduce basic relevant characteristics, leaving detailed discussion for Chapter 7.

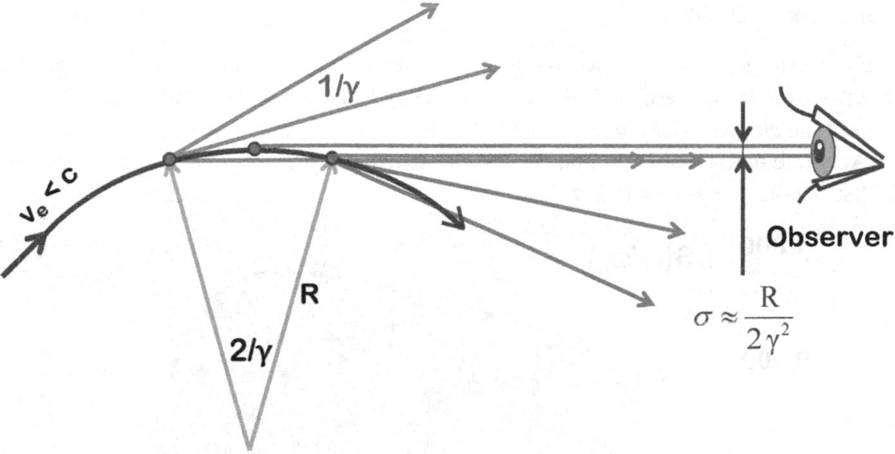

Figure 3.6 For illustration of emittance of a single photon.

3.3.1 EMITTANCE OF SINGLE RADIATED PHOTON

In order to discuss the ultimate brightness of the SR-based light sources, we require knowledge of the emittance of a single photon radiated from the curved beamline.

To answer this question we must first evaluate the size of the emitting region seen by the remote observer in the case of a single electron, as illustrated in Fig. 3.6. We first note that the angles of photons coming from the emitting region are spread over $\sigma' \approx 1/\gamma$. The size of the emitting region is given by the height of the arc segment and is thus equal to $\sigma \approx R/(2\gamma^2)$.

The estimate for the emittance of an SR photon emitted by a single electron can therefore be written as

$$\varepsilon_{ph} = \sigma\sigma' \quad \Rightarrow \quad \varepsilon_{ph} \approx \frac{R}{2\gamma^3} \tag{3.31}$$

(In a similar way as above, one can estimate beta function of photons as $\beta = \sigma/\sigma'$.)

Let's rewrite the photon emittance equation using the expression for photon wavelength

$$\omega_c = \frac{2\pi c}{\lambda_c} \approx \frac{c\gamma^3}{R} \quad \Rightarrow \quad \varepsilon_{ph} \approx \frac{\lambda_c}{4\pi} \tag{3.32}$$

We can see here that the emittance of synchrotron radiation is directly related to its wavelength. This is not a coincidence in a single example, but is actually an inherent property of photon radiation:

$$\varepsilon_{ph} = \frac{\lambda}{4\pi} \tag{3.33}$$

This, together with information about the SR spectrum in the next section, will bring us to discuss the brightness of SR light sources.

3.3.2 SR SPECTRUM

In all the estimations above, we assumed that the photons emitted are monoenergetic. It is not exactly the case, and in reality the energy of the photons will be distributed around the *characteristic frequency* of the SR photons ω_c.

Accurate mathematics, which we do not show here, predicts that the SR spectrum looks like the one shown in Fig. 3.7.

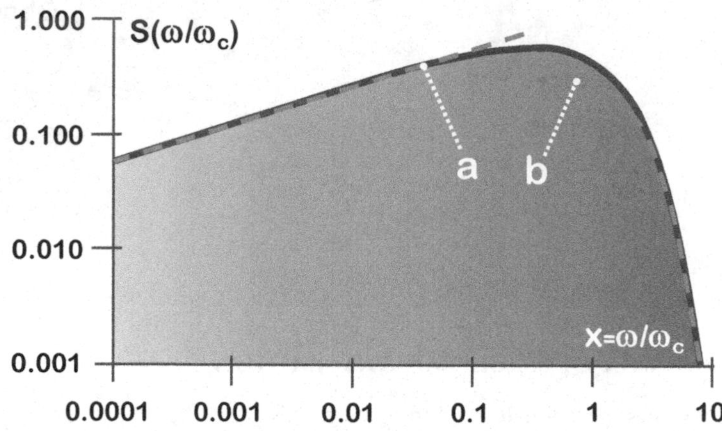

Figure 3.7 SR spectrum and its approximations for low (curve **a** — behaves as $4/3 \cdot x^{1/3}$) and high (curve **b** — behaves as $7/9 \cdot x^{1/2} e^{-x}$) energies.

We can indeed see that a large fraction of the photons will have energies close to ω_c. However, there is also a lower-energy tail, as well as some fraction of higher-energy photons.

It is also natural to expect that the photons' angular distribution will deviate from the $1/\gamma$ rule, and indeed, the lower-energy photons typically have larger angular spread.

3.3.3 BRIGHTNESS OR BRILLIANCE

Following discussion of the SR spectrum, we can introduce the notion of *bandwidth* — the interval of interest in the spectrum of photon frequencies. This bandwidth is denoted here BW and is expressed, typically, in %.

Let's assume that our photon beam is emitted from the area A_s and the emitted radiation has angular opening angles of $\Delta\Phi$ and $\Delta\psi$, as shown in Fig. 3.8.

The first concept to make note of is *flux*, which is expressed as the number of photons emitted per units of time and per unit of bandwidth:

$$\text{Flux} = \text{Photons}/\,(s \cdot BW) \tag{3.34}$$

The brilliance (or brightness) is then defined as flux per unit of emitting area and product of angles:

$$\text{Brilliance} = \text{Flux}/(A_s \cdot \Delta\Phi \cdot \Delta\psi)$$

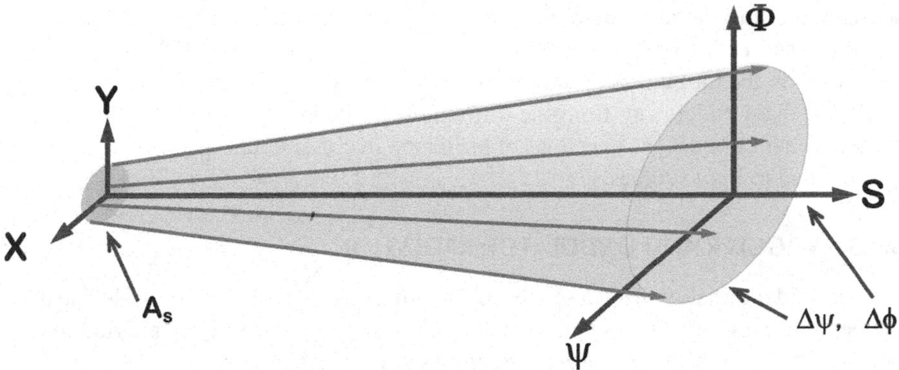

Figure 3.8 For illustration of brilliance or brightness. Here A_s is the emitting area and $\Delta\Psi$ and $\Delta\Phi$ are opening angles of emitted photons.

and is usually expressed in units of [Photons/$(s \cdot mm^2 \cdot mrad^2 \cdot BW)$].

In a typical case of Gaussian distributions, the definition of brilliance is based on the total effective sizes and divergences

$$\text{brilliance} = \frac{\text{flux}}{4\pi^2 \Sigma_x \Sigma_{x'} \Sigma_y \Sigma_{y'}} \tag{3.35}$$

where the total effective sizes include contributions from electrons as well as photons:

$$\Sigma_x = \sqrt{\sigma_{x,e}^2 + \sigma_{ph,e}^2} \qquad \sigma_x = \sqrt{\varepsilon_x \beta_x + (D_x \sigma_\varepsilon)^2} \tag{3.36}$$
$$\Sigma_{x'} = \sqrt{\sigma_{x',e}^2 + {\sigma'^2}_{ph,e}} \qquad \sigma_{x'} = \sqrt{\varepsilon_x / \beta_x + (D'_x \sigma_\varepsilon)^2}$$

and similarly for the other plane.

3.3.4 ULTIMATE BRIGHTNESS

As we have seen, brilliance is defined by the overall effective emittance, which convolves electron and photon distributions:

$$\varepsilon_{eff} = \sqrt{\sigma_e^2 + \sigma_{ph}^2}\ \sqrt{\sigma_{e'}^2 + \sigma_{ph'}^2} \tag{3.37}$$

Since the lowest photon emittance depends on the photon wavelength (Eq. 3.33), the smallest overall emittance will be obtained when:

$$\varepsilon_e = \sigma_e \sigma_{e'} \leq \varepsilon_{ph} \tag{3.38}$$

which corresponds to a *diffraction-limited* source.

In modern SR sources, the typical radiation of interest spans from 100 eV to 100 keV in terms of photon energy. Let's take an example of 12.4 keV of photon energy [2] which corresponds to $\varepsilon_{ph} \approx 8$ pm. For typical third-generation SR light

[2] *Wavelength* $\lambda \approx 1$ Å *corresponds to 12.4 keV photons.*

sources, the horizontal emittance ε_x is usually between 1 and 5 nm and the vertical emittance ε_y between 1 and 40 pm. Thus, such rings are close, in terms of its emittance, to the ultimate performance in the vertical plane. However, they are many orders of magnitude away from the diffraction-limited emittance in the horizontal plane. We will discuss the questions of brilliance in closer detail in Chapter 7, which is dedicated to light sources.

3.3.5 WIGGLER AND UNDULATOR RADIATION

Let's consider radiation from a sequence of bends, and in particular, let's assume that the bends are arranged in a sequence with $+ - + - +-$ polarity with a period of λ_u, so that the trajectory of the particle *wiggles* as shown in Fig. 3.9.

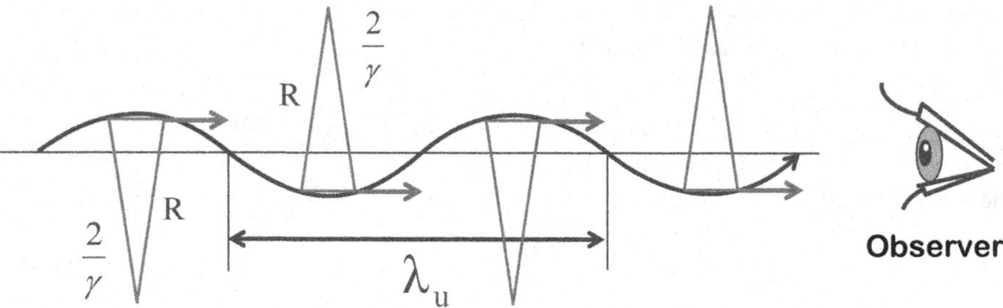

Figure 3.9 Radiation from sequence of bends.

An external observer will see photons emitted by the particle during its travel along the arc $2R/\gamma$. Let's now define the parameter K as the ratio of the wiggling period to the length of this arc:

$$K \sim \gamma \lambda_u / R \tag{3.39}$$

We can qualitatively see that if $2R/\gamma \ll \lambda_u/2$, then the radiation emitted at each wiggle is independent. This corresponds to $K \gg 1$ and is called the *wiggler regime*.

On the other hand, if $2R/\gamma \gg \lambda_u/2$, then we are in a regime where the entire wiggling trajectory contributes to radiation (therefore *interference* leads to *coherence* of radiation, which is explored in detail in further chapters). This corresponds to $K \ll 1$ and is called *the undulator regime.*

As we see, the *undulator parameter* $K \sim \gamma \lambda_u / R$ defines different regimes of synchrotron radiation: $K \gg 1$ is the wiggler regime, $K \ll 1$ is the undulator regime.

Fig. 3.10 shows the differences between radiation from a single bend and from a sequence of bends in wiggler and undulator regimes, respectively. We will consider this topic in more detail in the Chapters dedicated to light sources and to FELs, which are respectively 7 and 8.

3.3.6 SR QUANTUM REGIME

Let's define the parameter "*Upsilon*"as $\Upsilon = \hbar\omega_c/E$. The meaning of this parameter depends on the regimes of SR.

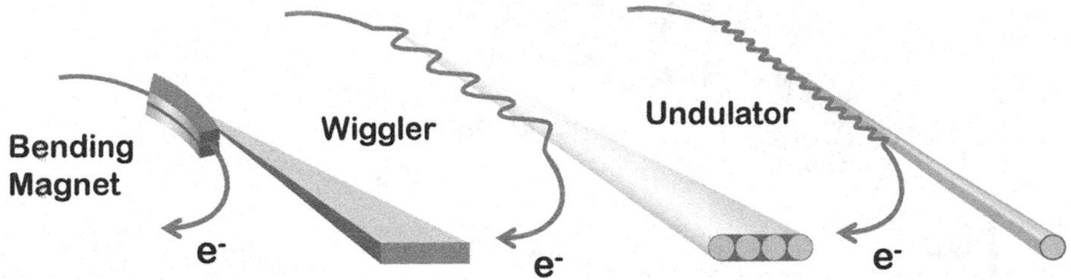

Figure 3.10 Wiggler and undulator radiation.

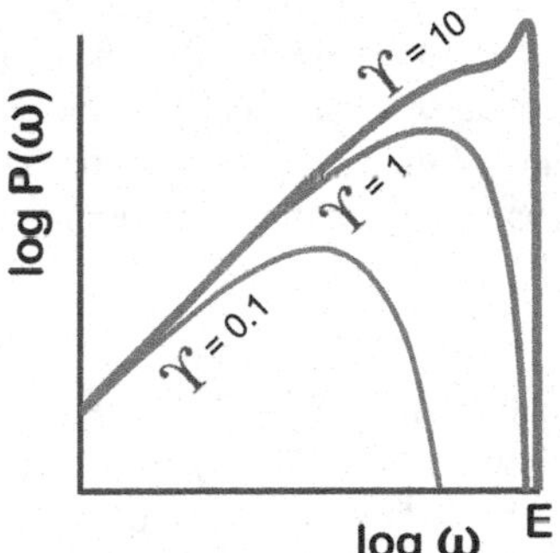

Figure 3.11 SR spectrum in classical and quantum regimes.

In classical synchrotron radiation regime, the energy of the emitted photon is much smaller than the energy of the particle. In quantum regime, the estimated photon energy approaches or exceeds the particle energy, and thus the SR formulae need to be appropriately adjusted.

When the parameter $\Upsilon \ll 1$, its physical meaning is the ratio of the characteristic photon energy to the energy of a single electron in the beam.

However, when $\Upsilon \sim 1$ and higher, the classical regime of synchrotron radiation is not applicable, and the quantum SR formulae of Sokolov–Ternov should be used. Such a situation may happen in particular in collision or highly relativistic focused beam, e.g., in linear colliders.

In a quantum regime, the shape of the SR spectrum changes, as there should not be a photon emitted that has energy larger than the energy of the initial particle.

The qualitative dependence of the SR spectrum in classical and quantum regimes is shown in Fig. 3.11.

Though the quantum SR is unlikely to occur in radiation from bends, it can happen in SR during beam collisions, as beams focused to tiny spots can produce enormous fields that cause the oncoming particles to radiate in a quantum regime.

3.4 LEP ENERGY INCREASE – INVENTION CASE STUDY

LEP was the electron-positron collider at CERN, and around the year 2000 it was running with the energy about 104 GeV per beam. The beam energy was limited by the energy losses due to synchrotron radiation, and by the amount of voltage from the installed superconducting RF cavities. The LEP team was looking for all possible ways to increase the beam energy, in pursuit of Higgs discovery.

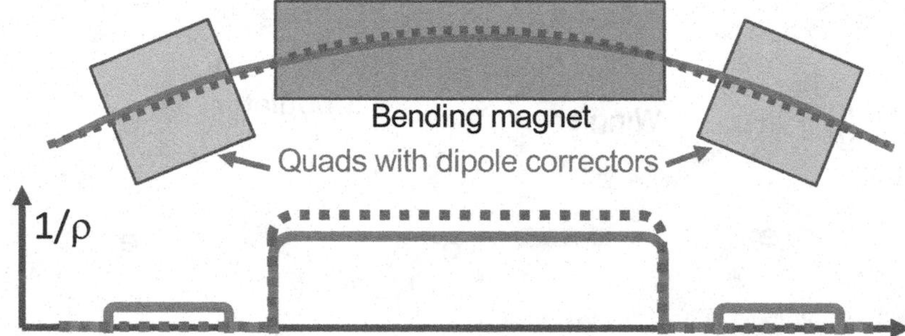

Figure 3.12 Use of dipole correctors of the quadrupoles allowed to reduce the bending field in the main dipoles, smooth the orbit, reduce synchrotron radiation, and achieve some energy increase in the electron-positron collider LEP. The bottom plot shows, qualitatively, the inverse bending radius along the beamline before (dashed line) and after (solid line) powering-up the dipole correctors of the quadrupoles.

LEP optical structure included all typical magnets: bending dipoles, quadrupoles with the dipole corrector coils, sextupoles, etc. An ingenious way was suggested[3] and implemented, allowing to achieve some increase of LEP energy without installing any more superconducting RF to compensate for SR energy losses, and without any other beamline modifications.

The suggested method involved powering-up the horizontal dipole correctors of the quadrupoles, with simultaneous reduction of the bending field in the dipoles. This allowed to smooth the trajectory, increase the average radius of the trajectory (see Fig. 3.12), thus reduce the SR energy losses, and get to a higher collision energy with the same installed RF.

This method allowed some noticeable increase of LEP collider energy, which eventually reached 209 GeV energy in the center of mass. And from TRIZ point of view, the creative use of the horizontal dipole correctors in the quadrupoles was a very clear application of the inventive principle of *using the resources (energy, fields, materials, contradictions, etc.) that already existed in a system.*

[3] P. Raimondi, CERN-OPEN-99-125, 1999.

EXERCISES

3.1. *Chapter materials review.*
A proton beam of E=50 TeV circulates in a 100-km perimeter ring. Estimate the synchrotron radiation energy loss per turn, the characteristic energy of the emitted photons and the cooling time.

3.2. *Chapter materials review.*
Describe how one needs to change the optics of third-generation SR sources in order to approach the diffraction-limited SR source, particularly in the horizontal plane.

3.3. *Chapter materials review.*
In a manner similar to how the equilibrium emittance was estimated in this chapter, derive the equilibrium energy spread of the beam.

3.4. *Mini-project.*
Define the approximate parameters (energy, sizes, fields in bending magnets) of a second-generation SR source aiming to achieve 10 keV of X-rays.

3.5. *Analyze inventions or discoveries using TRIZ and AS-TRIZ.* Analyze and describe scientific or technical inventions described in this chapter in terms of the TRIZ and AS-TRIZ approaches, identifying a contradiction and an inventive principle that were used (could have been used) for these inventions.

3.6. *Developing AS-TRIZ parameters and inventive principles.* Based on what you already know about accelerator science, discuss and suggest the possible additional parameters for the AS-TRIZ contradiction matrix, as well as the possible additional AS-TRIZ inventive principles.

3.7. *Practice in reinventing technical systems.* Synchrotron radiation emitted by the beam in bending magnets can hit the walls of the elliptical vacuum chamber, causing gas desorption, deterioration of vacuum, and reduction of the beam's lifetime. Suggest a way to modify the design of the vacuum chamber to mitigate this issue. Try to use the inventive principle of separating out the negative factor.

3.8. *Practice in reinventing technical systems.* The FODO beamline, with focusing and defocusing quadrupoles, can focus beams in all directions. A long-wavelength adiabatic deviation of the direction of the FODO beamline will make the beam follow the deviated beamline trajectory. The LEP energy increase discussed in Section 3.4 shows that the quadrupoles with dipole corrections help to steer the beam in a circular accelerator. In both of these examples, the beamlines have certain limited energy acceptance — the beam with an energy spread will have larger sizes and may experience losses of particles. Taking these examples to an extreme, suggest a way to focus and transport the beam with nearly 100% energy spread on a circular trajectory

3.9. *Practice in the art of back-of-the-envelope estimations.* Synchrotron radiation was first observed around 1947 in a General Electric 70 MeV synchrotron. It is known from historical photos that the entire vacuum chamber of this synchrotron could fit on a small dining table. Estimate the wavelength of the light that was observed. *(It is assumed that you can identify the most important effects playing roles in this task, can define the necessary parameters and set their values, and can get a numerical answer.)*

4 Synergies between Accelerators, Lasers and Plasma

In this chapter we will discuss the synergies between accelerators, lasers and plasma, connecting them via inventive principles.

Within the scope of our interests, the themes linking all three areas *(accelerators, lasers, plasma)*, include, among others, beam sources, laser beam generation, wave propagation (in structures and in plasma), laser propagation in plasma, plasma acceleration, radiation (synchrotron, betatron) and free electron lasers, as illustrated in Fig. 4.1.

Synergy — cross-fertilizing interaction of considered areas of physics.

Some of the themes that connect both accelerators and lasers are: focusing (chromaticity, aberrations, beam quality), cavities (RF and optical), laser-beam interaction (ponderomotive force), laser imprints (on e-beam in wigglers), cooling (e-, stochastic, optical stochastic, laser), Compton X-ray sources, chirped pulse amplification and bunch or pulse compression.

Accelerators' and plasma's connecting themes include: instabilities (plasma oscillation, beam instability, e-cloud, e-ion), beam–beam effects, plasma-focusing lenses, plasma mirrors, collision-less Landau damping (in plasma, in beams) and echo effects (in beams, in plasma).

Last, lasers' and plasma's themes include: gas lasers, optical parametric chirped pulse amplification and harmonic generation. In this chapter and in this entire text, we will touch on only some of the topics mentioned above.

In the following sections we will discuss accelerators, lasers and plasma in the order of how we interact with them in the real world: First, we *create* the beam/light pulse/plasma wave; then we prepare them for use, i.e., *energize* (accelerate, amplify, excite in plasma) them or *manipulate* (focus, compress, stretch, etc.) them; and, last, use or *interact* with them. This sequence is illustrated in Fig. 4.2.

The area of synergies between accelerators, lasers and plasma offers large number of invention examples that can be analyzed from TRIZ point of view. We will give, starting from this chapter, an example of invention case study, focusing on the step-by-step path to invention of the so-called Mak telescope.

4.1 CREATE

Let's discuss the topic of beam, light pulse and plasma wave creation in the *Create* — Energize — Manipulate — Interact sequence.

DOI: 10.1201/9781003326076-4

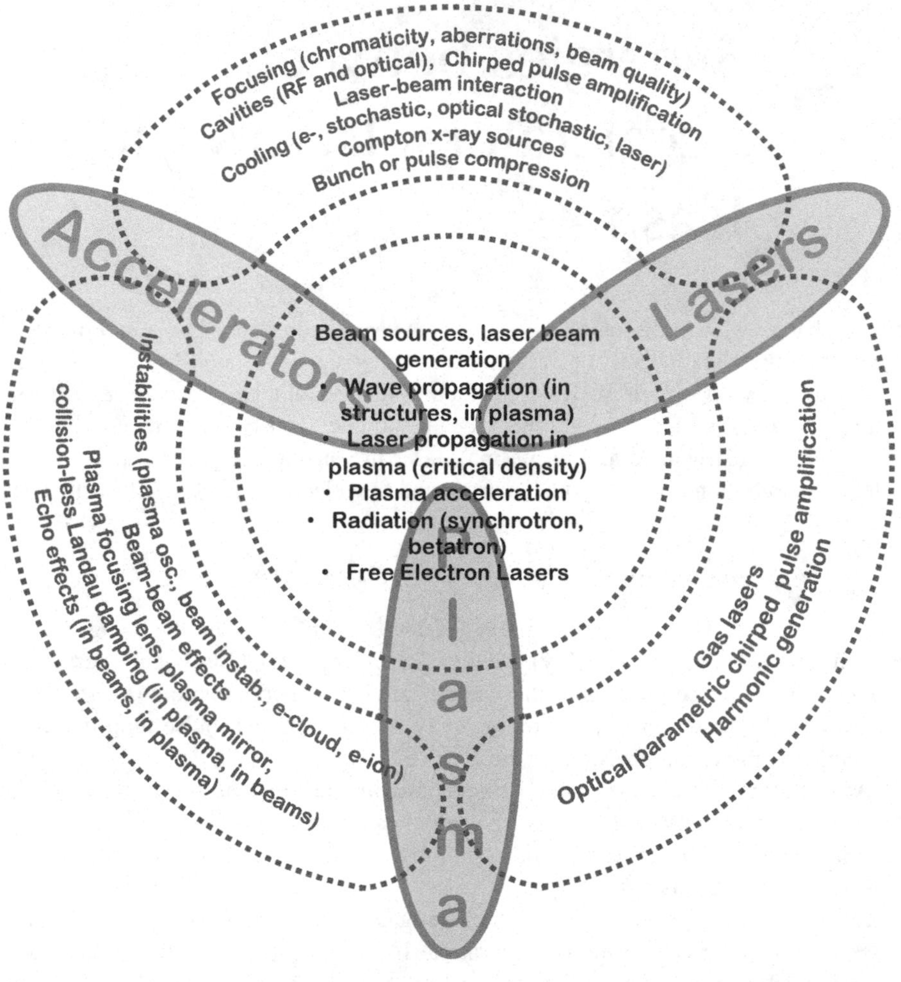

Figure 4.1 Synergies between accelerators, lasers and plasma.

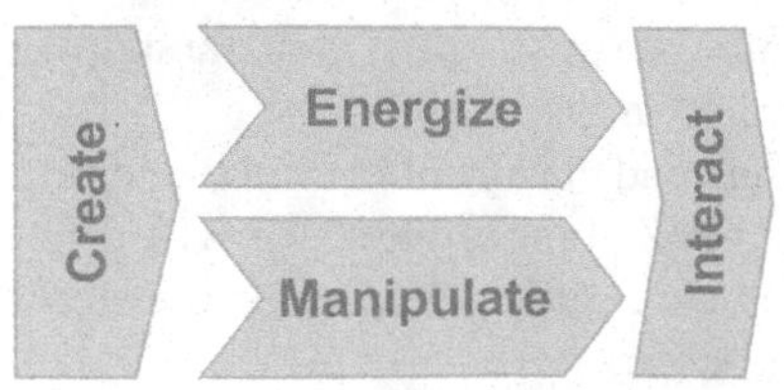

Figure 4.2 Discussion of synergies will follow this sequence.

Let's discuss beams, laser and plasma following the natural sequence: creating them, preparing for use, energizing (accelerating, amplifying, exciting waves in plasma), manipulating (focusing, compressing, stretching, etc.), and using them.

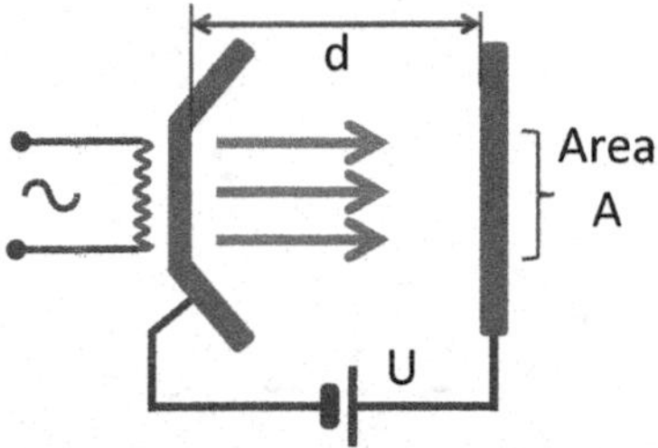

Figure 4.3 Thermal cathode e-gun.

Space charge effects limit the current in electron guns to $I = P \cdot U^{3/2}$ where the ideal Child-Langmuir microperveance P is

$$P \approx 2.33 \cdot \frac{A}{d^2} \left[\frac{\mu\mathrm{A}}{\mathrm{V}^{3/2}}\right]$$

4.1.1 BEAM SOURCES

We will begin our dialogue on particle sources by starting with *leptons* (electrons, positrons, muons, etc.) and later moving on to *hadrons* (protons, antiprotons, etc.) and ions.

The simplest form of an electron source is the thermal cathode gun (Fig. 4.3). According to the "three-halves power" law (or the Child-Langmuir law), the space charge effects of the non-relativistic accelerated electron beam often limit the current in electron guns.

$$I = P \cdot U^{3/2} \tag{4.1}$$

where I is the current, U is the cathode to anode voltage and the coefficient P is called *perveance*. The ideal Child-Langmuir perveance P is

$$P = \frac{4}{9}\varepsilon_0\sqrt{\frac{2e}{m}}\frac{A}{d^2} \tag{4.2}$$

where d is the gap between the cathode and anode and A is the cathode area. For electron beam $P \approx 2.33 \cdot 10^{-6} \cdot A/d^2$ [A/V$^{3/2}$]. It is often expressed in [μA/V$^{3/2}$] and then called *microperveance*.

A much more modern source of electrons is the laser-driven photocathode gun (illustrated in Fig. 4.4). Photo electron guns contain electrons that are generated by a laser field via the photoelectric effect. Such a gun usually consists of one-and-a-half RF cavities and a photocathode made either out of metal (e.g., copper) or from a special alloy. While a pure metal photocathode is robust, its *quantum efficiency* (the number of electrons per number of incident photons) is usually quite low, of the order of 10^{-4}. Photocathodes with alkali metals such as cesium can have a *quantum efficiency* of 10% or higher.

The photocathode guns driven by a laser pulse are also very suitable for production of short electron beams. The laser pulse is usually sent to the cathode at a small angle as shown in Fig. 4.4 to avoid interference with the accelerating electron beam.

As seen from Eq. 4.1, photoguns can provide a much higher pulsed electron current — due to a higher accelerating voltage — produced at the cathode by the RF structure. The typical accelerating voltage in guns ranges from around 30 MV/m in an L-band, 100 MV/m in an S-band and 200 MV/m in an X-band. [1]

[1] *RF frequencies bands: L: 1-2 GHz, S: 2-4 GHz, C: 4-8 GHz, X: 8-12 GHz.*

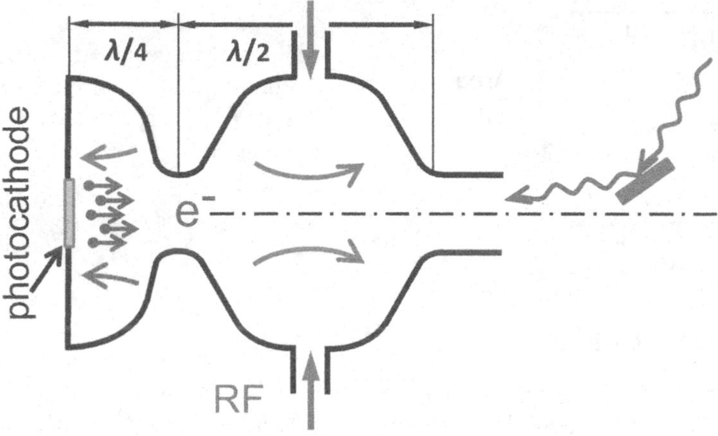

Figure 4.4 One-and-a-half-cell RF photocathode electron gun.

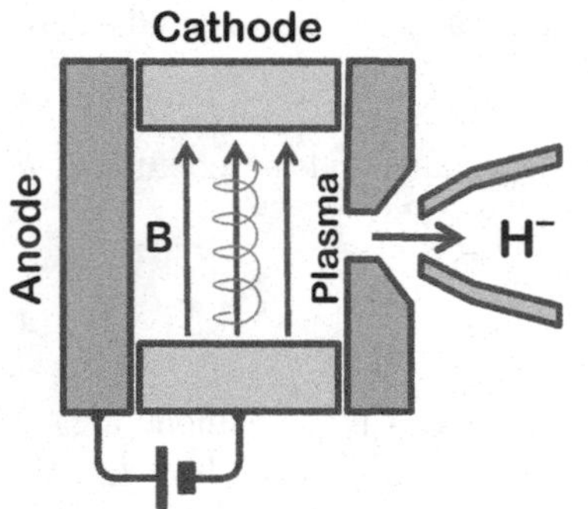

Figure 4.5 Surface-plasma Penning H^- ion source.

Adding small amount of cesium or other substances with low ionization potential acts as a catalysis and result in a considerable boost of negative ion production. Compare:

Hydrogen ionization energy: 13.6 eV
Cesium first ionization energy: 3.9 eV
Cesium acts as catalyst, or *intermediary*

Producing positrons usually requires creating e^+e^- pairs followed by separating the positrons. An electron or photon beam with a sufficient amount of energy is sent onto a target where the e^+e^- pairs will be produced. Separated positrons are then accelerated and sent to a damping ring, where their emittance will decrease due to radiation damping.

Ion or proton beams are produced by plasma-based ion sources — a large variety of source types exists. An example shown in Fig. 4.5 depicts a *Penning source* in which a magnetic trap is arranged in the cathode-anode area of an electron beam, ionizing gas via discharge. Ions of certain charges are extracted with the help of an electrode, and are then separated and sent for further acceleration and miscellaneous use. In particular for negative ion sources, *cesiation* (developed by V. Dudnikov) — the process when a small amount of Cs atoms is added into the gas — is often used to significantly enhance the emission of negative ions.

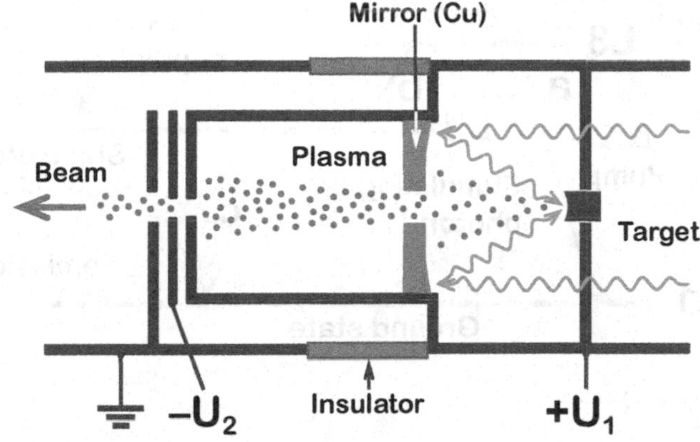

Figure 4.6 Conceptual diagram of the laser-driven ion source.

Plasma in ion sources can be created by various means such as ionization via laser (which was realized in the ion source[2] developed at CERN, shown in Fig. 4.6). In this example, the laser beam is focused by a metal mirror on a target, ionizing the target and producing plasma. The ions of plasma are then directed to the extraction electrodes via a hole in the mirror.

This brings us to the discussion of lasers.

4.1.2 LASERS

Lasers are a source of coherent light, and a laser diagram is shown in Fig. 4.7. The main laser components are the *gain medium* (which amplifies the light), the resonator (which gives optical feedback) and the pump source (which creates *population inversion*).

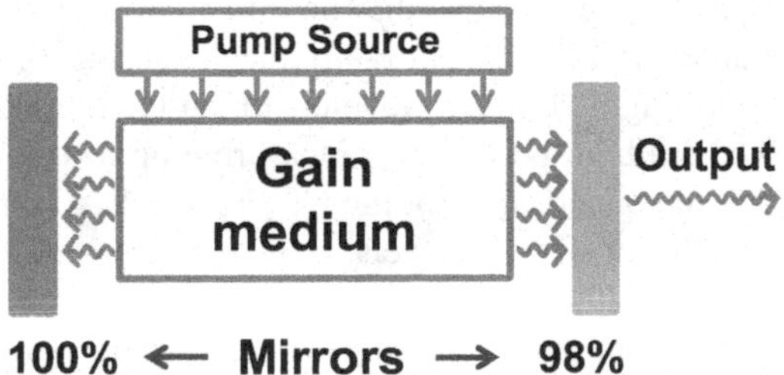

Figure 4.7 Laser diagram.

The gain medium contains atoms with specific conditions for the energy levels and for the lifetime of an excited state of the atom, suitable for arranging a *three-level*

[2]*Consider whether evolution from discharge ion source to laser-driven ion source can suggest a new inventive principle.*

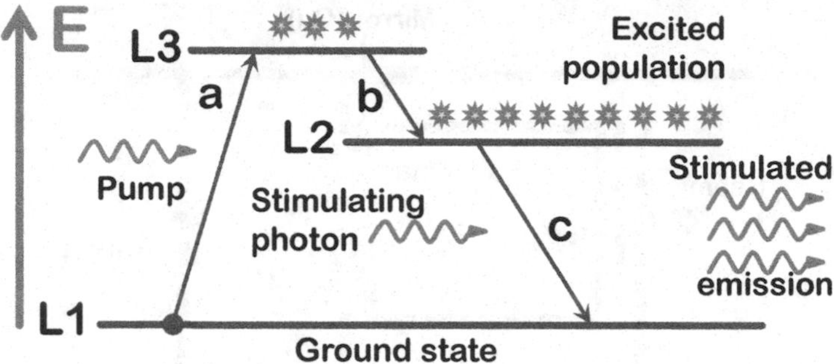

Figure 4.8 Three-level laser.

laser as described in the following paragraph. The pump gets the atom population from the ground state L_1 to the higher-energy level L_3. The excited population gets from L_3 to L_2 through non-radiative decay as the lifetime of the L_3 state is very short and all the population in state L_3 decays to state L_2. The lifetime of energy state L_2 is, in contrast, very long and therefore, a *population inversion* occurs with respect to the state L_1. Once the population inversion is obtained, stimulated emission can occur when an incident photon forces the population to drop from level L_2 to L_1 (as shown in Fig. 4.8), resulting in an optical gain.

As you can see from the above description, a minimum of three levels are needed to arrange a laser, but it is possible to create four and higher-level laser systems as well.

An advantage of four-level laser systems is that much less pumping power is needed for the creation of population inversion.

The first laser built was a three-level ruby laser emitting a 694-nm wavelength (shown conceptually in Fig. 4.9). The first mass-produced laser was a four-level He-Ne gas laser emitting a 694-nm wavelength. These first lasers were characterized by low emission power and very low *efficiency* — typically around 0.01–0.1%.

A CO_2 gas laser emitting a 10.6-μm wavelength, on the other hand, has an efficiency close to 30% and a high-power level measuring up to kW in CW. It is also interesting to note that the quantization of the CO_2 molecule's vibrational and rotational states enables this laser's system levels.

One more example of an efficient laser (at around 40%) is the *diode laser*. In this laser, the levels of the system are enabled by the quantization of energies of holes and electrons in the semiconductor diode. Due to the compactness of the diode laser, its output light has a very large divergence, low coherence and usually low power. However, the versatility of wavelength output and high efficiency makes this type of laser ideal for *pumping* — the excitation of gain medium in high-power [3] laser amplifiers. In the latter, the low power, low coherence and large divergence output

[3] *Nd:YAG — Neodymium-doped yttrium aluminum garnet: Nd:$Y_3Al_5O_{12}$. Yb:YAG — Ytterbium-doped YAG.*

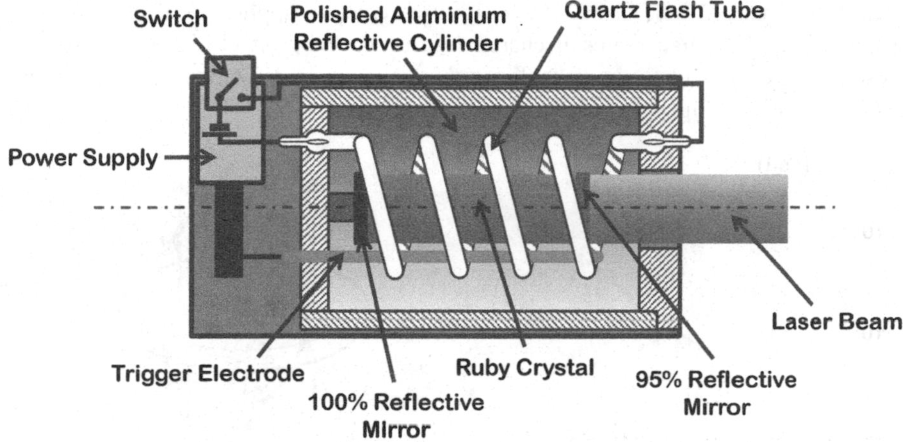

Figure 4.9 Conceptual diagram of a ruby laser. Quartz flash tube serves as the pump source and ruby crystal as the gain medium.

of the laser diode is amplified in a gain medium such as Nd:YAG, resulting in a high-power, high-coherence, high-efficiency laser beam as illustrated in Fig. 4.10. Note that pumping by the diode laser occurs at a shorter wavelength than the output radiation, in agreement with Fig. 4.8.

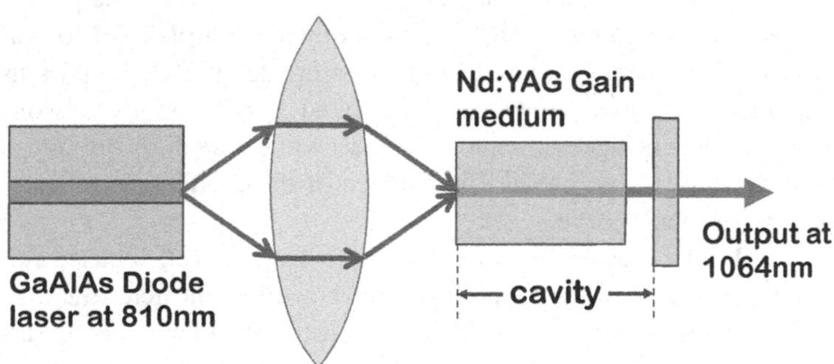

Figure 4.10 Diode laser in application for pumping of Nd:YAG laser.

The laser diode output can also be applied to fiber optics and fiber amplifiers, as we will discuss in the next section.

When used for laser pumping, many low-power diode lasers are usually assembled into arrays called "diode bars."

4.1.3 PLASMA GENERATION

To conclude our discussion regarding creation, let's briefly consider the process of how plasma is made. The most well-known method is discharge ionization in which plasma is produced by DC voltage or a pulse typically applied to a low pressure gas. The voltage U that creates discharge depends on the product of the gap g to gas pressure P, i.e. $(P \times g)$ — the behavior of this voltage is represented by the *Paschen curve* as shown in Fig. 4.11.

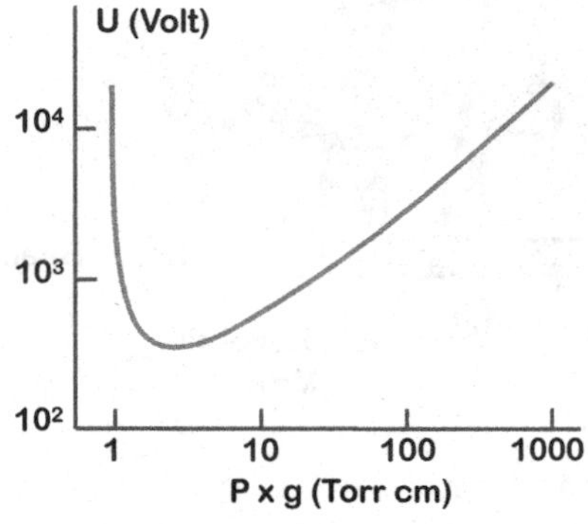

Figure 4.11 Paschen discharge curve for hydrogen gas.

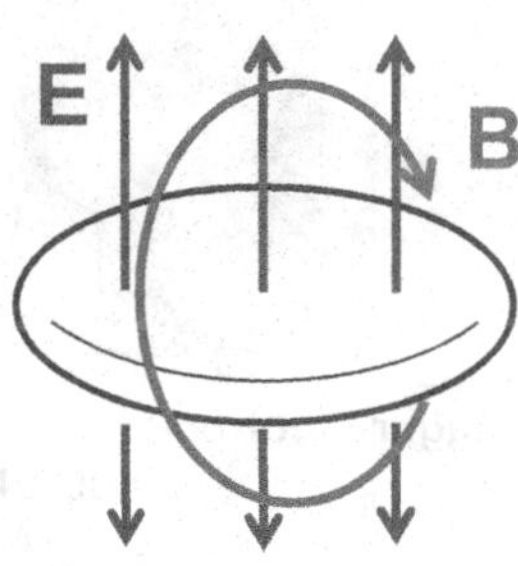

Figure 4.12 Fields of a relativistic electron bunch can produce field ionization of gas.

Several factors are important in sustaining conditions appropriate for a breakdown. On one side, the energy that the electron gains during its acceleration over the *mean free path* (before the next collision) should be larger than the first *ionization energy* of the gas molecules. A lower pressure (longer mean free path) [4] would therefore create preferable conditions for a breakdown. Multiple collisions and multiple electron ionizations are required in order to create suitable conditions for an *avalanche* and thus a breakdown. On the other hand, overly low $P \times g$ would mean that electrons would not have enough collisions while traveling through the gap, making an avalanche less probable. The balance of these effects is reflected in the behavior of the Paschen's curve.

Plasma can also be created by *field ionization*. Both a well-focused relativistic electron beam (see Fig. 4.12) and a laser can have a sufficient field to ionize gas. A gas or target can then be instantaneously ionized if the field level reaches an *atomic field* scale:

$$E_{Beam} \gg E_{Atomic} \quad \text{or} \quad E_{Laser} \gg E_{Atomic}$$

As we will discuss in Chapter 6, due to the *multi-photon ionization* and *tunneling* effects, much lower-level fields than atomic ones are often sufficient for ionization.

4.2 ENERGIZE

Next comes the process of *energizing* the beam, laser or plasma. In the case of a charged particle beam, this involves acceleration (electrostatic, betatron, acceleration in RF cavities and structures or plasma acceleration). In the case of a laser, it involves

[4] *Ionization energy of hydrogen is* ~ 13.6 *eV.*

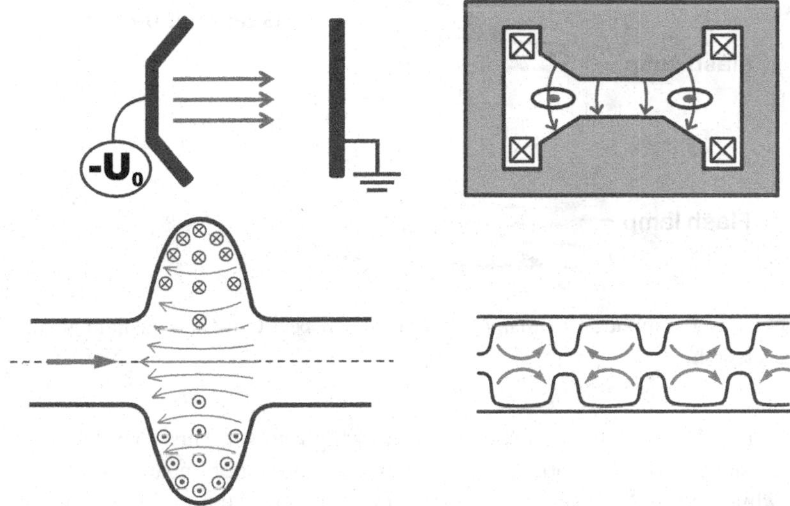

Figure 4.13 Electrostatic and betatron acceleration. RF cavity and RF structure.

amplification (standard, chirped pulse amplification, optical parametric chirped pulse amplification, etc.), and for plasma, excitation of waves by various means such as short pulse of a laser or beam.

4.2.1 BEAM ACCELERATION

The simplest and exceedingly widespread acceleration method is electrostatic acceleration (Fig. 4.13). Not so widely used now is betatron acceleration principle, which is based on Maxwell's Eq. 2.4.

Perhaps the most versatile acceleration method involves the use of resonators — also known as RF cavities. Cavities can be arranged in structures and made to be suitable for the acceleration of any types of particles from electrons (which almost immediately become relativistic) to protons or ions.

Often, acceleration is combined with focusing, either via magnetic quadrupoles inserted between acceleration sections or via an EM wave with a quadrupole component as in an *RFQ accelerating structure*.

4.2.2 LASER AMPLIFIERS

One of the possible principles of laser amplification is that a pumped gain medium of the *amplifier* amplifies light at the wavelength of the *oscillator* laser, which is made with the same material as the pumped gain medium of the amplifier. The diagram of such a laser amplifier is shown in Fig. 4.14.

The technical challenges caused by laser light amplification actually resulted in numerous inventions and breakthroughs, as discussed in Chapter 1.

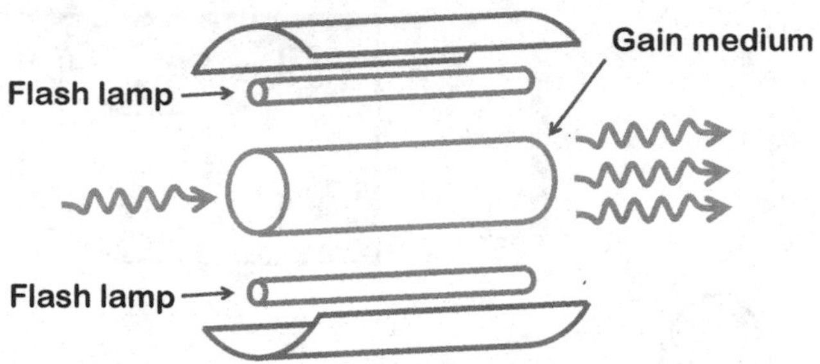

Figure 4.14 Laser amplifier. Flash lamp emits in broad spectrum. Gain medium amplifies selected wavelength.

In particular, lasers with ultra-short and ultra-high-power pulse have had their share of challenges. Ultra-short intense pulses can cause nonlinear effects in the medium, while the high-power pulses cause heating up of the amplifier medium. These issues limit the repetition rate, power and efficiency of laser systems. Some of the most powerful lasers fire just once every few hours!

4.2.3 LASER REPETITION RATE AND EFFICIENCY

Let's take a look at the challenges of high average power lasers from the point of view of AS-TRIZ. The TRIZ-formulated problem is as follows: as the intensity of the pulsed laser light increases, it takes much more time for the medium to cool down and be ready for its next use. The identified *contradiction* is between the parameter that needs to be improved, INTENSITY, and the parameter that gets worse, REPETITION RATE.

A general principle that can solve this contradiction can be looked up in nature (see Fig. 4.15) or taken from AS-TRIZ, where Principle 4, *Volume-to-surface ratio*, suggests changing this ratio in order to alter the object's characteristics (such as its cooling rate or its fields).

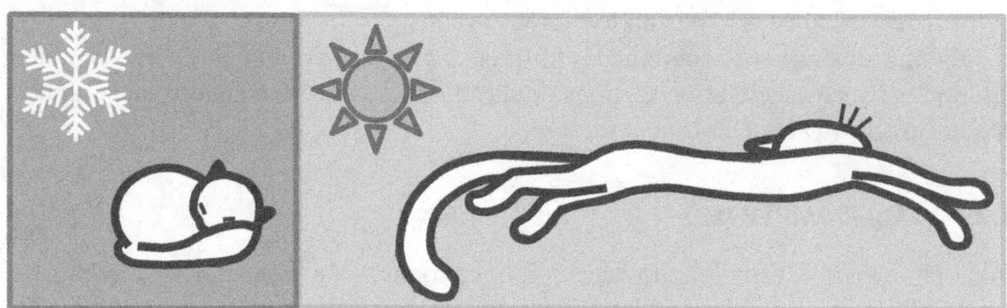

Figure 4.15 The cat intuitively knows the inventive principle of changing her surface-to-volume ratio depending on the external temperature.

4.2.4 FIBER LASERS AND SLAB LASERS

Fiber laser technology is a perfect illustration of the inventive principle of changing the surface-to-volume ratio.

Indeed, if the main technical issue with high average power laser amplifiers is cooling the active medium, then one would try to increase its surface, without changing its volume, to enhance thermal exchange with the environment. Fig. 4.16 clearly demonstrates that the typical geometry of the active medium (when its length L is similar to its radius R) is the most disadvantageous, in terms of cooling.

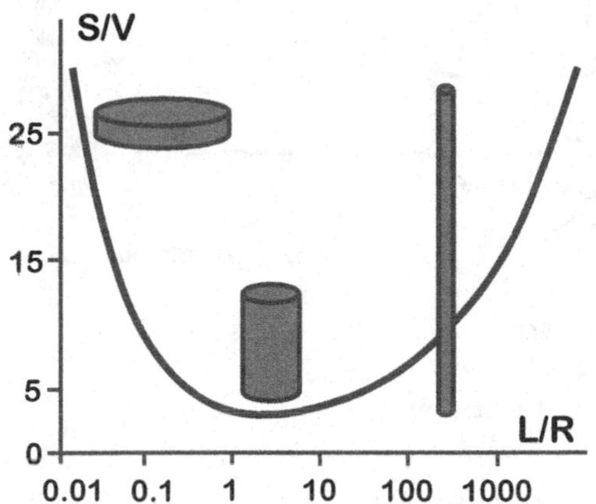

Figure 4.16 Volume-to-surface ratio S/V in units of $(2\pi V)^{1/3}$ vs. L/R.

Including the change of V/S into the principle "parameter change" connects it to fundamental symmetries, i.e. conservation laws of physics. E.g., Gauss theorem (divergence theorem) equates the total sources and sinks of a vectorial quantity, or the integral volume of its divergence, to the net flux of this vectorial quantity across the volume boundary.

On the other hand, either increasing L/R toward fiber geometry of the active medium or decreasing L/R toward slab geometry would both have advantages in terms of the surface-to-volume ratio S/V, thus making it easier to manage their temperature regimes.

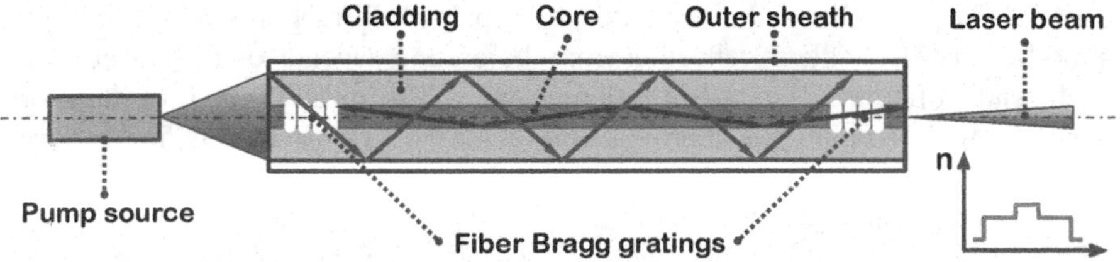

Figure 4.17 Schematic of a fiber laser and cross section of refractive index.

Correspondingly, the fiber laser and slab lasers are the technologies most suitable[5] for achieving higher average power laser pulses, higher repetition rate and higher wall-plug efficiency. The main components of a fiber laser are shown in Fig. 4.17.

[5] *Fiber laser technology uses the principle of larger surface-to-volume ratio.*

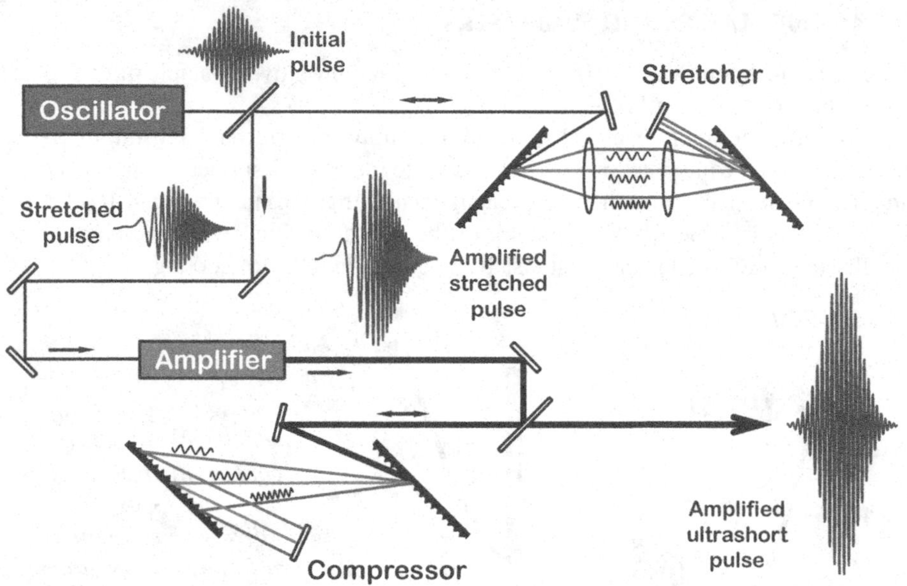

Figure 4.18 Schematics of CPA — chirped pulse amplification.

4.2.5 CPA – CHIRPED PULSE AMPLIFICATION

Invented for lasers by G. Mourou and D. Strickland, *chirped pulse amplification* (or CPA) is a process that stretches the short pulse in time, amplifies the now-longer laser pulse and then compresses the amplified pulse[6]. This multi-step process relieves the optical component of laser amplifiers from pulsed power, thus reducing nonlinear effects and avoiding material damage. The stretching and compressing of the pulses are based on the time–energy correlation property of the laser pulse.

The schematics of a laser stretcher and compressor are discussed in detail in Section 4.3.4. The stretcher and compressor use a pair of diffraction gratings and rely on the fact that the angle of reflection from the grating depends on the wavelength, which sends different colors along different paths. The spectral width of the short laser pulse with duration τ is approximately given by $\Delta f \sim 1/\tau$. This finite spread of the spectrum around the carrier frequency makes such laser pulses suitable for spatial–spectral manipulation.

The schematics of CPA are shown in Fig. 4.18. The initial short pulse is provided by a short-pulse *oscillator*. The first pair of gratings disperses the spectrum and stretches the pulse by a factor of about a thousand (for visibility, the longitudinal extent of the pulses in Fig. 4.18 is shown qualitatively). After stretching, the pulse is long and has a low peak power, which is thus safer for amplification. After passing

[6] *CPA technique was originally developed for radar, while chirped pulses could also be observed in nature, e.g., in the voices of bats.*

the power amplifier, the pulse is sent to the second pair of gratings, which reverses the dispersion of the first pair and compresses the pulse, producing a high-energy ultra-short laser pulse.

The invention of CPA was one of the factors that ultimately pushed laser technology to such peak power levels that these lasers became a possible competitor alongside particle accelerators.

Amplification of chirped pulses was used in radar and is now used in lasers — this trend from microwave to optical range can be taken as one of the generic principles of AS-TRIZ.

4.2.6 OPCPA – OPTICAL PARAMETRIC CPA

Another method of laser amplification is called OPCPA — *optical parametric* CPA. Its principle is based on nonlinear properties of crystals (typically barium borate or *BBO crystals*) which, being subjected to radiation of wavelength ω_s, generate radiation at two frequencies — ω_1 and ω_2 where, as energy is conserved, $\omega_1 + \omega_2 = \omega_s$ (as shown in Fig. 4.19).

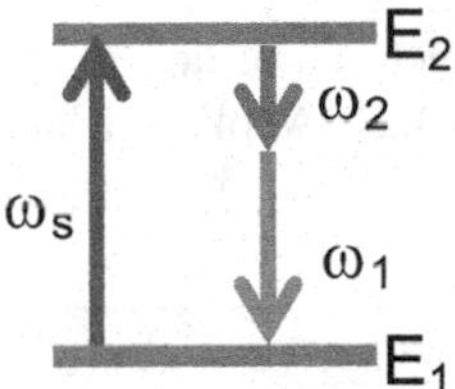

Figure 4.19 Optical parametric generation in nonlinear crystals.

In parametric process only the **real** *part of permeability is involved, and therefore no energy is left in the nonlinear crystal, and everything comes out in the form of light, which is beneficial for creating amplification systems with high average power.*

In *optical parametric amplification*, the input consists of two beams: the *pump* at ω_s and the *signal* at ω_1. The OPA output is the amplified ω_1 beam and weakened ω_s beam, plus an additional *idler* beam at ω_2.

The optical parametric amplifier system can be fed with a frequency-stretched *signal* pulse, as illustrated in Fig. 4.20. This makes the OPA system into a *chirped pulse* method known as OPCPA.

The OPCPA method is versatile; it can work from CW to femtosecond range in terms of pulse length, from UV to TeraHz in terms of light wavelength, and from mW to TW and PW in terms of the peak power.

The main advantage of OPCPA is that it works via a parametric process; i.e., no energy is left in the nonlinear crystal and everything comes out in the form of light. This is beneficial for high energy or high mean power systems since the thermal issues are eliminated.

4.2.7 PLASMA OSCILLATIONS

Jumping from lasers back to plasma topics, let's discuss the process of energizing the plasma, i.e., creating oscillations in plasma.

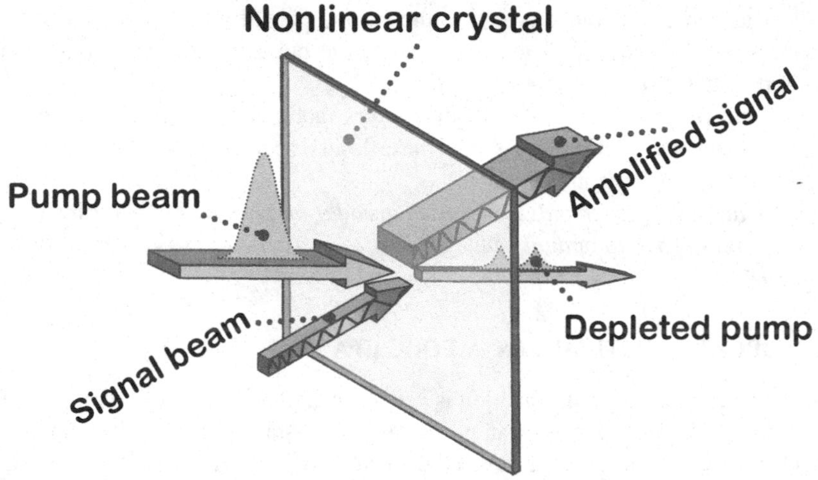

Figure 4.20 Optical parametric chirped pulse amplification — OPCPA.

Imagine that there is a region in plasma where electrons of density n shift with respect to the ions by a distance of x as shown in Fig. 4.21. Applying Gauss's law

$$\oint_{\partial\Omega} \mathbf{E} \cdot d\mathbf{S} = \frac{1}{\varepsilon_0} \int_{\Omega} \rho dV$$

will yield the value of an electric field produced by displaced charges

$$E = \frac{nex}{\varepsilon_0} \tag{4.3}$$

Writing an equation for the electrons' non-relativistic motion

$$F = m\frac{d^2x}{dt^2} = -eE = -\frac{ne^2x}{\varepsilon_0} \tag{4.4}$$

will then give us the expression for the oscillation frequency:

$$\omega_p^2 = \frac{ne^2}{\varepsilon_0 m} \tag{4.5}$$

Recalling the advice to express the end result in a form independent of the systems of units, we use

$$r_e = \frac{1}{4\pi\varepsilon_0} \frac{e^2}{m_e c^2}$$

to rewrite the oscillating frequency or the *plasma frequency* as

$$\omega_p^2 = 4\pi n c^2 r_e \tag{4.6}$$

We can also write a practical formula for $f_p = \omega_p/(2\pi)$:

$$f_p \approx 9000 n^{1/2} \quad \text{(Hz)} \tag{4.7}$$

where n is expressed in (cm^{-3}).

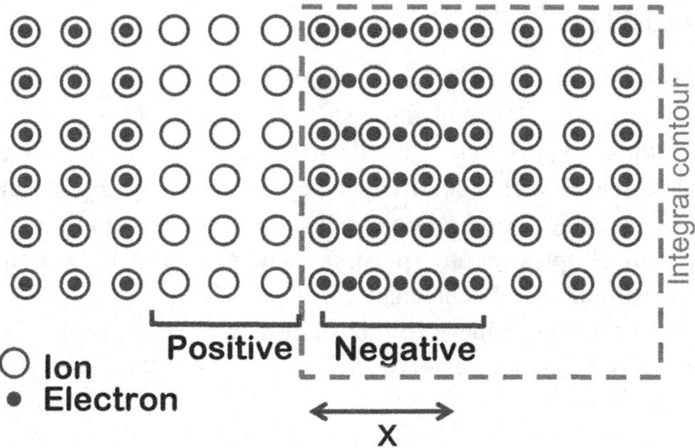

Figure 4.21 Plasma oscillations.

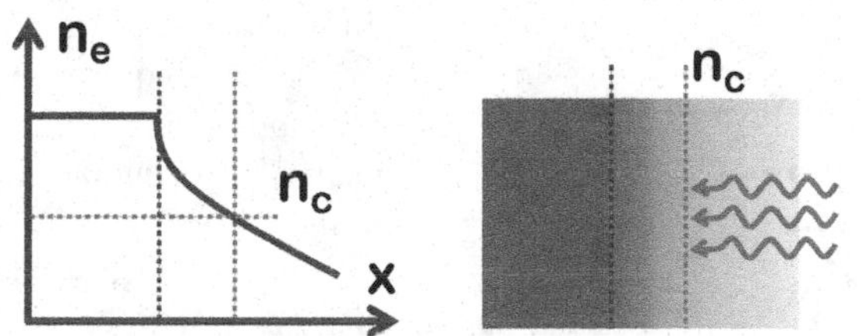

Figure 4.22 Laser penetration to plasma and critical density.

4.2.8 CRITICAL DENSITY AND SURFACE

Let's take this moment to qualitatively consider the process of laser plasma penetration and the corresponding factors of *critical density* and *critical surface.*

When a laser hits a target or a dense gas, the target surface or gas is heated and ionized, which forms plasma. Hot plasma then starts expanding into the vacuum, creating a gradient of plasma density as illustrated in Fig. 4.22, with a respective gradient of plasma frequency.

Qualitatively, if the plasma frequency ω_p in a particular layer of plasma is larger than the laser frequency ω, then the plasma electrons can move fast enough and can thus create electric currents that will screen the fields of the laser EM wave. Therefore, lasers can penetrate plasma only to the point when

$$\omega_p < \omega$$

The critical density is therefore

$$n_c = \frac{\omega^2}{4\pi c^2 r_e} \tag{4.8}$$

In other words, the laser beam of frequency ω cannot penetrate areas with $n > n_c$.

4.3 MANIPULATE

Let's consider some of the ways we can *manipulate* beams, laser pulses or plasma.

For particle beams the topics of interest include focusing (weak, strong, chromaticity, aberrations); compressing; cooling (electron, stochastic, optical stochastic, laser); phase plane exchange; transverse stability, etc. For laser beams we are interested in focusing; compression; phase locking; harmonic generation, etc. Plasma-manipulation topics include plasma focusing; Landau damping; and laser self-initiated focusing in the plasma channel.

We will touch on some of these in this section, and several other topics will be discussed in Chapter 12.

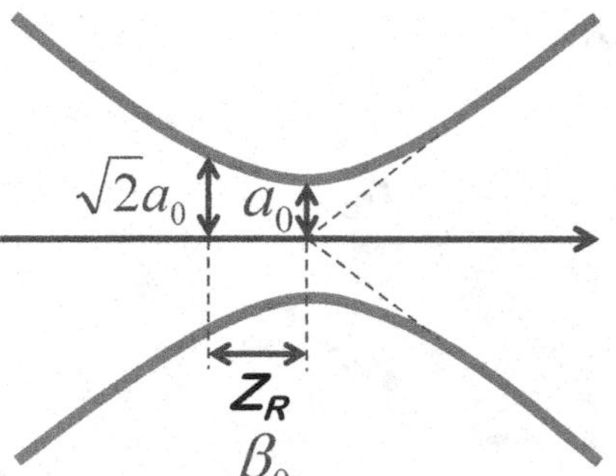

Different names are sometimes used for the same quantities in optics of light and in optics of charged particle beams. In particular, *Rayleigh length* in light optics is the same as *beta function* in the waist point in optics of charged particles.

Figure 4.23 Beam or light in the focus.

4.3.1 BEAM AND LASER FOCUSING

Optics of light and optics of charged particles have a lot of similarities. However, different names are sometimes used for the same quantities in beam and light optics. To illustrate this, consider a situation wherein the laser or particle beam is focused into a point so that its minimum size at the waist is equal to a_0 (also called w_0 in light optics). The distance from the waist point to the point where the beam size increases by $\sqrt{2}$ is called the *Rayleigh length* Z_R in light optics and is equivalent to the Twiss beta function β_0 at the location of the waist point in optics of charged particles (see Fig. 4.23).

4.3.2 WEAK AND STRONG FOCUSING

Analogies between light and beam optics — as well as mechanical analogies (in the spirit of TRIZ or synectics) — can help us build up an intuitive understanding of complex phenomena. Let's illustrate this in an example of weak versus strong focusing.

Weak focusing can be brought about by bending dipoles with the same gradient all along the perimeter of an accelerator (refer to Fig. 4.24). In this figure, the weak

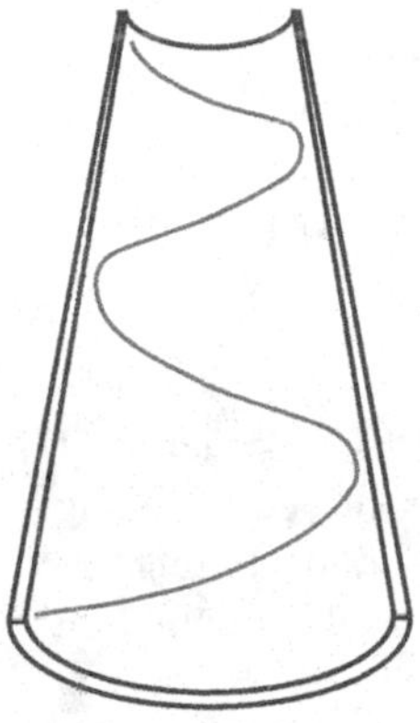

Weak focusing can be created by bending dipole magnets with the same gradient placed all along the perimeter of a circular accelerator. In this figure, the weak focusing is compared to the motion of a ball in a curved gutter.

Figure 4.24 Weak focusing.

focusing is compared to the motion of a ball in a gutter. Looking at Eqs. 2.19 and 2.20, one can conclude that there are conditions when the beam is focused simultaneously in the x and y planes (due to the presence of an additional focusing term in bends in the x plane). However, as follows from these equations or as Fig. 2.8 suggests, such focusing is weak — the spatial period of particle oscillation in this focusing field is of the order of the orbital circumference.

Weak focusing accelerators were mainly built in the early days. One of the disadvantages of weak focusing accelerators was the large transverse oscillations of particles, leading to wide apertures and correspondingly large and heavy magnets. For example, the 10 GeV weak focusing Synchrophasotron built in Dubna in 1957 (the biggest and the most powerful in its time), was registered in the Guinness Book of World Records for housing the heaviest magnet system, weighing 36,000 tons.

On the other hand, strong focusing can be generated via a sequence of focusing–defocusing quadrupoles, with the overall effect being equivalent to focusing, if certain conditions are satisfied. These conditions can be understood from the gutter analogy shown in Fig. 4.25 where the gutter is now bent, first of all, much more strongly, and second, it is bent up and down (see also Fig. 1.11). As clearly seen from the picture, stable motion is possible in this event only if the particle passes the areas of the downward-bent gutter near the center, in a way similar to what it would do in a FODO lattice. CERN's Proton Synchrotron, the first operating strong focusing proton accelerator, reached 24 GeV in 1959. It is constructed as a 200-m-diameter ring, with a magnet weight of 3,800 tons — weighing ten times less for twice the energy of Dubna's Synchrophasotron — resulting in a clear demonstration of the advantages of the strong focusing approach.

4.3.3 ABERRATIONS FOR LIGHT AND BEAM

Following Newton, we know that sunlight consists of a spectrum of colors and that each different color focuses differently — these are called chromatic aberrations. In

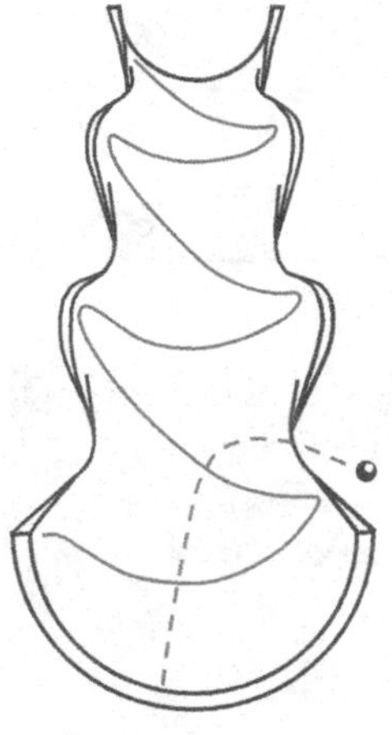

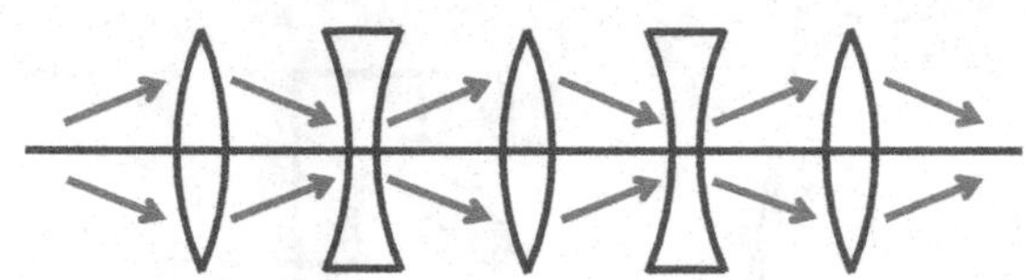

Strong focusing can be created by a sequence of focusing and defocusing lenses placed all along the perimeter of a circular accelerator. In this figure, the strong focusing is compared to the motion of a ball in a gutter which is periodically curved up and down, and can be also connected to the inventive principle of *system and anti-system.*

Figure 4.25 Strong focusing.

precise optical devices, such as photo cameras, the chromatic aberrations need to be compensated.

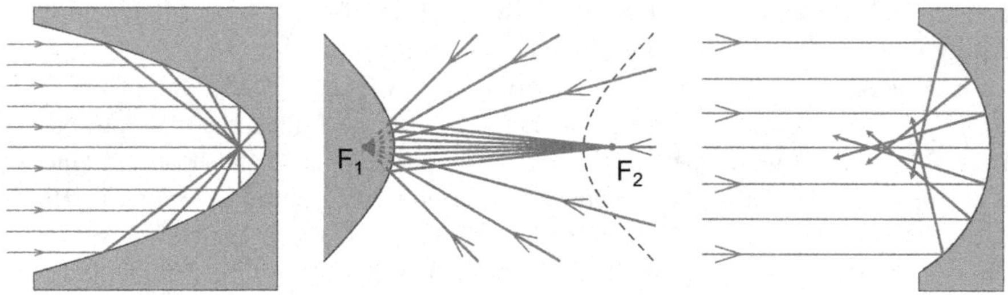

Figure 4.26 Mirror types, left to right — parabolic, hyperbolic, spherical.

Before starting to talk about those chromatic aberrations, it is useful to recall that a curved mirror, in contrast to a lens, does not produce chromatic aberrations. Different types of mirrors can be used for different purposes. To focus a parallel beam of light into a point, you need a parabolic mirror. Spherical mirror can also be used to focus a parallel beam, but it will introduce spherical aberrations. Hyperbolic mirror can be used to focus rays directed to a focal point F_2, into another focal point F_1 — see Fig. 4.26.

Plasma mirror can also be arranged, based on the critical surface phenomena described above — see Fig. 4.22 — and would illustrate the principle of *using the materials which are already destroyed.*

So, for light, one uses lenses made from different materials to compensate for chromatic aberrations, as illustrated in Fig. 4.27. In this example, using a strong focusing lens made from a low-dispersion *crown* glass coupled with a weak defocusing high-dispersion *flint* glass can correct the chromatic aberrations.

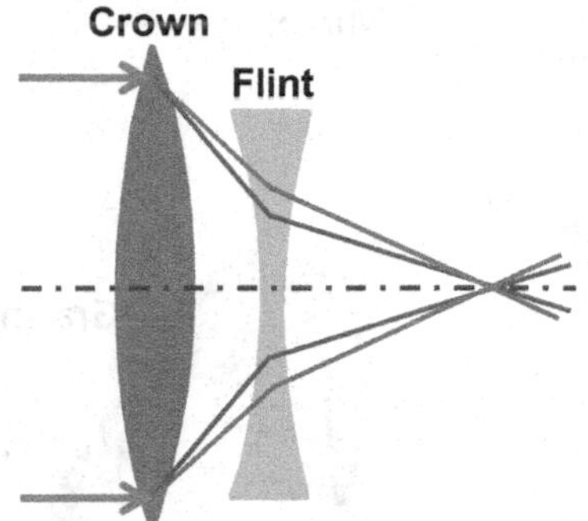

Compensation of chromatic aberration in this two-lens doublet is an illustration of inventive principle of *cancelling one harmful factor with another harmful factor.*

Figure 4.27 Lenses made from different glasses compensate chromatic aberrations.

For particle beams, chromatic aberrations are caused by the dependence of the focusing strength on the beam energy. There are no different materials and no specific Maxwell's equations in particle optics, therefore, other means have to be used for chromatic compensation in accelerators.

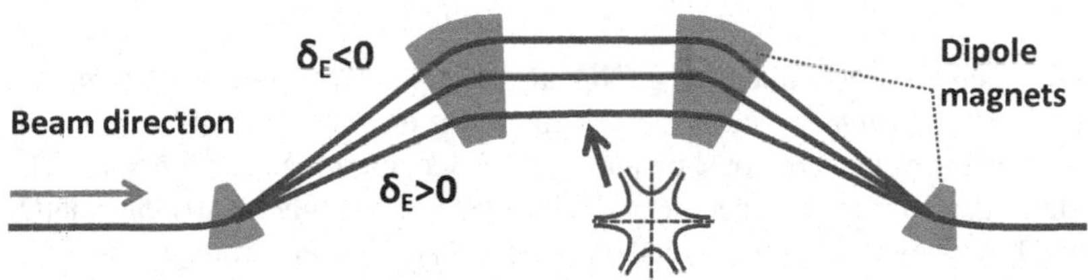

Figure 4.28 Compensation of chromatic aberration by inserting nonlinear sextupole magnets in a dispersive region.

Let's consider *chicane* arrangements of four bending dipoles as shown in Fig. 4.28. Such an arrangement creates a *dispersive* region, where orbit position depends on the energy. Let's insert a sextupole magnet with normalized strength S (or K') into the dispersive area

$$S = \frac{1}{2!B\rho}\frac{\delta^2 B}{\delta x^2} \tag{4.9}$$

The sextupole magnet produces the following effect (*kick*) on the angle of the beam trajectory

$$x' = x' + S\left(x^2 - y^2\right) \quad \text{and} \quad y' = y' - S\,2xy \tag{4.10}$$

In the dispersive region, one needs to replace x with $x+\eta\delta$ where η is the dispersion; thus, the sextupole kick will contain the energy-dependent focusing since the terms in the equation below correspond to a quadrupole with an effective gradient

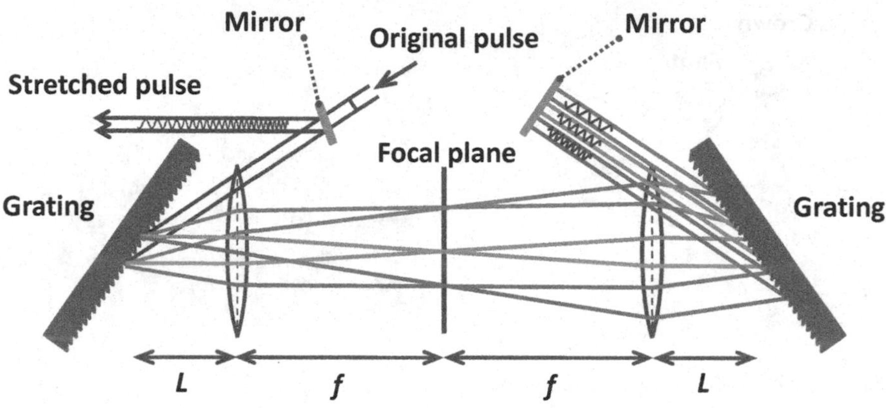

Figure 4.29 Laser pulse stretcher.

equal to $S \cdot \eta$:

$$x' \Rightarrow S(x+\eta\delta)^2 \Rightarrow 2S\,\eta\,x\delta + .. \tag{4.11}$$

$$y' \Rightarrow -S\,2\,(x+\eta\delta)\,y \Rightarrow -2S\,\eta\,y\delta + ..$$

Therefore, such an arrangement of nonlinear magnets placed in a dispersive region can be used for *chromatic correction* in the event of beam optics. One should also note that the terms that are not shown in Eq. 4.3.3, containing x^2 and δ^2, are unwanted additional aberrations — geometrical aberrations and higher-order chromatic aberrations. They often need to be corrected, either by special optic arrangements and appropriate placement of *sextupole pairs* or by higher-order magnets.

4.3.4 COMPRESSION OF BEAM AND LASER PULSES

Compression of laser pulses relies on the dependence of the light reflection angle from the grating on the light wavelength, in the case when the grating spacing is comparable to the wavelength. This produces an optical system where the path length through the system depends on the light's wavelength.

Energy-z correlation is used to compress/stretch either laser pulse or particle bunch — this is one more general principle of AS-TRIZ.

The combination of a pair of grated plates, lenses and mirrors can create a laser pulse stretcher — as realized in Fig. 4.29.

Equally, a laser pulse compressor can be arranged with a pair of gratings, as shown in Fig. 4.30.

The optical telescope placed inside the pulse stretcher is needed in order to provide a "negative distance" between the gratings.

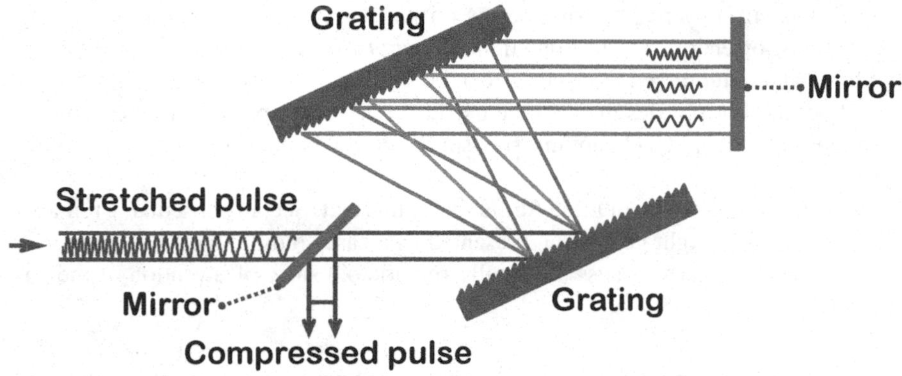

Figure 4.30 Laser pulse compressor.

Laser pulse compressor gratings are technological marvels. In order to stay below the damage threshold of the grating material, the size of the plates has to be around a meter in diameter for some of the highest power lasers. Taking into account that the typical space between grooves is around a micrometer, we can imagine the high levels of accuracy needed to produce such plates.

Charged particle beam compression is based on the same principle of the path length dependence on the particle's energy. Dispersion is created using dipole bending magnets and the initial beam is arranged to have an $E - z$ correlation, via its acceleration *off-crest* of RF voltage.

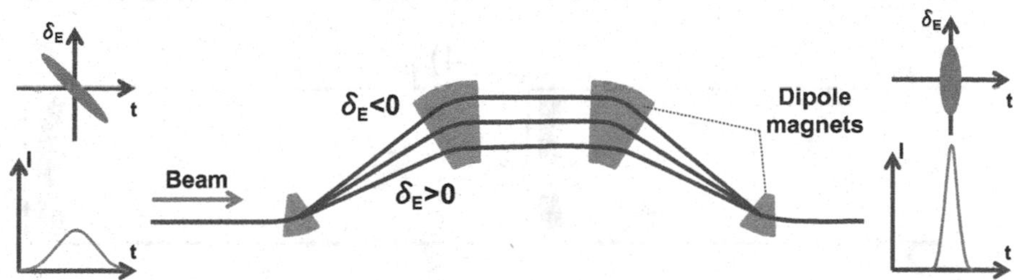

Figure 4.31 Bunch compressor.

Fig. 4.31 shows a bunch compressor and the phase space of the beam before and after compression.

4.4 INTERACT

We have reached the final step of the sequence of Create — Energize — Manipulate — *Interact.*

Once the beam is accelerated, it can be used in a variety of ways — from creating a radiation (synchrotron, betatron) source, a free electron laser, a collider, a spallation neutron source, to using beams for particle therapy, industrial or security applications, or energy applications in *accelerator-driven systems* (ADS). Equally, amplified and compressed laser pulses can be used in a variety of ways; in the context of accelerator physics they can be used as a driver for particle acceleration, as the main component of a Compton X-ray source or of a Photon collider, among other things.

We will touch on some of these further on in this text. See, in particular, a remarkable story of EUV light generation presented as a case study in the FEL Chapter 8. And in meantime, let's discuss an equally remarkable story of invention of the so-called *Mak telescope*.

4.5 CREATION OF MAK TELESCOPE – INVENTION CASE STUDY

Optical telescopes are known from around 1608. Refraction telescopes contain lenses and therefore suffer from chromatic aberrations. On the other hand, mirrors have no chromatic aberrations, and can be also used to build telescopes. The first reflecting telescope was invented by Newton in 1668 and its configuration is shown in the first picture in Fig. 4.32.

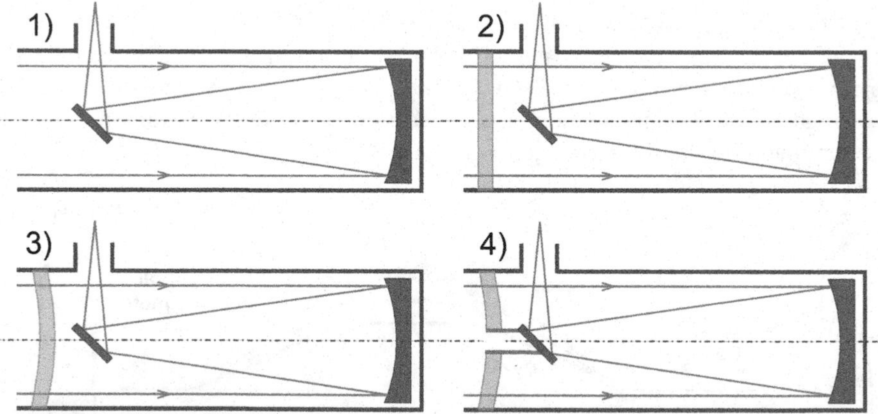

Figure 4.32 Steps toward Mak telescope invention. Starting from the Newton telescope configuration — step 1. Step 2 — add optical cover to protect the telescope and change the expensive parabolic mirror with a simple spherical mirror. Step 3 — change the optical cover into a meniscus lens, which will compensate spherical aberrations of the mirror. Step 4 — make a hole in the meniscus lens and connect the viewer mirror holders to that hole.

Maksutov — the inventor of the telescope now called *Mak* — was passionate about astronomical observations, and around 1940 was working on the task to create simple, inexpensive and robust reflector telescope for schools.

He noted the challenges of existing reflector telescopes, such as the Newton one — the open pipe, in particular, makes such design vulnerable and easy to damage, especially in school environment. Maksutov also noted that the viewport mirror holders are needed, which block part of light. And moreover, he noted that the main mirror needs to be parabolic, which is relatively hard to manufacture and thus expensive.

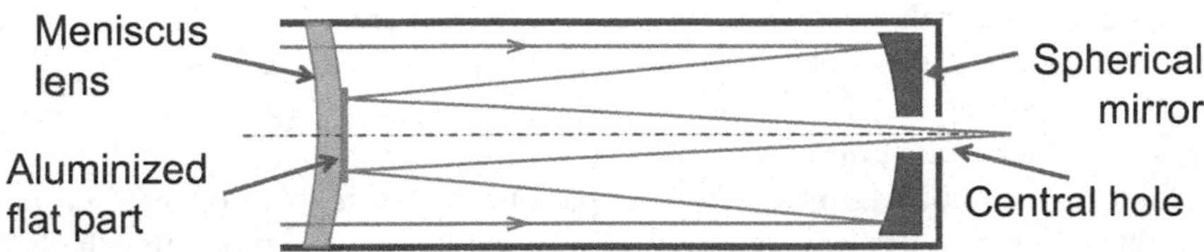

Figure 4.33 Final step of Mak telescope invention.

The process of re-inventing Mak telescope, in view of TRIZ approach, could look as follows. As the first step, to fix the telescope vulnerability due to its open pipe, Maksutov decided that a glass cover is needed — see step 2 in Fig. 4.32.

To reduce the production cost, he then suggested to use a spherical mirror instead of the parabolic one. This simplified the design and made it less expensive, but it introduced spherical aberrations.

Then the first Eureka came — he realized that he can make the glass cover in the shape of a weak meniscus lens (a lens with two spherical surfaces) that will exactly cancel the spherical aberration of the spherical mirror. From TRIZ viewpoint this is the principle of *cancelling one harmful factor with another harmful factor.*

The configuration then looked as the step 3 in Fig. 4.32, where the meniscus lens protected the telescope and compensated spherical aberration of the spherical mirror. The next challenge was that the viewport mirror placed inside required supporting holders, which complicated the design. Then the next Eureka came — he realized that one can make a hole in the meniscus lens in the center, to create holders for the mirror — step 4 in Fig. 4.32.

And yet another question came: why do we need this viewport mirror and holders? Can't we just use the meniscus lens? The final step of invention is shown in Fig. 4.33, where the meniscus lens protects the telescope, compensates for the spherical aberrations of the spherical mirror, and its central part covered with Aluminum also serves as a mirror to send the image to an eye.

Overall, the principles demonstrated in Mak telescope invention — the principle of compensating one harmful effect with another harmful effect, the principle of using energy and resources that we already have in the system — are principles that are often used in designs and inventions of systems for focusing of charged particles beams as well.

Invention of Mak telescope demonstrates the use of inventive principles of compensating one harmful effect with another harmful effect, and the principle of using energy and resources that we already have in the system.

EXERCISES

4.1. *Chapter materials review.*
A certain plasma's density is 10^{17} cm^{-3}. A laser light of which wavelength could still penetrate such a plasma? Also, estimate the corresponding plasma frequency.

4.2. *Chapter materials review.*
Taking into account a) Fermat's principle of the least amount of time for light propagation through an optical system and b) the observation that the diffraction angle should approach the law for reflection from a mirror as the wavelength becomes very short (and hence diffraction becomes less important), explain qualitatively why an optical telescope is needed inside the laser pulse stretcher, but not required in the laser pulse compressor.

4.3. *Analyze inventions or discoveries using TRIZ and AS-TRIZ.* Analyze and describe scientific or technical inventions described in this chapter in terms of the TRIZ and AS-TRIZ approaches, identifying a contradiction and an inventive principle that were used (could have been used) for these inventions.

4.4. *Developing AS-TRIZ parameters and inventive principles.* Based on what you already know about accelerator science, discuss and suggest the possible additional parameters for the AS-TRIZ contradiction matrix, as well as the possible additional AS-TRIZ inventive principles.

4.5. *Practice in reinventing technical systems.* Imagine that you need to measure the gas pressure inside of an electric light bulb very precisely. Can you suggest a method to do this? Breaking the bulb is not allowed. Hint: try to use one of the physical principles shown in illustrations to this Chapter.

4.6. *Practice in reinventing technical systems. Electron cloud* effects can arise in proton or positron storage rings, and can result in significant deterioration of the beam quality via a single bunch instability. The instability mechanism involves the circulating beam attracting electrons from the residual gas by positively charged beam bunches, with the residual gas electrons then bouncing in the vacuum chamber in resonance with the bunch frequency, leading to electron cloud density buildup and instability impacting the beam's quality. Suggest a way to mitigate electron cloud instability.

4.7. *Practice in the art of back-of-the-envelope estimations.* An electron gun immersed in a 0.3-T solenoidal magnetic field is intended to produce a round 2-mm- diameter beam for magnetized electron cooling of 1 MeV kinetic energy protons. Estimate the electron current. Discuss how would you increase the electron current. *(It is assumed that you can identify the most important effects playing roles in this task, can define the necessary parameters and set their values, and can get a numerical answer.)*

5 Conventional Acceleration

In this chapter we will discuss conventional acceleration. Beginning with a historical introduction, we will then touch on waveguides, resonant cavities fed by RF power generators, and conclude with an overview of the basics of linacs and longitudinal dynamics.

5.1 HISTORICAL INTRODUCTION

The "Livingston plot" presented in Fig. 1.6 of Chapter 1 oversees the development of accelerators over the past several decades. These days, we refer to them as conventional accelerators. These accelerators are also human-made, in contrast with, for example, a cosmic accelerator, which has the capacity to produce particles with energies exceeding 10^{20} eV, or lightning, which produces fairly minimal acceleration.

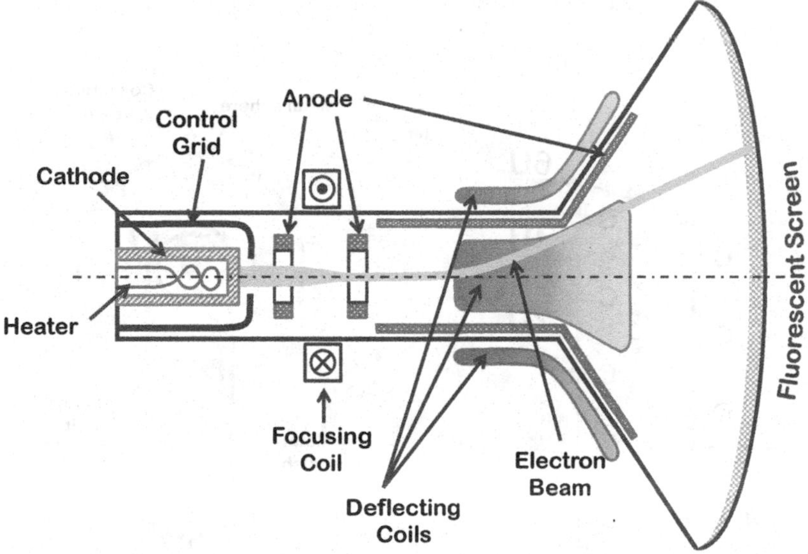

Figure 5.1 A cathode ray tube TV as an example of an accelerator.

To contrast the composition of older accelerators with that of a contemporary one, we will first look at an accelerator familiar to everyone — a cathode ray tube TV (Fig. 5.1), which is a common example of an accelerator with all subsystems present: a beam source, focusing and steering, an accelerating region and an interaction or a target area (i.e., a fluorescent screen).

DOI: 10.1201/9781003326076-5

5.1.1 ELECTROSTATIC ACCELERATORS

In an inspirational push in favor of accelerator development, Rutherford spoke at The Royal Society in 1928, lamenting, "I have long hoped for a source of positive particles more energetic than those emitted from natural radioactive substances."

The first accelerator developed was the Cockcroft–Walton generator which is based on a system of multiple rectifiers (Fig. 5.2). In this generator, voltage generated by the cascade circuit

$$U_{tot} = 2Un - \frac{2\pi I}{\omega C}\left(\frac{2}{3}n^3 + \frac{1}{4}n^2 + \frac{1}{12}n\right) \tag{5.1}$$

depends on the number of cells n, as well as generated current I, capacity C and frequency ω.

The Cockcroft–Walton accelerator helped to make Rutherford's dream a reality. In the first-ever transmutation experiment, the accelerated 700-keV protons were sent onto a lithium target, resulting in the production of helium.

Practical reasons (i.e., size of the device, performance of capacitors and diodes) limited the generated voltages and corresponding energies of the accelerated particles up to about $\sim 4\ MV$. Cockcroft–Walton accelerators can typically generate beam currents of several hundred mA with CW or pulsed particle beams of few μs pulse lengths.

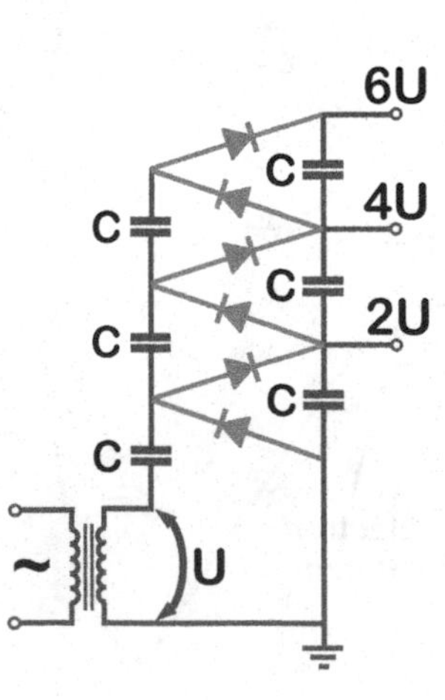

Figure 5.2 Cockcroft–Walton generator.

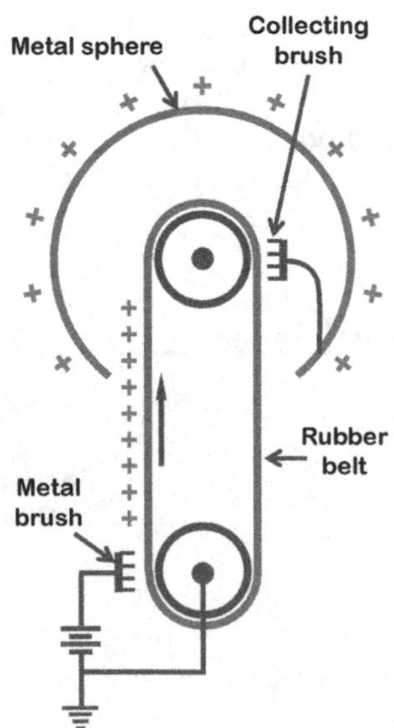

Figure 5.3 Van der Graaf accelerator.

Another example of an electrostatic accelerators is the "Van der Graaf" (see Fig. 5.3), where a metal brush deposits charges onto a rubber belt, carrying the charges into a metal sphere where they are later collected by another brush. The exchange of charges is performed with help from a discharge between the sharp tips of the needles of the brushes and the belt.

The ion source in the "Van der Graaf" accelerator is located inside of the metal sphere, and is charged to a high voltage. An electrical generator (mechanically connected to the upper axis of the rubber belt pulley) produces electrical power needed to operate the ion source or any other internal control electronics. The ions produced by the source are accelerated on their way down from the sphere.

With any electrostatic accelerator, it is difficult to achieve high energies due to limitations determined by the size of the vessels. In particular, the highest recorded energy created via the Van der Graaf accelerator was around ~25 MeV.

A version of the Van der Graaf accelerator that can produce twice higher energy of the particles is called a *tandem accelerator* (Fig. 5.4). In this case, negative ions are accelerated and, when they reach the charged sphere, they pass through a foil or gas target to perform their *charge exchange*. Now positive, ions continue to accelerate toward ground potential and thus reach twice the voltage of the charged sphere. An additional advantage of the tandem accelerator is having its ion source located at ground potential, which significantly simplifies its operation and maintenance.

Another version of the charge-carrying mechanism is realized in a *pelletron*, where instead of the rubber belt, metal pellets are connected by non-conductive links into a chain (see Fig. 5.5).

5.1.2 SYNCHROTRONS AND LINACS

Synchrotrons can accelerate particles to much higher energies than electrostatic devices can. Their name is derived from the process of *synchronous* change of the magnetic field of bending magnets according to the growing energy of the accelerated beam (see the quote at right).

Many modern accelerators are synchrotrons (e.g., the Large Hadron Collider at CERN is a synchrotron). These types of accelerators can reach very high energies, limited only (especially for electrons) by synchrotron radiation and the cost of their construction.

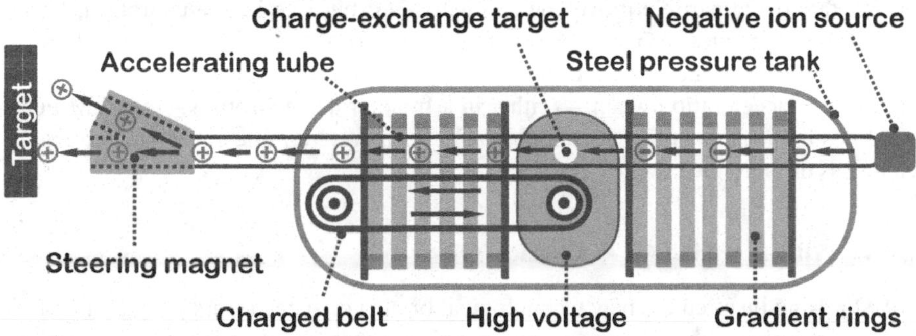

Figure 5.4 Tandem electrostatic accelerator.

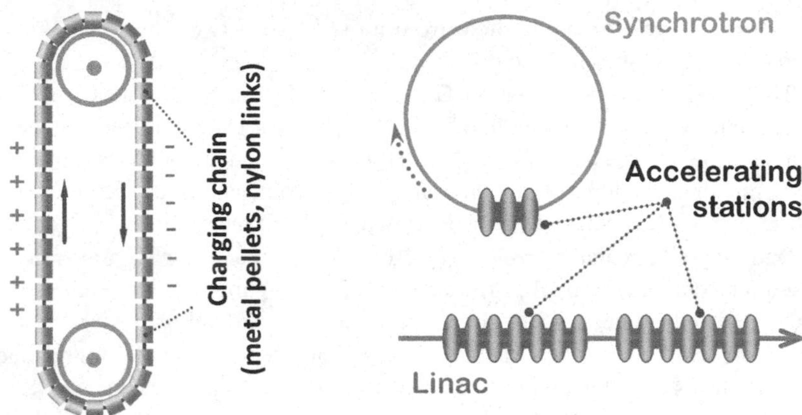

Figure 5.5 Pelletron charging mechanism. **Figure 5.6** Synchrotron and linac.

As we will momentarily explain, time-varying fields are the necessary conditions for acceleration of charged particles to high energies, and in particular for overcoming the limitations of the maximum achieved energy in the electrostatic accelerators.

Both linear and circular accelerators use EM fields oscillating in resonant cavities to achieve acceleration. In circular accelerators, particles follow an orbit guided by a magnetic field and return to the same accelerating cavity on every turn, while in linac accelerators the particles follow a straight path through a sequence of cavities, as shown in Fig. 5.6.

Powerful radio-frequency (RF) systems produce the required powerful electric fields in the resonant cavities. Accelerators of this kind progressed in large part due to the telecommunications industry, which drove the development of power systems with frequencies ranging from a few MHz to several GHz.

Recalling Maxwell's equation and its integral form

$$\nabla \times \mathbf{E} = -\frac{\partial \mathbf{B}}{\partial t} \qquad \text{or} \qquad \oint_{\partial\Sigma} \mathbf{E} \cdot d\ell = -\frac{d}{dt}\int_{\Sigma} \mathbf{B} \cdot d\mathbf{S}$$

we see that if the particle moves on an enclosed orbit, like in a synchrotron, finite acceleration (i.e., nonzero contour integral of **E**) would not be possible without a time-dependent magnetic field **B**. On the other hand, time-dependent magnetic flux can provide acceleration when — either in a linac or a synchrotron — the EM field oscillates in the resonant cavity and the particles receive a finite energy increment at each pass through the cavities.

5.1.3 WIDERÖE LINEAR ACCELERATOR

In 1925, Ising inspired the new technological branch of accelerators by realizing that limitations imposed by corona formation and discharge in electrostatic accelerators can be overcome with the use of alternating voltages. This was carried further by

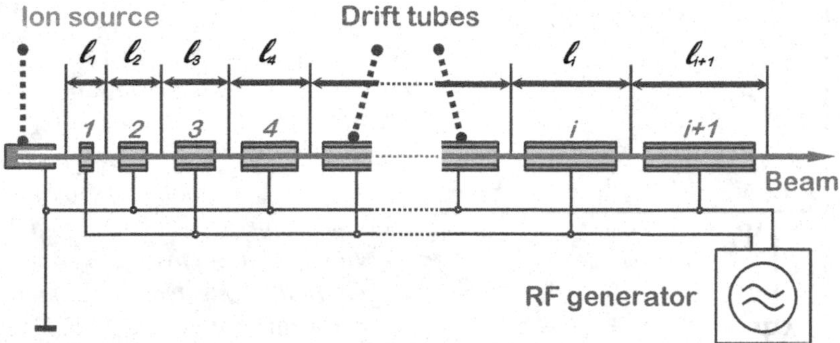

Figure 5.7 Wideröe linear accelerator.

Wideröe who, in 1928, performed the first successful test of the linac based on this principle.

The Wideröe linac has a series of drift tubes arranged along its beam axis and connected with an alternating high frequency RF voltage: $V(t) = V_{\max} \sin(\omega t)$ as shown in Fig. 5.7. During the first half of the RF period in the acceleration process, voltage applied to the first drift tube accelerates charged particles emerging from the ion source. Particles then enter the first drift tube and pass through it. Meanwhile, the direction of the RF field reverses without affecting the particles — seeing as the drift tube acts as a Faraday cage and shields the particles from external fields. When the particles reach the gap between the first and the second drift tubes, they accelerate and the process repeats in the following gaps as well.

The energy of the particle in Wideröe linac after passing the i-th drift tube is

$$E_i = iqV_{\max} \sin \Psi_0 \tag{5.2}$$

where q is the charge of the particle and Ψ_0 is average phase of the RF voltage that particles feel as they cross the gaps (see Fig. 5.8). As we can see, the energy is proportional to the number of stages i passed by the particle. Furthermore, the largest voltage to ground in the entire system never exceeds $V_{\max}$. This lets us reach high energies without using voltage levels, which can cause electrical breakdown.

The accelerating gaps between drift tubes in the Wideröe linac must increase in sync with the monotonically increasing velocity of the particle. Taking into account that the half-period of RF $\tau_{RF}/2$ should correspond to a particle passing with velocity v_i through one drift section, we write the distance between i-th and $(i+1)$-th gaps as

$$\ell_i = \frac{v_i \tau_{RF}}{2} = \frac{1}{f_{RF}} \sqrt{\frac{iqV_{\max} \sin \Psi_0}{2m}}$$

which we expanded using Eq. 5.2.

5.1.4 ALVAREZ DRIFT TUBE LINAC

The Alvarez linac is conceptually quite similar to the Wideröe linac. It differs in that, in an Alvarez linac, the accelerating voltage at individual drift tubes is created by an

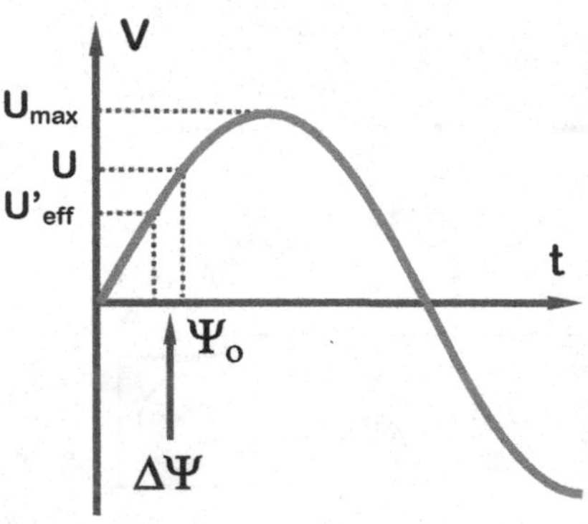

Figure 5.8 Voltage in Wideröe linac.

"Particles should be constrained to move in a circle of constant radius thus enabling the use of an annular ring of magnetic field...which would be varied in such a way that the radius of curvature remains constant as the particle gains energy through successive accelerations by an alternating electric field applied between coaxial hollow electrodes." — Mark Oliphant, 1943.

RF wave in a container (a tank made of a good conductor such as copper), in which the drift tubes are located (see Fig. 5.9). The drift tubes may have magnets installed inside to focus the beam during acceleration.

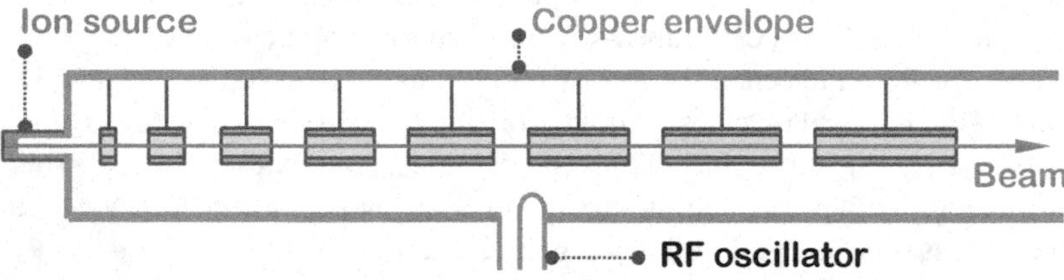

Figure 5.9 Alvarez drift tube linac.

The drift tube linacs are still used (particularly in hadrons), but they are being replaced by better-performing *RFQ-structures* (Fig. 5.10). Due to periodic transverse variations of their shape, such structures allow for the creation of not only accelerating fields, but also focusing fields. Recent progress via 3D computer simulations of RF fields and beam dynamics resulted in a widespread use of RFQs.

Most of the complicated linac designs discussed above are applicable to hadrons, which (in practically achievable accelerating gradients) become relativistic rather slowly, after a hundred or so meters of acceleration. Correspondingly, the size and shape of accelerating cavities need to vary along the hadron linac, so as to match their increasing velocity. For electrons, which have already become highly relativistic after a few tens of cm of acceleration, relatively simpler linac design is possible, in which cavities of the same size and shape are regularly placed along the acceleration path.

5.1.5 PHASE FOCUSING

As we have seen in the previous section, the energy transferred to particles in a drift tube linac depends on the voltage amplitude $V_{\max}$ and phase Ψ_0. We should note that a small deviation of the nominal voltage $V_{\max}$ would result in a particle velocity that

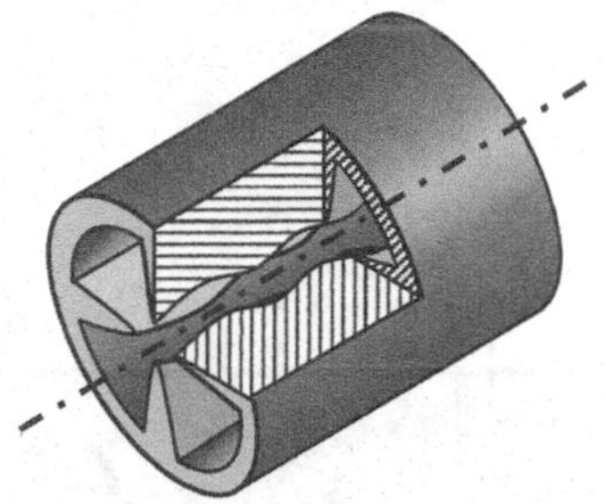

Figure 5.10 RFQ structure.

Drift tubes linacs in modern accelerators for protons or ions are now typically replaced by RFQ (radio-frequency quadrupole) structures, which combine acceleration with transverse focusing.

no longer matches the design velocity fixed by the length of drift sections. In this case, the particles would undergo a phase shift relative to the RF voltage and the synchronization of particle motion with respect to RF field will be eventually lost.

The system can be made to self-adjust to such small deviations if we use $\Psi_0 < \pi/2$ so that the effective accelerating voltage is $V_{eff} < V_{max}$, as shown in Fig. 5.8. In this instance, if a particle gains too much energy in the preceding stage and is traveling faster than the ideal particle and arrives at the next acceleration stage earlier, it will then feel the average RF phase $\Psi = \Psi_0 - \Delta\Psi$ and will be accelerated by the voltage

$$V'_{eff} = V_{max} \sin(\Psi_0 - \Delta\Psi) < V_{max} \sin\Psi_0 \tag{5.3}$$

which is below the ideal voltage. The particle will thus gain less energy and will slow down and return to the nominal velocity. The process will then repeat and the particles will continuously oscillate about the nominal phase Ψ_0, exhibiting the phenomena of *phase focusing*. If dissipation is present in the system (such as SR damping), these oscillations would eventually be forced to decay and the particles would gather around the synchronous phase. A similar effect of bunching can happen without damping, due to acceleration.

5.1.6 SYNCHROTRON OSCILLATIONS

The phase focusing of particles in accelerators manifests itself as the periodic longitudinal particle motion about the nominal phase, and is called *synchrotron oscillation*.

In circular accelerators, as the ideal particle encounters the RF voltage at exactly the nominal phase on each revolution, the RF frequency ω_{RF} must be an integer multiple of the revolution frequency ω_{rev}

$$h = \omega_{RF} / \omega_{rev} \tag{5.4}$$

where h is called the *harmonic number* of the ring: an integer chosen to allow matching the frequency of available RF generators.

For *relativistic* particles, the phase focusing is enabled by the dependence of the circular path on energy, as shown in Fig. 5.11. The stable point in this case is on

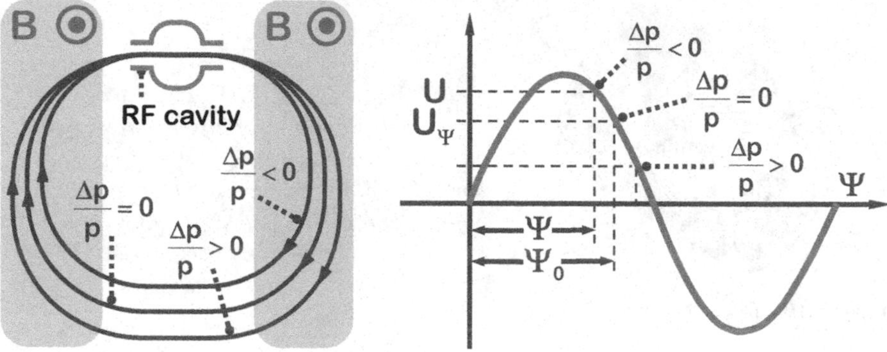

Figure 5.11 Synchrotron oscillations.

the declining slope of the sine wave, as particles with $\Delta p/p > 0$ will travel on the longer path, come to the RF cavity in the next turn somewhat later, and ultimately lower their acceleration. As the reader can already guess, for a particle that is not yet relativistic and whose velocity depends on energy, the situation is quite different — we will discuss this further later on in this chapter.

5.2 WAVEGUIDES

Prior to engaging in detailed discussion of RF accelerating cavities, let's introduce — via simple considerations and analogies — the basic properties of waveguides, as they provide important and intuitive understanding applicable for accelerating structures.

5.2.1 WAVES IN FREE SPACE

The velocity of an EM wave in a vacuum and in a medium is

$$\text{vacuum}: \quad v = c = \frac{1}{\sqrt{\varepsilon_0 \mu_0}}, \quad \text{medium}: \quad v = \frac{1}{\sqrt{\varepsilon_0 \varepsilon_r \mu_0 \mu_r}} \tag{5.5}$$

where ε_r is the dielectric constant and μ_r is the magnetic permeability of the medium.

The amplitudes of electric and magnetic fields in an EM wave in a vacuum are exactly the same if expressed in Gaussian-cgs units (demonstrating the naturalness of this system of units) and relates as $E = cB$ in SI units.

A plane EM wave with transverse electric and magnetic fields (TEM wave) propagating in free space in x-direction is shown in Fig. 5.12. The quantity

$$\mathbf{P} = (\mathbf{E} \times \mathbf{B}) / \mu_0 \tag{5.6}$$

is called the *Poynting vector* and equals to the local power flux.

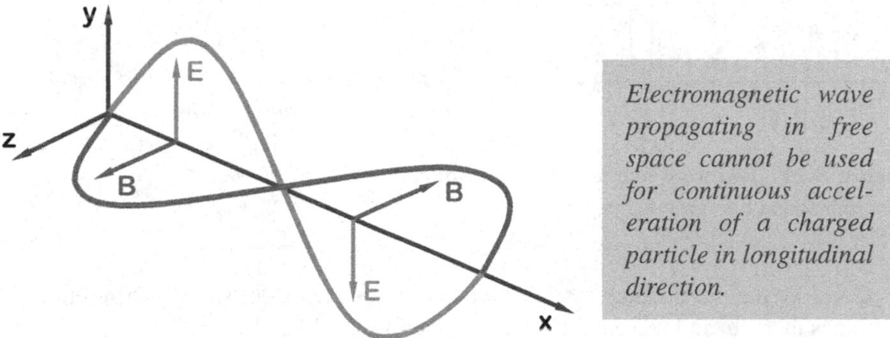

Electromagnetic wave propagating in free space cannot be used for continuous acceleration of a charged particle in longitudinal direction.

Figure 5.12 TEM wave in free space.

5.2.2 CONDUCTING SURFACES

In the derivations presented in this section, we discuss the behavior of EM waves bounded in metal boxes; therefore, we need to recall the boundary conditions of a wave at a perfectly conducting metallic surface.

On the surface of a perfect conductor, the tangential component of an electric field $E_{||}$ and the normal component of a magnetic field $B_{\perp}$ will vanish, as illustrated in Fig. 5.13.

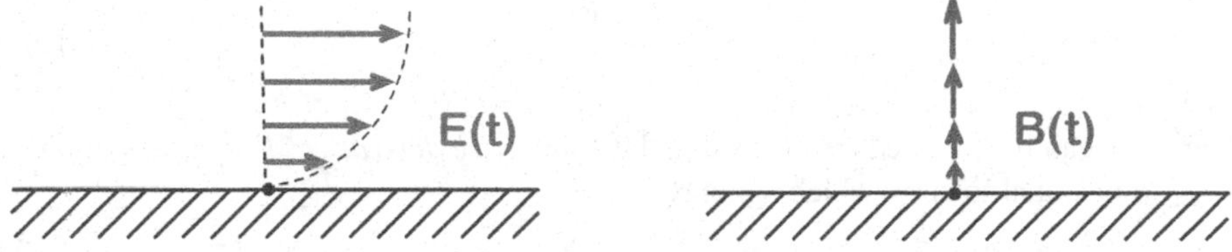

Figure 5.13 Boundary conditions on perfectly conducting surfaces.

A non-ideal surface with conductivity σ is characterized by *skin depth*; an EM wave entering a conductor is dampened to $1/e$ of its initial amplitude at the depth

$$\delta_S = \frac{1}{\sqrt{\pi f \ \mu_0 \mu_r \sigma}} \tag{5.7}$$

This allows us to introduce the notion of *surface resistance*

$$R_{surf} = \frac{1}{\sigma \, \delta_S} \ . \tag{5.8}$$

which plays an important role in determining performance of accelerating cavities.

5.2.3 GROUP VELOCITY

In preparation to discuss dispersion properties of waveguides and RF structures, let us recall the derivation of *group velocity* — the propagation velocity of energy (and information) in an EM wave.

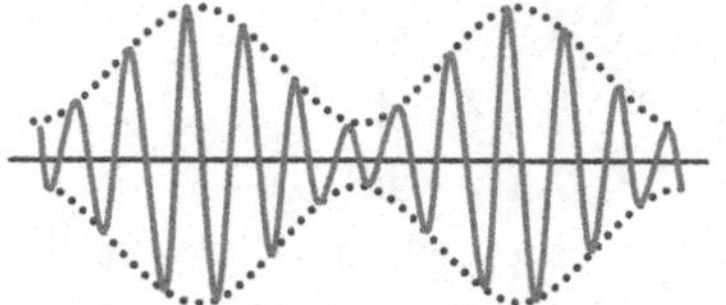

Figure 5.14 Two-wave interference for determination of group velocity.

Group velocity defines the propagation of energy (and information) in an EM wave and is always lower than c. Phase velocity can be higher or lower than the speed of light.

Consider the interference between two continuous waves of slightly different frequencies $\omega \pm d\omega$ and wavenumbers $k \pm dk$ (see Fig. 5.14):

$$\begin{aligned} E &= E_0 \sin\left[(k+dk)x-(\omega+d\omega)t\right] \\ &\qquad + E_0 \sin\left[(k-dk)x-(\omega-d\omega)t\right] \\ &= 2E_0 \sin\left[kx-\omega t\right]\cos\left[dk\,x-d\omega\,t\right] \\ &= 2E_0 f_1(x,t)\, f_2(x,t) \end{aligned} \tag{5.9}$$

The last line of Eq. 5.9 contains two functions, f_1 and f_2. The first function corresponds to a continuous wave with the mean wavenumber and frequency: $f_1(x,t) = \sin\left[kx-\omega t\right]$. In this wave, any given phase is propagated such that $kx-\omega t$ remains constant, which gives us the equation for the *phase velocity* of the wave:

$$v_p = \frac{\omega}{k} \tag{5.10}$$

The same can be obtained by requesting the *convective derivative*[1] $(\partial/\partial t + v_p \partial/\partial x)$ of f_1 to be equal to zero, which again results in

$$v_p = -\frac{\partial f_1(x,t)/\partial t}{\partial f_1(x,t)/\partial x} = \frac{\omega}{k}$$

The second function in Eq. 5.9 describes the evolution of the envelope of the pattern: $f_2(x,t) = \cos\left[dk\,x - d\omega\,t\right]$. Again, any point in the envelope propagates such that the quantity $dk\,x - d\omega\,t$ remains constant and therefore its velocity, i.e., the group velocity, is given by

$$v_g = -\frac{\partial f_2(x,t)/\partial t}{\partial f_2(x,t)/\partial x} = \frac{d\omega}{dk} \tag{5.11}$$

This prepares us for discussion of the notions of *dispersion* and *group* or *phase velocity* of a waveguide.

---

[1]*Convective derivative — the term originates from fluid mechanics — is the derivative taken with respect to a moving coordinate system.*

5.2.4 DISPERSION DIAGRAM FOR A WAVEGUIDE

We will start our conversation regarding wave propagation down a waveguide using two extreme cases.

First of all, if the wavelength λ of an EM wave in free space is much shorter than the transverse size a of the waveguide $\lambda \ll a$ (as shown in Fig. 5.15, case (i)), then the waveguide does not matter, and we expect the dispersion at large ω to approach the equation for free space (i.e., $\omega/c = k$). With the goal of deriving the dependence of frequency ω against wavenumber $k = 2\pi/\lambda$ in a waveguide, let us place a corresponding segment on a waveguide dispersion curve at a high frequency (Fig. 5.16, case (i)).

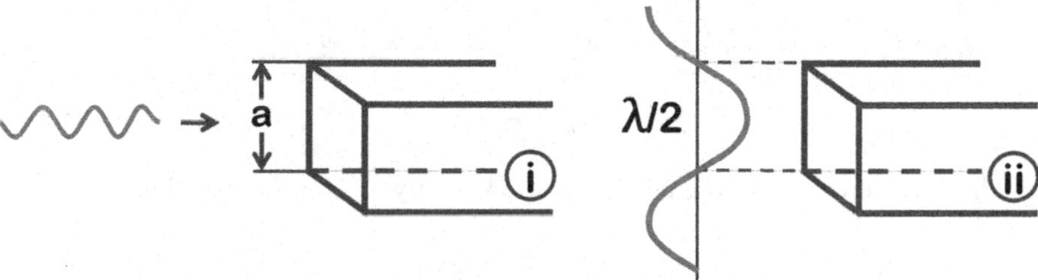

Figure 5.15 Waves in a waveguide, two extreme cases.

Another extreme case is shown in Fig. 5.15, (ii), when half of a wavelength in free space equals the waveguide transverse size. As can be seen from this diagram, the longest wavelength for which the boundary conditions at a perfectly conducting surface of the waveguide can still be satisfied, is given by $\lambda/2 \leq a$. This defines the cut-off parameters $\lambda_c = 2a$ or $\omega_c = \pi c/a$; that is, waves with wavelengths longer than λ_c cannot propagate in the waveguide.

As Fig. 5.15 (ii) suggests, the case of $\omega = \omega_c$ corresponds to an infinite wavelength in in the waveguide, or $k = 0$. We thus plot a corresponding point in Fig. 5.16.

In the intermediate region of frequencies, the dispersion curve should connect the point (ii) and region (i) in Fig. 5.16. Looking at the wave in a waveguide from a simple geometrical point of view (Fig. 5.17) and considering corresponding similar triangles, one can write

$$\left(\frac{\omega}{c}\right)^2 = k^2 + \left(\frac{\omega_c}{c}\right)^2 \tag{5.12}$$

thus describing dispersion of the wave in a rectangular waveguide, graphically shown in Fig. 5.18.

Eq. 5.12 suggests that, for any wavenumber k, the frequency is always greater than the cut-off frequency. Looking at the slope (i.e., derivative $d\omega/dk = v_g$) of the curve in Fig. 5.18 we can also observe that the longer the wavelength or lower the frequency, the slower the group velocity, and at the cut-off frequency no energy flows along the waveguide.

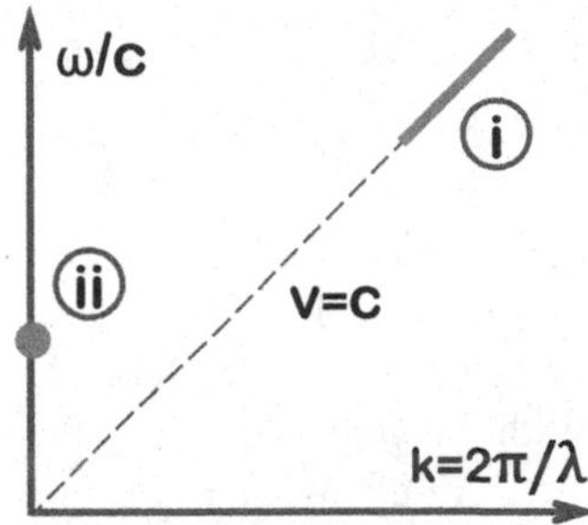

Figure 5.16 Dispersion of a waveguide, two extreme cases.

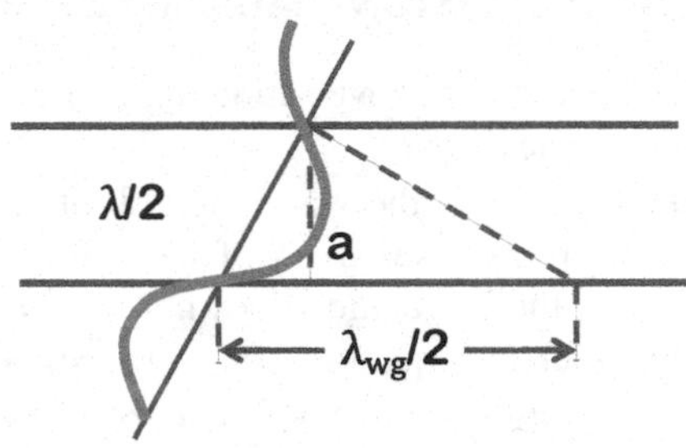

Figure 5.17 Intermediate case.

Eqs. 5.12 and 5.10 also help find that

$$v_p v_g = c^2 \tag{5.13}$$

which tells us that, in a waveguide, the phase velocity is always *larger* than the speed of light.

5.2.5 IRIS-LOADED STRUCTURES

We can conclude from the previous section that acceleration in a waveguide is not possible because the phase velocity of the wave exceeds that of light. Particles that travel slower than the wave would be periodically accelerating or decelerating, achieving zero net acceleration when averaged over a long time interval.

In order to make the acceleration possible, one needs to modify the waveguide to reduce the phase velocity to an appropriate value below the speed of light, so as to match the velocity of the particle.

Reduction of the phase velocity can be achieved by using iris-shaped screens installed into the waveguide with a constant step along the axes shown in Fig. 5.19.

The dispersion relation in a waveguide $\omega = c\sqrt{k_z^2 + (2\pi/\lambda_c)^2}$ changes due to installation of the irises. Qualitatively, behavior at low k resembles that of the waveguide curve, defined by the smallest diameter of the irises. At higher k, with the in-

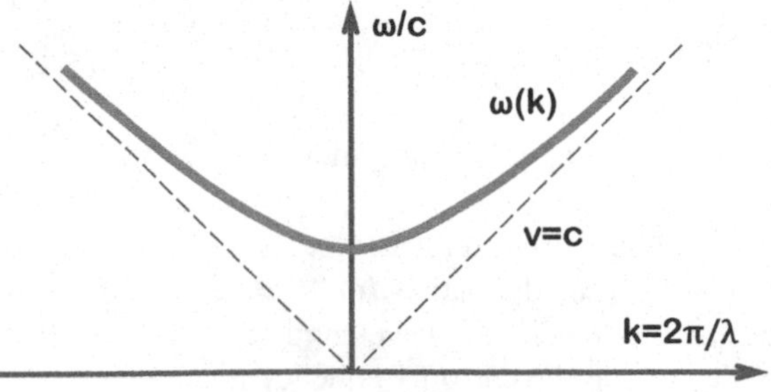

Figure 5.18 Dispersion of a waveguide.

Dispersion function of EM wave in a waveguide shows that phase velocity is always higher than the speed of light, and therefore acceleration of particles in waveguides is not possible.

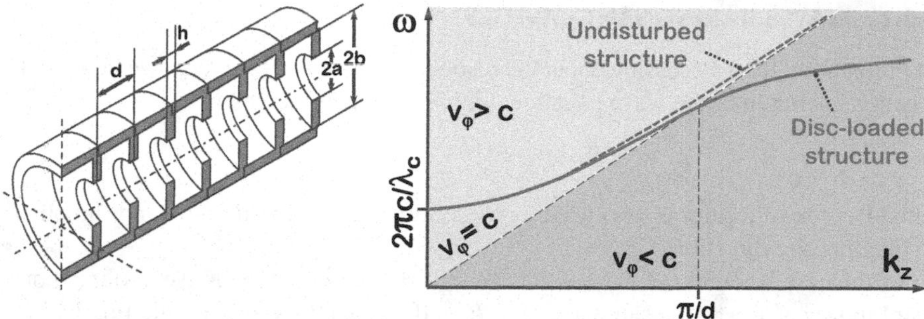

Figure 5.19 Iris-loaded accelerating structure.

Figure 5.20 Qualitative behavior of dispersion curve in iris-loaded structures.

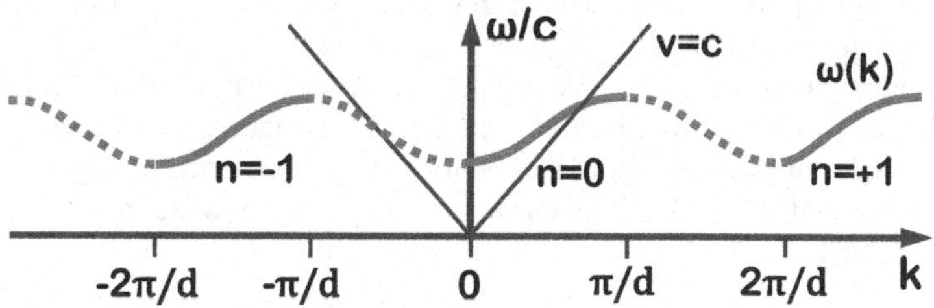

Figure 5.21 Extended dispersion diagram of an iris-loaded structure.

stallation of irises, the curve flattens off and crosses the boundary of $v_\varphi = c$ in the region of $k_z = \pi/d$ as illustrated in Fig. 5.20.

With a properly selected iris separation d and other parameters of the irises, the phase velocity of the iris-loaded structure can be set to any value, making it suitable for particle acceleration with arbitrary v/c.

An extended version of the dispersion diagram curve of an iris-loaded structure can also be constructed while noting that adding a multiple of $2\pi/d$ to the wavenumber k as

$$k_n = k_0 + \frac{2n\pi}{d}$$

would still satisfy the periodic boundary conditions at the irises. Therefore, the dispersion curve repeats itself at space harmonics corresponding to different integers, n, as shown in Fig. 5.21.

The first rising slope of the dispersion curve shown in Fig. 5.21 is usually used for acceleration.

5.3 CAVITIES

In this section, we will consider general properties of resonant cavities, their quality factors, shunt impedance, and will introduce the definition of the resonance modes.

5.3.1 WAVES IN RESONANT CAVITIES

In preparation for a discussion about the resonance modes in the cavity, we first recall a general solution of the wave equation, which can be written as

$$W(r,t) = Ae^{i(\omega t + k\cdot r)} + Be^{i(\omega t - k\cdot r)} \tag{5.14}$$

This describes the sum of two waves — one moving in one direction and the other in the opposite direction.

In the case wherein the wave is totally reflected from a conductive surface, both amplitudes need to be the same, i.e., $A = B$, and we can therefore rewrite Eq. 5.14 as

$$W(r,t) = Ae^{i\omega t}(e^{ik\cdot r} + e^{-ik\cdot r}) = 2A\cos(k\cdot r)e^{i\omega t} \tag{5.15}$$

This equation describes the field configuration with a static in time amplitude $2A\cos(k\cdot r)$, therefore corresponding to a standing wave.

Consider now that the waveguide we discussed earlier has a finite length ℓ and its entrance and exit are closed by two conducting surfaces, forming a resonance cavity. The resonant wavelengths of this cavity can be determined noting that a stable standing wave can form in this fully enclosed cavity if we assume that the following condition is satisfied

$$\ell = q\frac{\lambda_z}{2} \quad \text{with} \quad q = 0,\ 1,\ 2, \ldots \tag{5.16}$$

Therefore, only certain well-defined wavelengths λ are present in the cavity.

Near the resonant wavelength, the resonant cavity behaves like an oscillator with a high *quality factor* Q, allowing it to build up high voltages that can be used for particle acceleration. The cavities are often modeled as electrical oscillators, with their Q-value determined by losses of equivalent individual coils, capacitors and resistances of the circuit model.

5.3.2 PILL-BOX CAVITY

An enclosed section of a waveguide (either rectangular or cylindrical) forms the simplest RF cavity, called a *pill-box cavity*.

The conventionally accepted classification of the modes in pill-box cavities separates the cases of *transverse electric* or TE modes (zero electric field along the axis) and *transverse magnetic* or TM modes (zero magnetic field along the axis).

As Eq. 5.16 suggests, many modes can exist in a cavity, as defined by the corresponding dimension of the pill-box, and the integer number of the mode. The corresponding integer indexes are used to identify a particular mode. In a rectangular pill-box, a mode can be called TE_{klm} or TM_{klm} where the integer indexes indicate the number of half-wavelength variations across the corresponding dimension (x, y, z) of the cavity.

Cylindrical pill-box cavities are very common in accelerators. An example of a cylindrical pill-box cavity with holes for the beam is shown in Fig. 5.22. The classification of the modes in cylindrical cavities is very similar, TE_{klm} or TM_{klm}, but in this case the indexes refer to the polar coordinates (φ, r and z).

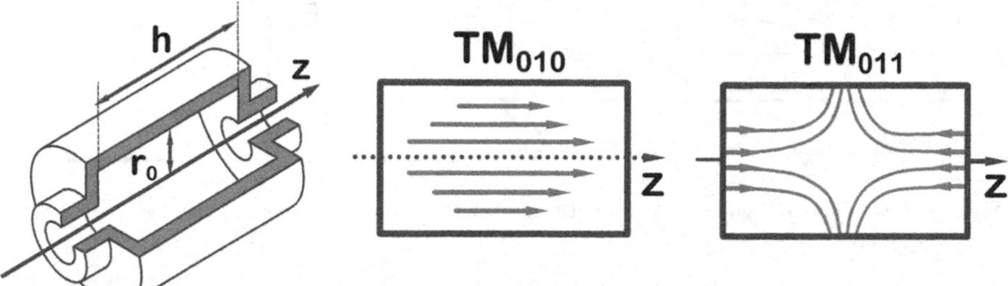

Figure 5.22 Cylindrical pill-box cavity.

Figure 5.23 Examples of pill-box cylindrical cavity modes with electric field lines shown.

Examples of modes in cylindrical pill-box cavities, with longitudinal electric fields and no variation over φ (TM_{0lm}), which are therefore suitable for use as an accelerating cavity, are shown in Fig. 5.23.

5.3.3 QUALITY FACTOR OF A RESONATOR

The quality factor of a resonator — Q — is defined as the ratio of the energy stored in the cavity to the energy dissipated per oscillating cycle, divided by 2π

$$Q = \frac{W_s}{W_d} = \omega \frac{W_s}{P_d} \tag{5.17}$$

Here, W_s is the energy stored in the cavity, W_p is the energy dissipated per cycle divided by 2π, P_d is the power dissipated in the cavity walls and ω is the frequency of the cavity.

The stored energy over the cavity volume is

$$W_s = \frac{\varepsilon_0}{2} \int |\mathbf{E}|^2 dv + \frac{\mu_0}{2} \int |\mathbf{H}|^2 dv \tag{5.18}$$

where the first integral is the energy stored in the E-field and the second integral corresponds to the energy[2] in the H-field. In an EM wave in space or a cavity, the energy oscillates back and forth between these two contributions.

The losses in the cavity are calculated by taking into account the finite conductivity σ of the cavity walls. Since the linear density of the current $\mathbf{j}$ along the walls of a perfect conductor can be written as

$$\mathbf{j} = \mathbf{n} \times \mathbf{H} \tag{5.19}$$

where $\mathbf{n}$ is the vector normal to the surface, we can therefore equate the power dissipated in the cavity walls to

$$P_d = \frac{R_{surf}}{2} \int_s |\mathbf{H}|^2 ds \tag{5.20}$$

[2] *In vacuum* $H = B/\mu_0$ *in SI units.*

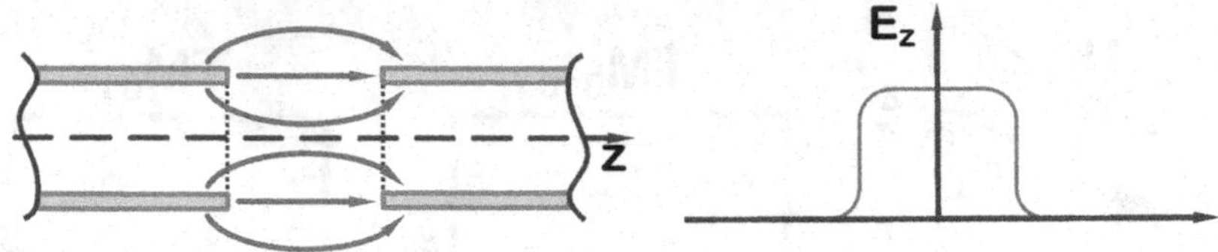

Figure 5.24 The RF gap — space between entrance and exit irises of cavity resonator in drift tube linac.

where the integral is taken over the inner surface of the conductor, and the surface resistance is given by Eq. 5.8.

5.3.4 SHUNT IMPEDANCE – R_S

The so-called *shunt impedance* R_s relates the accelerating voltage V to the power that needs to be fed into the cavity to compensate for the dissipation in the walls P_d.

The accelerating voltage along the path followed by the beam in an electric field E_z is

$$V = \int_{pass} |E_z(x,y,z)| d\ell \tag{5.21}$$

and is taken as peak-to-peak value. The shunt impedance is then defined as

$$R_s = \frac{V^2}{2P_d} \tag{5.22}$$

and is another important characteristic of an accelerating cavity.

5.3.5 ENERGY GAIN AND TRANSIT-TIME FACTOR

The energy gain of a particle as it travels a distance through the accelerating structure depends only on potential difference crossed by particle:

$$U = K\sqrt{P_{RF} l R_s} \tag{5.23}$$

where P_{RF} is the RF power supplied to the cavity, l is the length of the accelerating structure, R_s is the shunt impedance and K is a correction factor (typically ≈ 0.8).

In accelerators such as drift tube linacs, a so-called "transit-time factor" also plays a role and needs to be taken into account in order to evaluate the average energy gain.

Consider an accelerating gap corresponding to the space between drift tubes in a linac structure (Fig. 5.24). The accelerating field in this gap is uniform along the axis and depends sinusoidally on time $E_z = E_0 \cos(\omega t + \phi)$, where the phase ϕ refers to the particle in the middle of gap $z = 0$ at $t = 0$. The field varies as the particle traverses the gap, making the cavity less efficient and the resultant energy gain only a fraction of the peak voltage.

The *transit-time factor* Γ is the ratio of the energy actually given to a particle passing the cavity center at the peak field to the energy that would be received if the field were constant with time at its peak value. Taking into account that the energy gained over the gap G is

$$V = \int_{-G/2}^{+G/2} E_0 \cos(\omega t + \phi)\, dz = E_0 G \cos\phi \frac{\sin(\omega G/2\beta c)}{\omega G/2\beta c} \tag{5.24}$$

we write the following expression for the transit-time factor $\Gamma = \sin(\omega G/2\beta c)/(\omega G/2\beta c)$.

5.3.6 KILPATRICK LIMIT

The performance of any normal conducting accelerating structure depends on its susceptibility to RF breakdown (which can occur at very high fields). Empirically derived around 1950, the *Kilpatrick limit* expresses the relation between the accelerating frequency and maximum achievable accelerating field:

$$f\,[MHz] = 1.64\, E_k^2 e^{-8.5/E_k} \tag{5.25}$$

where E_k is expressed in [MV/m] and is depicted by the lower curve in Fig. 5.25.

Significant efforts and technological developments intended to improve surface quality and cleanness have resulted in a considerable increase of achievable accelerating gradients. In particular, Wang and Loew's empirical formula, devised in 1997, suggests the following behaviors:

$$E\,[MV/m] = 220 f^{1/3} \tag{5.26}$$

where f is expressed in [GHz] — shown by the upper curve in Fig. 5.25.

The $E \sim f^{1/3}$ dependence in Eq. 5.26 was, for a long time, an inspiration and a driving force for developing higher gradients acceleration at higher (multi-tens of GHz) frequencies. This dependence, however, eventually was not confirmed for the practical parameters of the accelerator RF pulses, where it was observed that in these regimes the maximum gradients appear to be rather independent of the frequency.

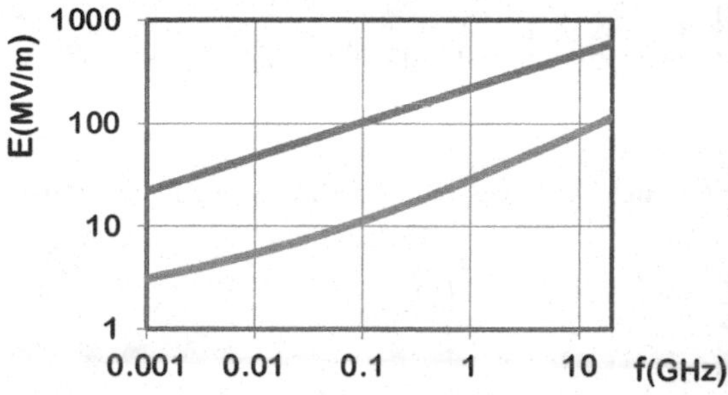

Figure 5.25 Breakdown Kilpatrick limit (lower curve) and Wang–Loew limit (upper curve).

Recent trends for achieving higher gradients involve cooling copper RF structures to liquid nitrogen temperatures, when copper hardens and can withstand higher fields without breakdowns, and its conductivity improves too, reducing the RF losses.

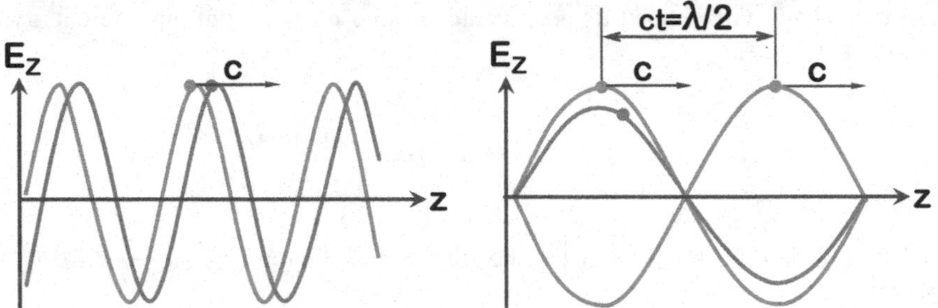

Figure 5.26 Acceleration in a traveling wave structure (left) and in a standing wave structure (right). The wave and particles' position in different moments of time are shown.

5.4 LONGITUDINAL DYNAMICS

In this section, we will discuss the basics of longitudinal dynamics in traveling and standing wave linacs, as well as in synchrotrons.

5.4.1 ACCELERATION IN RF STRUCTURES

Particle acceleration in linacs is achieved with RF structures, using EM modes with the electric field pointing in the longitudinal direction (the direction of the charged particle's motion). The RF electric field can be provided by either *traveling wave* structures or *standing wave* structures.

The acceleration conditions demand that the phase velocity of the *traveling wave* and the particle velocity be equal, so therefore disk-loaded structures are used to slow down the phase velocity of the electric field $v_p < c$ to achieve synchronism. In an appropriately synchronized traveling wave, the bunch of charge particles experience a constant electric field

$$E_z = E_0 \cos(\phi) \tag{5.27}$$

as illustrated in Fig. 5.26 (left plot).

In a *standing wave* structure, the electromagnetic field is the sum of two traveling waves running in opposite directions. Only the forward-traveling wave takes part in the acceleration process.

The electric field that the particle bunch observes in a standing wave is a varying function of time

$$E_z = E_0 \cos(\omega t + \phi) \sin(kz) \tag{5.28}$$

as illustrated in Fig. 5.26 (right plot).

The standing wave, despite its seeming disadvantage in comparison with the traveling wave (in terms of the average field seen by the bunch), is actually much more suitable in certain cases, such as for superconducting cavities.

5.4.2 LONGITUDINAL DYNAMICS IN A TRAVELING WAVE

Consider a particle moving in the E field of a traveling wave

$$E_z = E_0 \cos(\omega t - kz) \tag{5.29}$$

with a phase velocity $v_p = \omega/k$. The equations that describe the particle motion in the longitudinal plane in this field are

$$\frac{dp_z}{dt} = eE_0 \cos(\omega t - kz) \quad \text{and} \quad \frac{d\mathscr{E}}{dt} = eE_0 \dot{z} \cos(\omega t - kz) \tag{5.30}$$

We will define the synchronous particle as

$$\frac{dE_s}{dt} = eE_0 v_s \cos\varphi_s \tag{5.31}$$

and for any other particle, we will use, as coordinates, the deviations from the energy W and position u of the synchronous particle

$$\mathscr{E} = E_s + W \quad \text{and} \quad z = z_s + u \tag{5.32}$$

Then, after changing variables to

$$\varphi = kz - \omega t = \varphi_s - \frac{\omega}{v_s} u \tag{5.33}$$

we will obtain the system of equations for a particle motion in a traveling wave:

$$\frac{dW}{ds} = eE_0 \left[\cos\varphi - \cos\varphi_s\right] \quad , \quad \frac{d\varphi}{ds} = -\frac{\omega}{\beta_s^3 \gamma_s^3 c} \frac{W}{mc^2} \tag{5.34}$$

These describe the motion in the so-called "RF bucket" in a longitudinal phase space (φ , W) and feature stable enclosed trajectories as well as unstable trajectories. We will discuss the phase space trajectories and motion in the "RF bucket" in detail in the following section, after deriving similar equations for the case of acceleration in a synchrotron.

5.4.3 LONGITUDINAL DYNAMICS IN A SYNCHROTRON

Acceleration in a synchrotron is provided by the longitudinal electric fields generated in RF cavities placed on the orbit.

A particle in an RF cavity changes its energy according to the phase of the RF field in the cavity

$$\Delta E = eV(t) = eV_o \sin(\omega_{RF} t + \varphi_s) \tag{5.35}$$

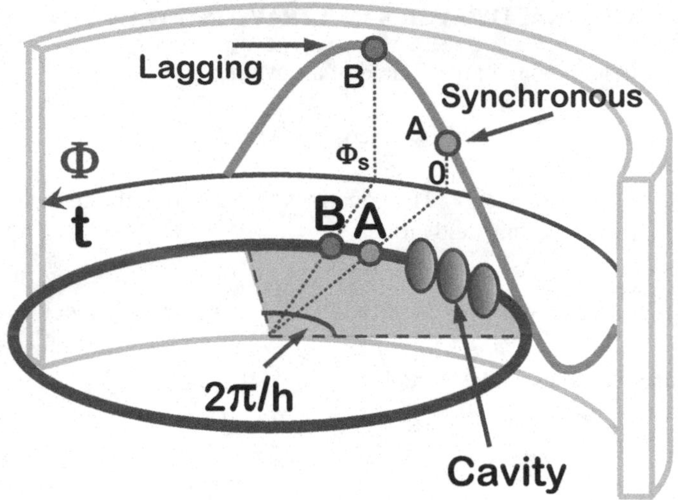

Figure 5.27 Synchronous and lagging particles in a synchrotron ring.

The *synchronous particle* (see Fig. 5.27) is the particle that arrives at the RF cavity when the voltage is such that it exactly compensates the average energy losses U_0

$$\Delta E = U_0 = eV_0 \sin(\varphi_s) \tag{5.36}$$

There are two points in time in the RF potential where the particle will get the correct energy from the RF wave: point P_1 and point P_2 (see Fig. 5.28), and, as we can guess, one is stable and the other is unstable.

In Section 5.1.6 we began our discussion of the dynamics of the particle arriving earlier or later than the synchronous particle, making a simplifying assumption that the particle is ultrarelativistic.

In the case of arbitrary energy, we need to take into account the dependence of the particle's time of flight around the ring of energy, which includes dependence of the circumference and particle velocity on energy.

The synchronous particle with a nominal energy E and velocity v travels around the nominal circumference C in time T so that $T = C/v$. Taking a logarithm and differentiating this expression will yield

$$\frac{dT}{T} = \frac{dC}{C} - \frac{dv}{v} \tag{5.37}$$

The first component of the above equation, dC/C, is expressed via the *momentum compaction factor* α_c, which connects the momentum deviation of the particle dp with the orbit length difference dC:

$$\frac{dC}{C} = \alpha_c \frac{dp}{p} \tag{5.38}$$

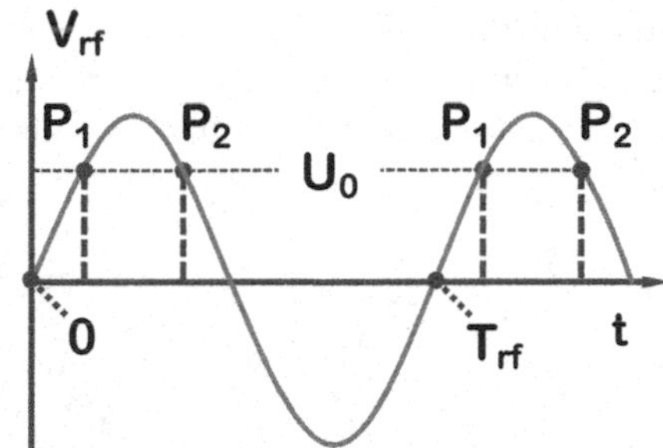

Figure 5.28 Motion in RF potential.

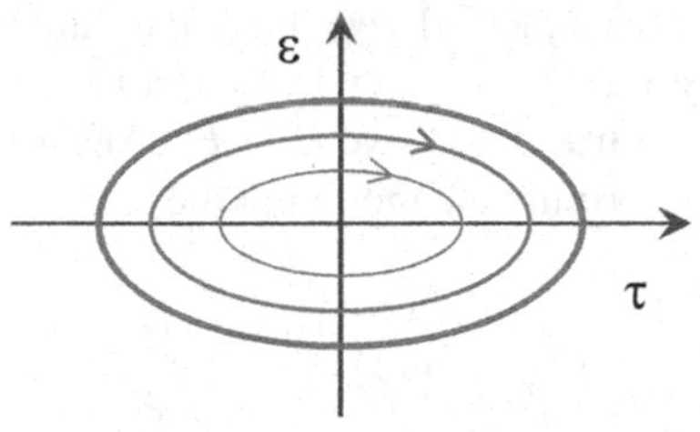

Figure 5.29 RF bucket trajectories in a linearized case are ellipses.

The momentum compaction factor α_c depends on the design of the focusing lattice of the ring and can be either positive (which is most common) or negative (particles with higher E travel over a shorter orbit — which seems counterintuitive, but is possible, although rare).

The second component of Eq. 5.37 can be expressed as

$$\frac{dv}{v} = \frac{d\beta}{\beta} = \frac{1}{\gamma^2}\frac{dp}{p} \tag{5.39}$$

Taking these two components together will yield

$$\frac{dT}{T} = \left(\alpha_c - \frac{1}{\gamma^2}\right)\frac{dp}{p} \sim \left(\alpha_c - \frac{1}{\gamma^2}\right)\frac{dE}{E} \tag{5.40}$$

which shows us that the time of flight depends on the energy deviation, on the relativistic factor γ and on the momentum compaction factor.

We now can conclude that if $\alpha_c - 1/\gamma^2 > 0$, the point P_2 in Fig. 5.28 is stable (in contrast to phase stability as described for a linac earlier). Indeed, according to this assumption, a particle with a higher energy has a longer flight time and therefore arrives later at the RF cavity, undergoes lower RF voltage (point P_2 is on a negative slope), thus gaining less energy, and so tends to return to a nominal energy. Similar deliberations can show that a particle with a lower energy will, in this case, gain more energy in a manifestation of the "principle of phase stability," which enables the capture of particles in the RF potential.

In the opposite case, $\alpha_c - 1/\gamma^2 < 0$, similar logic can lead us to a conclusion of stability of the point P_1 (as for a linac).

We have now arrived at the need to introduce the notion of *transition energy*. In rings with positive α_c passing, during acceleration, the energy corresponding to the gamma factor

$$\gamma_t = \frac{1}{\alpha_c^{1/2}} \tag{5.41}$$

corresponds to the moment of stability flipping from point P_1 to point P_2 on the RF slope. Preserving the quality of the accelerated beam requires a quick switch of the phase of the RF voltage at the moment of passing the transition energy.

Now, let's derive the longitudinal beam dynamics equations for particle acceleration in a synchrotron. We start by considering a particle moving in an electric field of a traveling wave $E_z = E_0 \cos(\omega t - kz)$ with a phase velocity $v_f = \omega/k$. Equations describing the motion in the longitudinal plane are

$$\frac{dp_z}{dt} = eE_0 \cos(\omega t - kz) \quad \text{and} \quad \frac{d\mathscr{E}}{dt} = eE_0 \dot{z} \cos(\omega t - kz) \tag{5.42}$$

We again define the synchronous particle as

$$\frac{dE_s}{dt} = eE_0 v_s \cos\phi_s \tag{5.43}$$

and use deviations from its energy and time to describe an arbitrary particle

$$\mathscr{E} = E_s + \varepsilon \quad \text{and} \quad t = t_s + \tau \tag{5.44}$$

Using these definitions, we obtain the first equation

$$\frac{d\varepsilon}{ds} = eE_0 \left[\cos(\omega\tau + \phi_s) - \cos\phi_s\right] \tag{5.45}$$

and using Eq. 5.40 to define the momentum compaction factor at high energy ($\gamma \gg 1$), we obtain the second equation

$$\frac{dT}{T} \sim \left(\alpha_c - \frac{1}{\gamma^2}\right) \frac{d\mathscr{E}}{\mathscr{E}} \quad \rightarrow \quad \frac{d\tau}{dt} \sim \frac{\alpha_c}{E_s} \frac{d\varepsilon}{dt} \tag{5.46}$$

which together describe an *RF bucket* in the longitudinal phase space with coordinates (τ, ε).

Rewriting these equations for the RF bucket for $\gamma \gg 1$ in terms of derivatives of time yields

$$\varepsilon' = \frac{qV_0}{L} \left[\sin(\varphi_s + \omega\tau) - \sin\varphi_s\right] \quad \text{and} \quad \tau' = \frac{\alpha_c}{E_s} \varepsilon \tag{5.47}$$

Linearizing these equations for the motion in the RF bucket gives us

$$\varepsilon' = \frac{e}{T_0} \frac{dV}{d\tau} \tau \quad \text{and} \quad \tau' = \frac{\alpha_c}{E_s} \varepsilon \tag{5.48}$$

which corresponds to the phase space motion with elliptical trajectories as in Fig. 5.29, with angular frequency defined as

$$\omega_s^2 = \frac{\alpha_c e \dot{V}}{T_0 E_0} \tag{5.49}$$

which is called *synchrotron frequency*.

In a realistic case of a practical accelerator design, we often cannot limit ourselves to a linear approximation and would need to consider the full nonlinear equations Eq. 5.47. We will explore these equations now, looking at them from a different

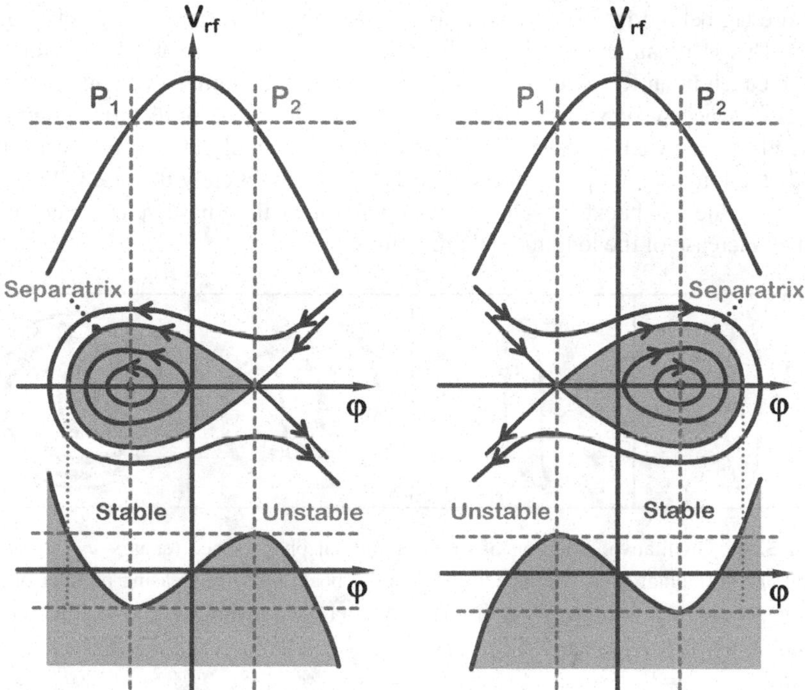

Figure 5.30 RF voltage and phase space and RF potential for cases below and above the transition energy.

perspective — via an analogy with classical mechanics. The equivalent equation can be rewritten as

$$\frac{d^2\varepsilon}{dt^2} = U\tau^2 \tag{5.50}$$

where U is an analog of a potential energy in an oscillator.

The shape of the potential and corresponding phase space trajectories are shown in Fig. 5.30 for two cases, below and above the transition energy. The potential energy analogy and its shape help to make sense of the behavior of the trajectories. They are ellipse-like in the vicinity of the stable point where the potential is parabolic, but become distorted and eventually unstable as they come closer to the *saddle point* of the potential where U flattens and passes through a local maximum, creating conditions for certain trajectories to take an arbitrarily long time for passing through the saddle point region. The trajectory that originates from the vicinity of the saddle point and separates the stable area from an unstable area is called the *separatrix*.

5.4.4 RF POTENTIAL – NONLINEARITY AND ADIABATICITY

The RF potential, as we can see in Eq. 5.50, is intrinsically nonlinear. Particles that are located close to the separatrix will have longer periods of oscillations and will therefore lag behind in their rotation in phase space with respect to particles in the center. This is illustrated in Fig. 5.31, which shows the qualitative behavior of a particle bunch in an RF potential (neglecting radiation damping) with an increasing number of synchrotron periods. We can see that while the beam distortions are negligible after one synchrotron period, after ten and especially after fifty synchrotron periods the longitudinal phase space distribution is completely distorted. The nonlinearity of the RF bucket results in *filamentation* of the phase space, causing an effective increase of the longitudinal emittance.

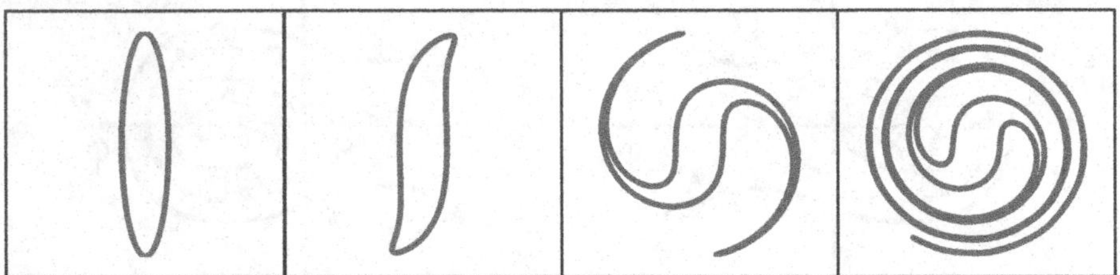

Figure 5.31 Qualitative evolution of the longitudinal phase space (energy vs. phase, for vertical and horizontal axes, correspondingly) of the beam for an increasing number of synchrotron periods. Left to right: initial distribution, after one synchrotron period, after ten and after fifty periods.

5.4.5 SYNCHROTRON TUNE AND BETATRON TUNE

Synchrotron oscillation in circular accelerators, as we saw in the previous sections, is a multi-turn phenomenon (the synchrotron frequency may correspond to many tens or hundreds of turns); i.e., such oscillations are very slow. The *synchrotron tune* is connected to the synchrotron frequency as follows

$$Q_S = \frac{2\pi T_{rev}}{\omega_S} \tag{5.51}$$

where T_{rev} is the period of revolution around the orbit. Typically, $Q_S \ll 1$.

In contrast, transverse betatron oscillations, in circular accelerators with strong focusing, are necessarily fast (there are usually many tens or hundreds of oscillations per revolution period). The *betatron tune* is defined as the number of oscillations around the ring or the ratio of the betatron frequency to the revolution frequency

$$Q = \frac{\mu}{2\pi} = \frac{1}{2\pi}\oint \frac{ds'}{\beta(s')}$$

where the integral is taken along the accelerator circumference. Typically, $Q \gg 1$.

The betatron Q is momentum dependent and this can link the two motions together. When synchrotron motion is thus coupled to betatron motion, this can manifest itself in the signals of *pick-up electrodes*, which measure the beam transverse

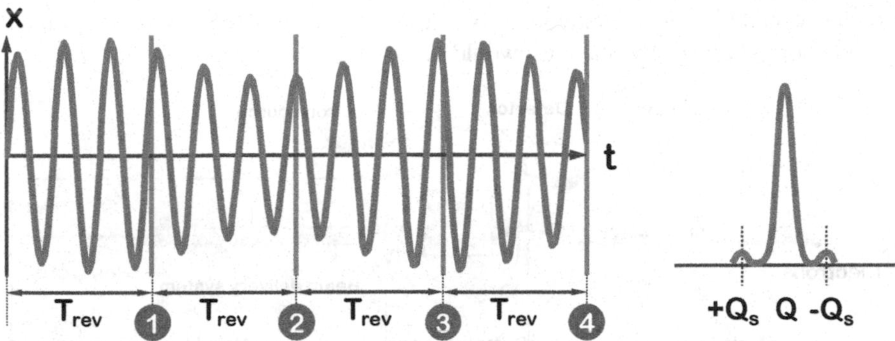

Figure 5.32 Betatron oscillations modulated by synchrotron motion (left) and a corresponding spectrum (right) with betatron tune and synchrotron sidebands.

Table 5.1
Operating frequencies and typical parameters for RF cavities

Warm cavities	**Gradient**	**Repetition rate**
S-band (3 GHz)	15–25 MV/m	50–300 Hz
C-band (5–6 GHz)	30–40 MV/m	<100 Hz
X-band (12 GHz)	100 MV/m	<100 Hz
Superconducting cavities	**Gradient**	**Repetition rate**
L-band (1.3 GHz)	< 35 MV/m	up to CW

oscillations as qualitatively shown in Fig. 5.32. As the main fast (betatron) signal is now modulated with a slow (synchrotron) component

$$x \propto \sin(2\pi Q t / T_{rev})(1 + \Delta \sin(2\pi Q_S t / T_{rev}))$$

the spectrum of transverse motion will now include *synchrotron sidebands* at $Q - Q_S$ and $Q + Q_S$ in addition to the main betatron frequency as is illustrated in Fig. 5.32.

5.4.6 ACCELERATOR TECHNOLOGIES AND APPLICATIONS

Table 5.1 shows typical accelerating gradients for RF cavities of various frequencies for linear accelerators.

As we can see, the achievable gradient generally increases with frequency, consistent with frequency dependences of Eqs. 5.25 and 5.26 (these relationships, however, haven't yet been demonstrated for frequencies higher than 12 GHz).

The practically achievable gradient is one of the main factors that define the size of accelerator-based facilities of various kinds. Let's consider a couple of examples.

In a *high-energy physics* application, a *linear collider* (Fig. 5.33) aiming at 500 GeV energy in the center of mass (CM), and built with L-band superconducting cavities, would be around 30 km long. A collider aiming at 3 TeV CM, built with an X-band normal conductive cavities, would be almost 50 km long.

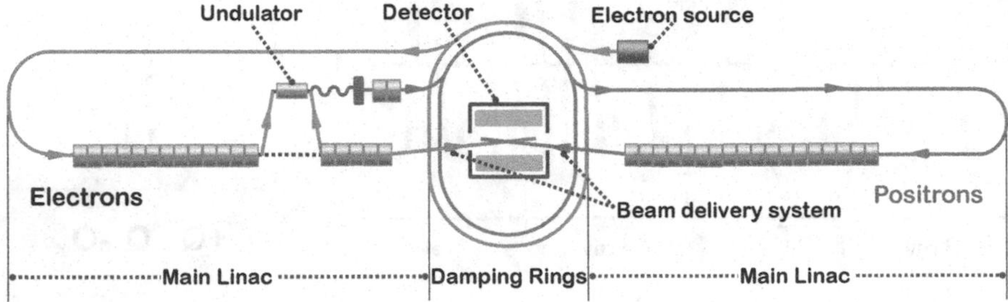

Figure 5.33 A generic linear collider.

While the linac length constitutes a major fraction of the length of a linear collider, in a *free electron laser*, the linac length is a noticeable fraction of the overall length (between around a quarter to a half). As an example, an FEL (see Fig. 5.34) with an electron beam energy of around 10 GeV built with S-band or C-band technology would be about a kilometer long.

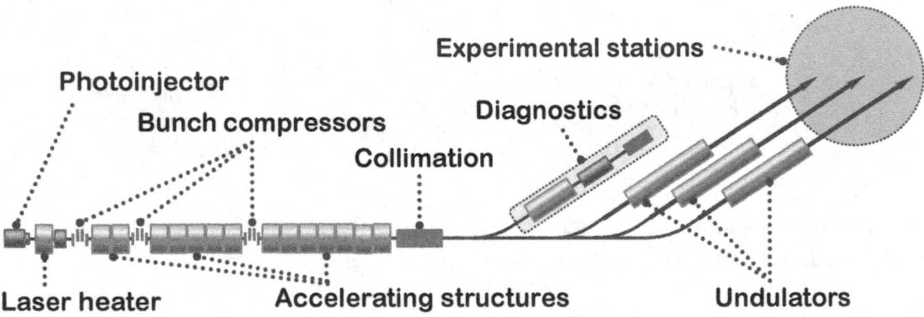

Figure 5.34 A generic free electron laser.

These examples demonstrate the motivations for developing new accelerator technologies that would make these and other accelerator-based facilities and instruments more compact. This brings us to the next chapter, where we will delve into methods of plasma acceleration.

5.5 FOCUSING IN DRIFT TUBE LINAC – INVENTION CASE STUDY

As we discussed earlier in this chapter, Alvarez linac drift tubes are usually arranged in a tank/vessel, made of good copper conductor, in which a cavity wave is induced (see Fig. 5.9), which excites the accelerating voltage between the tubes.

Such a linac may need to be quite long, and the conductor vessel can be large in diameter, so that it would not be possible to place the beam focusing elements outside of the vessel, and one would need to locate these focusing elements in between of the individual drift tube vessels. However, making short vessels, to include the focusing elements in between, would not be economical.

On the other hand, if we would use long drift tube vessels, we would have to include the beam focusing elements inside of the vessels, but it would appear that we do not have any space where to put them.

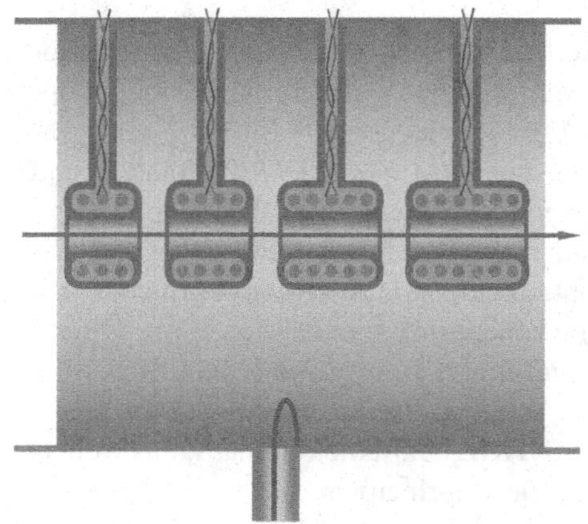

Figure 5.35 Drift tube linac with focusing.

In Alvarez linac the focusing elements (electromagnets or permanent magnets) can be placed inside of the drift tubes, demonstrating the application of the "nested doll" inventive principle, and also the principle of using resources (space, energy, fields, contradiction, etc.) that we already have in the system — in this case the space inside of the drift tubes.

This contradiction can be solved if we would apply the "nested doll" inventive principle, and also the principle of using resources (space, energy, fields, contradiction, etc.) that we already have in the system.

Indeed, we have the drift tubes, and the space inside of them, which is free of electric field, and that can be arranged to house electromagnets or permanent magnets, which can be configured to focus the beam. And we also already have the supports for the drift tubes that can be made hollow, to house the conductors for the focusing electromagnets, as shown in Fig. 5.35. This integration is also a clear example of the nested doll inventive principle.

Further evolution of this invention and integration of acceleration and focusing is an RFQ — radio frequency quadrupole structure (see Fig. 5.10), which combines RF acceleration and RF focusing and is now the leading choice for many modern hadron linacs.

EXERCISES

5.1. *Chapter materials review.*
Derive Eq. 5.34 for particle motion in the traveling wave.

5.2. *Chapter materials review.*
Discuss design approaches to beam optics that would result in achieving a negative value of the momentum compaction factor in a synchrotron.

5.3. *Mini-project.*
Define very approximate parameters (sizes, magnetic fields, parameters of RF system) of a 200-MeV rapid-cycling proton synchrotron capable of operating at a 10-Hz repetition rate. Assume injection at 1 MeV.

5.4. *Analyze inventions or discoveries using TRIZ and AS-TRIZ.* Analyze and describe scientific or technical inventions described in this chapter (e.g., tandem, RFQ, bunch and pulse compressors) in terms of the TRIZ and AS-TRIZ approaches, identifying a contradiction and an inventive principle that were used (could have been used) for these inventions.

5.5. *Developing AS-TRIZ parameters and inventive principles.* Based on what you already know about accelerator science, discuss and suggest the possible additional parameters for the AS-TRIZ contradiction matrix, as well as the possible additional AS-TRIZ inventive principles.

5.6. *Practice in reinventing technical systems.* A special type of cavity, a so-called open cavity, is used in the systems of RF pulse compression. The cavity looks like an oak barrel without a lid and a bottom. An RF wave can circulate in this cavity in the plane of its largest diameter, for a long time and without decay, in the case where the RF wavelength is much smaller than the cavity's diameter. Suggest a way to excite such an RF wave in this open cavity.

5.7. *Practice in reinventing technical systems.* Reinvent the way to deal with *Unidentified Flying Objects* in accelerators. The HERA electron-hadron collider in DESY suffered from difficult-to-identify events, attributed to UFOs — unidentified flying objects . The most popular hypothesis suggested that dust particles coming out of *ion pumps* were positively ionized and attracted to the circulating electron beams, creating an instability. Suggest a way to fix this problem.

5.8. *Practice in the art of back-of-the-envelope estimations.* LEP was a 27-km-long e+e- machine and, around the year 2000, was running with the energy of about 104 GeV per beam. The energy was limited by the energy losses due to synchrotron radiation and by the amount of voltage from the installed RF cavities. LEP optical structure included all typical magnets: bending dipoles, quadrupoles with corrector coils, and sextupoles. Assume that you cannot install more RF to compensate for SR energy losses, but can power up the dipole correctors to smooth the orbit and reduce SR losses. Estimate how much you can increase the energy of the LEP collider using this approach. *(It is assumed that you can identify the most important effects playing roles in this task, can define the necessary parameters and set their values, and can get a numerical answer.)*

6 Plasma Acceleration

Plasma acceleration is an emerging and promising field, whose rapid progress is enabled by developments in laser technology — particularly by the method of chirped pulse amplification. Plasma accelerators of electrons — the primary focus of this chapter — are the backbones of future compact light sources. Proton and ion plasma acceleration, briefly discussed here and in closer detail in Chapter 9, are a potential way to improve future medical accelerators.

The aim of this chapter, after a brief introduction and discussion of the motivations for pursuit of plasma acceleration, is to develop the framework that will help us to estimate the parameters of laser plasma-based light sources, so as to be prepared for Chapter 7.

6.1 MOTIVATIONS

The "Livingston plot," which depicts the energy of accelerated beams versus time (Fig. 1.6), illustrates the great history of accelerators and related inventions. It also shows the signs of saturation, highlighting the need for the next breakthrough in accelerator technology.

Traditionally, accelerating structures have been made from metal (normal conductive or super-conductive) and are typically limited in their accelerating gradient to $E_z < 100$ MeV/m. This limitation is imposed by the properties of the materials — since damage to the accelerating structure's walls (deterioration of their integrity) limits the gradient.

The "accelerating structures" produced on the fly in plasma by a laser pulse are, however, made from a material that is already "damaged" (plasma), and therefore do not exhibit the same limitations due to the material's properties.

Plasma acceleration was first proposed by T. Tajima and J. Dawson in 1979, which was, in fact, too early for laser and beam technologies to be ready to realize the proposed approach. Consequent parallel developments of laser and beam technologies — specifically those aimed at creating short, powerful pulses — created the new reality making the laser plasma acceleration the area with the highest degree of synergy between the physics of plasma, lasers and accelerators.

We should also recall (see margin notes) our discussion from Chapter 1 regarding the use of the AS-TRIZ approach to *post-facto* analyze the invention of plasma acceleration.

6.1.1 MAXIMUM FIELD IN PLASMA

By using plasma as an accelerating medium, we can remove the limitation of the accelerating gradient relating to the material's damage threshold. The maximum field in plasma will still be limited, but by other factors.

DOI: 10.1201/9781003326076-6

Let us look again at plasma oscillation, as illustrated in Fig. 4.21, and briefly recall that this diagram allowed us to estimate the plasma frequency ω_p. Assuming that a fraction of the charges is shifted by distance x, we selected an integral contour that enclosed the displaced charges, and we then equated, according to Maxwell's equations, the surface integral of electric field — $\mathbf{E} \cdot d\mathbf{S}$ to the volume integral of charge density — $\rho dV/\varepsilon_0$, to obtain the electric field created by the shifted charges $E = nex/\varepsilon_0$, which creates the restoring force. The equation of motion $F = md^2x/dt^2 = -eE = -ne^2x/\varepsilon_0$ then gave us the oscillation frequency $\omega_p^2 = ne^2/(\varepsilon_0 m)$, which, with use of $4\pi\varepsilon_0 r_e = e^2/(m_e c^2)$, we rewrote as $\omega_p^2 = 4\pi nc^2 r_e$ — the angular plasma frequency.

Plasma frequency in Hz is $f_p \sim 9000 n^{1/2}$ where plasma density n is in cm^{-3}
Maximum accelerating field in plasma $eE_{max} \approx 1\,\mathrm{GeV/cm} \cdot \left(n/10^{18}\mathrm{cm}^{-3}\right)^{1/2}$

Very similar calculations allow us to estimate the maximum accelerating field in plasma. Imagine that the plasma oscillation in Fig. 4.21 is excited by a charged object moving with velocity c. In the case where a total charge separation is achieved in plasma, the maximal field is estimated assuming

$$x \sim \lambda_p \sim \frac{c}{\omega_p} \tag{6.1}$$

which results in the following for the maximum field

$$E_{\max} \sim \frac{nec}{\varepsilon_0 \omega_p} = \frac{mc\omega_p}{e} \tag{6.2}$$

or equivalently

$$eE_{\max} \cong mc^2 \frac{\omega_p}{c} \tag{6.3}$$

We can use the practical formula $f_p \sim 9000 n^{1/2}$ where n is defined in cm^{-3} to obtain a formula for the maximum possible accelerating field in plasma:

$$eE_{max} \approx 1 \frac{\mathrm{eV}}{\mathrm{cm}} \cdot n^{1/2} \left(\mathrm{cm}^{-3}\right) \tag{6.4}$$

This means that 1 GeV/cm accelerating gradient can be achieved for plasma of $10^{18} cm^{-3}$ density.

Theoretical predictions, made back in 1979, of the principal feasibility of such large accelerating gradients were an essential driving force toward the development of plasma acceleration technology.

6.2 EARLY STEPS OF PLASMA ACCELERATION

We see that GeV/cm requires plasma with n=$10^{18} cm^{-3}$. The plasma wavelength

$$\lambda_p = \frac{c}{f_p} \quad \text{or} \quad \lambda_p \approx 0.1 mm \sqrt{\frac{10^{17} cm^{-3}}{n}} \tag{6.5}$$

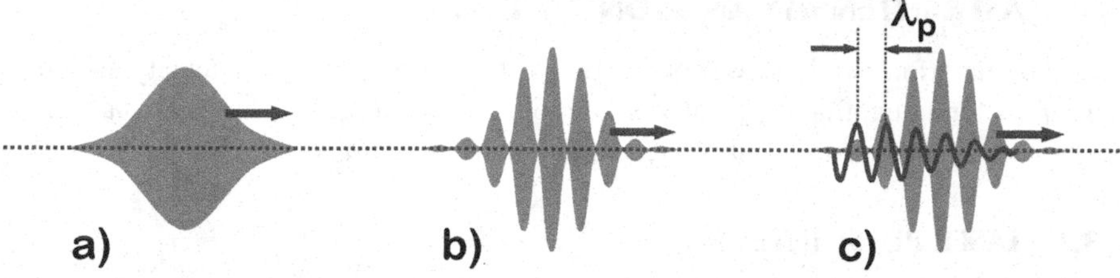

Figure 6.1 For illustration of plasma beat wave and self-modulated laser wakefield acceleration.

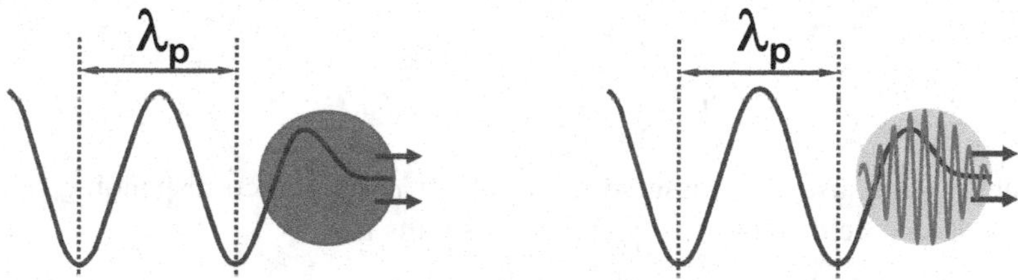

Figure 6.2 Plasma wakefield acceleration — PWFA.

Figure 6.3 Laser wakefield acceleration — LWFA.

corresponding to a plasma density of $10^{18} cm^{-3}$ is around $\lambda_p \approx 30\,\mu m$ (or around 100 fs). Thus, short sub-100-fs pulses are needed to excite plasma toward GeV/cm accelerating gradients.

In the absence of such short laser pulses, in the late 1970s and early 1980s, other methods of plasma excitation were suggested (by J.M. Dawson, 1979) such as the plasma beat wave accelerator (PBWA) and the self-modulated laser wakefield accelerator (SMLWFA); see Fig. 6.1.

In the PBWA, two laser pulses with envelopes as in Fig. 6.1a and frequencies differing by ω_p overlap to create a beating at the plasma's frequency as shown in Fig. 6.1b. This combined laser pulse is sent into plasma where it creates plasma excitation as shown in Fig. 6.1c.

In contrast to the previous method, in the SMLWFA, only a single laser pulse is sent into the plasma (Fig. 6.1a), where an instability (which we will not discuss here in detail) results in a self-modulation of the long laser pulse at λ_p (Fig. 6.1b), which again creates plasma excitation at wavelength λ_p (Fig 6.1c).

As a result of beam and laser technologies development, short sub-ps pulses of laser or beams became available and thus prompted rapid progress of plasma acceleration.

The plasma wakefield acceleration (PWFA) method uses a short, high-energy particle bunch to excite the plasma (Fig. 6.2). Similarly, a short laser pulse of high intensity can be used in a laser wakefield acceleration (LWFA) method (Fig. 6.3). In both of these cases, a high amplitude plasma wave is created, which can then be used for acceleration. We will discuss these methods in detail in the following sections.

6.3 LASER INTENSITY AND IONIZATION

Laser acceleration requires laser pulses of short duration and high intensity. In order to prepare for a quantitative discussion of the intensities required for plasma acceleration, let us introduce the basic concepts related to this subject.

6.3.1 LASER PULSE INTENSITY

Laser intensity (in a vacuum) is defined (in SI and Gaussian units respectively — recall that $\varepsilon_0 \approx 8.8 \cdot 10^{-12} A^2 s^4/(kg\, m^3)$) as

$$I = \frac{1}{2}\varepsilon_0 E_{\max}^2 c \quad \text{(SI)} \tag{6.6}$$

$$I = \frac{1}{8\pi} E_{\max}^2 c \quad \text{(Gaussian)} \tag{6.7}$$

The intensity I is usually measured in Watts per cm^2. The corresponding relation between electric field and intensity in practical units is

$$E_{\max}\left[\frac{V}{cm}\right] \cong 2.75 \times 10^9 \left(\frac{I}{10^{16} W/cm^2}\right)^{1/2} \tag{6.8}$$

Similarly, for the magnetic field:

$$B_{\max}\left[Gauss\right] \cong 9.2 \times 10^6 \left(\frac{I}{10^{16} W/cm^2}\right)^{1/2} \tag{6.9}$$

It is useful to remember that in an EM wave the field amplitudes of 300 V/cm and ≈1 Gauss are equivalent.

For example: a laser with 30 J energy in a 30-fs (10 μm)-long pulse corresponds to (assume rectangular distribution in space and time) a peak power of 10^{15} W and, if focused to a spot with a diameter of 3 μm as illustrated in Fig. 6.4, produces the intensity in the focus of 10^{22} W/cm^2. According to the equations given above, the fields in the focus of such a laser will approach 10,000 Mega Gauss.

6.3.2 ATOMIC INTENSITY

In order to develop a quantitative understanding of laser intensity values, it is best to compare the field of an intense laser with *atomic fields* — particularly with the field in a hydrogen atom.

The Bohr radius is given by:

$$a_B = \frac{\hbar^2}{me^2} = 5.3 \times 10^{-9} cm \tag{6.10}$$

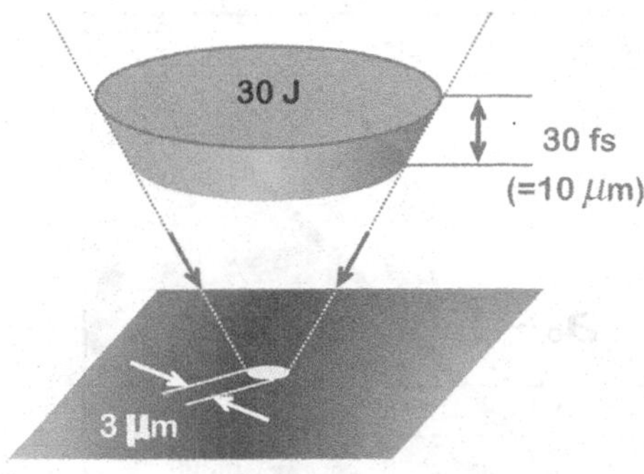

Contemporary high-power lasers are impressive. In the example shown, the peak power is 10^{15} W and the peak power density at the focus is $\sim 10^{22}$ W/cm^2. Laser of this power will instantly ionize any substance. Electrons carried along by the field of such a laser will rapidly become relativistic.

Figure 6.4 Laser focused to a tight spot.

The corresponding field is then defined as

$$E_a = \frac{e}{a_B^2} \qquad \text{(Gaussian units)} \tag{6.11}$$

$$= \frac{e}{4\pi\varepsilon_0 a_B^2} \quad \approx 5.1 \times 10^{11}\frac{V}{m} \quad \text{(SI)}$$

The corresponding atomic intensity is thus equal to

$$I_a = \frac{\varepsilon_0 c E_a^2}{2} \cong 3.51 \times 10^{16}\frac{W}{cm^2} \tag{6.12}$$

A laser with intensity higher than the above will ionize gas immediately. However, as we will show in the next sections, ionization can occur well below this threshold due to multi-photon effects or tunneling ionization.

6.3.3 PROGRESS IN LASER PEAK INTENSITY

Lasers able to produce peak intensity of atomic levels given by Eq. 6.12 were not available until the mid-1980s. This is illustrated in Fig. 6.5, which shows a qualitative overview of the progress in laser peak intensity throughout history.

The invention of the chirped pulse techniques — CPA and OPCPA — was a breakthrough in laser peak power, allowing reaching and exceeding atomic intensities (indicated by the line **b** in Fig. 6.5).

Fig. 6.5 also shows two other important intensity levels: the first one corresponds to the field ionization of hydrogen (line **a** in this figure), discussed in the next section, and the second one corresponds to the relativistic optics case (line **c**) when electrons become relativistic in the laser field (discussed in Section 6.3.6).

One more vital intensity limit relevant to Fig. 6.5 — but not shown as it would be significantly off-scale (around $2 \cdot 10^{29}$ W/cm^2) — is the *Schwinger intensity limit*, which corresponds to the case when the laser field can produce e^+e^- pairs from a vacuum (this will be discussed further in Section 6.3.8).

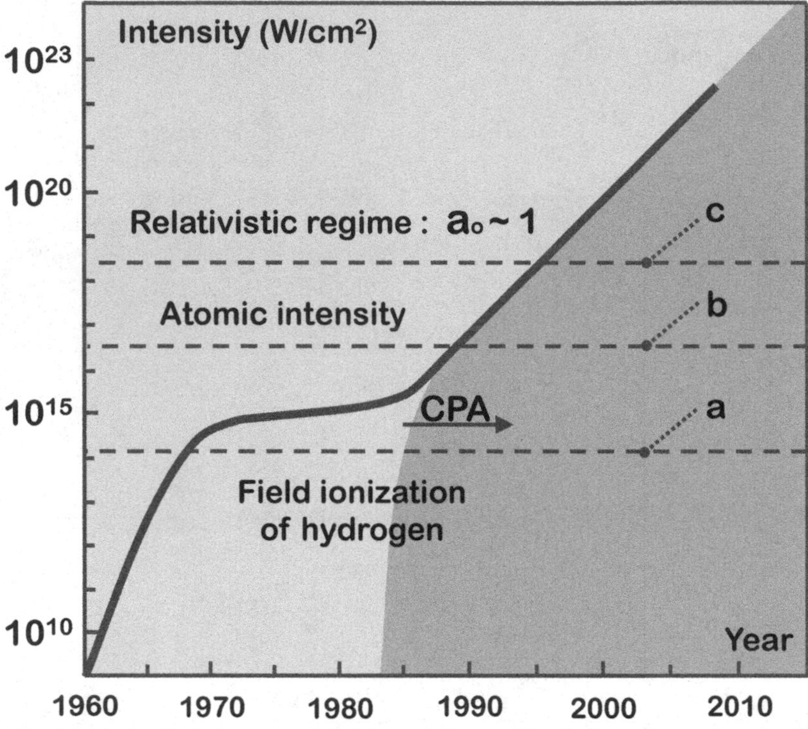

Figure 6.5 Qualitative overview of the progress in laser peak intensity.

6.3.4 TYPES OF IONIZATION

There are several types of ionization of interest, some of which are shown in Fig. 6.6, depicting the potential well of an electron in an atom.

In *direct ionization* (Fig. 6.6a), the photon transmits enough energy to an electron to overcome the potential barrier in one interaction. In a *multi-photon ionization* (Fig. 6.6b), the electron obtains the energy needed to overcome the potential barrier via the process of multi-photon absorption. *Tunneling ionization* (Fig. 6.6c) can occur when an electron quantum tunnels through the potential barrier.

The tunneling ionization, as presented in Fig. 6.6c, and with somewhat larger laser field intensity, will turn into another ionization mechanism called barrier suppression ionization.

6.3.5 BARRIER SUPPRESSION IONIZATION

Barrier suppression ionization (BSI) occurs when the laser field distorts the potential of an atom in such a way that the electron can freely escape the potential well.

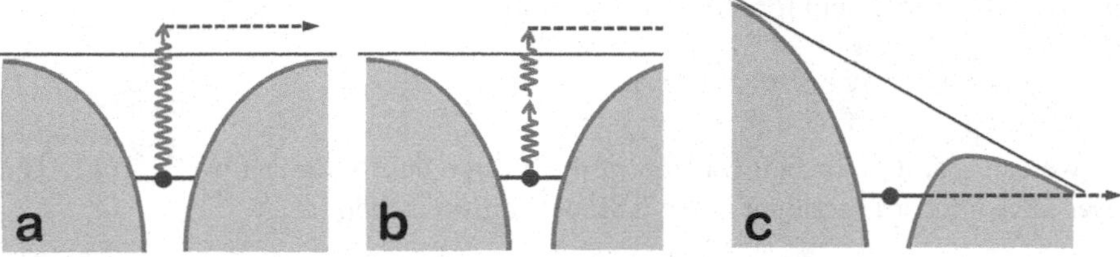

Figure 6.6 Types of ionization: (a) direct, (b) multi-photon, (c) tunneling.

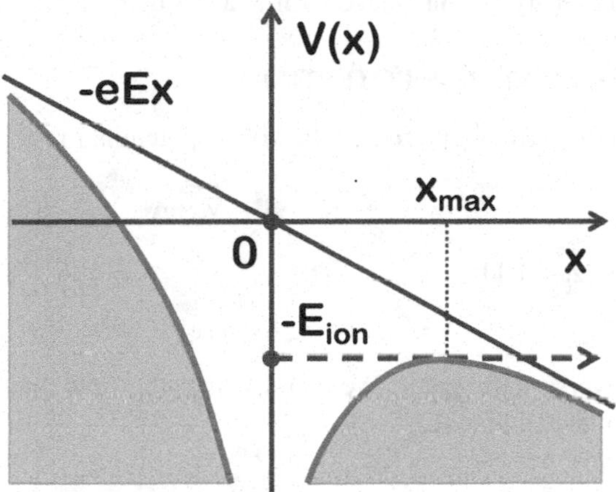

Figure 6.7 Barrier suppression ionization.

The Coulomb potential of a hydrogen atom distorted by a homogeneous field E can be written as (in Gaussian units):

$$V(x) = -\frac{e^2}{x} - eEx \tag{6.13}$$

The distorted potential is shown in Fig. 6.7. The position of the maximum of the potential on the right side of the plot is

$$x_{\max} = (e/E)^{1/2} \tag{6.14}$$

and the value of the potential at the maximum is

$$V(x_{\max}) = 2\left(e^3 E\right)^{1/2} \tag{6.15}$$

Equating the potential value at the maximum $V(x_{\max})$ to the hydrogen atom ionization potential E_{ion}

$$E_{ion} = \frac{e^2}{2a_B} \approx 13.6eV \tag{6.16}$$

gives us the critical field for the hydrogen atom

$$\varepsilon_c = \frac{e}{16a_B^2} = \frac{E_a}{16} \tag{6.17}$$

As we can see, it is a small fraction of the atomic field E_a given by Eq. 6.11. The laser intensity corresponding to the BSI mechanism is then

$$I_c = \frac{I_a}{256} \approx 1.4 \cdot 10^{14} W/cm^2 \tag{6.18}$$

which is more than two orders of magnitude lower than the atomic intensity I_a given by Eq. 6.12. The intensity I_c is indicated by line **a** in Fig. 6.5.

6.3.6 NORMALIZED VECTOR POTENTIAL

The laser field can be written in terms of the vector potential of the laser field **A** as

$$\mathbf{E} = -\frac{\partial \mathbf{A}}{c\partial t}\ , \qquad \mathbf{B} = \nabla \times \mathbf{A} \tag{6.19}$$

For a linearly polarized field:

$$\mathbf{A} = A_0 \cos(kz - \omega t)\, \mathbf{e}_\perp \tag{6.20}$$

where $\mathbf{e}_\perp$ is transverse unit vector. We can see that the field and vector potential amplitudes are connected via

$$E_0 = \frac{A_0 \omega}{c} \tag{6.21}$$

Comparing momentum gained by an electron e^- in one cycle of laser field

$$eE\,\Delta t \sim \frac{eE}{\omega} \tag{6.22}$$

with its rest mass $m_e c$, we can see that it is better to define the *normalized vector potential* as

$$\mathbf{a} = \frac{e\mathbf{A}}{m_e c^2} \tag{6.23}$$

with its amplitude given by

$$a_0 = \frac{eE_0}{m_e \omega c} \tag{6.24}$$

The amplitude a_0 will indicate if the electron motion in the laser field is relativistic: $a_0 \gg 1$, or nonrelativistic: $a_0 \ll 1$.

The normalized vector potential amplitude in practical units can be written as

$$a_0 \approx \left(\frac{I\,[W/cm^2]}{1.37 \cdot 10^{18}}\right)^{\frac{1}{2}} \cdot \lambda\,[\mu m] \tag{6.25}$$

where $\lambda = 2\pi c/\omega$ is the wavelength of the laser. For example, for a red laser with $\lambda = 0.65\ \mu$m, the value $a_0 = 1$ reached at intensity of $I \approx 3 \cdot 10^{18}\ W/cm^2$ (as indicated by line **c** in Fig. 6.5).

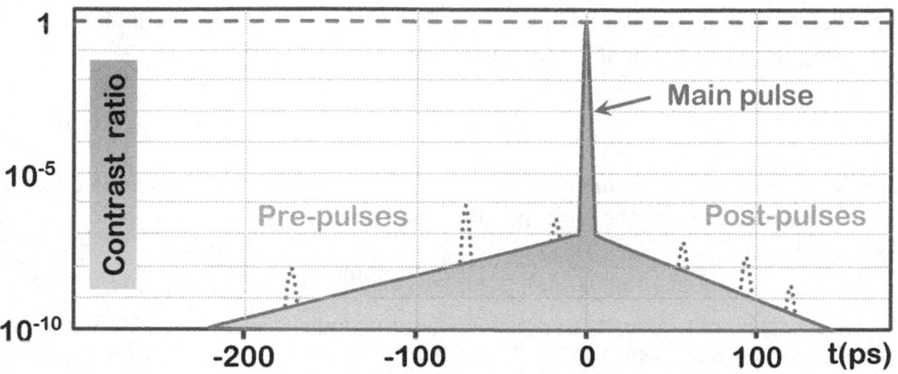

Figure 6.8 Qualitative temporal profile of a CPA-compressed laser pulse.

6.3.7 LASER CONTRAST RATIO

As we see in Fig. 6.5, different phenomena related to laser-matter interaction and plasma acceleration occur at significantly different intensities. This brings us to a dialogue regarding the temporal *contrast ratio* of a laser pulse.

The spatial contrast — the ratio of intensity at the laser focus to the intensity outside of the focus — is a standard concept intuitively known to everyone from everyday life.

For CPA-compressed pulses, which involve manipulations and exchanges between energy and longitudinal phase space coordinates, it is appropriate to introduce the notion of the temporal *contrast ratio* — a function of time given by the ratio of the peak laser intensity to the intensity in the front or back of the pulse.

A qualitative spatial profile of a CPA-compressed laser pulse is shown in Fig. 6.8 in terms of the contrast ratio, in logarithmic scale. It is typical that a short sub-ps pulse is accompanied by many tens of ps low-intensity pulses, as well as short pre-pulses or post-pulses, which are typically caused by nonlinear properties of the elements of the CPA system and non-ideal properties of the initial laser pulse.

The high contrast ratio is often the key parameter in plasma acceleration, as even a relatively low intensity can either ionize or destroy the target long before the arrival of the main high-intensity short pulse.

6.3.8 SCHWINGER INTENSITY LIMIT

A laser of high enough intensity can generate e^+e^- pairs from a vacuum. According to the time-energy uncertainty principle

$$\Delta E \, \Delta t \geq \frac{\hbar}{2} \qquad (6.26)$$

a virtual e^+e^- pair (thus $\Delta E \sim m_e c^2$) can appear for a short duration of time $\Delta t \sim \hbar/(m_e c^2)$. If an electric field E_S acting on the pair during Δt increases the momentum of e^+ or e^- by about mc (i.e., $\Delta t \cdot e E_S \sim mc$), the particles then become real and thus materialize from the vacuum thanks to the laser field. The corresponding laser field (ignoring factors of two in the estimations) is hence given by

$$E_S = \frac{m_e^2 c^3}{e\hbar} \approx 1.3 \cdot 10^{18} \ \ \mathrm{V/m} \tag{6.27}$$

and is called a *Schwinger limit* field — the scale above which the linear electrodynamics become invalid. The corresponding laser intensity is

$$I_S \approx 2 \cdot 10^{29} \ \ \mathrm{W/cm^2} \tag{6.28}$$

Reaching such laser intensity in practice would undoubtedly create a new scientific and technological breakthrough.

6.4 THE CONCEPT OF LASER ACCELERATION

We are now ready to discuss the concept of laser plasma acceleration — see Fig. 6.9 — wherein a powerful laser pulse enters gas (which can either be pre-ionized or not). We first note that the contrast ratio of the laser is not infinite, and so the ionization front starts in the gas at the front tail of the laser pulse, much in advance of the arrival of the main laser pulse. We then note that the main laser pulse needs to be of a length similar to or shorter than the plasma wavelength in order to excite the plasma efficiently.

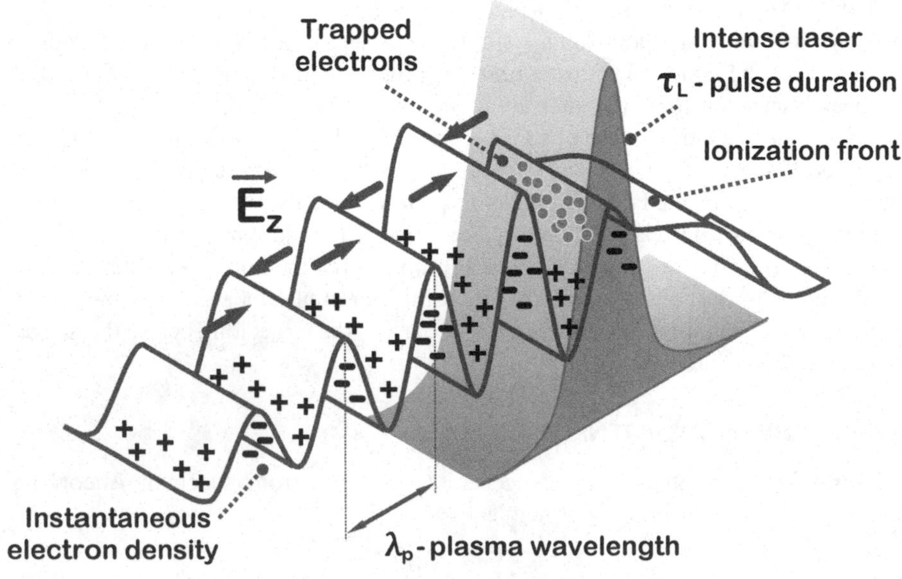

Figure 6.9 Laser acceleration — conceptually. Linear regime.

The electrons of the plasma can be trapped in the wave and then accelerated. Maximum acceleration can occur when the laser pulse causes total separation of the electrons and ion charges of the plasma; this regime is nonlinear, and the cavity that is formed in the plasma and can trap and accelerate electrons is called a *bubble.* Usually, electrons are trapped and accelerated in the first bubble. The mechanism of bubble formation in a strongly nonlinear approximation will be discussed in detail in the following section.

6.4.1 PONDEROMOTIVE FORCE

The formation of a bubble is the result of *ponderomotive force* that a laser pulse confined in space exerts on the plasma electrons.

We start from the assumption that the laser field E is homogeneous:

$$E = E_0 \cos(\omega t) \tag{6.29}$$

The corresponding transverse motion of electrons is

$$\ddot{y} = \frac{F}{m} = \frac{eE}{m} \quad \Rightarrow \quad y = -\frac{eE_0}{m\omega^2}\cos(\omega t) \tag{6.30}$$

We then assume that the field E has a gradient in transverse direction y and thus can be expressed as

$$E = E_0(y)\cos(\omega t) \approx E_0 \cos(\omega t) + y\frac{\partial E_0}{\partial y}\cos(\omega t) \tag{6.31}$$

We then find the force acting on an electron e^- averaged over time is

$$\langle F \rangle_t = \left\langle -\frac{eE_0}{m\omega^2}\cos(\omega t) \cdot \frac{\partial E_0}{\partial y}\cos(\omega t) \right\rangle_t \tag{6.32}$$

Replacing $\langle \cos^2 \rangle$ with $1/2$, we rewrite it as

$$\langle F \rangle_t = -\frac{e^2}{2m\omega^2}E_0\frac{\partial E_0}{\partial y} = -\frac{e^2}{4m\omega^2}\frac{\partial E_0^2}{\partial y} \tag{6.33}$$

We see from this equation that, since the intensity $I \propto E^2$, the ponderomotive force is proportional to the gradient of the laser intensity

$$\langle F \rangle_t \propto -e^2\frac{\partial I}{\partial y} \tag{6.34}$$

and that the direction of the force is such that the ponderomotive force pushes electrons out from the high intensity region.

We also note that the ponderomotive force is independent on the sign of the charge — positrons will be pushed out of the laser pulse just as well.

An alternative explanation of the ponderomotive force can also be suggested (see Fig. 6.10): the electrons oscillating in the time-varying laser field are pushed away more forcefully when they are in a higher intensity region — which on average repels the charged particles from the laser pulse's high intensity area.

Figure 6.10 For illustration of the mechanism of the ponderomotive force. Laser with transverse intensity gradient (left), qualitative picture of the electron motion (middle) and the intensity profile with direction of the force shown (right).

6.4.2 LASER PLASMA ACCELERATION IN NONLINEAR REGIME

Ponderomotive force plays a key role in the formation of the accelerating *bubble* in the nonlinear (also called *blow-out*) regime of laser plasma acceleration.

The ponderomotive force of a short (typically ~50 fs) and intense (typically $\sim 10^{18}$ W/cm^2) laser pulse expels plasma electrons while heavier ions stay at rest. The expelled electrons are immediately attracted back to the ions, forming the first bubble, as shown in Figs. 6.11 and 6.12, thus creating a plasma wave that trails behind the laser pulse. The gradient of the density of electrons creates spatial oscillation of the electric field within plasma (reaching ~100 GV/m) which can accelerate particles.

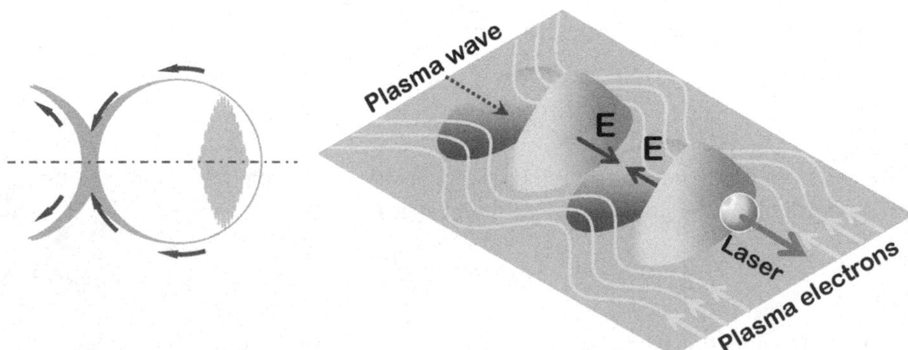

Figure 6.11 Plasma bubble formation.

Figure 6.12 Laser plasma acceleration in nonlinear regime — conceptually.

Having formed the first bubble, the electrons continue their oscillations around the ions, but their motions quickly become incoherent and so the second and subsequent bubbles gradually become smaller. In a sense, only the first bubble (and sometimes the second) is useful for acceleration.

6.4.3 WAVE BREAKING

The high accelerating gradient in the plasma is useful only if a particle beam can be injected into the bubble. Luckily, self-injection of background plasma electrons into the plasma bubble can occur through the *wave breaking* phenomenon.

Wave breaking transpires when, within the nonlinear regime, certain particles outrun the wave, as is represented in Fig. 6.13 in the analogy with ocean waves.

Figure 6.13 Wave breaking concept — the wave nonlinearity gradually rises from left to right.

Other methods of getting particles into the bubble include injection of an external electron beam (challenging if the bunches are short) and various other methods that create an electron bunch inside of the bubble in the right place and at the right time. These typically involve using multiple laser pulses and mixes of gases with different ionization potentials.

6.4.4 IMPORTANCE OF LASER GUIDANCE

As a laser pulse travels through the gas or plasma, several competing effects are taking place.

The most notable is *diffraction*; indeed, a laser beam focused to a size of several tens of μm will diffract very fast. Other effects include *dephasing* — the gradual separation of the accelerating beam (which quickly become relativistic) from the laser (which propagates in gas or plasma slower than the speed of light) — and also *depletion* — the gradual decrease of the laser intensity.

Additional peculiar effects include *longitudinal compression* of the laser pulse by plasma waves; *self-focusing*, in particular due to the relativistic effect (the electrons of plasma at the axis become relativistic and have higher masses, affecting the plasma refraction coefficient); and *ionization-caused diffraction* (gas on the axis where intensity is higher will be ionized first, affecting diffraction).

A possible solution that could solve some of the issues listed above involves creating a channel within the plasma where a special density profile $n(r)$ will be formed to assist guiding the laser pulse for a significant distance.

This solution was realized in practice in the form of a capillary discharge channel developed at Oxford University by S. Hooker (ca. 2006). A schematic of the capillary channel is shown in Fig. 6.14. In this example, a sub-mm hole is created in a sapphire block, hydrogen gas is delivered to the capillary via side holes, and discharge electrodes act to pre-form plasma with a density profile featuring minimum on the axis (as gas near cold walls of the capillary has higher density according to $P = nkT = conts$) — such a density profile is suitable for refraction-assisted laser guidance.

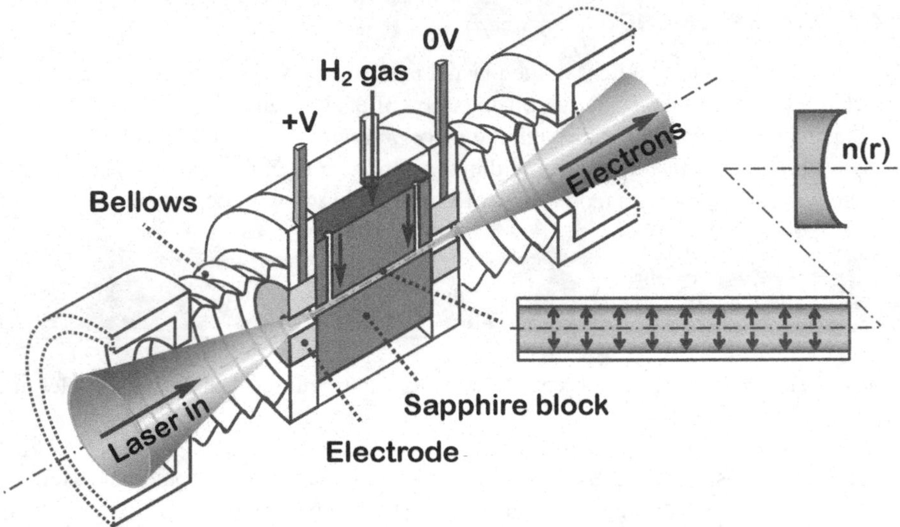

Figure 6.14 Capillary channel technique of laser plasma acceleration.

The capillary channel technique was essential in exceeding the GeV barrier in laser plasma acceleration for the first time ever (W. Leemans et al., 2006), creating a mono-energetic 1 GeV beam after accelerating in just 3 cm of plasma.

6.5 BETATRON RADIATION SOURCES

Beams accelerated in a laser-formed plasma bubble can oscillate, which generates synchrotron (betatron) radiation. Strong radial electric fields within plasma bubbles are responsible for the electrons experiencing transverse oscillations. This can generate bright betatron radiation in an extensive range of photon energy (around 1 - 100 keV). In this section we will estimate the expected parameters of radiation produced by a laser plasma source.

6.5.1 TRANSVERSE FIELDS IN THE BUBBLE

Transverse oscillations of the accelerating electron beam in the plasma bubble are caused by a transverse focusing force that is produced by ions. We can assume that the ions are heavy and are stationary within the bubble. The ions produce a focusing force that can be determined using

$$\oint \mathbf{E} \cdot d\mathbf{S} = 4\pi \int \rho dV \quad \text{(Gaussian units)} \tag{6.35}$$

and by assuming cylindrical symmetry. The focusing force is therefore

$$eE = 2\pi n e^2 r \tag{6.36}$$

An electron with relativistic factor γ will oscillate in this field as

$$\frac{d^2r}{ds^2} = \frac{2\pi ne^2 r}{\gamma mc^2} = \frac{\omega_p^2}{2\gamma c^2} r \tag{6.37}$$

The period of oscillation is thus given by

$$\lambda = \sqrt{2\gamma}\, \lambda_p \tag{6.38}$$

(and is changing during electron acceleration).

6.5.2 ESTIMATIONS OF BETATRON RADIATION PARAMETERS

In Chapter 3 we estimated the characteristics of synchrotron radiation, in particular the energy loss per unit length:

$$\frac{dW}{ds} = \frac{2}{3}\frac{e^2\gamma^4}{R^2} \tag{6.39}$$

the characteristic frequency of photons:

$$\omega_c = \frac{3}{2}\frac{c\gamma^3}{R} \tag{6.40}$$

and the number of photons emitted per unit length:

$$\frac{dN}{ds} = \frac{\alpha\gamma}{R} \tag{6.41}$$

Fig. 6.15 shows a qualitative representation of the evolution of the plasma bubble and oscillation of the accelerating beam. We assume that the beams are self-injected into the bubble due to the wave breaking phenomena when two beamlets overshoot and enter the bubble symmetrically from the top and bottom (as shown in Fig. 6.15a). Their initial transverse velocity forces the beamlets to oscillate in the focusing field of the ions. The beamlets then continue to simultaneously accelerate while exhibiting transverse oscillations.

Let's assume that the amplitude of beam oscillations in the plasma bubble is equal to r_b and the period of oscillations is λ. The radius of the curvature of the beam trajectory in this case is equal to:

$$R = \frac{\lambda^2}{4\pi^2 r_b} \tag{6.42}$$

Substituting the period of oscillation given by Eq. 6.38, we obtain for the radius of the curvature

$$R = \frac{\gamma\lambda_p^2}{2\pi^2 r_b} \tag{6.43}$$

Substituting this into Eq. 6.40, we get an estimation of the radiation wavelength for the laser plasma betatron source:

$$\lambda_c = \frac{1}{3\pi}\frac{\lambda_p^2}{r_\beta}\frac{1}{\gamma^2} \tag{6.44}$$

Using Eq. 6.41 together with Eq. 6.43 we can also estimate the number of photons N_γ emitted per λ:

$$N_\gamma \approx \sqrt{2\gamma}\; 2\pi^2 \,\alpha\, \frac{r_b}{\lambda_p} \tag{6.45}$$

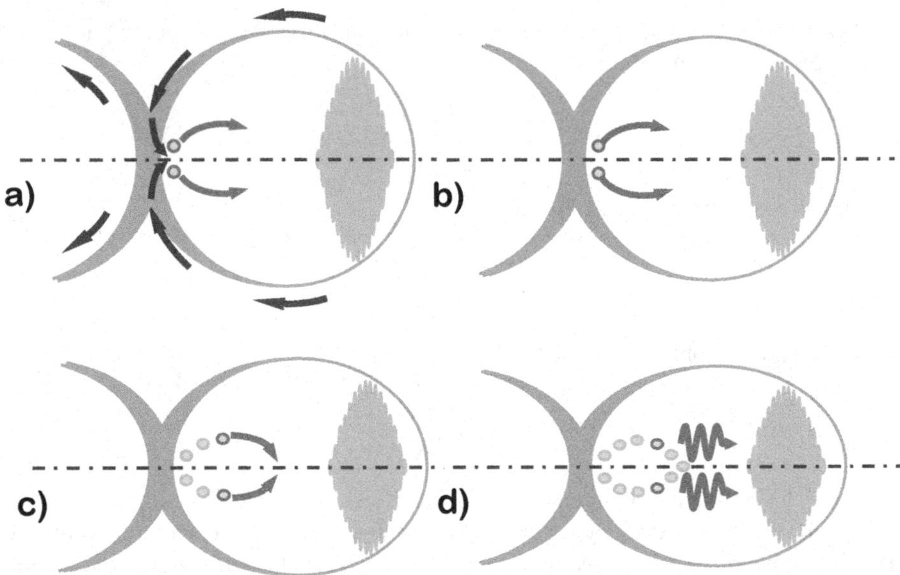

Figure 6.15 Laser plasma betatron source — conceptually. Wave breaking and self-injection — (a). Oscillation of accelerating electron beams in the plasma bubble — (b)-(c), sequential time moments. Betatron radiation produced by oscillating beams — (d).

Let us consider a practical example of a beam accelerated in the bubble characterized by $\lambda_p = 0.03$ mm, to up to 1 GeV ($\gamma = 2 \cdot 10^3$). The synchrotron radiation is most notable at higher energies, thus we will ignore lower-energy radiation occurring during acceleration. The oscillation amplitude r_b can only be guessed very approximately and usually needs to be obtained via careful simulations. In a very rough approximation, r_b is around 1%–10% of the bubble size (which is $\sim \lambda_p$). We therefore assume that $r_b = 0.001$ mm. Substituting this into the above equations, we obtain $\lambda_c = 0.025$ nm (or $\sim$ 50 keV) and N_γ per λ is $\sim$ 0.3 per each accelerated electron. Considering that the accelerating bunch can carry tens of pC to nC charge, we can conclude that such a light source can generate many hard X-ray photons.

6.6 GLIMPSE INTO THE FUTURE

Plasma acceleration is a technique that opens new opportunities for creating scientific, technological and medical instruments. In this section we will give a brief review of the progress made in laser acceleration to date, and, after comparing the evolution of plasma accelerators to the evolution of computers in the latter half of the 20th century, we will take a glimpse into the future.

6.6.1 LASER PLASMA ACCELERATION – RAPID PROGRESS

The last decade has yielded rapid progress in the field of laser plasma acceleration. The pace of research and development in this area received a significant boost in 2004 when the first quasi-monoenergetic beam was generated.[1]

In 2006, Oxford and Berkeley teams[2] broke the GeV barrier in laser plasma acceleration and demonstrated quasi-monoenergetic properties of this accelerated beam. Demonstration of these promising properties further increased the research momentum, and applications of the accelerated beam started to be developed, based on generation of radiation in conventional as well as plasma wigglers (betatron radiation).

The first use of laser plasma-produced betatron radiation for biological imaging was reported[3] in 2011. Multi-GeV laser plasma acceleration[4] was mastered to produce around 4 GeV beams, and most recently around 7.8 GeV beams[5]. Even further progress is certainly expected.

6.6.2 COMPACT RADIATION SOURCES

Laser-driven plasma accelerators can already generate electron beams with several GeV of energy, ~10 fs bunch duration and ~10–100 pC of charge per bunch.

These parameters make laser plasma technology potentially suitable for creating compact radiation sources (see Fig. 6.16). As the beams in these sources are created by a laser in the first place, such sources would also have the advantage of automatic synchronization of accelerated electron beams and generated X-rays with the initial laser pulses. Such synchronization is a powerful asset for *time-resolved studies*.

The far-from-desirable parameters of laser plasma light sources so far include the repetition rate, wall-plug efficiency and beam qualities (energy spread and emittance).

The repetition rate and efficiency are presently limited by laser technology and are the subjects of active research by many groups. Various promising ideas that are beyond the scope of this book have been suggested and are being developed.

---

[1] S. Mangles et al., Nature, v. 431, p. 535-538, (2004).

[2] W. Leemans et al., Nature Physics, v. 2, p. 696-699, (2006).

[3] S. Kneip et al., Applied Physics Letters, 99, 093701, (2011).

[4] W. Leemans et al., Phys. Rev. Letters, 113, 245002, (2014).

[5] A. Gonsalves et al., Phys. Rev. Letters, 122 084801, (2019).

The beam quality, which is a focus of significant attention, is gradually improving, and eventually the laser plasma accelerated beam might be suitable for the generation of coherent radiation in a free electron laser application.

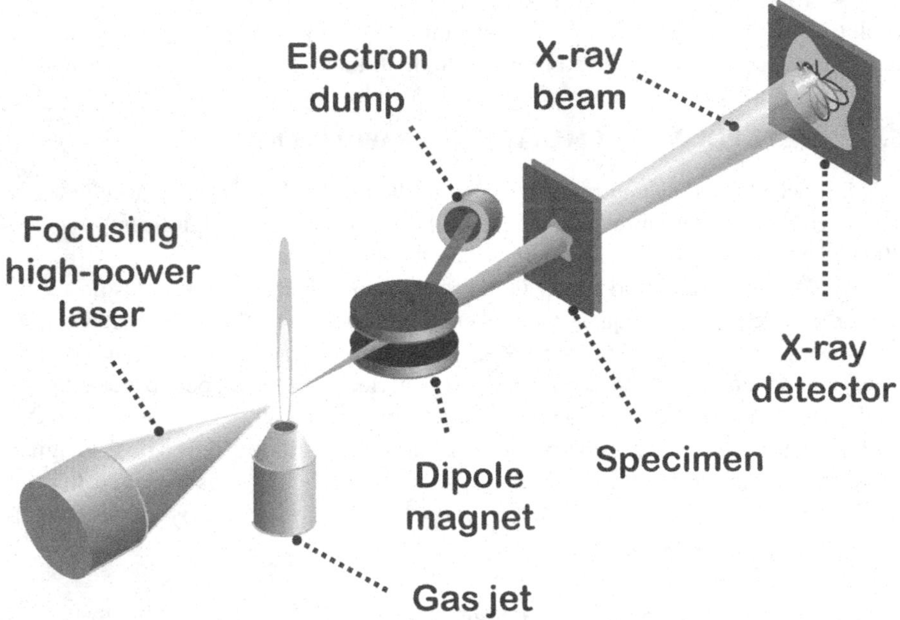

Figure 6.16 Laser plasma betatron radiation light source — conceptually.

Modern synchrotron-based light sources are large machines with perimeters of several hundred meters. The linac-based free electron lasers can be around a kilometer or more in length. Both of these types of light sources operate with electron beams of a few to about 10 GeV.

Despite the fact that similar electron energies can be reached in a much more compact laser plasma accelerator, it is unlikely that laser plasma-based light sources would entirely replace conventional light sources in the foreseeable future. It is more probable that the types of light sources will evolve in a similar manner to that of computers.

6.6.3 EVOLUTION OF COMPUTERS AND LIGHT SOURCES

Early computers were large, bulky and slow. Development of compact personal computers started in the early 1980s, but was not accepted immediately: "IBM bringing out a personal computer would be like teaching an elephant to tap dance," as newspapers mocked around 1981. Still, as we know today, a plethora of large computers and super-computers co-exist with a variety of personal and compact computers, from laptops to mobile phones and smart watches (Fig. 6.17).

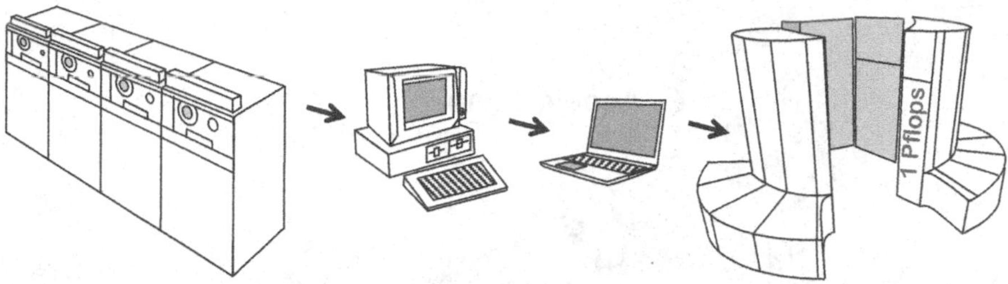

Figure 6.17 Computers' evolution.

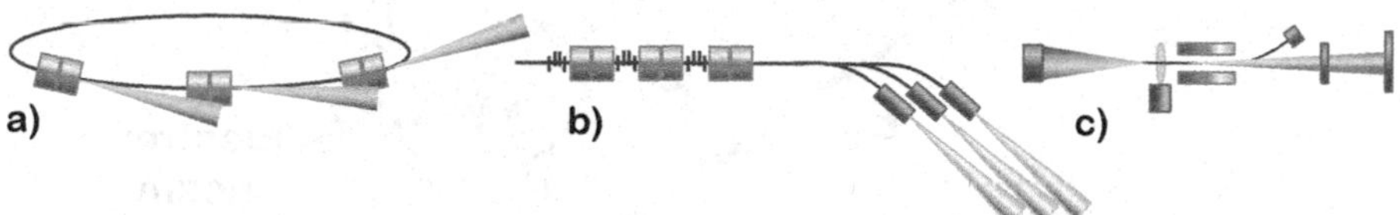

Figure 6.18 Light sources' evolution.

The expected future evolution of light sources may follow a similar pattern. The synchrotron-based light sources and FELs — (a) and (b) in Fig. 6.18 — will eventually be joined by compact plasma-based light sources — (c) in Fig. 6.18 — as a result of intense research, commercialization, and work with users, the industry and economists — efforts of all of which will result in a change of the paradigm.

All types of light sources will then continue to co-exist and national-scale facilities will be complemented by a variety of compact plasma acceleration-based light sources.

6.7 PLASMA ACCELERATION AIMING AT TEV

While application of plasma acceleration to compact light sources is practically within reach, the application of plasma acceleration technology to high-energy physics discovery machines is significantly further away. In this section, we will briefly review some of the primary challenges to plasma acceleration on the way to TeV energy.

6.7.1 MULTI-STAGE LASER PLASMA ACCELERATION

In laser plasma acceleration, the laser pulse propagating through a medium (plasma) has $v < c$ and the accelerating electrons that quickly become relativistic will soon dephase from the plasma wave. Consequently, if we are aiming at multi-tens of GeV or TeV acceleration of electrons, many stages of acceleration will be necessary.

The length of a single stage can be estimated by taking into account that the group velocity of a laser pulse is given by

$$v_g = \sqrt{1 - \omega_p^2/\omega^2} \qquad (6.46)$$

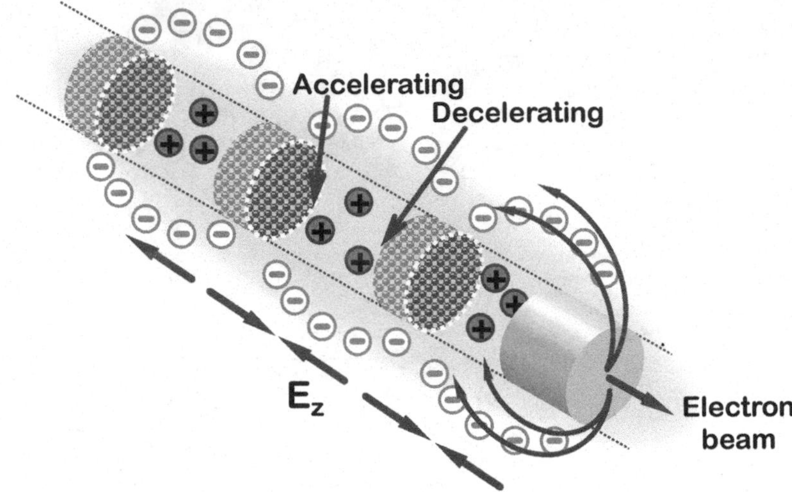

Figure 6.19 Beam-driven plasma acceleration — conceptually.

and that the dephasing occurs when an electron outruns the wave by a half of a period. For a relativistic electron the dephasing time t_d is thus given by

$$(c - v_g)\, t_d = \lambda_p/2 \tag{6.47}$$

We then substitute the expression for the group velocity and get the estimate for the dephasing length:

$$L_d \approx \lambda_p\, \omega^2/\omega_p^2 \tag{6.48}$$

For example, for a laser with a wavelength of 1 μm and $\lambda_p = 30$ μm, the dephasing length is $L_d \approx 30$ mm. With an accelerating gradient (at the corresponding plasma density) of 1 GeV/cm, a single stage could yield around 3 GeV. Acceleration to a TeV would thus require several hundred stages. Preservation of beam qualities during multi-stage acceleration is the research area that promises significant advances in the near future.

6.7.2 BEAM-DRIVEN PLASMA ACCELERATION

Plasma can be excited not with a laser pulse, but with a short intense bunch of charged particles (e.g., electrons). In this case, the bubble will be formed due to the bunch's field, and will have a very similar shape and properties to the laser example.

This *beam-driven acceleration* approach (Fig. 6.19) has an advantage in that the driver beam has $v = c$, and thus dephasing of the witness beam from the driver is no longer an issue. Another advantage of this method is that the driver beam can carry much more energy than a laser pulse.

These advantages manifested themselves via a much higher beam energy achieved in beam-driven acceleration — the maximum final energy obtained so far is around

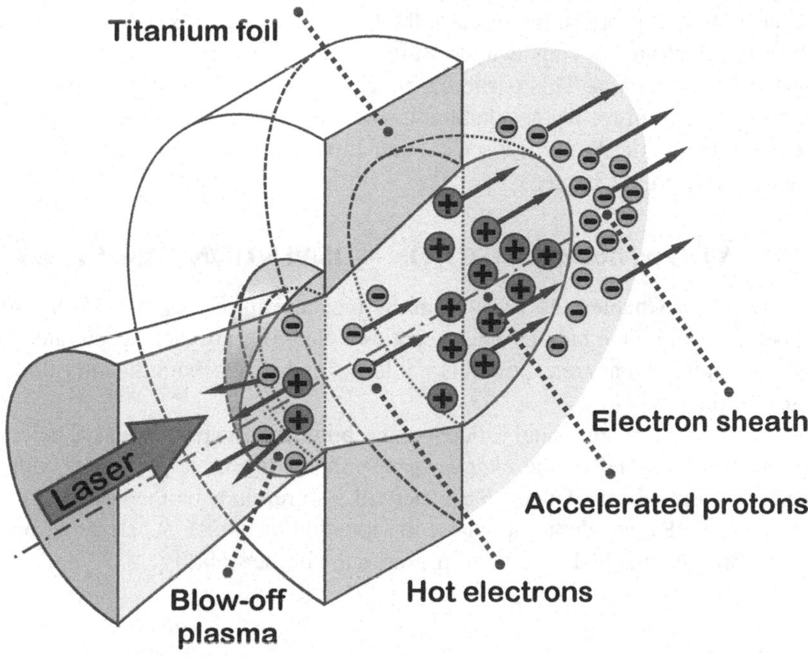

Figure 6.20 Sheath laser plasma acceleration of protons or ions.

80 GeV with an initial beam energy of 42 GeV (which acted in this SLAC linac energy doubling experiment[6] both as a witness as well as a driver).

6.8 LASER-PLASMA AND PROTONS

So far, we have only discussed acceleration of electrons. However, plasma acceleration of protons and ions is possible and is being actively developed. We will touch here on this subject only briefly, and leave the details for Chapter 9.

The major reason why laser plasma acceleration of protons should be developed is its potential value to proton therapy. Conventional proton therapy systems — which require around 250 MeV protons and include beam sources (cyclotron or synchrotron), beamlines and especially the beam delivery gantries — are large and expensive. A desire to create compact laser plasma acceleration proton therapy systems is one of the main motivations for development of plasma acceleration of protons.

Proton and ion laser plasma acceleration is a rapidly developing area, and it is not yet quite settled and ready for the textbooks. Different models, with different assumptions and simplifications, are created to explain and predict beam properties in different regimes of acceleration. One of the models, illustrated in Fig. 6.20, is sheath

[6] I. Blumenfeld et al., Nature, 2007.

laser acceleration of protons. In this case, the laser heats and ionizes a foil, creating a sheath of hot electrons moving away from the foil, which in turn pull and accelerate the ions from the plasma. This particular mechanism, unfortunately, creates a very wide energy spectrum for the accelerated beam. Moreover, its scaling with the laser power is not favorable. We will discuss other more suitable mechanisms of proton plasma acceleration in Chapter 9.

6.9 LWFA DOWNRAMP INJECTION – INVENTION CASE STUDY

As we saw in this Chapter, the wave-breaking mechanism results in capturing some plasma electrons into the bubble. These captured electrons are then accelerated. The number of electrons that are captured is not large, and may be significantly fluctuating pulse-to-pulse.

There is a way to significantly increase the number of captured and accelerated electrons, which remind us the *change of parameter* inventive principle, with parameter being the plasma density. This method also reminds us the critical gamma transition during RF acceleration, where the location of the RF accelerating bucket had to be abruptly changed in order to preserve the phase stability.

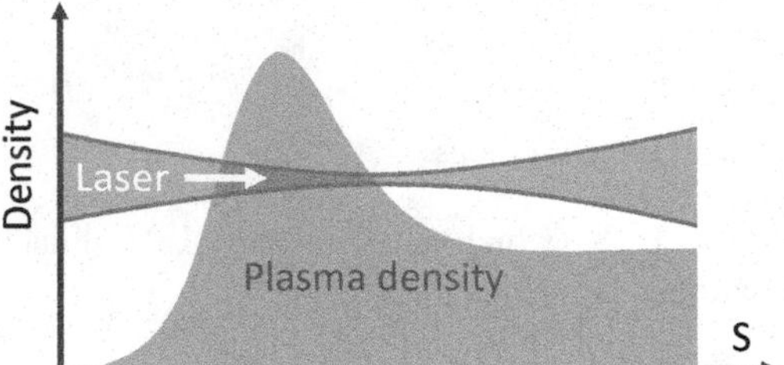

Downramp injection, when the laser beam is focused into the area of decreasing plasma density, reminds us the parameter change inventive principle.

Figure 6.21 Downramp injection.

This method is the so-called downramp injection[7] — achieved when the laser beam is focused to the area, where the plasma density is decreasing along the beamline, as shown in Fig. 6.21.

Since the bubble size is proportional to the plasma wavelength, which in its turn depends on the plasma density, in this downramp injection case, the length of the bubble will grow for a certain distance along the beamline. Plasma electrons, which were on the boundary of the trapping area, will now appear inside of the plasma bubble — therefore more electrons will be trapped and accelerated, fluctuations will reduce, and injection will become more predictable. A constant-density region placed after the downramp will allow to accelerate these electrons to a high-energy.

[7] Bulanov PRE 1998; H. Suk PRL 2001; Geddes PRL 2008; Gonsalves Nature Physics 2011

EXERCISES

6.1. *Chapter materials review.*
What laser intensity (in W/cm^2) would correspond to a normalized vector potential of $a_0 = 10$, and what are the maximum values of the electric and magnetic fields in the laser wave for a ruby laser or for a CO_2 laser?

6.2. *Chapter materials review.*
Assume that we get a 1 GeV electron bunch from a laser plasma accelerator and would like to create an undulator from plasma using the focusing force of the plasma's ions. Suggest the plasma parameters and the amplitude of the beam oscillation that would correspond to the radiation at the boundary of the undulator regime, i.e., with the undulator parameter equaling $K = 1$.

6.3. *Mini-project.*
Assume that the LHC proton beam of $E = 7$ TeV is going to be used as a driver for the plasma acceleration of electrons. Select the parameters for the plasma and estimate to what bunch length you would need to compress the LHC beam so that it could be used for plasma acceleration. Define, roughly, the parameters of the corresponding bunch compressor.

6.4. *Analyze inventions or discoveries using TRIZ and AS-TRIZ.* Analyze and describe scientific or technical inventions described in this chapter in terms of the TRIZ and AS-TRIZ approaches, identifying a contradiction and an inventive principle that were used (could have been used) for these inventions.

6.5. *Developing AS-TRIZ parameters and inventive principles.* Based on what you already know about accelerator science, discuss and suggest the possible additional parameters for the AS-TRIZ contradiction matrix, as well as the possible additional AS-TRIZ inventive principles.

6.6. *Practice in reinventing technical systems.* One physics lab was studying plasma discharge. It was arranged in a cylindrical vessel that was placed horizontally. The team observed that the plasma of the discharge, being hotter and thus having a lower density than the surrounding gas, would move upward, disturbing the experiment. In order to keep the plasma in place, the team installed a solenoid around the plasma vessel. However, when the intensity of the plasma discharge was increased, it was observed that the solenoid could not keep the plasma in place anymore. One member of the team then suggested rebuilding the installation and using a much stronger solenoid. However, the head of the lab said that there is a much simpler way! And so, they made quick adjustments to the installation to fix the problem and keep the plasma centered. Try to recreate the solution to this challenge. Hint: one can use a device that can be found in any household.

6.7. *Practice in the art of back-of-the-envelope estimations.* In an electron-positron collider, the electron beam with an energy of 250 GeV is circulating in a storage ring with a 100-km perimeter. The average current of the circulating electron beam is 1 A. Estimate the total power needed to feed the RF cavities in such an accelerator. *(It is assumed that you can identify the most important effects playing roles in this task, can define the necessary parameters and set their values, and can get a numerical answer.)*

7 Light Sources

In this chapter, we will focus on light sources — first synchrotron radiation light sources and then Compton light sources. For the former, we will base our observations on results derived in Chapter 3, and will then introduce the necessary formalism for the latter.

7.1 SR PROPERTIES AND HISTORY

We know from courses on electrodynamics that electromagnetic radiation is emitted by charged particles when they are accelerated. In the particular case when the relativistic particles move on a curved trajectory — i.e., when they are accelerated radially (when the velocity is perpendicular to the acceleration vector) — the emitted electromagnetic radiation is called *synchrotron radiation*.

This SR was at first considered a nuisance, as it causes energy losses in the accelerated particles. Its unique properties, however, were eventually shown to have the potential to pave the path to a new and important type of scientific instrument: the SR sources.

7.1.1 ELECTROMAGNETIC SPECTRUM

The usefulness of SR and Compton sources is due to the fact that they can cover a large range of the electromagnetic spectrum (Fig. 7.1), from infrared (IR) and ultraviolet (UV) to vacuum ultraviolet (VUV: the wavelength range strongly absorbed in air — thus the name), and on to hard X-rays and γ-rays.

The EM spectrum shown in Fig. 7.1 also indicates a particular range of wavelengths — the so-called *water window* is defined as the range between the K-absorption edges of oxygen (0.53 keV) and carbon (0.28 keV) — see Fig. 7.2. In this range of soft X-ray energies, water is relatively transparent, which simplifies SR experiments when biological samples need to be used as water solutions.

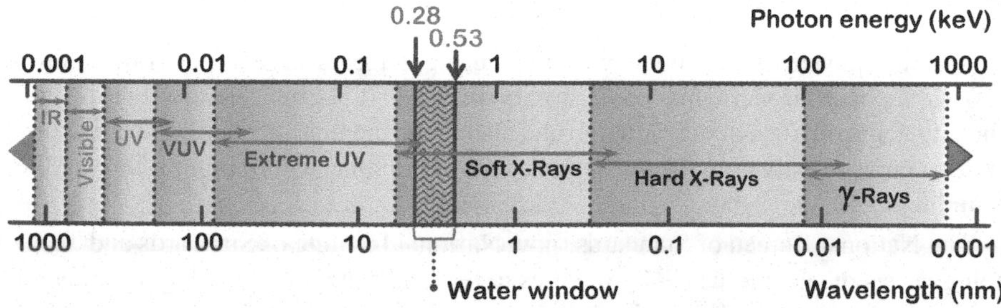

Figure 7.1 Electromagnetic spectrum covered by SR and Compton sources.

DOI: 10.1201/9781003326076-7

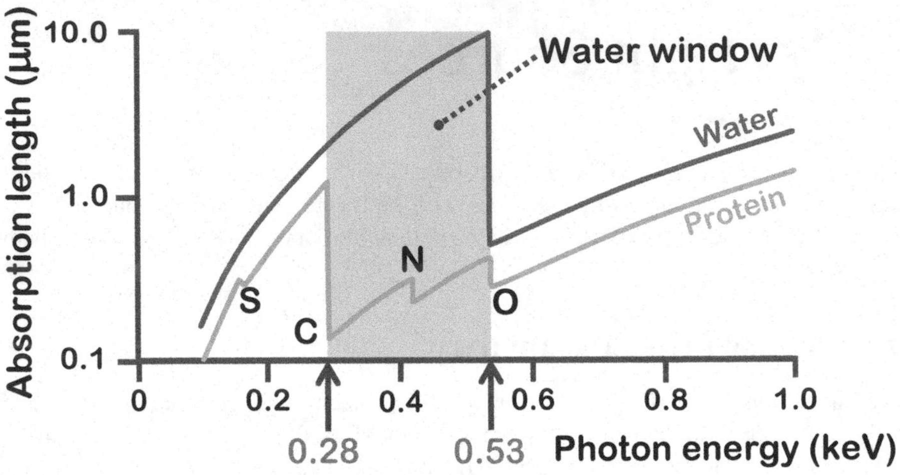

Figure 7.2 Photon attenuation in water in comparison with a typical protein.

The success and widespread use of SR and Compton sources is thanks to their flexible spectral parameters. This allows one to select the photon energy required for an experiment by adjusting the energy of the beam or laser, or the field strength of wigglers or undulators (useful, for example, when trying to fit within the range of the water window).

7.1.2 BRIEF HISTORY OF SYNCHROTRON RADIATION

In 1944, D. Ivanenko and I. Pomeranchuk predicted that the maximum energy of electrons in a betatron is limited due to energy losses caused by radiation of relativistic electrons.

This radiation was first observed around 1947, by accident, in a General Electric 70 MeV synchrotron (this gave the name *synchrotron* to the observed radiation), in the *visible* spectrum. It is interesting to note that earlier deliberate attempts to find this radiation in a betatron had failed, as researchers looked for the radiation in a microwave range where the betatron walls were opaque.

The first physics experiments with SR were conducted in 1956 at Cornell, on a 320 MeV synchrotron. In this run, D. Tomboulian and P. Hartman studied the spectral and angular properties of the radiation and also made the first soft X-ray spectroscopy experiments, investigating the transparency of beryllium and aluminum foils near the K and L edges.

The National Bureau of Standards (now National Institute of Standards and Technology) was the next to use SR properties to their advantage, modifying a section of a vacuum chamber of a 180 MeV electron synchrotron to enable access to SR. Soon, it was apparent that the era of SR light sources had begun.

7.2 EVOLUTION AND PARAMETERS OF SR SOURCES

A large demand for new scientific instruments stimulated enormous advances in SR light source technology. In just a few decades, several generations of SR sources technology have been developed, each exhibiting an improvement with every evolutionary step.

7.2.1 GENERATIONS OF SYNCHROTRON RADIATION SOURCES

It is now a tradition to distinguish the many generations of SR sources according to the following classifications. The first-generation light sources are the accelerators built for high-energy physics, nuclear physics or other purposes, which used for synchrotron radiation experiments parasitically.

A large demand for SR experiments resulted in the construction of dedicated accelerators, creating the second-generation of purpose-built synchrotron light sources. The SRS at Daresbury, England, was the first dedicated machine (operated between 1981 and 2008).

The second-generation light sources employed SR emitted from bending magnets. Advances in accelerator science and technology, inspired by a demand from SR users, quickly created an opportunity for the next technological breakthrough: the third-generation light sources — accelerators optimized for high brilliance due to low electron beam emittance and the use of *insertion devices* (wigglers and undulators). Examples of such SR sources include the European Synchrotron Radiation Facility (ESRF) in France, the Diamond light source in the UK and many others.

The third-generation of SR sources is presently the most widespread. There are several tens of such machines around the world and the number is growing, following demand in the field of science. The brightness of the third-generation machines is several orders of magnitude higher than that of the previous generation (see Fig. 7.3) and exceeds — by about ten orders of magnitude — the brightness of the sources available in the beginning of the 20th century.

The fourth-generation light source was brought to fruition via the *free electron laser* idea, which was developed in the 1970s by John Madey. An essential part of the FEL, the undulator (invented in 1947 by Ilya Ginzburg) provided a significant cornerstone for the foundation of FEL technology development.

The fourth-generation light sources enjoy all the latest developments of accelerator science: they are linac-based with low-emittance photo-injectors, assisted with several bunch compressors that help to obtain the ultra-short bunches. Examples of such FELs include: FLASH (Germany), LCLS (USA), and SACLA (Japan). Transition to the fourth-generation, which are coherent sources, is also an interesting illustration of the other way around inventive principle; see Fig. 7.4.

The next generation of SR light sources will inevitably arrive, and very soon. The exact design and underlying technology are not yet determined, and in fact several ideas are in competition, one being the *ultimate storage rings* discussed in Chapter 3 and another being related to plasma acceleration light sources discussed in Chapter 6, among others.

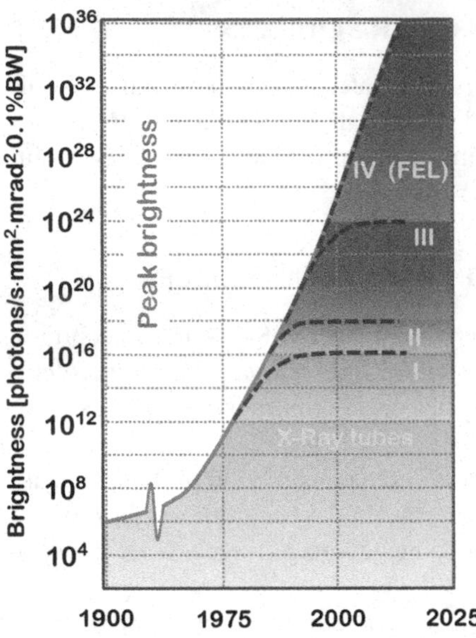

Figure 7.3 Generations of SR sources.

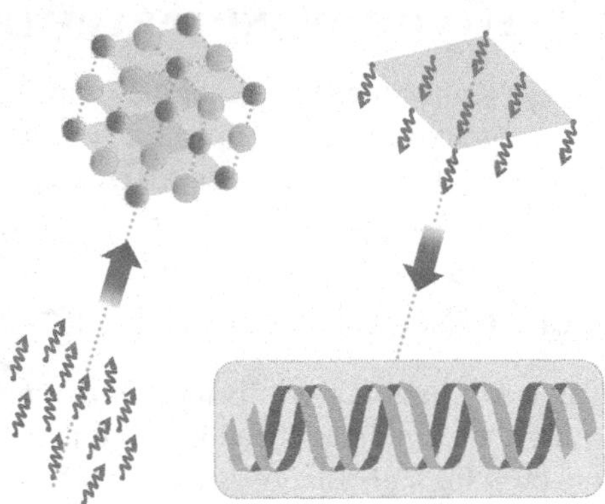

Figure 7.4 From 3rd to 4th generation and the inventive principle of the other way around — the era of studies of crystalline structures by incoherent photon beams in the 3rd generation is replaced by the era of studying non-crystalline structures by coherent (i.e., crystal-like) photon beams in the 4th generation.

7.2.2 BASIC SR PROPERTIES AND PARAMETERS OF SR SOURCES

Without repeating the formulae, which can be found in Chapter 3, let us briefly recall our approach for deriving important parameters of SR sources and of the synchrotron radiation itself.

Recall that a simple back-of-the-envelope treatment of SR was possible when the amount of radiation *left behind* was determined from simple geometrical consideration and the volume integral over the field squared gave us the energy lost per unit of length. This led us immediately to the estimation of the cooling time and, after considering the quantum (statistical) character of radiation, we came up with an estimate for the equilibrium horizontal emittance of the beam (with vertical equilibrium emittance determined by the coupling of the ring).

Knowledge of equilibrium emittances, as well as knowledge of the single photon emittance that we estimated on the way, firstly led us to an estimation of the SR flux and then of the brightness of SR sources.

High-intensity photon flux (defined as the number of photons per second and per spectral bandwidth) allows for either rapid experiments or the use of weakly scattering objects or crystals. The brightness of a light source (also called, interchangeably, brilliance or spectral brightness defined in units of $photons/(s \cdot mm^2 \cdot mrad^2 \cdot BW)$) is determined as the number of photons emitted per second per surface area and per solid angle and per fraction of a spectral bandwidth. The high brilliance of SR sources is enabled by low emittance (and thus a low emitting area) and a low divergence of radiation emitted by an ultra-relativistic beam.

SR covers a broad light spectrum, from microwaves to hard X-rays, hence allowing a variety of experiments. Experiments that require precise photon energies can benefit either from the use of a monochromator or from the ability to adjust the emission wavelength of the insertion devices. Polarization of SR can be either linear or circular, as required by the experiment, and depends on the design of insertion devices.

The coherence properties of SR sources vary — from partial coherence in third-generation sources to full temporal coherence in early FELs and full temporal and spatial coherence in the latest designs of pre-seeded FELs. We will discuss some of the corresponding methods in Chapter 10.

The temporal structure of SR radiation in third- and fourth-generations span many orders of magnitude, starting from tens of ps SR flashes in ring-based sources to ultra-short tens of fs flashes in FELs. The sources are designed to have high spatial and temporal stability of the emitted radiation, to sub-micron levels in space, and tens of fs levels in time.

Discussing the dependences of SR properties on parameters in Chapter 3 led us to the concept of the ultimate brightness of a *diffraction-limited* SR source — the *ultimate storage ring* where the equilibrium horizontal emittance is at least as small as the emittance of radiated photons.

The tendencies to increase the ring perimeter and tighten the focusing (increased segmentation of the ring), are demonstrated by the developing plans of upgrades of many third-generation SR sources, as well as by the emerging designs of ultimate performance storage rings.

Reaching the ultimate parameters, as Eq. 3.30 suggests, may require increasing the perimeter of the ring and decreasing the average of η^2/β_x i.e., employing tighter focusing, which can be achieved by shortening the distances between focusing elements.

7.3 SR SOURCE LAYOUTS AND EXPERIMENTS

The modern third-generation light sources are state-of-the-art facilities that provide multiple X-ray beamlines for a variety of experiments (Fig. 7.5).

In this section, we will look into a generic layout of such sources, and will briefly touch on their experimental capabilities.

7.3.1 LAYOUT OF A SYNCHROTRON RADIATION SOURCE

A schematic of a generic third-generation synchrotron radiation source is shown in Fig. 7.6. Electrons are typically generated in an RF gun and accelerated in a *linac* (usually to a few hundred MeV), further accelerated to the required energy (of a few GeV) in a *booster*, and then injected into the storage ring where the circulating electrons emit an intense beam of synchrotron radiation.

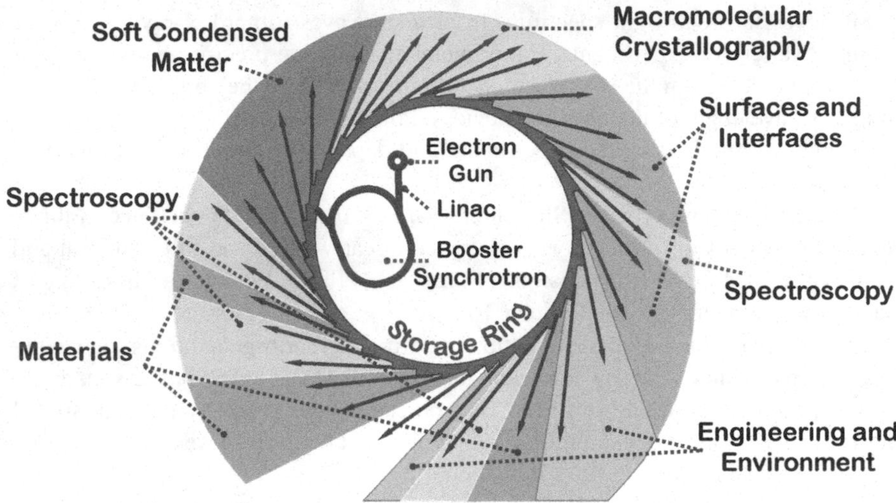

Figure 7.5 Generic SR light source with multiple X-ray beamlines and showing typical allocation of beamlines to experiments.

The optics of the storage ring are arranged in such a way so they have many empty drift sections where *insertion devices* (ID) — wigglers and undulators — can be installed. Each of these IDs will direct light into a corresponding X-ray beamline, which can then be tailored to a particular type of experiment (life science, materials, etc.). The typical number of X-ray beamlines is a couple of dozen, as shown in Fig. 7.5.

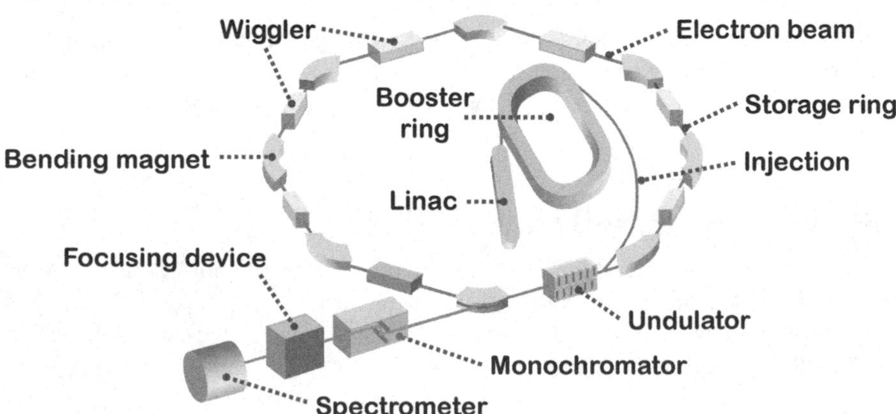

Figure 7.6 Schematics of a generic third-generation SR light source.

The second-generation sources use radiation emitted in bending magnets, which emit a continuous spectrum characterized by critical energy ε_c, which can be estimated as ε_c(keV) = 0.665 B(T)E^2 GeV. For example, for $B = 1.4$ T and $E = 3$ GeV the critical energy is $\varepsilon_c = 8.4$ keV.

The third-generation employs insertion devices (undulators and wigglers). Either of these devices is a periodic array of magnetic poles that provide a sinusoidal magnetic field **B** on axis: $\mathbf{B} = (0, B_0 \sin(k_u z), 0)$ where $k_u = 2\pi/\lambda_u$. The maximal radius of the curvature of the orbit in the sinusoidal field R defines two distinct regimes (see Section 3.3.5). If $2R/\gamma \ll \lambda_u/2$ (parameter $K \sim \gamma\lambda_u/R \gg 1$), then the radiation emitted at each period of sin-like field is independent; this corresponds to a wiggler regime (similar to radiation from a sequence of bends). The insertion devices working in a wiggler regime are usually used to achieve high photon energies and flux.

The insertion devices can be placed outside of the beam vacuum chamber or can be made as *in-vacuum devices* — either undulators or wigglers — as needed to achieve maximum field on the beam orbit and therefore maximum radiation from the insertion devices. The insertion devices can be made with permanent magnets, electromagnets or superconducting magnets, and the gaps of the insertion devices can be made adjustable so that they can be wide open during beam injection, and can be narrowed during normal operation.

In contrast, the undulator regime corresponds to the case $2R/\gamma \gg \lambda_u/2$ (or $K \ll 1$), which means that the entire wiggling trajectory will contribute to radiation. Undulators are typically used to generate high brilliance radiation in a quasi-monochromatic spectrum — the bandwidth of undulator radiation is inversely proportional to the number of undulator periods N_u and can be estimated as $\Delta f/f \sim 1/N_u$. The precise definition and meaning of parameter K, as well as the wavelength of undulator radiation, will be discussed in the next chapter, in relation to FELs.

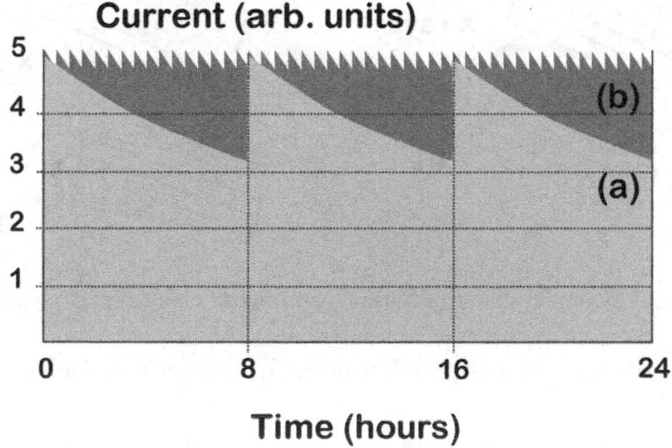

Figure 7.7 Current in SR light source without (a) and with (b) top-up injection mode.

To a large extent, the scientific performance of third-generation SR sources depends on their stability. The current in the storage ring, decaying between injection cycles (due to the *Touschek effect*: intrabeam scattering resulting in a change of particle momentum and its consequent loss on the energy acceptance aperture) forces the power of emitted SR to change, affecting the temperature regime and stability of the ring and of the X-rays' beamlines. The *top-off injection* (also called *top-up*) is the operation regime (see Fig. 7.7) that keeps the beam current in the ring almost constant, improving the stability significantly. In this regime, a small amount of current is injected into the ring much more frequently than in the standard regime.

The top-up injection brings advantages to the light sources — no issue with beam lifetime anymore, since the beam can be replaced frequently, and stable temperature conditions to X-ray optics of the users' beamlines. However, top-up injection also brings challenges. Since users will desire continuous operation of the beamlines, the shutters, that can be closed to protect the X-ray beamlines, need to remain open during injection. Similarly, the gaps of the insertion devices need to remain in the nominal narrow position. Both these factors will create additional challenges for the ring dynamic aperture and for the injection. We will discuss this in connection with one of invention study case further below.

7.3.2 EXPERIMENTS USING SR

Synchrotron radiation allows for a wide array of experiments, ranging from utilization of phenomena in X-ray scattering, X-ray absorption or X-ray fluorescence, to various advanced methods that enhance the resolution of obtained images (e.g., relying on X-ray absorption near the atomic spectral edges of particular elements contained in the studied samples). Already, a variety of imaging methods allow for the use of SR in biological, chemical, medical and material studies, and in many other areas of science and technology.

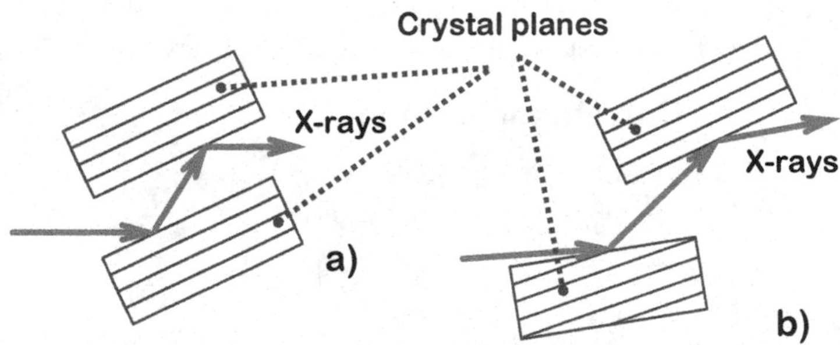

Figure 7.8 Crystal monochromator of X-rays. Symmetric case (a) and asymmetric case (b).

SR experiments often require an X-ray beam with a well-defined wavelength. *Monochromatization* is typically performed by the crystal monochromators. A variety of configurations of crystal monochromators are possible — Fig. 7.8 shows two particular arrangements. In both of these cases, the geometry is selected in such a way that the desired X-ray wavelength λ corresponds to *Bragg conditions*

$$n\lambda = 2d\sin\theta \tag{7.1}$$

which correspond to the maximal reflection of X-rays from the crystal. In the above equation, n is an integer, d is the distance between the crystal planes and θ is the angle between X-rays and scattering crystal planes. The monochromator plates are usually made from crystals of Si or Ge. The symmetrical configuration shown in

Fig. 7.8a is standard, while the asymmetrical one (Fig. 7.8b) allows for the increase in angular resolution — thus narrowing the resulting energy spread of the X-ray beam.

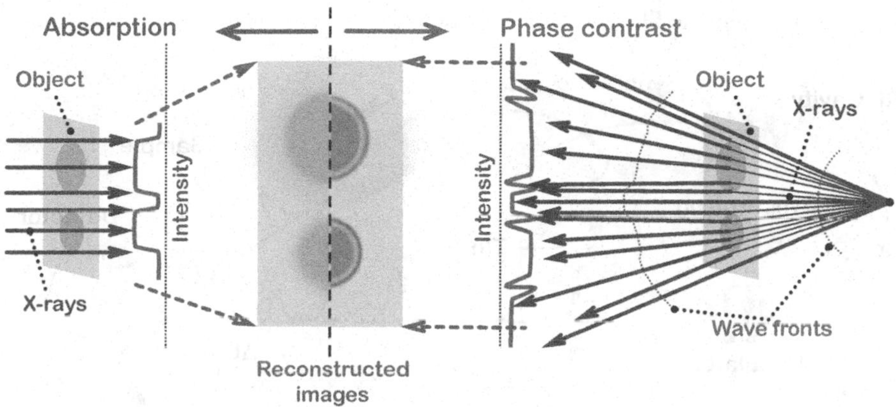

Figure 7.9 Absorption (left) and phase contrast (right) X-ray imaging and comparison of reconstructed image (middle).

Due to the small size of the area that emits X-rays, the SR light sources can utilize an advanced technique called *phase contrast imaging* — shown in Fig. 7.9 in comparison with standard absorption imaging.

Phase contrast imaging is particularly appropriate for studies of biological objects where the density difference, and thus absorption difference between different tissues, is minimal, which complicates the goal of achieving high-resolution images relying on absorption (left part of Fig. 7.9). However, benefiting from the point-like nature of the emitting source, one can increase the distance between the object and the detector plane, and rely instead on refraction of X-rays caused by the density variations in the object. The consequent interference pattern on the detector plane will have much sharper features, thus reconstructing images with better resolution and contrast (right side of Fig. 7.9).

The phase contrast imaging technique is especially beneficial for laser plasma betatron light sources (see Chapter 6), as the emitting areas can have sizes below a micrometer. The relatively low average brightness of such sources would then be compensated by higher spatial resolutions, which are additionally enhanced by extremely short temporal durations of X-ray flash.

To conclude this section, let us recapitulate the capabilities of the modern SR sources by referring to a well-known example. In 1952, DNA structure was studied in R. Franklin and R. Gosling's experiments using an X-ray tube that had a brilliance of around $10^8\left(\mathrm{ph/\,sec\,/mm^2/mrad^2/0.1\,BW}\right)$. At that time, the duration of exposure needed to acquire the necessary statistics was typically as long as one day (around 10^5 sec). The modern third-generation light sources with a brilliance of the order of 10^{20} can provide the same exposure in just 100 ns.

The much higher brightness is not the only advantage of modern SR sources. Engaging lasers in combination with SR sources creates a completely new type of experiment — a *pump-probe* configuration, in which a laser pulse synchronized with the beam revolution in an SR source excites the object just before the main X-ray pulse arrives from the SR source (as illustrated in Fig. 7.10).

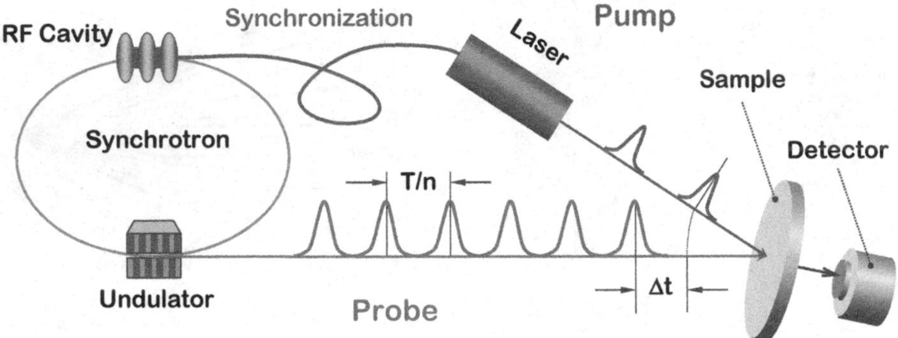

Figure 7.10 Pump-probe experiment arrangement. Here T and n are revolution period and number of bunches in the SR ring, Δt is time delay between the pump laser pulse and SR probe pulse. RF-laser synchronization is achieved with fiber optics connection.

The specific feature of this kind of experiment is the ability to vary the time delay Δt between the pump and probe pulses. The pump-probe experiments are therefore in particular useful for studies of ultra-fast phenomena such as spin dynamics in metals, structural molecular dynamics of proteins, photosynthesis, ultrafast photo-switching and many others.

7.4 COMPTON AND THOMSON SCATTERING OF PHOTONS

Thomson and Compton processes describe the scattering of an EM wave or photon on a charged particle. The Compton scattering, in particular, can be very useful for creating compact X-ray sources — enabled by the development of electron accelerators and laser technologies.

In this section — after reviewing the basic formalism of the Thomson and Compton processes — we will discuss the typical design and characteristics of Compton X-ray sources.

7.4.1 THOMSON SCATTERING

The elastic scattering of an electromagnetic plane wave by an electron at rest (or low energy E) with mass m_e and charge q, is a process known as *Thomson scattering*.

The total cross section of a classical Thomson scattering is given by the following equation:

$$\sigma_{Th} = \frac{8\pi}{3} r_e^2 \approx 0.665 \cdot 10^{-28} \; [m^2] \tag{7.2}$$

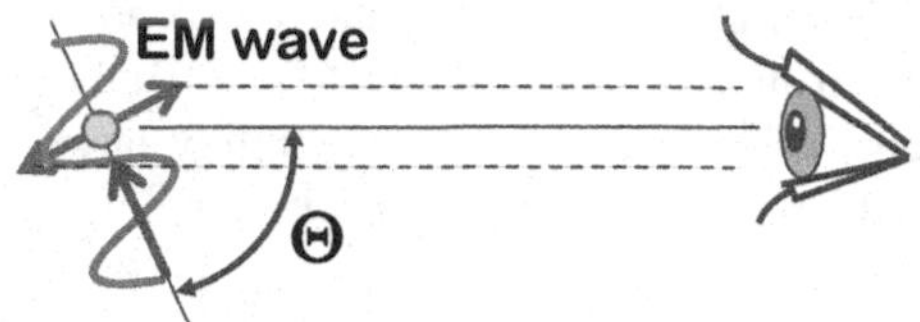

Figure 7.11 Thomson scattering.

Thomson scattering — elastic scattering of an EM plane wave by an electron at rest. Compton interaction — collision between electron and photon with a substantial energy exchange.

And the differential cross section, illustrated in Fig. 7.11, is equal to

$$\frac{d\sigma}{d\Omega} = \frac{1}{2} r_e^2 \left(1 + \cos^2\theta\right) \tag{7.3}$$

Thomson scattering is an approximation of an elastic process — the energies of the particle and photon are the same before and after the scattering (i.e., the recoil of the electron can be neglected, in contrast to the Compton scattering).

7.4.2 COMPTON SCATTERING

Compton scattering describes the inelastic process where we can no longer neglect the transfer of energy between the particle and the photon.

We are, in particular, interested in the instance when a collision between a high-energy electron and a low-energy photon results in a substantial fraction of the electron energy being transferred to the photon. In the laboratory reference frame, this manifests as backscattering of the photon with a significant energy boost; this process is known as *Compton backscattering* (or inverse Compton scattering), as illustrated in Fig. 7.12.

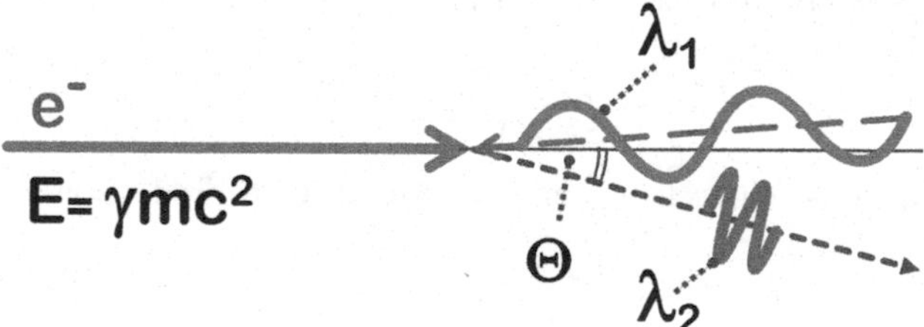

Figure 7.12 Compton backscattering. Initial photon with wavelength λ_1 and after scattering with λ_2.

In the approximation of small angles and relativistic electrons (see Fig. 7.12), the wavelength of the photon after Compton backscattering is described by the following equation:

$$\lambda_2 = \lambda_1 \left(1 + \theta^2\gamma^2\right) / \left(4\gamma^2\right) \tag{7.4}$$

As we can see from Eq. 7.4, in the case of relativistic electrons, the photon gains considerable energy after interaction: its wavelength is shortened by the factor of $4\gamma^2$.

Let's consider two examples in the case of green light with λ_1= 532 nm (corresponding photon energy is 2.33 eV). If the total electron energy is 5.11 MeV (γ = 10) then λ_2 = 1.33 nm (equivalent to 0.93 keV energy of the photons). For a slightly larger energy of electrons of 18.6 MeV (γ = 36.5) the scattered wavelength would reach an angstrom: λ_2 = 0.1 nm (or 12.4 keV).

Green laser (532 nm) scattered from an 18.6 MeV electron beam turns into X-rays with 0.1 nm wavelength.

7.4.3 COMPTON SCATTERING CHARACTERISTICS

The total Compton cross section can be approximated by the following equation:

$$\sigma_{tot} = \frac{8\pi\, r_e^2}{3} \tag{7.5}$$

which shows that the total Compton scattering cross section is very close to the Thomson one.

We can evaluate the rate of emitted X-rays as the product of the Thomson cross section and the *luminosity* $\mathscr{L}$ characterizing interaction between the electron and laser beams. Assuming head-on collision of the beams, we write for the luminosity:

$$\mathscr{L} = \frac{N_e N_\gamma f}{2\pi\, \sigma_x \sigma_y} \tag{7.6}$$

In the above, N_e and N_γ are the numbers of electrons and photons in the colliding bunches and f is the repetition frequency of collisions. The beam sizes in Eq. 7.6 are the convolution of electron and photon beam sizes:

$$\sigma = \sqrt{\sigma_e^2 + \sigma_\gamma^2} \tag{7.7}$$

The above equations help us to estimate the rate of X-ray production as

$$\frac{d\mathscr{N}_\gamma}{dt} = \sigma_{tot}\, \mathscr{L} \tag{7.8}$$

In relativistic approximation $\gamma \gg 1$, the expression for the final frequency of Compton scattered photons is given by the following equation

$$\omega_f \approx \frac{2\gamma^2 \omega_i (1 - \cos\phi_1)}{1 + (\gamma\phi_2)^2 + 2\gamma\frac{\omega_i}{m_e}(1 - \cos\phi_1)} \tag{7.9}$$

where the angles are defined in Fig. 7.13 and which identifies the following characteristics of Compton scattering. There is a clear dependence between scattered photon energy and its angle — this can be useful for the selection of monochromatic

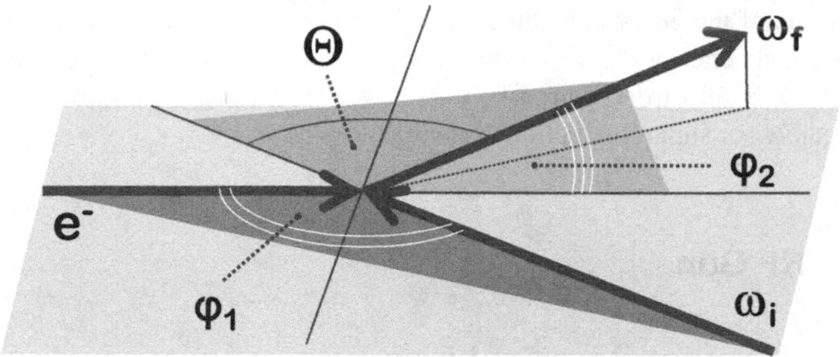

Figure 7.13 Compton scattering — definition of frequencies and angles.

beam with help of collimation. The majority of the X-ray flux is emitted into a cone with an opening angle of $4/\gamma$. The photons of maximum energy $\omega_c = 4\,\omega_i \gamma^2$ come from a head-on collision ($\varphi_1 = \pi$), while the photons of half-maximum energy come from a $\varphi_1 = \pi/2$ collision.

The equations and dependencies defined above aid us in estimating basic parameters of Compton-based light sources. Let us now consider the designs of such light sources and their typical characteristics.

7.5 COMPTON LIGHT SOURCES

A generic Compton light source based on a linear accelerator is shown in Fig. 7.14.

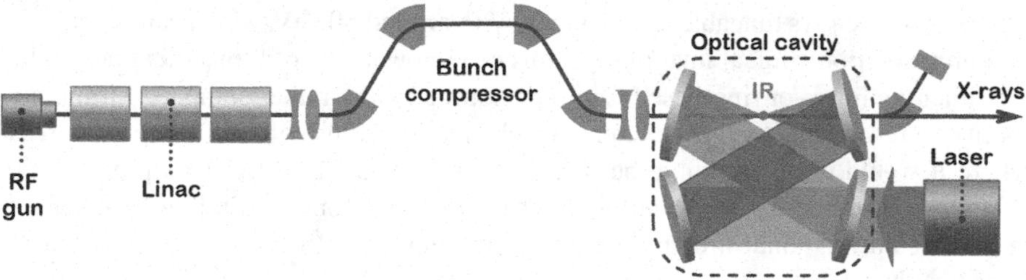

Figure 7.14 Generic Compton source of linac type.

In this design, a train of electron bunches produced by a photo RF gun is accelerated in a linac up to a few tens of MeV (or several hundred MeV, depending on the application). It is then compressed longitudinally and sent for collision with laser bunches accumulated in the laser cavity. The laser pumps the cavity through a semi-transparent mirror and the laser intensity buildup in the cavity can exceed the intensity in the single laser pulse by more than a hundred-fold, increasing the brightness of the linac-based Compton source.

Further enhancement of the brightness of the linac based Compton source can be achieved by employing the method of *energy recovery*. The Compton cross section of electron-photon interaction is rather low and therefore, in typical configurations, the majority of electrons will pass through the laser pulses without interacting. The major fraction of the electron beam can therefore be decelerated, after Compton interaction, and its energy recovered, reducing the required RF power and increasing the current of the electron beam. The energy recovery-based Compton sources are mostly suitable for superconducting linac technology.

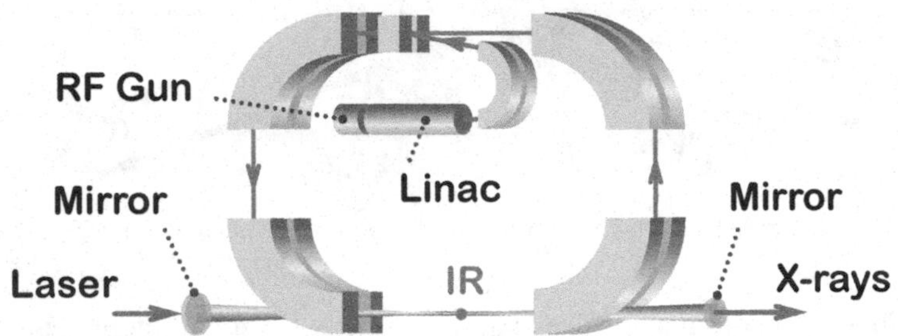

Figure 7.15 Generic Compton light source based on electron storage ring.

Another type of Compton source is one based on an electron storage ring, as illustrated in Fig. 7.15. In this case, a short linac injects an electron beam into a compact ring, where — due to SR — the electron beam emittances are cooled, thus helping to achieve a higher luminosity of electron-laser interaction. The laser cavity is typically located around or inside one of the straight sections of the ring with the interaction region (IR) in the center.

Parameter ranges of existing or planned Compton sources allow for their application in a variety of areas. In particular, 10–30 MeV accelerator produces (with a typical laser) X-rays tuneable from a few keV to around 50 keV, which can be applied to high resolution clinical imaging systems or various types of biomedical research.

Either the linac- or ring-based Compton source can be rather compact, fitting in a room of a few meters by a few meters, which enables the use of such sources in areas where it would not be possible before. For example, a THOMX[1] Compton source is being considered for use in cultural heritage applications, and might be installed in a museum for nondestructive studies of precious paintings without the need to transport them.

Compton sources aimed at larger energies of X-ray photons, toward the 1–5 MeV range, have their particular niche. The phenomenon of nuclear resonance fluorescence helps to create imaging instruments with excellent isotopic sensitivity; therefore such Compton sources can also assist in nuclear-waste management.

The Compton sources based on a linac, mentioned above, require superconducting RF (SRF) technology to achieve maximal brightness. Compton sources are compact and therefore may be suitable for small organizations, labs, hospitals or universities. However, most of the modern SRF cavities with frequencies around 1 GHz require operation at 2 K temperature. There is, nonetheless, a contradiction here as the cryogenics system, in particular for 2 K, can be bulky and expensive.

[1] A. Variola et al., THOMX Conceptual Design Report, LAL RT 09/28, SOLEIL/SOU-RA-2678, 2010.

A possible solution for this issue is to revert to lower frequency SRF cavities, which operate at 4 K temperatures, resulting in a significant reduction in complexity, size and cost of the cryogenic system. The practical frequency range of SRF cavities that work at 4 K is around 200-500 MHz. The larger transverse size and lower accelerating gradient of the lower frequency cavities can sometimes be compensated by modifying their design, e.g., use of spoke cavities.

7.6 HYBRID MULTI-BEND ACHROMAT – INVENTION CASE STUDY

Let's look at synchrotron light sources' pursuit for smallest beam emittance, and at the TRIZ example of inventing a better conveyor and connect them together.

The equilibrium beam emittance in synchrotron light sources is defined (see Eqs. 3.28 and 3.30) by the so-called curly-H function $\mathcal{H}$:

$$\mathcal{H} = \frac{\eta^2 + \left(\beta_x \eta' - \beta_x' \eta /2\right)^2}{\beta_x} \tag{7.10}$$

The equilibrium emittance is proportional to the average value of the curly $\mathcal{H}$ function along the ring. If we can optimize the dispersion η and beta function β_x in such a way as to minimize the average value of curly $\mathcal{H}$ for a regular optical cell of SR light source, we can reduce the beam emittance and thus increase the brightness.

Evolution of the types of SR sources' optics exhibits the tendency to increase the number of bends per cell, and replace simple FODO optics with achromats (double-, triple- or multi-bend), which allowed to create space for insertion devices, as well as to minimize the equilibrium emittance — see Fig. 7.16.

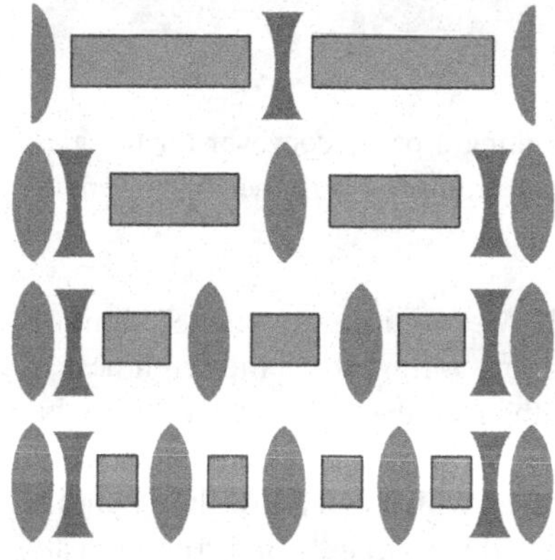

Figure 7.16 FODO to MBA evolution.

Light sources evolved from use of FODO optics to DBA — double bend achromat, then to TBA — triple bend achromat, and eventually to MBA — multi bend achromat. This allowed to create straight sections for the insertion devices, but also resulted in increase of the number of bends per cell with corresponding reduction of curly-H function $\mathcal{H}$*, better optimization of optics, minimization of equilibrium emittance and maximization of the brightness.*

We can ask a question — is it possible to improve MBA optics even further?

Before discussing the answer to this question, let's consider a classical TRIZ example of inventing a better conveyor for glass manufacturing. Imagine, that we need to transport hot glass, just from the oven, on a conveyor. Large rollers of the conveyor will cause deformation of the glass sheet and will impact its quality. To improve the glass quality, one can make the conveyor rollers smaller, and use many more of them. But this will increase the cost of the conveyor.

We can ask a similar question in this case — is it possible to improve the glass quality without increasing the cost?

Considering the trend, that in order to improve quality, one can make rollers smaller, and use more of them, we can conclude that for maximum glass quality the rollers need to be of *zero size*. So, they should be of atomic size! This means that we can use a metal with low melting temperature, for example tin, to create a liquid metal conveyor — as illustrated in Fig. 7.17.

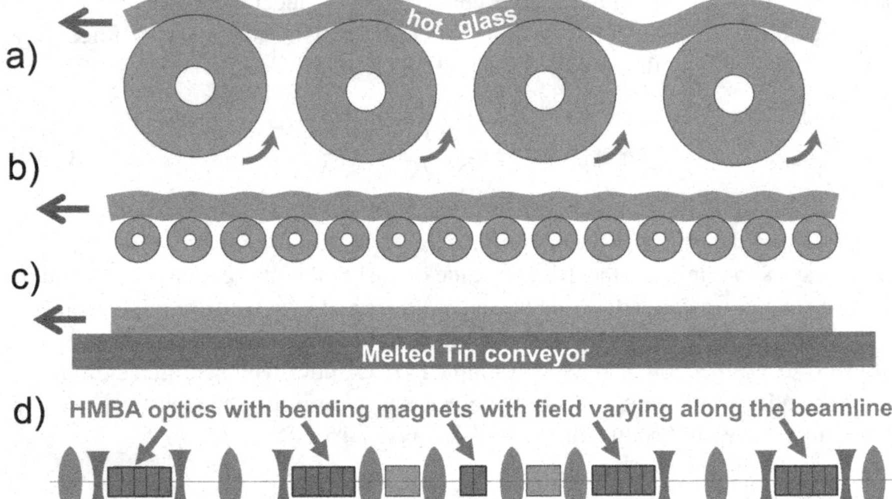

Figure 7.17 An analogy between TRIZ task of inventing a better conveyor for hot glass (steps a-b-c), and invention of the Hybrid Multi Bend Achromat (step d), where the bending field is varying along the beamline.

Applying this inventive approach to FODO-MBA evolution, we will see similarities — the increase of the number of bending magnets will improve the brightness of the light source, but would increase its cost. And then we can ask ourselves — what is the equivalent of using the atom-sized rollers in the case of the ultimately better MBA cell? This would be zero-size bending magnets, which can allow their bending fields and the optics in and around them to be adjusted individually and continuously.

The practical implementation of this approach would be — instead of going to even larger number of smaller bending magnets — to use bending magnets with field varying along the magnet, for better optimization of optics. This invention is called Hybrid-MBA[2], and is the best so far optics design approach, which allowed to achieve world-record low emittance at the ESRF-EBS synchrotron light source.

[2]P. Raimondi et al., PRAB, 24, 110701 (2021)

EXERCISES

7.1. *Chapter materials review.*
Describe a method to create a monochromatic X-ray beam in Compton sources.

7.2. *Chapter materials review.*
In laser plasma acceleration, the final energy of an accelerated electron beam is 1 GeV. The wavelength of the laser used for laser plasma acceleration is 800 nm. Part of the same laser pulse is redirected with mirrors to collide head-on with the accelerated electron beam. Estimate the energy of photons created in such a Compton source, as well as the angular spread of the photons.

7.3. *Mini-project.*
Select a desired photon energy of a laser plasma acceleration betatron X-ray source (e.g., from 1 to 100 keV) and devise a consistent set of basic parameters describing the source. Discuss the justifications for selecting particular values of certain parameters (for plasma or laser, etc.). Estimate the brightness of the source.

7.4. *Mini-project.*
Select a desired photon energy of a Compton X-ray source (e.g., from 1 keV to 10 MeV) and devise a consistent set of basic parameters describing the source. Discuss the reasons for selecting particular values of certain parameters (for electron beam, laser, etc.). Estimate the brightness of the source.

7.5. *Analyze inventions or discoveries using TRIZ and AS-TRIZ.* A liquid anode X-ray tube is a contemporary technology that increases the photon flux. Analyze this technology in terms of the TRIZ and AS-TRIZ approach, identifying a contradiction and a general inventive principle that was used (could have been used) in this invention.

7.6. *Developing AS-TRIZ parameters and inventive principles.* Based on what you already know about accelerator science, discuss and suggest the possible additional parameters for the AS-TRIZ contradiction matrix, as well as the possible additional AS-TRIZ inventive principles.

7.7. *Practice in reinventing technical systems.* A hydrostatic level system (HLS) is often used for the alignment of synchrotron light sources. An HLS consists of many vessels of 10-15 cm height filled with water and connected by pipes. The pipes running from the bottoms of the vessels provide a connection to water, and the pipes running from the tops of the vessels provide a connection to the air. In an ideal case, the water levels in all the sensors are the same (with respect to the gravitational equipotential plane) and thus measuring these levels relative to the vessel (e.g., with capacitor sensors) would allow one to measure the misalignments of the SR source magnets to which the vessels are attached. However, there is a challenge: water has a very large thermal expansion coefficient (0.0002 at 20°C). If the height of the water column in the sensors is just 0.1 m, then variations of the temperature along the accelerator will give a measurement error of 20 μm/degree. Now imagine that you need to align the accelerator to better than 1 μm but you cannot stabilize the temperature to the required level. Suggest a way to change the design of the water level alignment system to solve this problem.

8 Free Electron Lasers

In this chapter we will continue to build upon the results of the previous chapters — particularly Chapter 3 (Synchrotron Radiation) and Chapter 7 (Light Sources) — and will discuss the present reigning champion among the X-ray light sources: the free electron laser.

We will begin with a brief, historical introduction. Then, we will recall the properties of radiation from a sequence of bends, wigglers and undulators, and then discuss how their radiation spectra compare. Next, we will follow up on undulator resonance conditions and microbunching. Finally, we will discuss the precise physical meaning and exact definition of the undulator parameter K, which was introduced in earlier chapters as a qualitative factor.

Following this introduction of basic FEL concepts, we will discuss FEL designs and parameters, as well as possible future advances in the evolution of FELs.

8.1 FEL CONCEPTUALLY

In a third-generation light source, the phase relationship between the radiation emitted by each electron is random and therefore the spatial and temporal coherence of the radiation is poor. Even if an undulator were inserted into the ring, the electrons would emit radiation *incoherently*.

In contrast to the third-generation light sources, operation of a free electron laser relies on *microbunching* of the beam caused by interaction of the radiation with the beam, i.e., the beam interacts with itself via the radiation it emits.

Microbunching in FEL happens primarily at the resonant wavelength determined by the undulator parameter.

The conceptual mechanism of energy exchange between the beam and EM wave (FEL light) is the following. The FEL undulator creates a wiggling beam trajectory, so that particle will have transverse velocity, parallel to the electric field of the EM wave. In synchronous conditions the transverse electric field of the EM wave is directed against the particle transverse velocity, therefore energy is removed from the particles and thus fed into the EM wave.

Once the electron beam is microbunched, each microbunch emits radiation as a single particle of a large charge, in phase with each other — i.e., *coherently*, as illustrated in Fig. 8.1. Correspondingly, the radiation power and brightness of FEL will scale as N_e^2 and not as N_e as in third-generation sources, thus giving an enormous boost in performance.

Radiation coming out from FEL undulator will be emitted in a narrow band of frequencies with relative spread of $\sim 1/N_u$ — as illustrated in Fig. 8.2. This comes from the observation that FEL radiation will look like a signal with N_u sin-like periods and

DOI: 10.1201/9781003326076-8

Figure 8.1 Random a) and coherent b) addition of radiation fields from electrons placed randomly or assembled in microbunches.

with zero before and after, which power spectrum is

$$I(\omega) \propto \left(\frac{\sin^2 \xi}{\xi^2}\right) \quad \text{with} \quad \xi = (\omega - \omega_0)T/2 = \pi N_u \frac{\omega - \omega_0}{\omega_0} \tag{8.1}$$

where ξ is dimensionless detuning. We will discuss this spectrum shape and its derivative further below, in the section describing the FEL gain.

We will now consider the relevant FEL phenomena and characteristics step by step, but before that, we will discuss the FEL invention path.

8.2 FEL HISTORY – INVENTION CASE STUDY

The FEL concept, as well as the term itself, was suggested by John Madey in the early 1970s during his work at Stanford University. His research benefited from the earlier work by Hanz Motz who, in 1953, built an undulator (which was proposed in Vitaly Ginzburg's 1947 theoretical paper, wherein the undulator was described as a device generating electromagnetic radiation via relativistic electrons).

In 1971, John Madey wrote his first FEL-related published journal entry on the subject of stimulated emissions of radiation in a periodic magnetic field. This was shortly followed by a patent on FEL filed in 1972 — but that's a whole 'nother story. The FEL created by Madey used a 43 MeV electron beam to create radiation with wavelengths of 3.4 μm in a 5 m helical undulator, with a 3.2 cm period and a field of 0.24 T.

Following Madey's pioneering work, many FELs have been created all around the world — their wavelengths gradually shortening as the technology matured. The most modern FELs have recently reached the Angstrom range, creating

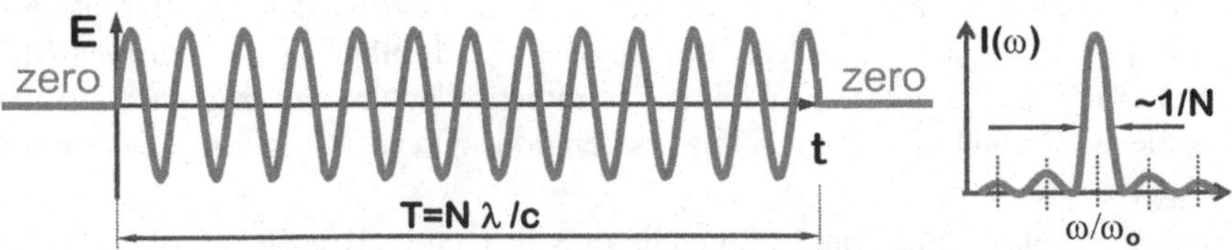

Figure 8.2 Sin-like signal coming from an undulator with N periods, and its spectrum.

unsurpassed possibilities for discovery science, bio-medical studies and technology-aimed research.

It would be educational and appropriate, for the purpose of our book, to look at the FEL invention path, and attempt to trace possible post-factum applied invention principles. Let's look at this history again step by step.

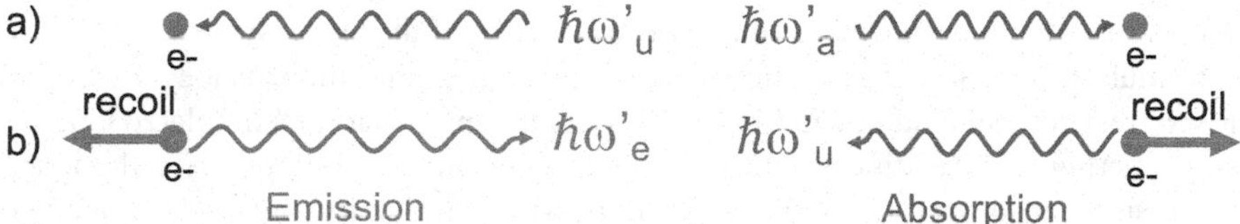

Figure 8.3 Quantum approach to FEL analysis.

In the first analysis of the FEL[1] Madey used the term *stimulated bremsstrahlung* and *stimulated Compton scattering*, referring to the observation that the undulator's magnetic field is seen by the relativistic electron as an EM wave. He then considered two effects — an electron can either scatter an undulator "photon" in the forward direction and loose momentum (emission), or it can scatter a laser photon in the backward direction and gain momentum (absorption), as illustrated in Fig. 8.3.

Observing that the electron mass and recoil play a role in the photon-electron interaction, one can write the following relations between the photon energies in these processes:

$$\hbar\omega'_e < \hbar\omega'_u \quad \text{and} \quad \hbar\omega'_a > \hbar\omega'_u \tag{8.2}$$

Correspondingly, the emission and absorption of a photon of a given frequency requires slightly different "undulator photon" energies, and hence different electron energies. The probability curves for emission and absorption (which should look similar to the spectrum in Fig. 8.2) are therefore slightly shifted in energy.

The above observation results in the following "*Madey's Theorem*": the "Gain Curve" i.e. the rate of (emission – absorption) is the derivative of the spontaneous emission curve — as illustrated in Fig. 8.4.

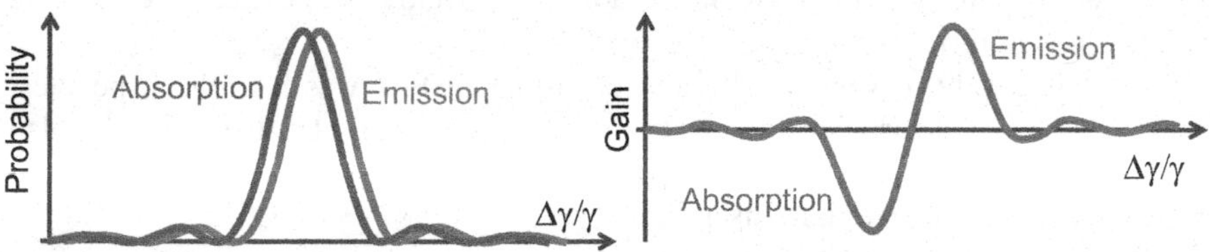

Figure 8.4 Illustration of the Madey's theorem for derivation of the gain curve.

Let's discuss the term that Madey used in the first analysis of FEL — "stimulated Compton scattering." What does "stimulated" really mean in this case? It is important

[1] J.M.J. Madey, J. Appl. Phys., 42, 1906 (1971).

to discuss and understand the difference between use of this term for lasers (see, e.g., Fig. 4.8, which is illustrating a three-level laser), and for FELs.

In FEL descriptions the term "stimulated" refers to the assumption that probability of photon-electron collision will be influenced by the presence of the scattered radiation which exists in the system.

I.e., it is not referred to an individual stimulating act, as in the three-level laser. In FEL "stimulated" is referred to an averaged behavior of the entire system.

Stimulated Compton scattering was quite popular at that time. Madey's work is closely related (as he quoted in his 1971 paper) to a proposal for Stimulated Compton scattering of a relativistic electron beam from infrared microwave radiation[2], where an RF resonator with electron beam passing through it would be surrounded by mirrors for creating the cavity for IR radiation.

The main technical difference of Madey's proposal with respect to this earlier work was the use of a static magnetic field, rather than an electromagnetic one — Madey described various undulator-style configurations for periodic magnetic fields.

While the first analysis of FEL was performed from quantum physics standpoint, the Planck constant $\hbar$ cancelled out from the final results, and it was soon shown that FEL can be described from completely classical approach, and that its operation is similar to electron beam devices developed earlier.

The path to FEL invention is really fascinating. The first analysis of FEL relied on quantum approach, however the Planck constant $\hbar$ disappeared in the final results and many peers expressed doubts whether FEL is really a "true" laser. And indeed, soon enough a fully classical picture of FEL operation was developed (R.Palmer 1972, Hopf et al., Colson, 1976).

Moreover, while at first the physical principles of FEL operation were considered different to those of earlier devices, a connection was made with earlier theoretical work, showing that the FEL did indeed operate according to the same principles as earlier electron beam devices (Kroll et al., 1978), operating on the physical principle of electrons bunching on the scale of the radiation wavelength and therefore emitting radiation coherently.

Correspondingly, it eventually became clear that the FEL was essentially the latest in a long series of electron beam devices that generate coherent radiation, and it also followed a natural invention evolution path, with clearly identifiable steps, which we are particularly interested in to discuss.

A klystron (see Fig. 13.2) is the first device to mention. It was invented by Sigurd and Russell Varian, who while working on IOT (see Fig. 13.1) in 1937, added a second cavity resonator for signal input to the inductive output tube. This input resonator acted as a pair of inductive grids to alternately "bunch" and release packets of electrons down the drift space of the tube, so the electron beam would be composed

[2]R.H. Pantell et al., IEEE J. Quantum Electr. QE-4, 905 (1968).

of electrons traveling at different velocities. This "velocity modulation" of the beam translated into the same sort of amplitude variation at the output resonator, where energy was extracted from the beam. The bunched beam would excite EM wave in the output resonator coherently.

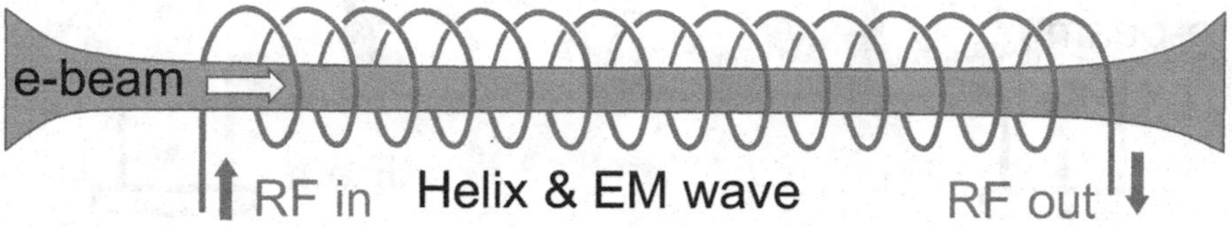

Figure 8.5 Traveling Wave Tube electron beam and helix

While in klystron the interaction of RF with electron beam happens in two distinct points along the beam (location of the input and output resonators), in the Traveling Wave Tube (TWT) developed by Kompfner and Pierce in 1947-1950, the EM wave — beam interaction was expanded into a long region. To provide beam-wave synchronization, a helix was placed around the beam, to slow down the EM wave and ensure synchronism with the electron beam; see Fig. 8.5.

Invention of an undulator was an important milestone. In 1947 V. Ginzburg suggested[3] use of periodic magnetic structures for generation of radiation of relativistic electrons.

In 1951 H. Motz made a proposal[4] for producing coherent millimeter waves from a pre-bunched electron beam. Motz also shown (in 1959), using the same analysis as for TWT, that undulator can be used to amplify radiation and in later experiments with a 3-5 MeV electron beam demonstrated semi-coherent emission of bunched beam in 6-8 mm radiation band.

Perhaps the most interesting device on the path toward FEL is so-called Ubitron (stands for Undulating Beam InTeRactiON) invented[5] in 1957 by R.M. Phillips.

Ubitron was a "fast wave" structure based on a new interaction mechanism — undulating electron beam immersed into EM wave, and was both a microwave tube and a non-relativistic FEL amplifier; see Fig. 8.6. As R.M. Phillips described[6], the invention came in 1957 accidentally, when he was trying to explain why an X-band periodically focused coupled cavity TWT oscillated, while a solenoid focused version did not. The most apparent difference between the two was behavior of the electron beam — one wiggled and the other spiraled. As R.M. Phillips recalled later, out of this investigation of ways of coupling an RF wave to an undulating electron beam, came the idea to couple the TE_{01} mode by allowing the wave to slip through

[3]V. Ginzburg, C.R. Acad. Sci. USSR, v.56, 145, 253, 1947.

[4]H. Motz, J. Appl. Phys. 22, 527 (1951).

[5]R.M. Phillips, IRE Trans. on Electron Devices, vol. 7, no. 4, pp. 231-241, Oct 1960.

[6]R.M. Phillips, Nucl. Instr. Meth. A272, 1 (1988).

the beam such that the electric field would reverse the direction at the same moment the electron velocity reversed — we will see below that the same mechanism of synchronization applies to FEL.

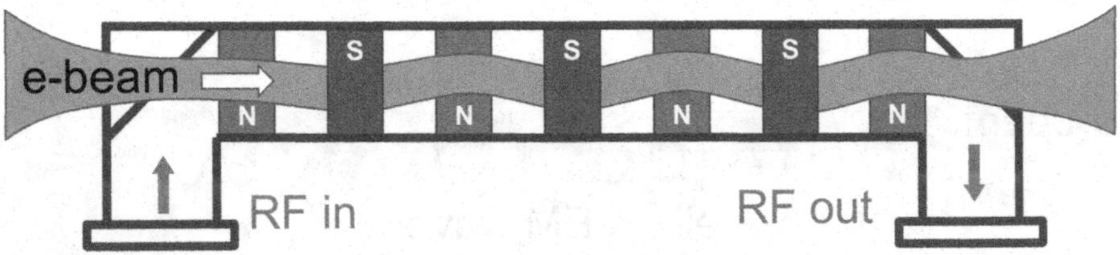

Figure 8.6 Ubitron — Undulating Beam InTeRactiON — is non-relativistic FEL amplifier.

Ubitron was setting records of RF power a decade earlier before the term "free electron laser" was coined and FEL operation demonstrated. Ubitron devices ranged from S-band (3 GHz) to the most successful V-band (54 GHz, $\lambda \sim 5$ mm) version operated with 70 kV hollow electron beam and producing 150 kW output power.

Coming back to Madey's FEL — the first demonstration[7] of FEL amplifier came in 1976 when 24 MeV beam with 70 mA peak current in 5 m helical undulator with 3.2 cm period generated radiation at 10.6 μm wavelength with peak gain of 7% per pass. The first FEL oscillator was demonstrated a year later, when 43 MeV beam with 2.6 A peak current generated radiation at 3.4 μm wavelength with peak power of 7 kW — a spectacular start of the FEL era.

Looking back at the path to FEL invention — Thomson and Compton scattering in around 1920, Klystron in 1937, Traveling Wave Tube and undulator in 1947–1950, Ubitron in 1957–1960, and Laser in 1958 — one can indeed see that the FEL was essentially the latest in a long series of inventions of electron beam devices that generate coherent radiation.

The path from klystron to Traveling Wave Tube, then to Ubitron and then to FEL is an illustration of application of several inventive principles — "continuity of useful action," "going to another direction" and "going from microwave to optical."

Connecting the FEL invention path to TRIZ inventive principles, one can observe that the step from klystron to TWT was an application of the "continuity of useful action" principle — add an extended cavity (helix with traveling EM wave) to make interaction of the beam with EM wave continuous.

The step from TWT to ubitron was application of the inventive principle "going to another direction" — adding wiggling motion that connected EM wave with the largest source of energy — longitudinal motion.

[7]L.R. Elias et al., Phys. Rev. Lett. 36, 717, (1976)

Finally, the step from ubitron to FEL was an illustration of the principle or trend of "going from microwave to optical" — the trend which can also be seen from comparing the radar technology with laser CPA technique.

8.3 SR FROM BENDS, WIGGLERS AND UNDULATORS

We started our discussion of FELs in Chapter 3, where we approached radiation from wigglers and undulators and compared them to radiation from bending magnets. We recall that, for relativistic electrons with $\gamma \gg 1$, the emitted photons go into $1/\gamma$ cone and, if the radius of the curvature of the trajectory in the magnetic field is R, then the external observer will see the photons emitted during the particle travel along the arc $2R/\gamma$. This allowed us to estimate the characteristic frequency[8] of SR as $\omega_c = 1.5c\gamma^3/R$.

We also recall that the extra factor of γ^2 appears in this formula for ω_c due to the difference between the speed of photons c and the speed of particles v, estimated as $(1-v/c) = 1/(2\gamma^2)$. Knowledge of the characteristic frequency allows us to determine the spectral characteristic of the SR emitted from the bending magnets.

8.3.1 RADIATION FROM SEQUENCE OF BENDS

Assume that a set of bending magnets are arranged in a sequence with +–+– polarity with period λ_u, so that the particle trajectory through this sequence of magnets wiggles as illustrated in Fig. 8.7.

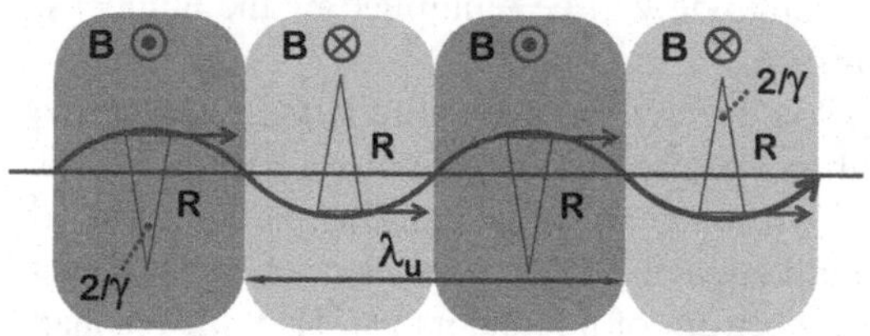

Figure 8.7 Trajectory and radiation in a sequence of bending magnets.

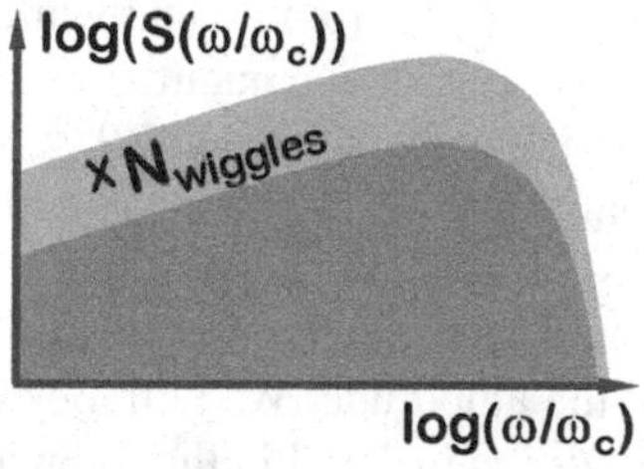

Figure 8.8 Wiggler (top) and bending magnet (bottom) SR spectra.

If the length of the emitting region[9] (that a remote observer can see) is much less than the length of an individual bend, i.e., $2R/\gamma \ll \lambda_u/2$, then the radiation emitted in each bend is independent. Such an arrangement of bends is called a wiggler (and corresponds to $K \gg 1$ where $K \sim \gamma\,\lambda_u/R$, and where λ_u/R can be noted as being approximately equal to the maximal angle of the trajectory).

In a wiggler configuration, the spectrum of emitted SR is thus expected to resemble to the spectrum from the bends — the spectrum shape will be similar to

---

[8] *Recall that γ^3 dependence of ω_c is due to the length of the emitting arc ($\propto 1/\gamma$) and due to the photon and particle velocity difference ($v - c \propto 1/\gamma^2$).*

[9] *Precise definition of K follows in just a couple of pages.*

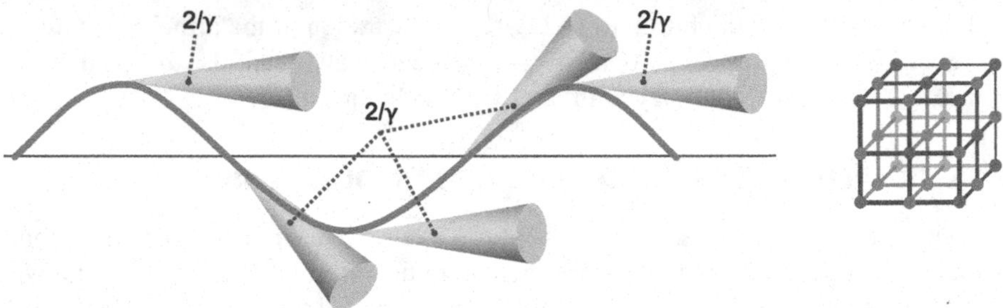

Figure 8.9 Radiation from wiggler, regime of $K \gg 1$.

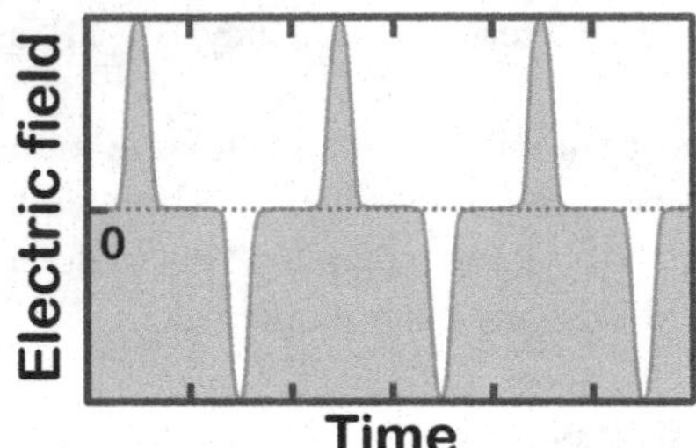

Figure 8.10 Time profile of radiation observed from wiggler.

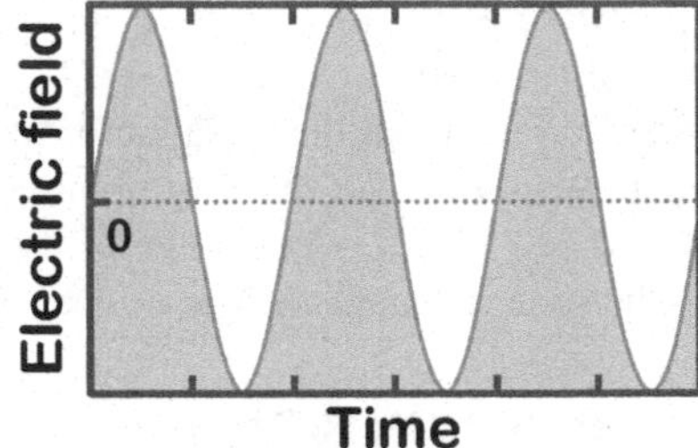

Figure 8.11 Time profile of radiation observed from undulator.

the spectrum from a bend while the amplitude will be multiplied by the number of wiggles, as shown qualitatively in Fig. 8.8.

The opposite regime $2R/\gamma \gg \lambda_u/2$ is different — the entire wiggling trajectory contributes to radiation. It is logical to assume that there is a wavelength in this $K \ll 1$ undulator regime that would be resonant with the emitting particle. Indeed, it will emit radiation at this wavelength at each wiggle, allowing coherent buildup of the amplitude. We will look at the spectrum of undulators and also this resonant condition in detail in the following sections.

8.3.2 SR SPECTRA FROM WIGGLER AND UNDULATOR

Despite the similarity of the overall shape of the SR spectra of bend magnets and wigglers, the external observer will note an important difference in the details of the spectra.

The fields emitted from wigglers and detected by the external observer manifest themselves as periodic signals — short flashes repeating with a period corresponding to the time of flight between wiggler periods, as illustrated in Figs 8.9 and 8.10. Given the periodic nature of radiation emitted by wigglers, the spectrum of SR wiggler radiation should consist of harmonics defined by the wiggler period corrected by the factor $(1-v/c) = 1/(2\gamma^2)$, which takes into account the relative velocity of particles and radiation (illustrated in Fig. 8.12 at left). The relative width of each peak in the wiggler spectrum corresponds to the number of wiggles N_w, i.e., $\Delta\lambda/\lambda \approx 1/N_w$.

As the the entire trajectory contributes to radiation emitted from the undulator, the time structure of the observed radiation is periodic and continuous, as shown in Figs 8.13 and 8.11.

The spectrum corresponding to the observed undulator radiation will thus contain just one harmonic at a wavelength close to $\lambda_u/(2\gamma^2)$. We will clarify this statement in the next section.

8.3.3 MOTION AND RADIATION IN SINE-LIKE FIELD

In the example that we considered in Fig. 8.7, we assumed that the segmented field of the wiggler is uniform and sharply changes its sign during the transitions between segments. This may correspond to zero-aperture magnets but, in practice, is not possible.

A much better way to estimate the field of a wiggler or undulator is to assume that the field is sine-like, i.e.,

$$B_y(z) = B_0 \sin(k_u z) \tag{8.3}$$

where $k_u = 2\pi/\lambda_u$.

Let us consider a trajectory through such a sine-like field and let us parameterize it in such a way that the maximum angle of the trajectory is equal to K/γ, as shown in Fig. 8.14.

The trajectory parametrization can therefore be written as

$$x = \frac{K}{\gamma}\frac{\lambda_u}{2\pi}\sin\left(\frac{2\pi z}{\lambda_u}\right) \quad \text{and} \quad x' = \frac{K}{\gamma}\cos\left(\frac{2\pi z}{\lambda_u}\right) \tag{8.4}$$

This shows us that if $K < 1$, then the trajectory angle is always less than $1/\gamma$ and the external observer will be able to see the emitted fields without interruptions; ultimately, the entire trajectory contributes to radiation.

Let's now connect the maximum field in the undulator B_0 with parameter K. The bending radius is connected to the trajectory curvature as $d^2x/dz^2 = 1/R$ which gives

$$K = \frac{\lambda_u \gamma}{2\pi R} \tag{8.5}$$

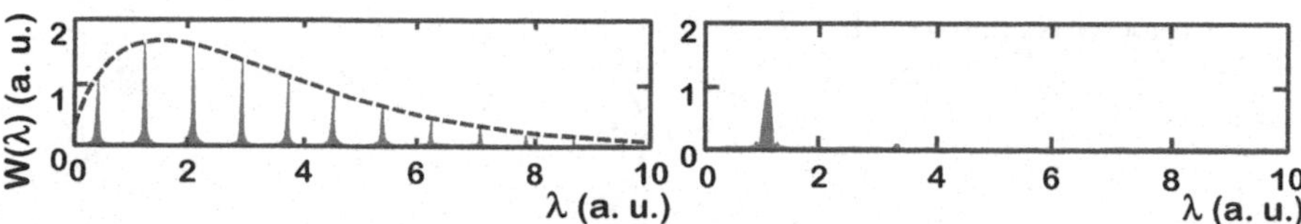

Figure 8.12 Spectrum from wiggler (left) and undulator (right), qualitative comparison. Dashed line on the left spectrum corresponds to the spectrum from bends of the same strength. Horizontal axis is in units of $\lambda_u/(2\gamma^2)$.

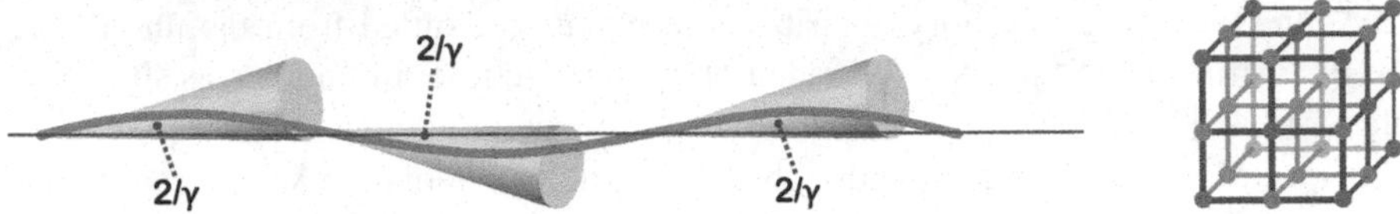

Figure 8.13 Radiation from undulator, with $K \ll 1$.

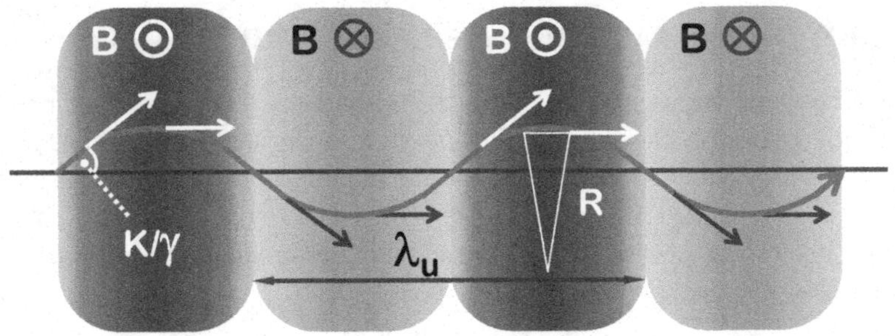

Figure 8.14 Trajectory and radiation in sine-like field.

and by substituting the expression for the bending radius in the magnetic field $R = pc/(eB_0)$ we obtain

$$K = \frac{\lambda_u e B_0}{2\pi mc^2} \tag{8.6}$$

This finally give us the precise definition of the undulator parameter.

We are now ready to discuss the basics of FEL operation.

8.4 BASICS OF FEL OPERATION

We will now consider all the basic relevant phenomena — resonance condition, energy exchange and microbunching — step by step.

8.4.1 AVERAGE LONGITUDINAL VELOCITY IN AN UNDULATOR

The average longitudinal velocity in an undulator is an important parameter that determines the resonant wavelength. If the particle moves in a free space, its longitudinal velocity is approximated as

$$v_{z0} = \beta c \approx c\left(1 - \frac{1}{2\gamma^2}\right) \tag{8.7}$$

For the sine-like trajectory that we parametrized above, the transverse velocity is given by

$$v_x = \beta c \frac{K}{\gamma} \sin\left(\frac{2\pi z}{\lambda_u}\right) \tag{8.8}$$

In the second-order approximation, the longitudinal velocity can be written as follows

$$v_z \approx \beta c \left(1 - \frac{1}{2}\frac{v_x^2}{\beta^2 c^2}\right) \tag{8.9}$$

and therefore the average longitudinal velocity can be expressed as

$$\langle v_z \rangle \approx c\left(1 - \frac{1}{2\gamma^2}\left(1 + \frac{K^2}{2}\right)\right) \tag{8.10}$$

We can see that, in comparison with free space, there is an additional longitudinal *retardation*, which is due to the transverse velocity in an undulator. This retardation is equal to

$$c\left(\frac{K^2}{4\gamma^2}\right) \tag{8.11}$$

and is determined by the undulator parameter.

8.4.2 PARTICLE AND FIELD ENERGY EXCHANGE

Energy exchange between an EM wave and electron depends on the electric field and velocity:

$$\frac{dW}{dt} = e\mathbf{E}\cdot\mathbf{v} \tag{8.12}$$

If the electrons have only the longitudinal velocity (depicted in the left image of Fig. 8.15) no energy can be transferred between the electrons and the EM wave, as in this case $e\mathbf{E}\cdot\mathbf{v} = 0$.

On the other hand, an electron beam with sine-like trajectory as in an undulator overlaid with an EM wave (right plot in Fig. 8.15) can exhibit an energy exchange between the EM wave and electrons as in this case

$$e\,\mathbf{E}\cdot\mathbf{v} \neq 0 \tag{8.13}$$

because $\mathbf{v}_\perp \neq 0$. Therefore, if electrons have a transverse velocity, energy can be transferred between electrons and EM wave — which is the principle FEL relies upon.

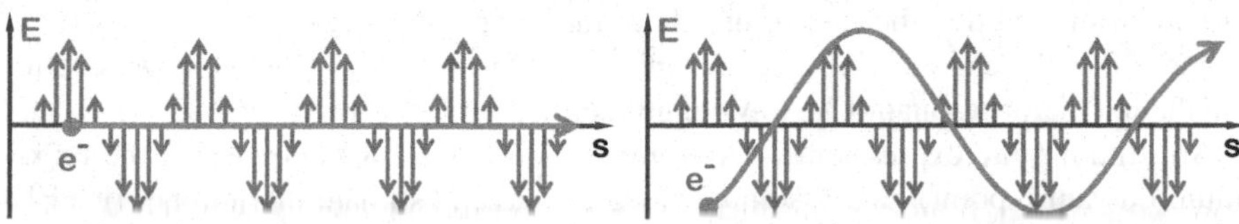

Figure 8.15 EM wave and particle trajectory — straight (left) and wiggling (right) in an undulator.

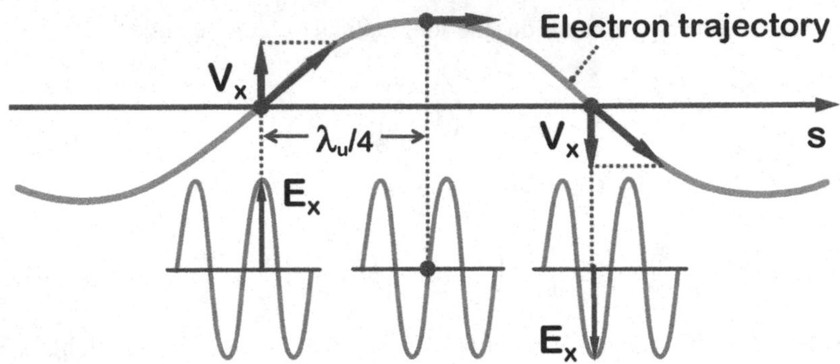

Figure 8.16 EM wave-particle resonance condition of energy transfer.

8.4.3 RESONANCE CONDITION

For certain λ of an EM wave, a resonant energy transfer between electrons and the EM wave can occur — as illustrated in Fig. 8.16.

The necessary condition for the resonant energy transfer is that the EM wave slips forward with respect to an electron by a $\lambda/2$ per half period of electron trajectory, i.e.,

$$\lambda = \lambda_u \left(1 - \langle v_z \rangle / c\right) \tag{8.14}$$

Taking into account the average velocity in an undulator defined by Eq. 8.10, we therefore obtain, for the resonant EM wavelength:

$$\lambda = \frac{\lambda_u}{2\gamma^2}\left(1 + \frac{K^2}{2}\right) \tag{8.15}$$

We note that, in the undulator case when $K \ll 1$, the resonance wavelength is very close to the relativistically transformed undulator period $\lambda_u/(2\gamma^2)$.

We should also note that slippage by $3(\lambda/2)$, $5(\lambda/2)$, $7(\lambda/2)$ and so on would also be in resonance, which may result in generation of odd, higher harmonics.

8.4.4 NUMBER OF PHOTONS EMITTED

Earlier we estimated the rate of energy loss as $dW/dS \approx r_e \gamma^4 mc^2/R^2$, which was defined only by comparison of the field velocity with the speed of light, and therefore should be still valid in the case of undulator radiation.

Instead of the characteristic wavelength of synchrotron radiation we now need to use the undulator resonant EM wavelength defined by Eq. 8.15.

Substituting the expression for K from Eq. 8.6 (where R is defined at the maximum curvature point), and assuming $K \ll 1$, we can estimate the number of FEL photons with wavelength λ emitted per electron per one oscillation over the undulator wavelength λ_u:

$$N_{\text{photons}} = \pi \alpha K^2$$

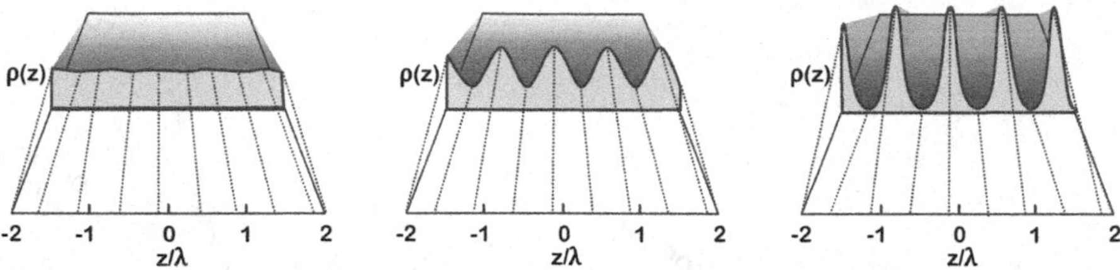

Figure 8.17 Microbunching. Density of the beam along the longitudinal coordinate for the initial noise (left), intermediate regime of microbunching (middle) and saturated microbunching (right).

For arbitrary K the estimate will change as $N_{\text{photons}} = \pi\alpha K^2/(1+K^2)$, and for large K will be equal to just $\pi\alpha$ per one oscillation over λ_u — we see that in all cases the number of photons per oscillation is usually very small, and does not exceed 2.2%.

8.4.5 MICROBUNCHING CONCEPTUALLY

The interaction of particles with the resonant EM wave that we defined in the previous section can create energy modulation in the particle beam.

As particles move along the curved sine-like trajectory, this energy modulation results in different routes taken over different trajectories, depending on the particles' energy. Different path lengths can in turn create density modulations along the beam.

An initial EM wave of resonant wavelength can be external (*seeding*) or can emerge from the noise that is always present in the beam. The latter corresponds to the *self amplified spontaneous emission* (*SASE*) process, which is illustrated in Fig. 8.17. Here the corresponding harmonics from the noise of the initial distribution evolve through the linear regime into saturation regime where complete modulation of density eventually occurs. The resulting microbunches emit coherently at wavelength λ with radiated power $P \sim N^2$.

8.5 FEL TYPES

The two major kinds of FELs are single pass and multi pass. While multi-pass FELs were built primarily in the earlier days of FEL technology, single-pass devices are dominating the arena today.

8.5.1 MULTI-PASS FEL

Multi-pass FEL is similar to a standard laser: mirrors help to build up the optical amplitude on the resonance harmonic while the electron beam trajectory is arranged to pass through the undulator while avoiding mirrors, as shown in Fig. 8.18.

Similar to a normal laser, a single-pass FEL stores the radiation in a cavity. Such systems usually have low gain and therefore many reflections of radiation off of

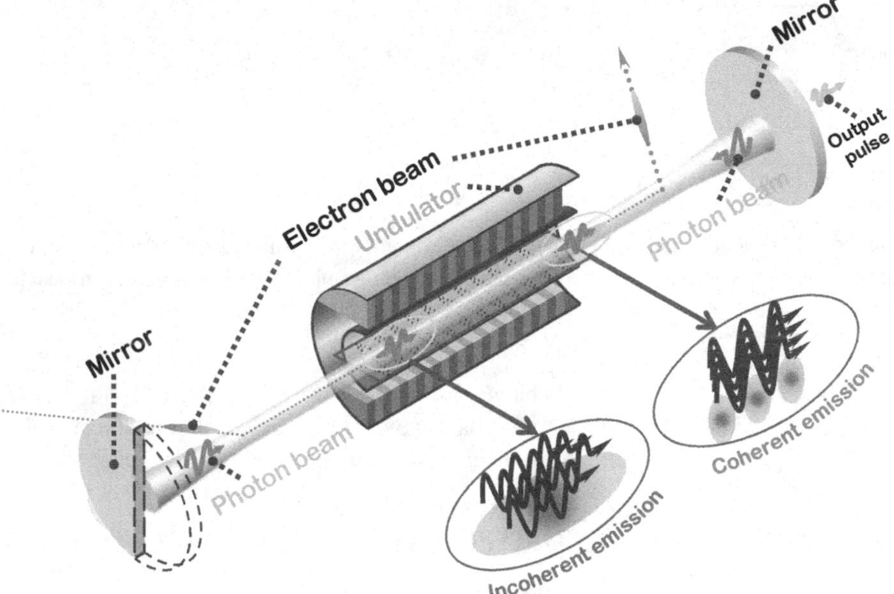

Figure 8.18 Multi-pass FEL.

mirrors are needed to create sufficient amplitude. The multi-pass design is suitable in particular for the light sources used in visible or near-visible ranges of radiation.

8.5.2 SINGLE-PASS FEL

In a single-pass FEL, the radiation has to grow within a single passage of the beam through the undulator (Fig. 8.19). Such FELs can be either seeded or the SASE type.

The need to use single-pass systems is actually a necessity dictated by an absence of good mirrors in the X-ray spectral region. The single-pass system has to be a high-gain system, which puts extreme constraints on the quality of the electron beam, as well as on the accuracy of the undulator. A typical permanent magnet undulator is shown conceptually in Fig. 8.20.

8.6 MICROBUNCHING AND GAIN

The term FEL contains the word *laser*; however, FEL can be explained entirely using the approach of classical electrodynamics. In this section, we will consider microbunching in detail and then discuss the FEL gain factor.

Interaction of the radiation emitted in an undulator with the electron bunch itself can, in certain conditions, be sufficiently strong to generate a significant modulation of the electrons' energy in the beam.

In this case, the energy change of the particles will occur due to the coupling between the transverse (typically horizontal, for planar undiulator) oscillation of the

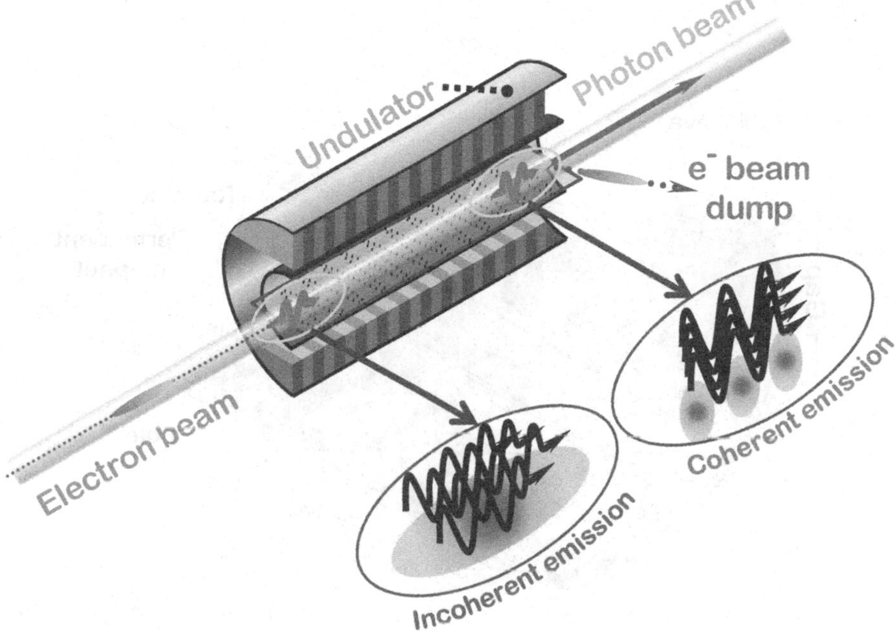

Figure 8.19 Single-pass FEL.

electron in the undulator and the transverse (thus also horizontal) component of the electric field of the emitted EM plane wave.

The energy variation in the beam can then create microbunching due to different path length over curved trajectory, which can amplify the EM wave in certain conditions.

In planar undulator the trajectory radius of curvature oscillates, and moreover, the planar undulator "wave" is a superposition of two waves with different polarization, only one of which will interact with the beam, and the second will average out. On the other hand, in a helical undulator the radius of trajectory is constant, and there is only one wave interacting with the beam. Therefore, while helical undulator is more difficult to manufacture, it is simpler for considerations of FEL beam interaction.

8.6.1 MICROBUNCHING IN HELICAL UNDULATOR

Let's consider for simplicity a helical undulator magnet with magnetic field

$$\mathbf{B} = B_0 \left(\mathbf{e}_x \cos\left(k_u z\right) + \mathbf{e}_y \sin\left(k_u z\right)\right)$$

where $\mathbf{e}_x$ and $\mathbf{e}_y$ are unit vectors in two transverse directions. We will use the same approach for parametrization of the transverse velocity via K

$$\frac{\mathbf{v}_\perp}{c} = -\frac{K}{\gamma} \left(\mathbf{e}_x \cos\left(k_u z\right) + \mathbf{e}_y \sin\left(k_u z\right)\right)$$

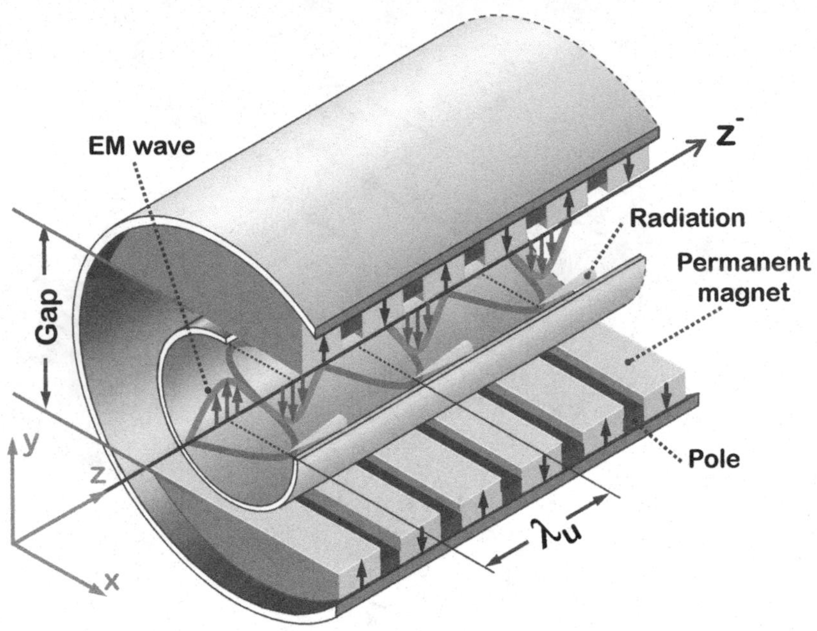

Figure 8.20 Radiation in an FEL undulator composed of permanent magnets.

which ensures that the maximum angle of the trajectory is equal to K/γ and K is defined by the same Eq. 8.6. Since the undulator does not change the energy and velocity, we can write $v_\perp^2 + v_z^2 = \beta^2 c^2$ and can thus define the average transverse velocity in the helical undulator $\langle v_\perp \rangle^2 = c^2 K^2/\gamma^2$. Therefore, for the case of helical undulator we have the following expression for the average horizontal velocity

$$\langle v_z \rangle = c\left(1 - \frac{1}{2\gamma^2} - \frac{K^2}{2\gamma^2}\right) \tag{8.16}$$

Note that there is a factor of two difference with respect to the planar undulator (see Eq. 8.10) in the last term.

Let's now express the electric field of radiation as (where $k = 2\pi/\lambda = \omega/c$)

$$\mathbf{E} = E_0 \left(\mathbf{e}_x \sin(kz - \omega t + \phi_0) + \mathbf{e}_y \cos(kz - \omega t + \phi_0)\right)$$

The change of particle energy over time in terms of relativistic factor is

$$\dot{\gamma} = \frac{e}{mc}\frac{\mathbf{v}_\perp}{c} \cdot \mathbf{E} \quad \text{which can then be expressed as} \quad \dot{\gamma} = -\frac{eE_0 K}{\gamma mc}\sin(\Phi)$$

where $\Phi = (k + k_u)z - \omega t + \phi_0$ is called *ponderomotive* phase. Note that if $(k + k_u)z = \omega t$ then $\Phi = \phi_0$ corresponding to the case of optimal energy transfer, when the energy will be equal to the resonant one $\gamma = \gamma_r$ (or slowly changing).

We can now derive a similar condition for the interference or synchronism that we discussed earlier using a visual picture. For that let's make a time derivative of the above and replace the $\dot{z}$ with $\langle v_z \rangle$ and replace ω and k using $k = 2\pi/\lambda = \omega/c$. We get $\langle v_z \rangle (k + k_u) = ck$ and using the expression for the average longitudinal velocity and assuming that the gamma factor is large, we can derive

$$\lambda = \frac{\lambda_u}{2\gamma_r^2}(1 + K^2) \tag{8.17}$$

Note a factor of two difference inside of the brackets in comparison with Eq. 8.15 — this is due to helical undulator considered in this example.

Let's now consider the above equations near the resonance and define

$$\eta = (\gamma - \gamma_r)/\gamma_r$$

This gives us two equations — the first one for the energy:

$$\dot{\eta} = -\frac{eE_0 K}{\gamma_r^2 mc}\sin(\Phi) \tag{8.18}$$

while the second one, for the phase, can be derived from:

$$\dot{\Phi} = (k + k_u)\langle v_z \rangle - \omega \quad \text{which gives} \quad \dot{\Phi} = 2\pi\left((1/\lambda + 1/\lambda_u)\langle v_z \rangle - c/\lambda\right)$$

and assuming λ is at the resonance, we get the following simple equation:

$$\dot{\Phi} = \frac{4\pi c}{\lambda_u}\eta \tag{8.19}$$

These two equations can be written as the second order differential equation for Φ

$$\ddot{\Phi} = -\Omega^2 \sin(\Phi) \tag{8.20}$$

where for the helical undulator case $\Omega = (4\pi e E_0 K/(\gamma_r^2 m))^{1/2}$. The above equation represents the motion of a simple pendulum and is the same as equation for motion in a synchrotron discussed in earlier chapters. It is called the FEL-pendulum equation, which describes the interaction of particles with radiation in an FEL.

The equations Eq. 8.18 or Eq. 8.19 describe the familiar motion in a separatrix, where the period of oscillations depends on the initial conditions, particularly whether the particle is infinitely close to the separatrix, in which case the period becomes infinitely long. Examples of solutions to this equation for different initial conditions are shown in Figs 8.21 and 8.22. Note that if the initial energy of the electron beam is exactly on the resonance, there will be no average energy change after oscillations in the FEL potential. However, if the initial energy is offset, the final energy can be lower, and therefore the EM wave amplitude can grow.

Small signal gain can be obtained by averaging the energy loss/gain over all phases, and dividing by the radiation beam intensity.

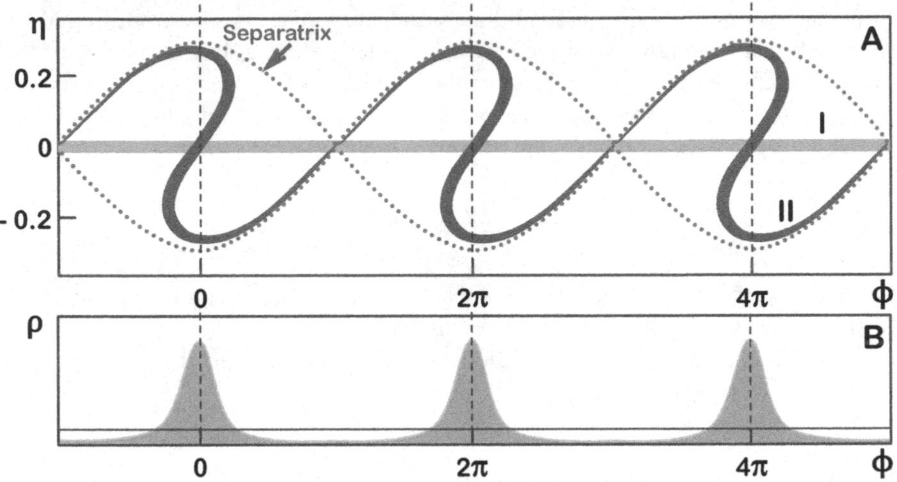

Figure 8.21 Illustrating solutions of FEL-pendulum equation and microbunching for different initial conditions. The initial beam (I) is on-energy and after half a synchrotron period is bunched (II), demonstrating a symmetrical profile of beam density (B) and no net change of the beam energy.

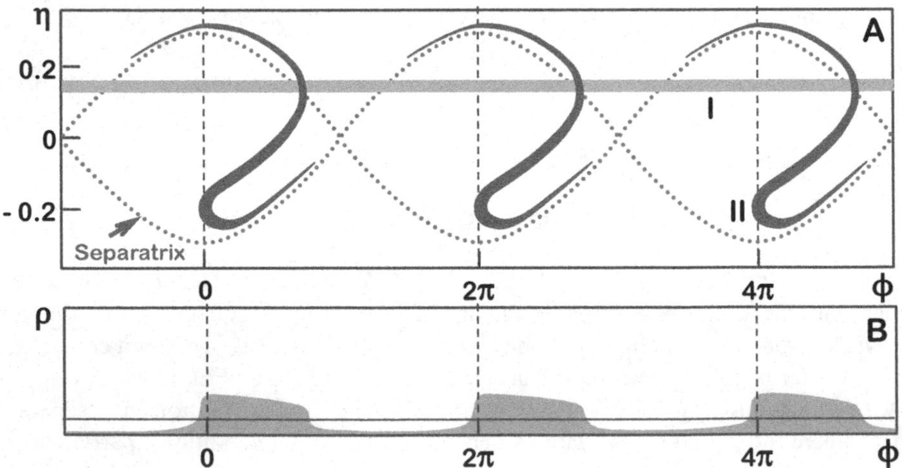

Figure 8.22 Illustrating solutions of FEL-pendulum equation and microbunching for different initial conditions. The initial beam (I) is slightly off-energy and after half a synchrotron period is bunched (II), demonstrating a profile of beam density (B) with sharp edge in the distribution, and also nonzero change of the average beam energy.

8.6.2 FEL LOW-GAIN CURVE

The FEL pendulum equations can be solved in the following approximations: the radiation amplitude is small and one can keep only the first order terms in E_0; the gain is small; i.e., the energy change of the electrons is low ($\Delta\gamma \ll \gamma$); the radiation wavelength is very close to the fundamental undulator radiation wavelength and one can average all quantities over one undulator period to remove fast oscillations.

In the EM wave-particle interaction process described by the FEL pendulum equations, each electron will gain or lose energy depending on the relative phase between the transverse oscillation in the undulator and the phase of the radiation plane wave.

By averaging the particle beam distributions represented by solutions of FEL pendulum equations illustrated in Figs 8.21 and 8.22 over the initial phases of the electrons, one can obtain the average energy variation in the beam.

As there is a balance and conservation of energy, the variation of the electron energy is equivalent to a variation of the generated EM wave's energy in the FEL. This allows to define the FEL gain G as a relative change of the wave's energy during one passage of the undulator, equaling to the change of the energy of all electrons involved in the interaction:

$$G = \frac{\Delta E_{tot}}{W_0^L} = -m_e c^2 \frac{N}{W_0^L} < \Delta\gamma >_\varphi \tag{8.21}$$

where W_0^L is the initial energy of the wave over the entire length of the undulator.

Going back to the planar undulator we give the following formula the gain function for reference and without derivations

$$G(\xi) = -\frac{\pi^2 r_e \hat{K}^2 N_u^3 \lambda_u^2 n_e}{\gamma_r^3} \cdot \frac{d}{d\xi}\left(\frac{\sin^2 \xi}{\xi^2}\right) \tag{8.22}$$

where ξ is defined in the same way as in Eq. 8.1, i.e., $\xi = \pi N_u(\omega - \omega_0)/\omega_0$ and n_e is the number of electrons per unit volume.

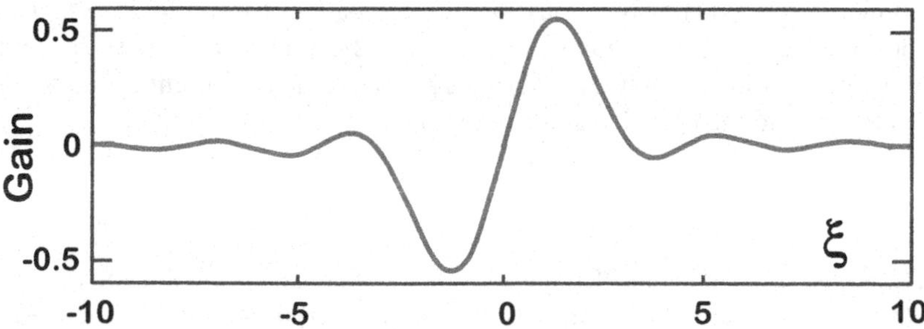

Figure 8.23 FEL normalized low-gain curve.

The normalized low-signal and low-gain FEL curve is shown in Fig. 8.23 and, as we see, is proportional to $d/d\xi(\text{sinc}^2(\xi))$ (where $\text{sinc}(x) = \sin(x)/x$), and is the same as the one discussed earlier in connection to Madey's gain curve analysis.

Please note that Eq. 8.22 contains the so-called modified undulator parameter $\hat{K}$. For helical undulator $\hat{K} = K$, where K is defined by Eq. 8.6, and for planar undulator

$$\hat{K} \equiv K \cdot (J_0(\varkappa) - J_1(\varkappa)) \quad \text{where} \quad \varkappa = \frac{K^2}{4 + 2K^2}$$

where J_0 and J_1 are Bessel functions, and the correction term in brackets is often called JJ, so that the modified parameter is written as $\hat{K} = K \cdot JJ$. The Bessel functions appear in the solutions due to averaging of sin-like functions in the planar undulator, where, for example, the radius of trajectory curvature vary along the trajectory. For $K \leq 1$ one can use the approximation $JJ(\varkappa) \approx 1 - \varkappa^2/4 - \varkappa/2$.

The positive gain in Fig. 8.23 corresponds to an amplification of the EM wave (a standard FEL case) while the negative gain corresponds to an acceleration of the beam and decrease of the EM wave's amplitude. The latter case relates to the so-called *inverse FEL*.

Note that while Eq. 8.22 is defined in terms of ξ, which depends on radiation frequency difference $(\omega - \omega_0)$, it is more convenient to define it in terms of the energy offset $(\gamma - \gamma_r)$. To convert the expression, one can rewrite the Eq. 8.15 via ω and then differentiate it

$$\omega = \frac{2ck_u}{1 + K^2/2} \cdot \gamma^2 \quad => \quad \frac{d\omega}{d\gamma} = \frac{2ck_u}{1 + K^2/2} \cdot 2\gamma = \frac{2\omega}{\gamma}$$

which allows to define the low signal gain Eq. 8.22 in terms of the energy offset η, using the following relation

$$\xi = \pi N_u \frac{(\omega - \omega_0)}{\omega_0} = -2\pi N_u \eta \quad \text{where} \quad \eta = \frac{(\gamma - \gamma_r)}{\gamma_r} \tag{8.23}$$

Behavior of the gain curve shown in Fig. 8.23 suggests that for optimal FEL operation the initial electron energy should be offset to a point of the maximum gain, and that saturation will occur when the energy is shifted to the point of the maximum absorption. The broad maximum of the gain curve in Fig. 8.23 is located at ξ between 1.3 and 1.6 — which, based on Eq. 8.23, allows to estimate the optimal beam energy detuning roughly as $0.25/N_u$. Therefore, the maximum energy transfer from an electron to the radiation can be estimated as

$$\frac{(\Delta\gamma)_{\max}}{\gamma} \approx \frac{1}{2N_u}$$

and in this case, during beam-FEL interaction, the beam energy will evolve from the point of maximum gain to the point of maximum absorption on the gain curve. The maximum peak power that can be extracted from the beam is then

$$P_{\text{laser}} \approx \frac{1}{2N_u} IE$$

where I is the peak beam current and E is the beam energy.

8.6.3 HIGH-GAIN FELS

Most modern FELs, especially those aimed at hard X-rays, are high-gain systems. In these FELs, the gain is so large that the EM wave amplitude changes within a single pass in the undulator, and therefore our estimations determined in the previous section need to be revised.

A detailed analysis — which takes into account the wave equation with driving terms determined by the oscillating current density of the beam — predicts an exponential growth of the radiation power of

$$P(s) \propto \exp(s/L_g) \tag{8.24}$$

where L_g is the gain length. The exponential growth continues until saturation is reached, at which point the emitted power starts to oscillate as illustrated in Fig. 8.24.

High gain FEL formulae and equations can be parametrized via the so-called FEL parameter or Pierce parameter ρ defined as

$$\rho = \left[\frac{I}{\gamma^3 I_A} \frac{\lambda_u^2}{2\pi\sigma_x\sigma_y} \frac{(K \cdot JJ)^2}{32\pi} \right]^{1/3} \tag{8.25}$$

where I is the peak beam current, σ_x and σ_y are transverse sizes of the electron beam, and $I_A = mc^3/e \approx 17$ kA is Alfven current.

The high gain 1-D model then gives the following approximate dependencies:

$$\text{for the FEL gain length} \qquad L_g \approx \frac{1}{4\pi\sqrt{3}} \frac{\lambda_u}{\rho} \tag{8.26}$$

$$\text{for the FEL length to saturation} \qquad L_s \approx \frac{\lambda_u}{\rho} \tag{8.27}$$

$$\text{for the FEL radiaton peak power} \qquad P_r \approx 1.6 \cdot \rho P_b \tag{8.28}$$

where P_b is the peak beam power $P_b = EI/e$.

As we see from above, the FEL length to saturation is about 20 times the gain length, and the FEL peak power is a small fraction of the beam peak power, since the parameter ρ is typically $\ll 1$.

Analysis of a high-gain curve, such as the one shown in Fig. 8.24, and of the approximate relations given above, helps to determine an optimal length of the undulator for a particular FEL design.

8.7 FEL DESIGNS AND PROPERTIES

We will now review typical accelerator parameters, requirements and radiation characteristics of modern FELs.

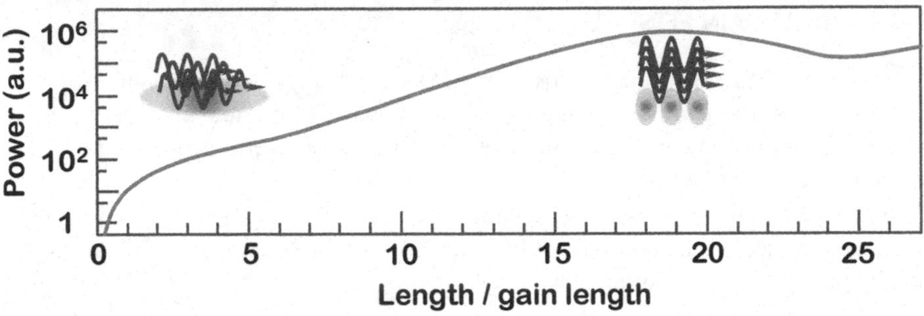

Figure 8.24 High-gain FELs, typical behavior of the emitted power — exponential growth eventually turned into saturation.

8.7.1 FEL BEAM EMITTANCE REQUIREMENTS

As we have determined in Chapter 3, the emittance of a synchrotron radiation photon beam is given by Eq. 3.33 and is equal to $\varepsilon_{ph} = \lambda/4\pi$. It is intuitive to assume that efficient generation of FEL radiation (the term, *lasing*, is often used) requires a fair amount of overlap between the electron and the photon beam.

The geometrical emittance of an electron beam needed for efficient lasing, therefore, needs to be smaller than the one for the photon beam. Thus,

$$\varepsilon \leq \frac{\lambda}{4\pi} \tag{8.29}$$

or in terms of the normalized emittance: $\varepsilon_N \leq \gamma\lambda/(4\pi)$. As an example, for $\lambda = 0.2$ nm and $\gamma = 3 \cdot 10^4$ ($\approx$ 15 GeV), the required normalized emittance is ≤ 0.5 mm·mrad, which results in a necessity to use a very bright electron source. We can also note that the requirement for geometrical emittance can be eased for higher-energy electron beams since, during acceleration, the geometrical emittance decreases in inverse proportion to the beam energy.

The width of the gain curve (see Fig. 8.23) also suggests that for efficient lasing the energy spread in the electron beam must be less than

$$\frac{\sigma_\gamma}{\gamma} < \frac{1}{2N} \tag{8.30}$$

As we discussed above, radiation slips with respect to the electron beam by λ for every λ_u, and it then also useful to define the so-called cooperation length

$$\ell_c \approx \lambda \frac{L_g}{\lambda_u} \tag{8.31}$$

which is the the length over which electrons within the bunch can "communicate" with each other (i.e., how far ahead the radiation emitted by an electron goes by the time it travels through L_g). The cooperation length is also important for defining the length of the slice that determines the beam emittance requirements.

Table 8.1
FEL and laser comparison

	LASER	FEL
Characteristics	Source of narrow, monochromatic and coherent light beams	Source of narrow, monochromatic and coherent light beams
Configurations	Oscillator or amplifier	Oscillator or amplifier
First demonstration	1960	1977
Laser media	Solids, liquids, gases	Vacuum with electron beam in periodic magnetic field
Energy storage	Potential energy of electrons	Kinetic energy of electrons
Energy pump	Light or applied electric current	Electron accelerator
Theoretical basis	Quantum mechanics	Relativistic mechanics and electrodynamics
Wavelength definition	Energy levels of laser medium	Electron energy, magnetic field strength and period

An important clarification of the above emittance requirement relates to the concept of *slice emittance*. Even if the undulator is 100 m long, for any reasonable undulator period (e.g., $\lambda_u \approx 1$ cm) and wavelength (assume $\lambda \approx 0.1$ nm), the total slippage will be around 1 μm. A typical electron bunch is usually much longer. Therefore, only a small longitudinal fraction of the bunch contributes to a particular spatial portion of generated radiation. The requirement for the emittance defined above in Eq. 8.29, as well as for the energy spread defined in Eq. 8.30, is thus applicable to the slice with length ℓ_c that is generating this portion of the radiation.

8.7.2 FEL AND LASER COMPARISON

It can be useful to compare the properties and features of FELs and lasers side by side — see Table 8.1. Amongst many similarities, the most notable difference is in their theoretical foundations: while the laser is a purely quantum device, an FEL is entirely classical.

8.7.3 FEL RADIATION PROPERTIES

Due to the coherent nature of radiation, FELs can provide peak brilliance of around 8–10 orders of magnitudes larger than that of storage ring light sources (see Fig. 8.25).

Since the repetition rates of the typical FELs is still below that of storage rings, the average brilliance of FELs exceeds that of third-generation light sources typically by about 2–4 orders of magnitude.

The FELs are also unsurpassed in temporal resolution, exceeding the third-generation by 2–4 orders of magnitude, able to reach femtosecond resolutions in

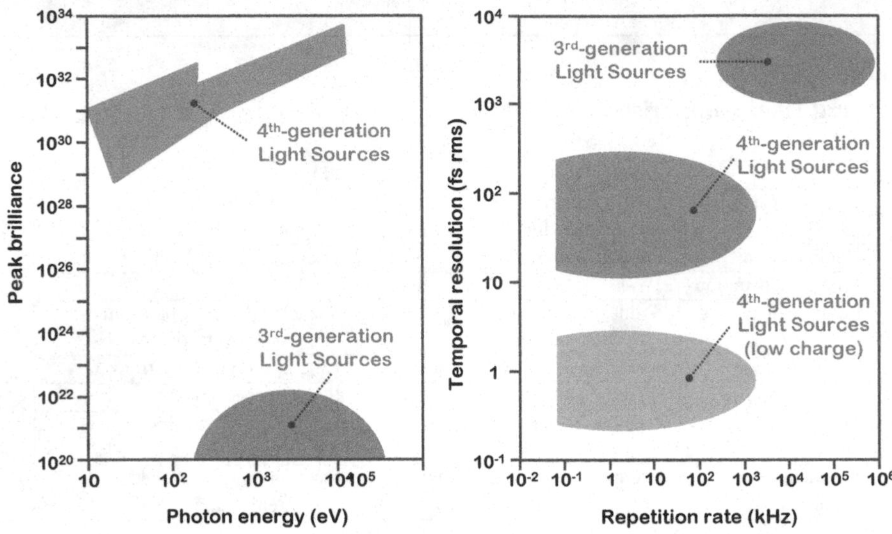

Figure 8.25 Peak brilliance (left) and temporal resolution (right) of typical FEL in comparison with third-generation SR sources. The peak brilliance in the left plot is defined in units of [photons/(s $\cdot$ mm^2 $\cdot$ mrad^2 $\cdot$ 0.1% BW)].

special configurations of low-charge sliced beams (the methods of ultra-short bunch generation are discussed in more detail in Chapter 12).

8.7.4 TYPICAL FEL DESIGN AND ACCELERATOR CHALLENGES

Modern high-brightness FELs aimed at hard X-ray ranges are sophisticated machines. A typical layout of such an FEL is shown in Fig. 5.34. A high-brightness laser-driven photo RF gun, operating from a kHz to potentially MHz range, can power a linear accelerator, a normal conducting (NC) or a superconducting (SC). Typically, two bunch compressors, at intermediate and final energies, are used to achieve the shortest possible beam length. The accelerated and compressed electron beam is sent into an undulator or switched between different undulators.

Practical constraints (size, cost) split the FEL designs into two families — NC and SC FELs, which operate under two distinct sets of parameters. Normal conducting FELs can have the highest electron beam energy and therefore the shortest X-ray wavelength; however, they are pulsed and usually have rather low (sub-kHz) repetition rates.

Superconducting FELs, on the other hand, can be either pulsed (but with long pulses and many bunches in a single pulse) or are continuously operating (CW) with repetition rates of bunches up to MHz. They, however, are generally limited in electron energy and cannot reach the shortest wavelength (see Table 8.2).

Accelerator physics challenges in FELs start from the source — a very low emittance gun is needed, seeing as the normalized emittance cannot be improved in the

Table 8.2
NC and SC FEL, typical parameters

Characteristics	NC-FEL	SC-FEL
Linac energy	6–15 GeV	2–4 (8 – 14) GeV
Linac frequency	S, C, X band	L band
Repetition frequency	50–200 Hz	kHz–MHz
Operation mode	pulsed	long pulse to CW
Minimum wavelength	0.1 nm	0.3–1 nm

linac. Acceleration and compression through the linac needs to be performed to keep the emittance low while managing the collective effects of high-brightness beams (which is not trivial, particularly at a low energy). Compression of the bunch to a tens of fs range is challenged by *coherent synchrotron radiation* (CSR), which will be discussed in closer detail in Chapter 12.

Another group of challenges related to FELs is simultaneously connected to both the physics and technologies of accelerators and lasers. First, this refers to the synchronization of conventional lasers and FEL electron sources for pump-probe experiments. Second, conventional lasers are essential to applications in FELs for the seeding of radiation in order to ensure improved temporal coherence of generated X-rays, as well as for *high harmonic generation* (HHG) (discussed further in Chapter 12).

8.8 BEYOND THE FOURTH-GENERATION LIGHT SOURCES

Let us summarize the properties of third- and fourth-generation light sources and outline prospects for future generations.

The third generation (storage rings) and FEL have complementary properties. The storage ring synchrotron light sources are extremely stable, can serve several tens of user beamlines, and their radiation can approach full transverse coherence for ultimate diffraction-limited storage rings.

The fourth-generation (FELs) sources have high brightness, short pulses and full transverse coherence. However, FELs can serve only a few beamlines at a time. Hard X-ray high brightness FELs are machines that require considerable resources for their construction and can usually be realized only as national-scale facilities.

New technological solutions are required if we are to build more economical and compact radiation sources.

Progress with laser plasma accelerators in the last several years has opened up new possibilities to work toward compact betatron radiation sources and has also created an aspiration in the community to use this technology to drive compact FELs.

The first generation of undulator radiation in the soft X-ray range driven by an LWFA-produced beam has already been obtained. In this first experiment, a laser

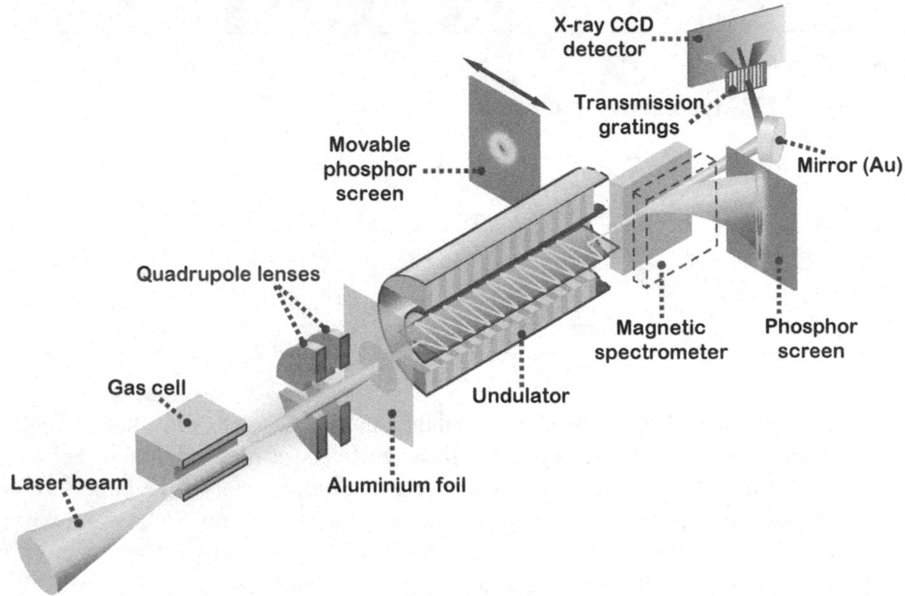

Figure 8.26 Generic layout of a compact light source driven by an LPWA.

plasma wakefield accelerator produced 55–75 MeV electron bunches, which were then sent to an undulator to generate visible to IR-range synchrotron radiation.[10]

A generic layout of a compact light source based on laser plasma acceleration and an undulator is shown in Fig. 8.26. Such compact sources are often referred to as *table-top* although at present time a more appropriate term would be "room-sized."

Laser plasma wakefield accelerators demonstrated the possibility of generating a GeV beam with promising electron beam qualities, including a normalized emittance of the order of 1 mm mrad and an energy spread of close to 1% for the entire bunch.

Recall that lasing in FEL requires the beam slice to have appropriately small emittances; these characteristics are already within reach for laser plasma acceleration and will likely be achieved with relatively modest improvements on what has presently been obtained. The beam energy spread will, however, require more noticeable improvements from the presently achieved values of a few percent for the entire beam to around a few hundredth of a percent for a radiation-generating slice.

Furthermore, an FEL based on laser plasma acceleration will require significant improvement to the stability of LPWA beams; at present, the repeatability of its beam parameters is poor and important characteristics of beams can often exhibit nearly 100% pulse-to-pulse fluctuations.

We should also note that an election beam generated in a plasma bubble typically has inconvenient ratios between its size and its angular spread. While the beam size

[10]H.P. Schlenvoigt et al., Nature Phys. 4, 130 (2008).

Figure 8.27 For illustration of filamentation. An intact paper sheet (left) may have very low volume; however, when crumpled (middle picture) it will have its effective volume increased by orders of magnitude in comparison with the case if it would be tightly rolled.

may be of the order of a micron, the angular spread can reach several milliradians. Even if the calculated emittance of such a beam is very small, its usefulness strongly depends on the ability to quickly capture this beam into an appropriately designed focusing channel, in order to avoid effective emittance growth due to filamentation — nonlinear wrapping-around in the beam phase space as illustrated (in a *synectics* approach of using analogies) by a paper sheet that increases its effective volume by orders of magnitude after being crumpled — see Fig. 8.27.

Preservation of the emittance of an LPWA beam during its transport from the plasma bubble into the undulator would require properly catching the accelerated beam into a focusing system. The focusing distances of the first lenses of the catching optics need to be comparable to the focusing in the plasma bubble, and the focusing distances of further lenses should gradually increase to match the focusing strength of the undulator beamline. An alternative approach would be to create an undulator inside of the plasma — a concept which, if proven feasible, may create light sources of ultimate compactness.

The challenges of making the plasma acceleration FEL are being steadily and successfully addressed. It has been reported that an LWFA-based FEL with ~500 MeV beam achieved[11] lasing at 27 nm, demonstrating a 100-fold gain the the last third undulator, manifesting the exponential gain regime.

A similarly remarkable breakthrough has been reported[12] in beam-driven plasma acceleration FEL, where the electron bunch, produced by conventional accelerator and then boosted in energy by PWFA, kept its qualities sufficient for lasing and

[11] Wentao Wang et al., "Free-electron lasing at 27 nanometres based on a laser wakefield accelerator," 516, Nature, Vol 595, p. 516, (2021).

[12] R. Pompili et al., "Free-electron lasing with compact beam-driven plasma wakefield accelerator," Nature, Vol 605, p. 659, (2022)

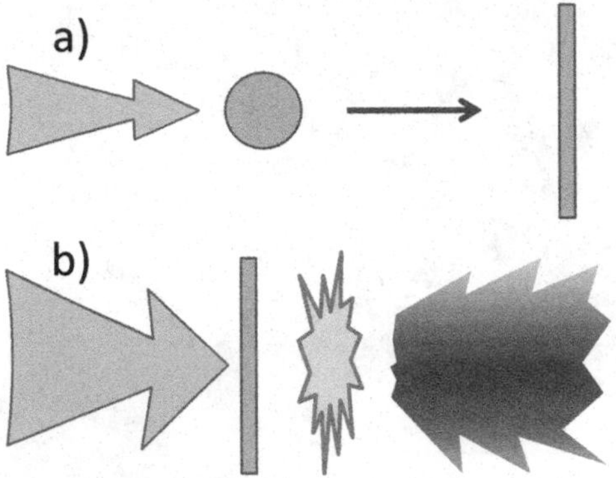

Figure 8.28 Tin droplet EUV light generation. Step a) laser-pre pulse hits round tin droplet, which then converts to a flat pancake-shape droplet. Step b) main laser pulse converts flat droplet to hot plasma for EUV generation.

Development of tin droplet pre-pulse-based EUV generation illustrates application of TRIZ principles of Volume to Surface ratio, principles of preliminary action, intermediary, and the principle of using the energy and resources that already exist in the system.

demonstrated the exponential growth of radiation intensity over six consecutive undulators.

Theses breakthrough examples are surely just the start, and we will see many more plasma acceleration compact FELs in the near future.

8.9 EUV LIGHT GENERATION – INVENTION CASE STUDY

The topics of synergies between lasers and plasma, discussed in the Chapter 4, can be illustrated by the following story of EUV light generation for semiconductor industry. There is a natural connection of this story and this area in general with accelerators and FELs, which may emerge in the nearest future.

Generation of EUV (Extreme Ultra Violet) light with 13.5 nm wavelength is very important for semiconductor industry, in particular for the lithography process. This example is extremely relevant for our book, since it connects to both lasers and plasma, and illustrates the links and bridges to several inventive principles.

Production of 13.5 nm EUV light was achieved with powerful laser pulses focused on liquid tin (Sn) droplets, which would then be converted to very hot plasma. Plasma emits radiation in a wide range, including EUV. Multi-layer optics selects radiation near the 13.5 nm range[13], feeding the semiconductor production areas.

The approach described above was not very efficient, and the main challenge was due to the fact that the liquid tin droplets were round. Different layers of tin sphere would receive different amount of energy from the laser pulse. Only part of tin mass would produce hot enough plasma that generated useful radiation, resulting in low-energy conversion efficiency (laser energy to 13.5 nm radiation energy). Other parts

[13]V.Y. Banine et al., J. Phys. D: Appl. Phys., v. 44, 253001, (2011).

of tin droplet would produce the clusters of molecules and ions that fly downstream, damaging the precious and delicate multi-layer optics.

The path to resolving this challenge is a remarkable example that can illustrate application of TRIZ principles (even though the authors may not followed them consciously). Let's now reproduce this path. One could have started to look at this problem by asking a question — is the droplet shape optimal? Laser interacts with the droplet primarily via its surface, therefore, remembering the volume to surface inventive principle, one could have asked the next question — what would be a more optimal shape of the tin droplet?

As the first step, remembering the plot in Fig. 4.16, and the Volume/Surface inventive principle, we can conclude that the optimal droplet shape would be a flat slab or a pancake. The laser pulse will then interact with a much larger surface. The next question would then be the following — how to create a pancake shape out of the initially round tin droplet?

For the next step, we can get a hint from the inventive principle of preliminary action, and be also inspired by the principle of using resources that we already have in the system, and therefore conclude that we need to send a laser pre-pulse to the round tin droplet, to give it a kick that will convert it to a pancake, and only after that conversion to send the main laser pulse.

This was indeed the approach that was used to successfully modify[14] the EUV generation. The laser pre-pulse focused on the spherical tin droplet would create a hydrodynamic motion of liquid tin and at some proper moment its shape will convert to a flat pancake. The main laser pulse will then be focused on the flat pancake tin droplet, which would absorb all the energy from the main laser pulse, and thus efficiently convert to a very hot plasma, emitting the 13.5 nm EUV radiation.

Connection of this story to accelerators is obvious — FELs can be designed to produce 13.5 nm or a shorter wavelength, as required for the semiconductor industry. The present challenge of using accelerators for this purpose is the system's compactness, reliability and the wall-plug efficiency. However, it is likely that all these challenges can be solved in a couple of decades.

[14]Oscar O. Versolato, Plasma Sources Sci. Technol., v. 28, 083001, (2019).

EXERCISES

8.1. *Chapter materials review.*
Evaluate the slice emittance requirements for an FEL based on a 1 GeV electron beam. Discuss the factors affecting the requirements for the slice energy spread of this FEL's beam.

8.2. *Chapter materials review.*
Discuss the phenomenon of filamentation in plasma acceleration in connection to the TRIZ inventive principle of preliminary action.

8.3. *Mini-project.*
Define, very approximately, the main parameters (energy, length, undulator field and step size) of a linac-based FEL aimed at 15 KeV X-ray energy.

8.4. *Analyze inventions or discoveries using TRIZ and AS-TRIZ.*
Analyze and describe scientific or technical inventions described in this chapter in terms of the TRIZ and AS-TRIZ approaches, identifying a contradiction and an inventive principle that were used (could have been used) for these inventions.

8.5. *Developing AS-TRIZ parameters and inventive principles.*
Based on what you already know about accelerator science, discuss and suggest the possible additional parameters for the AS-TRIZ contradiction matrix, as well as the possible additional AS-TRIZ inventive principles.

8.6. *Practice in reinventing technical systems.*
FEL radiation takes away only a small fraction of the beam energy. Still, at some point along the undulator, the beam energy offset may become noticeable and the resonant condition can be violated, resulting in reduced efficiency of radiation generation. Suggest a way to modify the system to prevent the loss of synchronism when the beam energy is decreasing along the undulator.

8.7. *Practice in reinventing technical systems.*
An FEL requires a high-peak current, i.e., short bunches. Beams coming from a photoinjector are typically long, and bunch compression is required, which is typically performed with a sequence of magnetic chicanes together with an energy-position correlated chirp introduced in the bunch in an RF accelerating section. Suggest an alternative way to create an energy-position chirp in an electron bunch, using ideas and techniques discussed in this Chapter 8.

8.8. *Practice in the art of back-of-the-envelope estimations.*
A free-electron laser with energy recovery is fed from an accelerator built with dual-axis cavities, as shown in Fig. 15.4. The FEL is designed to generate radiation with a wavelength 13.5 nm. Estimate the energy of the electron beam at the final energy of deceleration, just before the beam is directed to the beam dump. *(It is assumed that you can identify the most important effects playing roles in this task, can define the necessary parameters and set their values, and can get a numerical answer.)*

9 Proton and Ion Laser Plasma Acceleration

In Chapter 6, we primarily discussed the plasma acceleration of electrons. There are strong motivations, however, for development of plasma acceleration of protons and ions, as further advances in this area could improve current ways of treating tumors.

The advantages of using protons and ions (in comparison to using electrons and X-rays) are associated with the phenomenon of *Bragg peak* — we will start our discussion by looking at this phenomenon in detail.

Regarding the application of accelerators for therapy, it is important to know how either the electrons or protons can affect the living cells. We will therefore briefly overview, in this chapter, how the mechanisms of DNA respond to radiation.

The conventional accelerator systems used for particle beam therapy for tumors typically require protons with around 250 MeV of energy. The accelerator is typically a synchrotron or a cyclotron (so far the most popular). A beam therapy facility usually includes beam delivery gantries (which send the beams in a selected target volume), as well as collimators and degraders. We will review the design and functionality of conventional beam-therapy facilities in a later section.

A conventional accelerator system for particle beam therapy, with a beam source and several gantries, may require several thousands of square meters of space, making these respective facilities large and expensive to maintain. On the other hand, plasma acceleration of protons and ions is one of the most promising means by which we can make beam therapy more affordable and more accessible to patients. After we discuss the motivations to apply plasma acceleration to this area, we will briefly overview the present understanding of the different regimes and mechanisms of proton/ion plasma acceleration.

We would like to once again note that proton/ion laser plasma acceleration is a rapidly developing area and, moreover, it involves rather complicated physics. There are various regimes of acceleration identified and several mechanisms that explain the experimental results. The different mechanisms of acceleration compete, and often more than one mechanism acts at the same time. The mechanisms of acceleration are often just approximate models that represent a gradual improvement of our understanding of proton/ion plasma acceleration. The state of this rapidly developing area makes it hardly suitable for textbooks just yet. We will therefore grant only a cursory look at an overview of the presently identified acceleration mechanisms and their scaling rules.

We will conclude this chapter with a glimpse into the future, and, as it already became a tradition for this second edition, with a consideration of an invention case study, related in this case to an ingenious and inventive approach to simulations of laser plasma acceleration.

DOI: 10.1201/9781003326076-9

9.1 BRAGG PEAK

The treatment of tumors with the help of X-rays or proton/ion beams is based on the delivery of energy to malignant cells of the target volume, intended to prevent and eradicate the growth of these unwanted cells. We will review the biological effects of radiation in the next section, and will now compare the effects of the photons with the effects of charged particles.

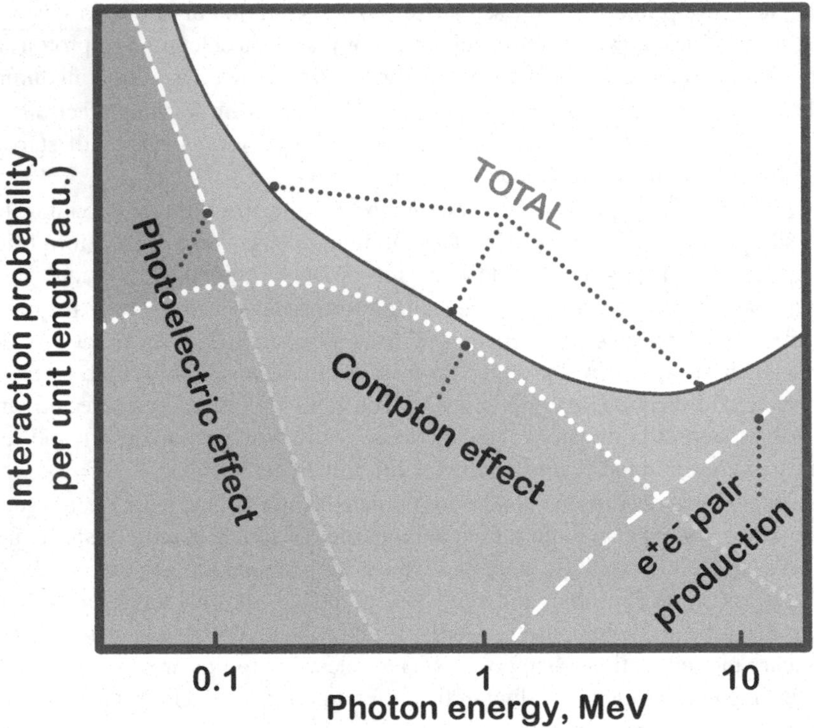

Figure 9.1 Photon matter interaction, qualitatively.

Photons penetrating through a medium will lose their energy due to several factors. An incident photon can interact with an atom and get absorbed, causing a *photoelectron* to be ejected from the atom. The effect is dominant at lower energies (as illustrated in Fig. 9.1). An incident photon can also lose part of its energy via the Compton effect — the inelastic collision of photons with the electrons of the atoms. Finally, if the energy of the photon is sufficient, it can create an e^+e^- pair. This effect has a threshold character and manifests itself for photons with energies greater than about an MeV.

The photons that are used in radiation therapy are usually produced from electron beams of several MeV, converted on a solid target to photons via the process of *bremsstrahlung*. As interaction between the photon beam and medium results in a gradual decrease of intensity with the depth of the absorbing medium, the so-called *stopping power* $S(E)$ is gradually decreasing too, as presented in Fig. 9.2. The stopping power is defined as $-dE/dx$ — the energy loss per unit length taken with the minus sign.

In contrast to photons, charged particles lose their energy in matter primarily through a Coulomb interaction with the outer-shell electrons of the absorber's atoms. Excitation and ionization of atoms result in a gradual slowdown of the particle. A slower moving particle will interact with an atom for longer time, resulting in a larger energy transfer. Therefore, the charged particles will have an increased energy loss per unit length at the end of their passage through the medium. Qualitatively, this is the origin of the *Bragg peak* — the peak in energy loss that occurs just before the particles come to a complete stop.

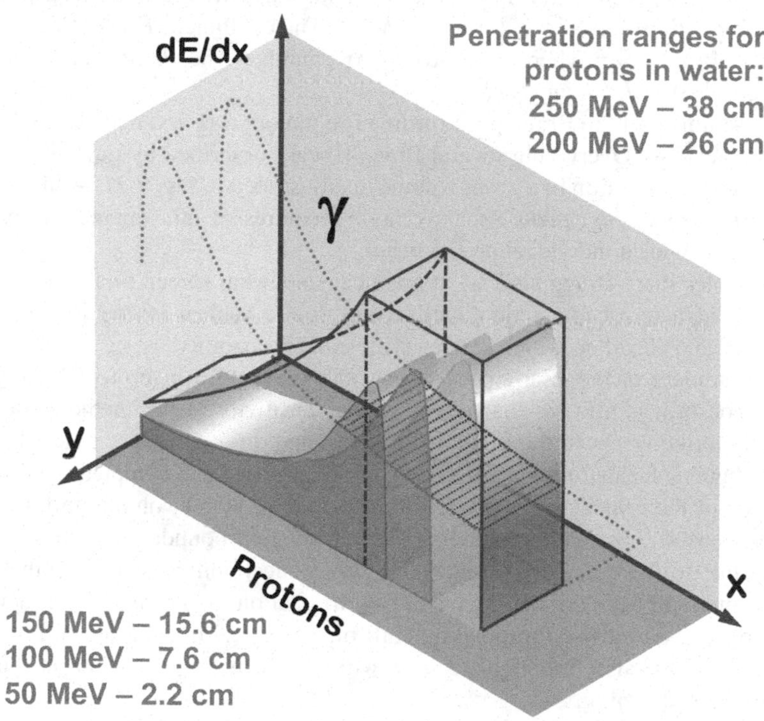

Figure 9.2 Absorption of photons (dotted lines) in comparison with absorption of protons in media. Overlaying multiple Bragg peaks creates a near uniform dose distribution in a certain target volume. The ranges of protons penetration in water are also shown in the figure for different proton energies.

The above-mentioned phenomenon is named after William Bragg, who discovered it in 1903. The Bragg peak is the reason why protons are better for treatment of tumors than X-rays, in some cases, as protons localize the deposited dose in the destined target volume (see Fig. 9.2), this minimizes the impact on healthy tissues — especially in cases where the target volume is located close to critical organs.

Quantitatively, the Bragg peak can be explained by the following formula for the mean energy loss of moderately relativistic heavy particles (*Bethe equation*):

$$\langle -\frac{dE}{dx} \rangle \approx Kz^2 \frac{Z}{A} \frac{1}{\beta^2} \left(\frac{1}{2} \ln \frac{2m_e c^2 \beta^2 \gamma^2 W_{max}}{I^2} - \beta^2 \right) \tag{9.1}$$

Here, z is the charge number of an incident particle, Z and A are the charge number and atomic mass of the absorber, respectively, and β and γ are the relativistic factors of the incident particle. The parameters under the logarithm are: I — the mean excitation energy of the atom's electron, and $W_{\max}$ — the maximum energy transfer in a single collision. For a particle with mass M, the latter is defined as $W_{\max} = 2m_e c^2 \beta^2 \gamma^2 / (1 + 2\gamma m_e / M + (m_e / M)^2)$. The coefficient K is defined as follows: $K = 4\pi N_A r_e^2 m_e c^2$ where N_A is Avogadro's number. The coefficient K approximately equals 0.3 $MeV \cdot cm^2/mol$.

The usefulness of the Bragg peak for treating tumors was first realized by Robert R. Wilson in 1946. Overlaying several Bragg's peaks described by Eq. 9.1 creates a uniform dose distribution in a given volume (as illustrated in Fig. 9.2) — this is often called a *spreadout Bragg peak*. Such overlaying requires an adjustment to the energy and intensity of each individual proton beam.

The ideally sharp Bragg peak is, in practice, somewhat spread — first due to the statistical character of interaction, and second due to nuclear interactions between the protons and absorber, which happen with some probability.

The sharpness of the Bragg peak is an enabling feature of proton therapy, but simultaneously it is a factor that increases the sensitivity of the method to errors, especially to the errors in the predicted depth range. In a particular case when the target volume is located near a critical organ, ideally, one can completely eliminate irradiation of the critical organ while filling the entire target volume uniformly. In practice, however, one cannot obtain a sharp irradiation boundary of the irradiated volume, due to the necessity to allow for some uncertainties of the depth range. Possible motion of the critical organs during irradiation — as well as shrinkage of the tumor (and possibly corresponding shift of the critical organs) as the treatment progresses — are also important factors, which need to be taken into account in proton therapy planning.

The above-mentioned sensitivity to errors places a particularly strong requirement on the energy of protons in the cases when plasma acceleration is used. The beam needs to have a well-defined energy. This can be ensured either via predictable plasma acceleration or by an appropriate energy-selection system.

9.2 DNA RESPONSE TO RADIATION

The underlying mechanism of X-ray and proton cancer therapy is the ability to prevent the replication of malignant cancer cells. This control is achieved by damaging the *DNA* of cancerous cells. Let's now consider the basic facts about DNA's response to radiation.

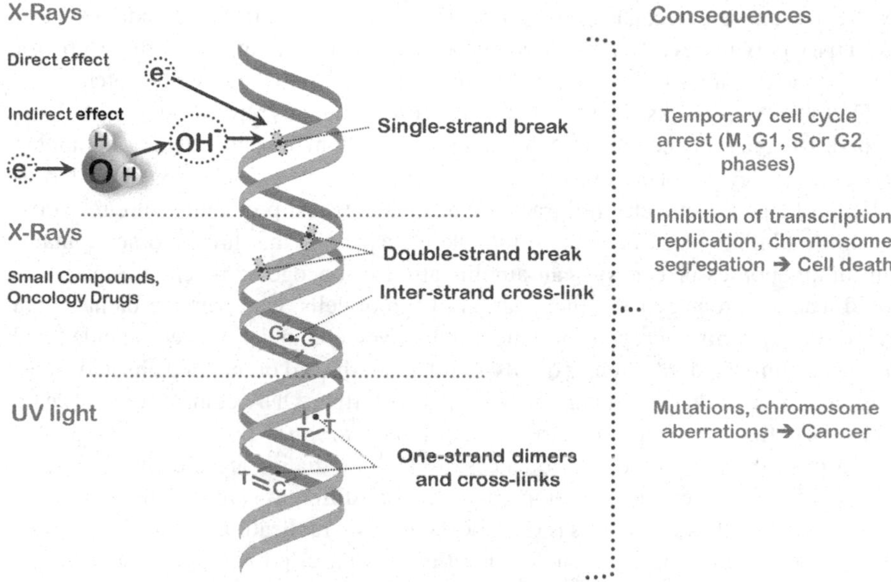

Figure 9.3 Radiation effects on DNA.

The effect of ionizing radiation on DNA can produce many outcomes, from neutral to negative (stimulating growth of cancer cells), or positive (eliminating cancer cells) depending on the initial state of the irradiated object, the radiation dose and the use of *concomitant agents* — pharmaceutical substances used prior or together with irradiation.

In some cases, radiation simply does not cause DNA damage, but passes through the cell without any side effects. With an increased dose, the radiation starts to physically affect the DNA. The effects are distinguished as direct and indirect.

Direct radiation effects take place when X-rays create ions and corresponding electrons (e-) that physically break the nucleotide pairs of the DNA (see Fig. 9.3).

Nevertheless, in the majority of cases, X-rays act indirectly when they induce water radiolysis and consequent production of OH- hydroxide — highly active free radicals — that then form hydrogen peroxide and cause single-strand DNA breaks.

We should also note that the effects described above are also very similar to the case of irradiation by protons, taking into account that, in this instance, the electrons and *bremsstrahlung* photons are produced when the protons lose their energy to electrons of the absorbing medium.

Double-strand breaks represent the most detrimental damage produced in DNA by ionizing radiation.

Several minor DNA defects — such as one-strand dimer formations (caused by UV light — see bottom part of Fig. 9.3) or single-strand breaks could be corrected by the internal cell mechanism, *base excision repair*, which involves endonucleases and DNA polymerase. Nevertheless, some errors can also appear at this step (for example, wrong nucleotide(s) insertion could cause a future coding mismatch).

Double-strand breaks are more difficult to repair and thus represent the most detrimental damage produced in DNA by ionizing radiation. When a cluster of complex DNA damage events occur together, for example from dense ionization at the terminal end of an energetic electron track, there is the highest probability of either mis-repair or failure to repair damage. A loss of entire parts of the chromosomes containing tumor-suppressor genes — an amplification of oncogenic-potential regions — could lead to carcinogenesis, the creation of tumor cells. Unrepairable damage can trigger the *apoptosis mechanism*, where defective cells undergo "cell suicide" and are, thus, eliminated. Mis-repair of DNA with cross-linked or mutated chromosomes may not be immediately lethal, but will lead to further DNA damage or cell death when the cell next attempts division — *mitotic catastrophe*.

Various examples of the consequences of DNA irradiation are shown on the right side of Fig. 9.3. They include temporary arrest at different stages of the cell cycle (mitosis (M), both check-points (G1 and G2) or DNA replication/synthesis (S)). The irradiation effects also may include inhibition of transcription and a variety of cell division abnormalities, all of which could result in a defective cell death.

The positive effects of ionizing radiation are caused by the same mechanisms, since what could be detrimental for a normal cell could also be destructive for a malignant one.

In order to further enhance the destructive effect of X-rays on malignant cells, some concomitant agents[1] such as *small molecule oncology drugs* could be involved. The upper limit for a small molecule's weight is approximately 900 daltons[2], which allows rapid diffusion across cell membranes so that they can reach intracellular sites of action.

The effects of irradiation and the use of concomitant agents can lead to an increased number of difficult-to-repair inter-strand cross links, as shown in Fig. 9.3. The apoptosis mechanism, launched by the strong synergetic effect of these two techniques, leads to a more efficient elimination of malignant cells.

9.3 CONVENTIONAL PROTON THERAPY FACILITIES

The design of conventional proton therapy facilities is dictated by several factors, such as the size of the proton source, the number of treatment beamlines and the size of the gantry.

A generic layout of a proton therapy facility is shown in Fig. 9.4. Typically, either a cyclotron or a synchrotron is used as a source for a proton facility. In the event that

[1] In molecular biology, a small molecule drug (or a small compound) is a low molecular weight organic compound that may help to regulate biological processes.

[2] Dalton (Da) is the standard unit defined as one twelfth of the mass of an unbound neutral atom of C_{12} in its nuclear and electronic ground state. In other words, one Da is approximately equal to a mass of a proton or neutron.

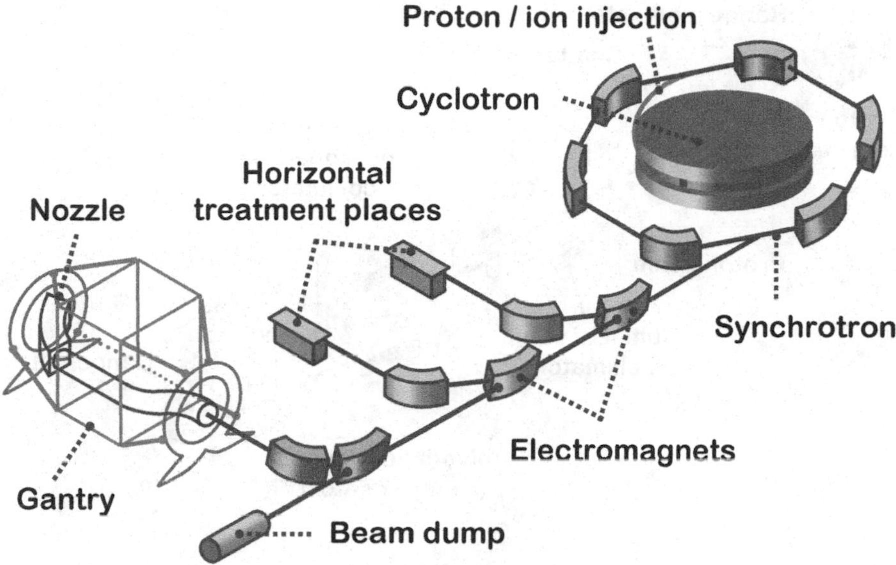

Figure 9.4 Generic proton or heavy-ion therapy facility.

ions are used for therapy (e.g., carbon, which gives certain advantages), a cyclotron can be used to inject the beam into a synchrotron.

The beamline distribution system then directs the accelerated beam into the treatment rooms. The beam can have a fixed location in the room or can be brought into the room via a *gantry*, which allows flexibility in the direction of the arriving beam.

A typical facility would usually have several treatment rooms with fixed beams and one or more rooms with gantries. The gantry is a complicated mechanical device that needs to rotate the entire section of a beamline while maintaining its precise (usually sub-mm) alignment. The energy of the protons or ions and the achievable strength of the bending magnets of the gantries would typically make the gantry on the order of ten meters in size and a hundred of tons in weight. The use of superconducting magnets in the gantries is possible and would typically reduce the weight of the gantry by around a factor of four.

The overall size of the proton/ion treatment facilities is defined by the size of the accelerator and beamlines, but, most notably, by the size and the height of the gantries. The cost of the components and the size of the necessary building are the prevailing limiting factors — there are presently several tens of facilities operating worldwide, while the generally accepted estimation states that there should be one proton treatment facility for every 10 million people.

9.3.1 BEAM GENERATION AND HANDLING AT PROTON FACILITIES

The typical elements of the proton therapy system are illustrated in Fig. 9.5.

First, the proton beam coming from the source passes through a scatterer, which increases the divergence of the beam. The transverse beam shape is then adjusted by

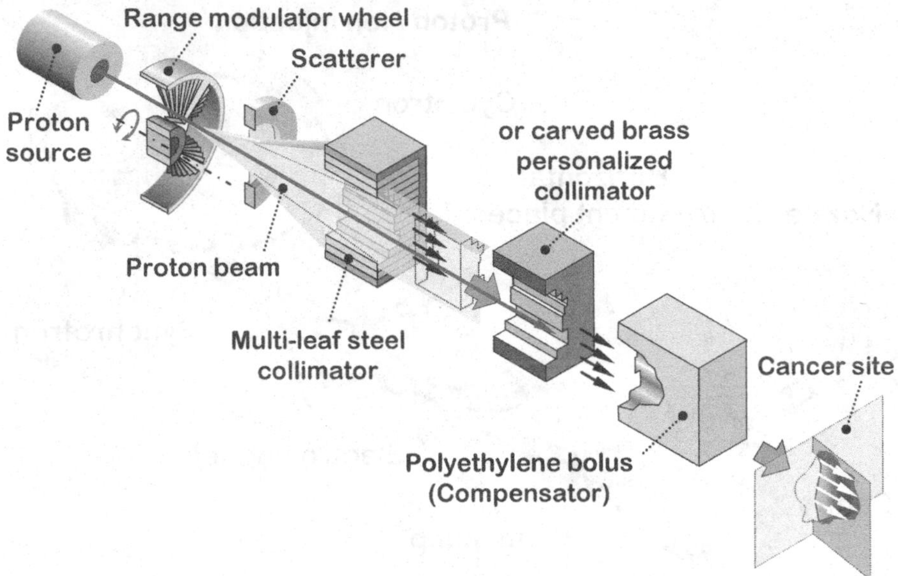

Figure 9.5 The elements of the proton therapy beamline.

a multi-leaf steel collimator to match the shape of the target volume. The leaf of the collimator can be mechanically controlled on a sufficiently short time scale. Instead of an adjustable multi-leaf collimator, a fixed-shape brass collimator can be used, tailored for a particular case.

Figure 9.6 Pencil beam scanning.

X-ray or a proton therapy system fed by a cyclotron require use of multi-leaf collimators — mechanical devices that need to move frequently in a radiation environment — which are often responsible for more than 50% failures of the entire therapy system. Proton synchrotrons, which offer smaller beam size and energy adjustability, eliminate the need for such collimators and allow pencil beam scanning.

Another element of the system is called the bolus, or range-shifter, and serves to compensate for the depth difference of different regions of the target volume. The compensator is usually made from polyethylene and reduces the energy of the beam to fit the required penetration depth at each point.

The elements as described above are, in particular, suitable for the beam coming from standard cyclotrons, the energy of which usually cannot be changed easily. In the case of synchrotrons, where the beam energy can be finely scanned, the *pencil beam scanning* approach can be employed, as shown in Fig. 9.6. In this case, the beam should be small enough so that one could "paint" the desired target volume transversely, and simultaneously control the depth by appropriately adjusting the beam energy.

9.3.2 BEAM INJECTORS IN PROTON FACILITIES

So far, the most widely used accelerator for proton therapy facilities has been a cyclotron.

A standard cyclotron is shown in Fig. 9.7. A constant magnetic field (often arranged to decrease with the radius to ensure transverse focusing) houses the electrodes (dees) where oscillating voltage is applied.

The standard cyclotron is intended for CW operation, and this is in fact one of its main advantages. The frequency of accelerating voltage ω_0 and the magnetic field B_0 are constant and are connected via the equation for the time of flight around the orbit

$$\omega_0 = \frac{v}{R} = \frac{qB_0}{m\gamma} \tag{9.2}$$

where m is the mass at rest and γ is the relativistic factor that can change during acceleration.

Since the condition $v = \omega_0 R$ must be valid up to the moment of extraction, the final energy of the cyclotron is fixed by its geometry and cannot be changed. Applications of the standard cyclotron to proton therapy thus have to rely on the use of range-shifters to adjust the penetration depth.

In addition to the fixed final energy, standard cyclotrons also suffer from the relativistic effects. Once protons accelerate to several tens of MeV, the perfect relation between the revolution time and frequency of the field starts to break.

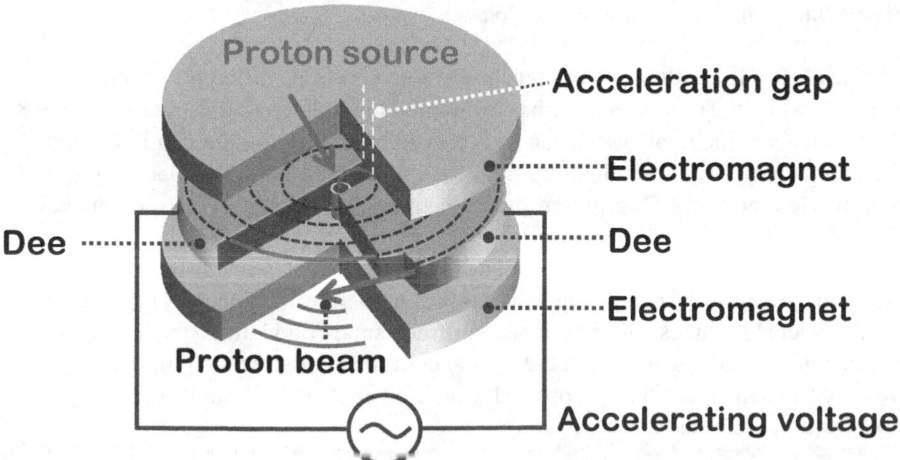

Figure 9.7 Schematic of a cyclotron.

In contrast to a standard cyclotron, a *synchrocyclotron* can have variable energies of the accelerated beam and it can also achieve higher energies — several hundreds of MeV.

In the synchrocyclotron, the relativistic effects are compensated by continuously decreasing the frequency of the accelerating voltage ω during acceleration so that

$$\omega = \frac{\omega_0}{\gamma(t)} \tag{9.3}$$

The time-varying accelerating frequency also means that only a bunch of a certain length can be in sync with the field — the synchrocyclotron therefore cannot accelerate CW current but can only produce pulsed beams.

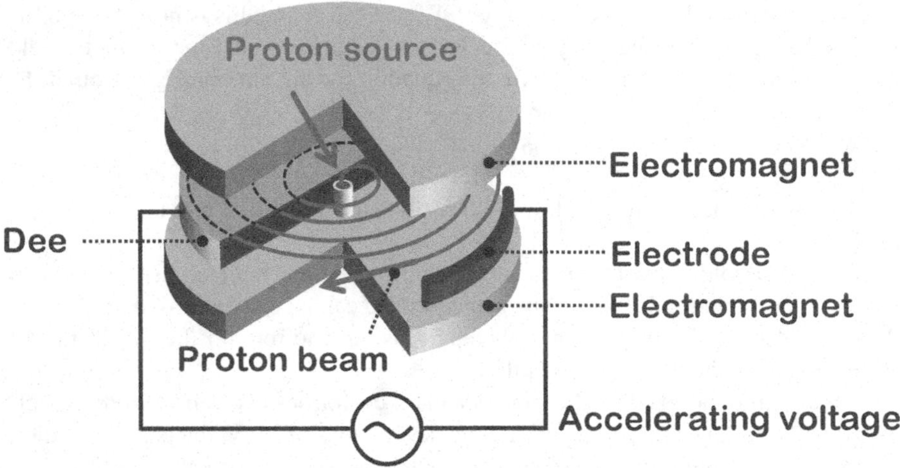

Figure 9.8 Schematics of a synchrocyclotron.

The electrode configuration in synchrocyclotrons is also different — only one dee remains, while the other electrode has a modified open shape as illustrated in Fig. 9.8. Adjustments to the final energy can be achieved by modifying the relations between the magnetic field and accelerating frequency, making it possible to accelerate protons to GeV energies. The pulsed mode of operating synchrocyclotrons, however, limits their *duty factor* and thus limits their intensity.

The variability of the final energy of a synchrocyclotron is an advantage that keeps attracting attention to their potential use as proton therapy machines. A notable recent development[3] includes a design of an iron-free[4] superconducting synchrocyclotron, where dual nested solenoids are used to cancel the external fields of the device. The iron-free design makes the synchrocyclotron light and compact, and particularly suitable for mounting directly on a gantry.

---

[3]A. Radovinsky et al., MIT report PSFC/RR-13-9, 2013.

[4]*Compare this with Section 10.2.2 and analyze this design from the TRIZ point of view.*

Another type of cyclotron — the *isochronous cyclotron* — compensates the relativistic effects by allowing the magnetic field to increase with the radius of the orbit as

$$B = \gamma(R)\, B_0 \tag{9.4}$$

To increase the field with a radius in an isochronous cyclotron, shims attached to the poles can be used in such a way that the azimuthal fraction of the shims increase with the radius, as illustrated in Fig. 9.9.

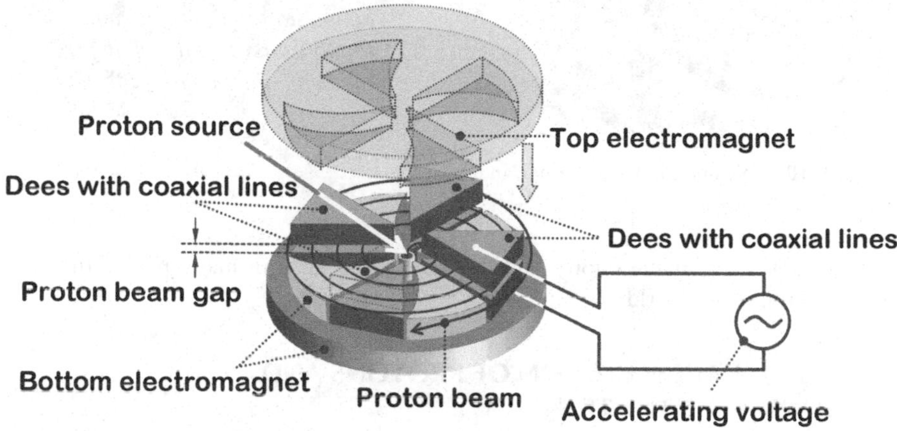

Figure 9.9 Schematics of an isochronous cyclotron.

The field increasing with the radius contradicts the requirements for weak focusing, and therefore strong focusing is arranged in isochronous cyclotrons by employing a spiral shape for the pole shims (often called *flutter* configuration), so that the regions of the field transition shown in Fig. 9.10 would introduce edge focusing on the beam.

The isochronous cyclotrons, as their acceleration frequency is constant, are suitable for CW operation. Proton energies of around a GeV can be achieved by compensating for the relativistic effects. The final energy of isochronous cyclotrons is fixed, as its adjustments require modification of the poles.

CW operation and the ability to reach higher energy made isochronous cyclotrons very popular for proton therapy applications. Superconducting isochronous cyclotrons have also been developed, with the goal to increase the compactness of proton therapy devices.

Synchrotrons are naturally also a possible choice as an accelerator for use in proton therapy. Their advantage is that they can provide variable energies; however, there are also a number of disadvantages. First of all, the size: synchrotrons are larger than cyclotrons. Second, synchrotrons are pulsed machines, while proton therapy benefits from a slow dose delivery. Therefore, slow extraction methods, such as those based on excitation of nonlinear resonances, may need to be used. Overall,

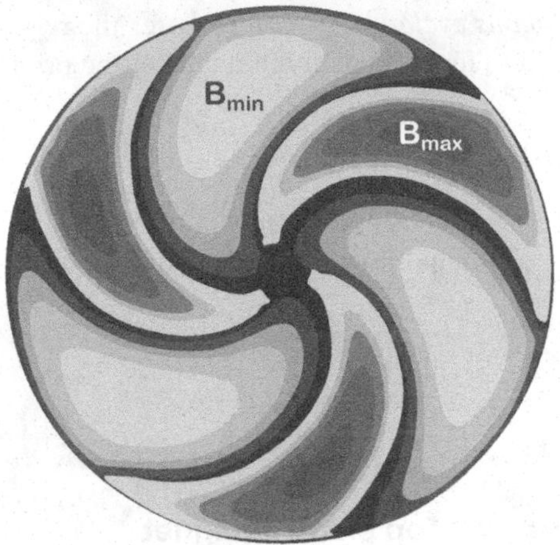

Figure 9.10 Example of a field profile in an isochronous cyclotron.

Spiral looking shape of magnetic field in isochronous cyclotrons allows to increase the average field with radius, to compensate for relativistic increase of proton mass, and also allows to create additional transverse focusing for the beam due to edge effects.

synchrotrons are a viable choice and they are being used in many proton therapy facilities around the world.

9.4 PLASMA ACCELERATION OF PROTONS AND IONS – MOTIVATION

The motivation for developing laser plasma acceleration techniques for beam therapy can be defined by the following factors. On the one hand, there are advantages to using protons instead of X-rays for certain tumors, particularly in pediatric cases, where use of protons significantly reduces the probability of tumor recurrence or side effects due to lower dose for non-target tissues. On the other hand, widespread use of proton therapy facilities is limited by their overall cost.

Proton therapy systems require 250 MeV beams (or above 330–350 MeV when protons are also used for diagnostic imaging). Such systems, especially the beam delivery gantries, are large and expensive (see Fig. 9.4). The cost and size arguments are even more pronounced for heavy ion therapy systems based on the use of carbon ions, which create certain therapeutic advantages.

Meanwhile, laser plasma acceleration has demonstrated rapid progress, delivering several GeV in energy of quasi monoenergetic electron beams and more than 100 MeV proton beams. The progress is due to advances in lasers, where a CPA laser beam of a few hundred TW or around a PW (corresponding, e.g., to 400 J energy in 400 ns or 30 J in 30 fs), when focused to a 5 μm spot, can create an intensity on the order of 10^{25} W/m^2 — which is suitable for proton plasma acceleration. Such lasers, while still bulky and expensive, are rapidly improving and will become more efficient and compact in the future.

Ultimately, the desire to create compact laser plasma acceleration proton therapy systems is one of the main motivations for developing plasma acceleration of protons.

9.5 REGIMES OF PROTON LASER PLASMA ACCELERATION

In this section, we will very briefly describe several different regimes (mechanisms) of laser-driven proton acceleration.

We will start with a discussion of an already classical mechanism called *sheath acceleration* or *target normal sheath acceleration* (TNSA). As we will see, this classical method provides protons with too large an energy spread and poor scaling with laser power.

We will then follow up on more recent and promising mechanisms that rely on *radiation pressure acceleration*, specifically the mechanisms of *hole-boring* and *light-sail acceleration*. The latter has favorable scaling with laser power and promising perspectives for achieving near-monoenergetic beams.

We will briefly touch upon other mechanisms that have been identified, such as *shock acceleration* and the *relativistic transparency regime* — also called the *break-out afterburner*. Various academic review papers[5–7] can be consulted to gain deeper insights into the discussed areas.

Once more, we would like to stress that the mentioned mechanisms of plasma acceleration represent a gradual improvement of our understanding of this very complicated phenomenon. Different mechanisms compete, and often more than one mechanism is active in a particular case. As we strive to sharpen our understanding of these mechanisms, further significant progress in this area will be inevitable.

9.5.1 SHEATH ACCELERATION REGIME

Sheath acceleration (TNSA — target normal sheath acceleration regime) is illustrated conceptually in Fig. 9.11. Here, the laser pulse is focused on a thin metal foil that creates plasma. The plasma electrons quickly become relativistically hot and leave the foil, creating a sheath of charge, which then pulls out the ions and protons from the plasma.

The surface of the foil is typically contaminated (which is, in this case, useful) by a thin layer of hydrogen that becomes the source of protons that are most readily accelerated. Acceleration of protons and ions happens on both sides of the foil. Still, the side opposite the laser irradiation side exhibits sharper boundaries of the electron sheath, providing higher energies and better beam quality of the accelerated protons.

Experimental investigations of the TNSA mechanism using lasers with intensities above 10^{19} W/cm^2 demonstrated[8] that multi-Mev ion acceleration from the rear surface of thin foils is possible.

The scaling rules for the TNSA mechanism can be derived by taking into account that a Debye sheath will form from hot electrons on the rear of the foil, and the potential difference U through this sheath will be of the order of the electron temperature

[5]A. Macchi et al., Rev. Mod. Physics, 85, 751 (2013).

[6]H. Daido et al., Rep. Prog. Phys. 75, 056401 (2012).

[7]M. Borghesi et al., Fusion Science and Technology, 49, 412 (2006).

[8]Maksimchuk et al., PRL, 84, 4108 (2000); Snavely et al., PRL, 85, 2945 (2000).

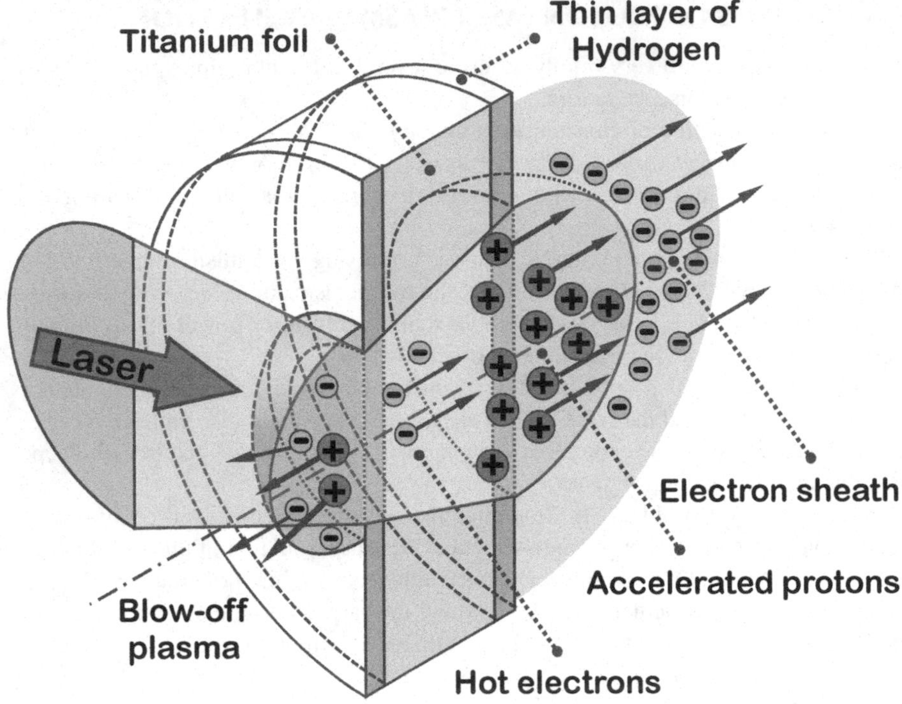

Figure 9.11 Sheath laser acceleration of protons.

T_e multiplied by the Boltzman constant: $U \approx k_B\, T_h$. Considering that the temperature of electrons is related to laser intensity I and wavelength λ as $T_e \propto \sqrt{I\lambda^2}$, one can obtain a scaling for the maximal energy of accelerating protons as

$$W_{\text{max}} \approx k_B\, T_e \propto \sqrt{I\lambda^2} \tag{9.5}$$

In the above equation, we have assumed in the first approximation that the maximum energy of the protons is equal to the value of the potential difference U. In fact, energies several times larger than that are expected, taking into account that the sheath is expanding and protons are "surfing" on the expanding potential.

In some of the first experiments,[9] proton energies close to 20 MeV were obtained. The spectrum of protons was, however, very broad, and there were not many protons at the large energy side of the spectrum. Qualitative behavior of the TNSA proton spectrum is shown in Fig. 9.12. Even though later experiments demonstrated that the proton energies close to 100 MeV were possible (see review[10]), these disadvantageous qualities of the spectrum remained.

---

[9]E.L. Clark et al., PRL, 84, p.670. (2000)

[10]M. Borghesi et al., Plasma Phys. Control. Fus., 50, 124040 (2008).

Some of the characteristic properties of TNSA-produced beams include a large divergence (of several degrees) combined with micron-scale beam size. Such a beam may formally have a low emittance; however, it will quickly filament and the emittance will increase if the beam is not captured in an appropriate focusing system. The TNSA mechanism can produce 10^{11}–10^{13} protons per shot, but not many of those protons will be at the high-energy edge of the spectrum.

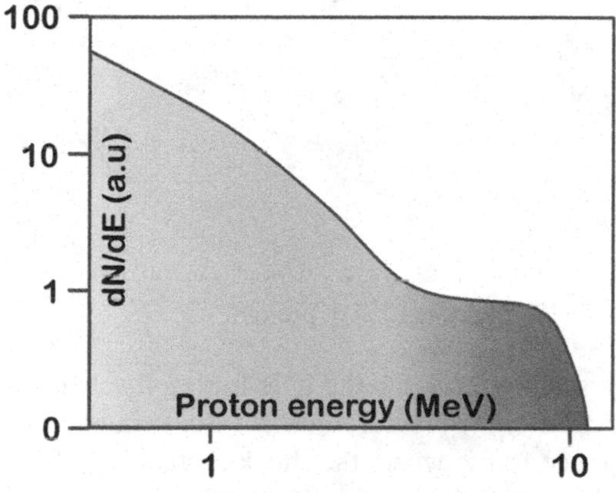

Figure 9.12 TNSA spectrum, qualitative behavior.

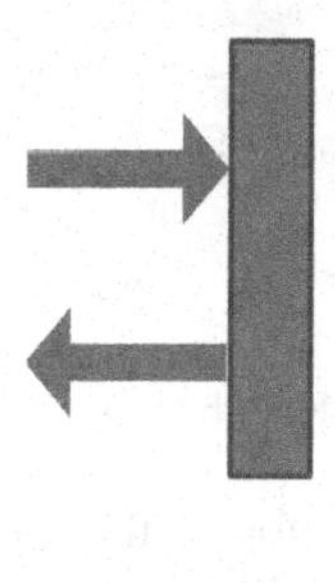

Figure 9.13 Radiation pressure acceleration concept.

Scaling rules predicted by TNSA models allow us to make projections (with large uncertainties) toward reaching a proton energy of around 200 MeV. These projections have shown that the required laser intensities are on the order of 10^{21} W/cm^2.

Various ways to improve TNSA have been considered, starting from the brute force method to increase the laser intensity, to more subtle methods involving the enhancement of the laser energy transfer to electrons and increase of the electron density. Reduction of foil thickness and reduced mass targets, enhanced coupling by a conically shaped target and use of nano-particle structured targets, among others, have all been used with some degree of success.

Despite the mentioned improvements, the TNSA mechanism still has a major disadvantage due to the shape of the spectrum of accelerated particles — the number of particles at the high end of the spectrum remains very low.

9.5.2 HOLE-BORING RADIATION PRESSURE ACCELERATION

The radiation pressure acceleration mechanism is based on the effects equivalent to *radiation pressure*, which light exhibits on a mirror when it reflects from its surface, as illustrated in Fig. 9.13. The radiation pressure for perfect reflection is proportional to the laser intensity I_L as

$$P_L = \frac{2\, I_L}{c} \qquad (9.6)$$

Similar pressure can be applied to a thin foil, upon whose surface plasma quickly forms. The radiation pressure effect is transmitted into the plasma by electrons via the ponderomotive force. Displaced electrons produce space charge that creates a steady pressure, which in its turn transfers the effect to the ions.

Two versions of radiation pressure mechanisms have received distinct names — the *hole-boring* and the *light-sail* mechanism.

In the case of hole-boring, the space charge due to electrons acts on ions that are pushed into the overdense plasma, initially compressing the foil and then pushing a region of the foil forward, as illustrated conceptually in Fig. 9.14.

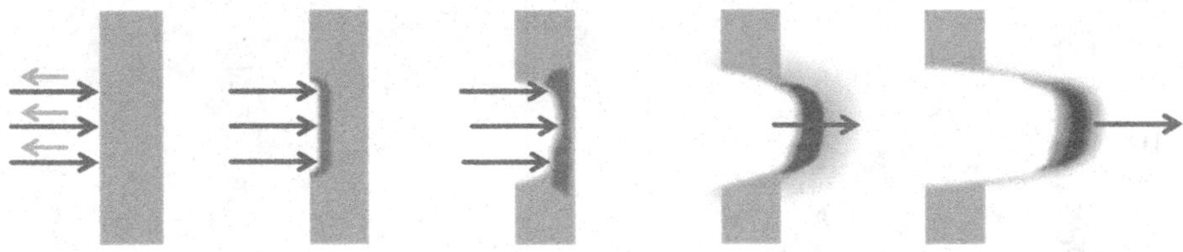

Figure 9.14 Hole-boring radiation pressure laser acceleration of protons.

The derivation of the approximate scaling rules for the hole-boring mechanism can be obtained by including the radiation light pressure into the fluid equation of motion, and transferring it to the reference frame where the shock is stationary (i.e., where the time derivatives are zero). This will yield $\rho u^2 = P_L$ and therefore the velocity of ions can be estimated as

$$u \approx \sqrt{\frac{I_L}{\rho c}} \tag{9.7}$$

Consider an example of a laser with $I_L = 10^{21}$ W/cm^2 shining on an aluminum target with $\rho = 2.7\,10^3$ kg/m^3. The resulting velocity of ions in this case is $u \sim 0.01c$ or about 3 mm/ps. The energy of the protons associated with this shock is 60 keV. By "bouncing" the stationary protons of the shock front, the protons can gain twice the velocity or four times the energy. Reducing the density of the foil could lead to a further increase of the energy of accelerated ions — the latter can be achieved by using gas targets.

The most attractive feature of the hole-boring radiation pressure mechanism is that the resulting proton beam has been demonstrated[11] to have a nearly monochromatic peak at the maximum energy.

9.5.3 LIGHT-SAIL RADIATION PRESSURE ACCELERATION

In the *light-sail*[12] regime, the radiation pressure mechanism is taken to the extreme when the foil is so light that it starts to accelerate immediately as a whole, as shown schematically in Fig. 9.15.

---

[11]C.A. Palmer et al., Phys. Rev. Lett., 106, 014801 (2011).

[12]T. Esirkepov et al. Phys. Rev. Lett., 92, 175003 (2004).

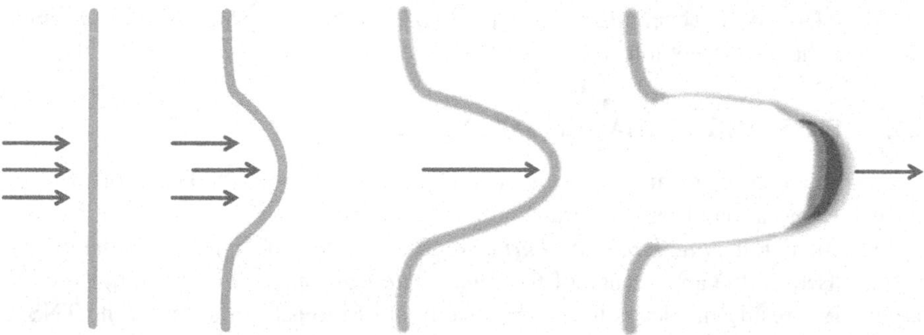

Figure 9.15 Light-sail radiation pressure laser acceleration of protons.

In a simple model of light-sail acceleration, one can assume that an area A is illuminated with an intensity I_L and if the light reflection is perfect, the force acting on this region of the foil is given by

$$F = 2A\frac{I_L}{\rho c} = m\frac{dv}{dt} = Ad\rho\frac{v_i}{t} \tag{9.8}$$

where m is the mass of this segment of the foil, $m = A\ d\ \rho$, with d being the foil thickness and ρ the density. The velocity of ions will therefore grow proportionally to the laser intensity and laser pulse duration:

$$v_i \approx \frac{2I_L\tau}{\eta\ c} \tag{9.9}$$

where $\eta = \rho\ d$ — the areal density and τ is the laser pulse duration. As we can see, the energy scaling in the light-sail regime is more favorable than in the hole-boring case, as in the non-relativistic case of light-sail acceleration the velocity of ions is proportional to I_L and the energy is thus proportional to I_L^2.

The light-sail radiation pressure mechanism cannot be completely decoupled from the competing TNSA mechanism. The electron heating — which was ignored in the simplified picture above — may cause foil deterioration. A possible improvement for the latter involves use of a circularly polarized laser, which reduces the effects of TNSA and foil heating.

The scaling of a light-sail mechanism for 10 PW pulses with laser intensity of 10^{22} W/cm^2 predicts[13] that GeV proton beams with good near-monochromatic spectral characteristics can be produced.

A Rayleigh–Taylor instability of the shape of the foil can develop during acceleration, resulting in deterioration of the resulted spectrum. However, this instability has been shown[14] to be stabilized by simultaneous acceleration of multi-species ions.

[13]B. Qiao et al., Phys. Rev. Lett., 102, 145002 (2009).

[14]B. Qiao et al., PRL, 105, 1555002 (2010), T. Pu Yu et al., PRL, 105, 065002 (2010).

Overall, the proton energy scaling as $(I_L \tau/\eta)^2$ for the light-sail radiation pressure regime is the most favorable and promising.

9.5.4 EMERGING MECHANISMS OF ACCELERATION

Various other mechanisms have been suggested that describe the behavior of laser plasma acceleration of ions in certain parameter ranges.

In particular, the *break-out afterburner* regime has been described[15] as the mechanism based on the appearance of the *relativistic transparency* of an initially solid target. As a result, this leads to the enhancement of ion acceleration via the TNSA mechanism.

Shock acceleration is a mechanism[16] based on the appearance of a high Mach number electrostatic shock in the overdense plasma created by the sharp front of the laser pulse. The propagating electrostatic shock will then reflect the plasma ions to the doubled velocity of the shock, resulting in an appearance of monochromatic proton peaks observed in experiments using overdense gas jet targets.

As we can see, while the number of forthcoming theoretical and experimental tasks and questions is still very high in this area, waiting for an inquisitive mind, there are mechanisms of proton plasma accelerations which, given the rate of the laser technology progress, can already yield the practically useful techniques for creating beam therapy facilities based on proton plasma acceleration.

9.6 GLIMPSE INTO THE FUTURE

Creating a more compact and affordable design of a proton therapy facility is a formidable task, one that is attracting the attention of many research teams worldwide. The plasma acceleration community, together with laser and conventional acceleration communities, are joining forces to solve this task.

Techniques of beam control and energy selection, along with methods of capturing divergent and chromatic beams, all developed alongside conventional accelerators and combined with novel opportunities enabled by modern and future lasers, should help us to find a way to create a viable design and a prototype system that can be used efficiently in practice.

In Chapter 12, we will briefly review various methods of beam control and manipulation. Some of these methods can be particularly applicable to the design of a plasma proton acceleration facility.

9.7 BOOSTED FRAME LWFA – INVENTION CASE STUDY

Plasma accelerator calculations take a lot of CPU time. Simulations need to include all important 3D effects and will need to propagate, through the simulation grid, the

---

[15] Yin et al., Phys. Rev. Lett. 107, 045003 (2011).

[16] L. Silva et al., PRL 92, 015002 (2004); D. Haberberger et al., Nature Phys., 8, 95 (2012).

laser or beam driver pulses, which can have tiny features of the order of laser wavelength, i.e. sub-micron, over long distances comparable to a fraction of a meter. The main challenge is coming, therefore, from the large ratios of characteristic lengths that need to be resolved in simulations — sub-micron features of the laser pulse and plasma details, and close to a meter length of the plasma channel. Correspondingly, the grids for numerical computation have to be dense and large (as illustrated in Fig. 9.16), and the required computing resources are enormous. This makes the accurate computations to be very difficult even at supercomputers.

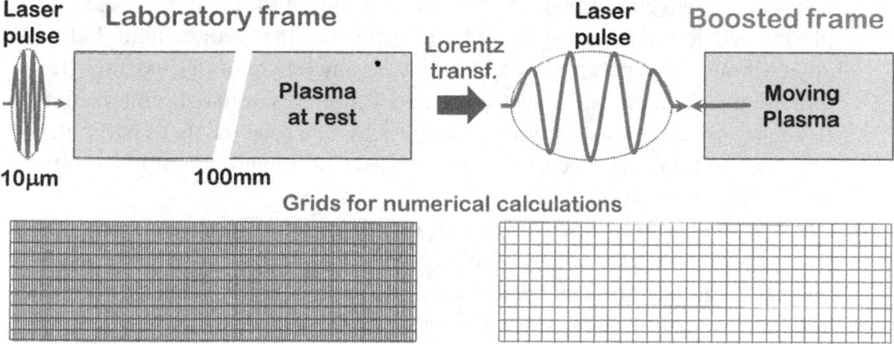

Figure 9.16 Boosted frame LWFA simulations approach.

A change of the reference frame is a common inventive principle in physics and in this case it can create really wonderful results. One can show[17] that there exists a relativistic reference frame, where the scale of the laser pulse length and the scale of the plasma channel length are comparable — as illustrated in Fig. 9.16. Correspondingly, the numerical grid does not need to be as dense, in this reference frame, and accurate computations are now much more feasible.

This approach of involving Lorentz relativistic transformation into an appropriate reference frame is called *boosted frame* plasma simulations. It allowed, for example, to perform simulations[18] of 10 GeV stages of 1 TeV electron-position LWFA collider — such simulations would not be possible without of the boosting frame approach.

Retrospectively, this very impactful approach can be connected to the inventive principle of "parameter change," where the relevant parameter is the corresponding relativistic factor of the boosted reference frame.

[17] J.-L. Vay, Phys. Rev. Lett. 98, 130405, (2007).

[18] J.-L. Vay et al., Physics of Plasmas 18, 123103 (2011).

EXERCISES

9.1. *Chapter materials review.*
Discuss the advantages and challenges of particle beam therapy in comparison to X-ray therapy.

9.2. *Chapter materials review.*
Discuss the key requirements for the laser and the target that may result in 200 MeV mono-energetic beams of protons in laser plasma acceleration.

9.3. *Mini-project.*
Discuss and develop a plan to create a 250-MeV proton source based on plasma acceleration, aiming for it to be applied in the medical field. Select approximate laser parameters and target parameters, and discuss their requirements. Discuss and select a method for energy monochromatization or energy collimation/selection. Describe why you selected these particular values of certain parameters (for target or laser, collimation or monochromatization system, etc.).

9.4. *Analyze inventions or discoveries using TRIZ and AS-TRIZ.*
Analyze and describe scientific or technical inventions described in this chapter (e.g., isochronous cyclotron) in terms of the TRIZ and AS-TRIZ approaches, identifying a contradiction and an inventive principle that were used (could have been used) for these inventions.

9.5. *Developing AS-TRIZ parameters and inventive principles.*
Based on what you already know about accelerator science, discuss and suggest the possible additional parameters for the AS-TRIZ contradiction matrix, as well as the possible additional AS-TRIZ inventive principles.

9.6. *Practice in reinventing technical systems.*
A FODO optical beamline system of focusing and defocusing magnetic quadrupoles can transport the beams of charged particles — for example, ions — stably and for a long distance. In an analogy with FODO, suggest a way to keep a charged ion, or a cloud of ions (that can be used as a target for laser plasma acceleration or as a *qubit* for a *quantum computer*), stationary and stably in a fixed location. Hint: apply the inventive principle *the other way around.*

9.7. *Practice in reinventing technical systems.*
A patient's motion during radiation treatment is one of the challenges that get in the way of providing accurate dose deliveries only to malignant tissues. Suggest a possible alternative way for providing the radiation dose to mitigate this challenge.

9.8. *Practice in the art of back-of-the-envelope estimations.*
Imagine that the beam size and angular spread of the ion beam produced by TNSA plasma acceleration are 10 μm and 0.1 rad and that the energy spread is 10%. Estimate where you would need to place the focusing lens with respect to the TNSA target to capture and focus such a beam. *(It is assumed that you can identify the most important effects playing roles in this task, can define the necessary parameters and set their values, and can get a numerical answer.)*

10 Beam Cooling and Final Focusing

We will now examine several topics related to beam cooling, phase space manipulation and final focusing of beams — the term usually applied to focusing beams before interaction point in colliders.

10.1 BEAM COOLING

Beam cooling methods are usually necessary for antiparticles, e.g., antiprotons $\bar{p}$, as they are produced on a target very "hot" — with large emittance and energy spread, or for intense bunches of protons or ions. The cooling methods are intended to decrease beam emittances as well as to counteract emittance growth due to effects such as intrabeam scattering.

10.1.1 ELECTRON AND STOCHASTIC BEAM COOLING

Electron cooling, invented by G.I. Budker and realized at Novosibirsk, consists of creating a region along the orbit of the beam where a "cold" electron beam will co-propagate together with a "hot" antiproton beam (see Fig. 10.1) at the same velocity. Energy exchange between the beams will eventually result in cooling of the $\bar{p}$ beam.

Another method of cooling the antiproton beams was proposed by S. van der Meer and realized at CERN. In this approach, a beam particle's betatron oscillation is detected by a pick-up electrode, and then the signal is amplified and sent via a short path across the ring onto a kicker (see Fig. 10.2). This will apply a kick to the same particle, resulting in a reduction of its oscillations. This method is called *stochastic cooling* and was indispensable for ensuring the discovery of W and Z bosons (Carlo Rubbia and Simon van der Meer, Nobel Prize in Physics, 1984).

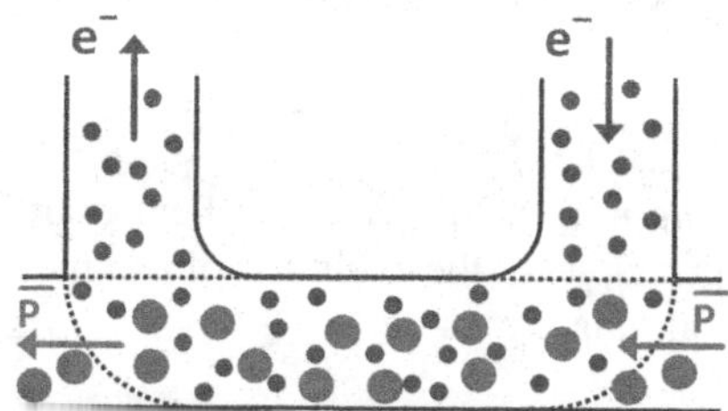

Figure 10.1 Electron cooling.

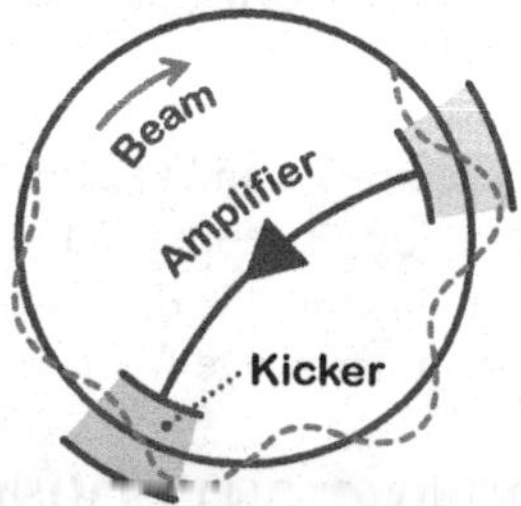

Figure 10.2 Stochastic cooling.

DOI: 10.1201/9781003326076-10

As it should be apparent from the description of stochastic cooling, this method can clearly have drawbacks in the case of a large number of particles in the beam, as the pick-up electrode would see signals from many particles, which would then smear individual contributions.

10.1.2 OPTICAL STOCHASTIC COOLING

"Standard" stochastic cooling entails sampling of the beam by electrostatic pick-up electrodes (see Fig. 10.3), and therefore its cooling rate is limited by the system bandwidth, which is defined by pick-up length. An extension of the method, *optical stochastic cooling*, uses optical pick-ups and optical amplifiers as shown in Fig. 10.4, resulting in potential increases of bandwidth and of the cooling rate by several orders of magnitude.

Figure 10.3 Standard stochastic cooling.

Figure 10.4 Optical stochastic cooling.

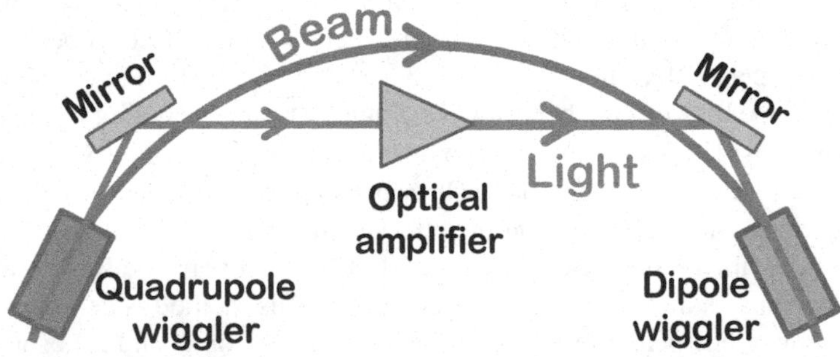

Figure 10.5 Layout of optical stochastic cooling system in an accelerator.

In optical stochastic cooling, quadrupole wigglers and dipole wigglers play the role of pick-up electrodes and kickers as shown in Fig. 10.5.

Beam cooling methods, and stochastic cooling in particular, are excellent examples that demonstrate the discussed AS-TRIZ principle — the evolution of technical systems from microwave frequencies into optical range.

10.1.3 IONISATION COOLING

The three main methods of beam cooling (apart from Synchrotron Radiation-based cooling) are shown conceptually in Figs. 10.1, 10.2 and 10.5.

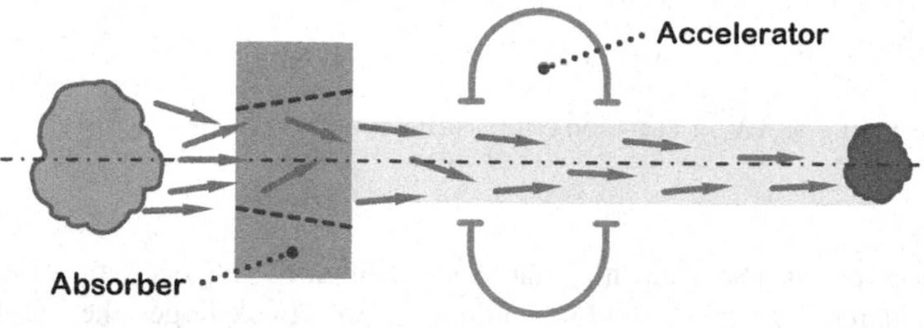

Figure 10.6 Ionization cooling concept.

In addition to electron and stochastic cooling, which we have already discussed, we add *ionization cooling* — which is based on subjecting particles to ionization losses in an absorber, and then restoring the longitudinal component of the particles' momentum as illustrated in Fig. 10.6.

Stochastic and electron cooling are well-developed techniques. In particular, stochastic cooling was essential for the discovery of W and Z Bosons (C. Rubbia and S. van der Meer, 1984 Nobel Prize in Physics). Electron cooling was most recently used for improving the operation of the Tevatron collider.

Conversely, ionization cooling is a concept still under development. Conceptually simple, it is quite challenging technologically. However, this technique might be the only way to reduce the emittance of short-lived particles such as muons.

10.1.4 COOLING RATES ESTIMATE

Let's estimate the cooling rates for stochastic and electron cooling.

For stochastic cooling estimates, let's assume that there are N particles uniformly distributed around the ring, that the revolution time around the ring is T, and that the pick-up and amplifier electronics has a bandwidth W. Correspondingly, an off-axis particle shown in Fig. 10.2 gives a signal with a duration of

$$T_s \approx \frac{1}{W}$$

where the index of T_s stands for "sample" and indicates the smallest fraction of the beam that can be observed:

$$N_s = \frac{NT_s}{T} = \frac{N}{WT}$$

Let's now consider a test particle with a transverse coordinate x to which an idealized stochastic cooling system will apply a correction $-\lambda x$ after one turn. Other particles in the sample will also contribute; therefore, the corrected coordinate of the test particle after one turn will be equal to

$$x_c = x - \lambda x - \sum_{S'} \lambda x_i$$

where the summation S' covers all other particles in the sample. We can rewrite it as

$$x_c = x - \lambda N_s \langle x \rangle_s \equiv x - g \langle x \rangle_s$$

where the gain $g \equiv \lambda N_s$. Let's then express the evolution of $\delta \langle x^2 \rangle_s = \langle x_c^2 - x^2 \rangle_s$ as

$$\delta \langle x^2 \rangle_s = -(2g - g^2) \langle x \rangle_s^2$$

We can see from the above equation that the maximum cooling decrement is achieved when the gain $g = 1$, or $\lambda_{\text{opt}} = 1/N_s$, and if the gain is twice larger, then cooling will turn into heating. In this case of optimal gain, the stochastic cooling decrement will be equal to

$$\frac{1}{\tau} \approx \frac{W}{N} \tag{10.1}$$

For example, if the bandwidth W is 10 MHz and the total number of particles is 10^7, the cooling time will be around a second. We can also see from the above expression that moving from MHz and GHz bandwidth range into THz range, as is possible for *optical stochastic cooling*, will drastically shorten the cooling time or cool a much denser beam.

For electron cooling estimates, let's consider the extreme case when the temperature of the electron beam with density n_e is ultimately low. This can be achieved in the case when the electron beam is guided by a homogeneous magnetic field and the transverse temperature of electrons is excluded from consideration, since the Larmor circles of electrons are much smaller than the average distance between electrons $1/n_e^{1/3}$ and collisions are adiabatic in comparison with Larmor motion. In this case of strong magnetization, the longitudinal temperature of electron beam will be defined by the initial temperature of the cathode and by the mutual repulsion of electrons:

$$T_{||} \approx \frac{T_c^2}{2E_k} + e^2 n_e^{1/3} \tag{10.2}$$

where E_k is electron energy after acceleration and T_c is the cathode temperature. The first term is typically much smaller than the second one, and the spreads of longitudinal velocities in the electron beam can be estimated as

$$v_e \approx (e^2 n_e^{1/3}/m)^{1/2}$$

In this case, the maximum friction force experienced by a heavy charged particle moving with a velocity of around v_e should approach

$$F_{\max} \approx e^2 n_e^{2/3} \tag{10.3}$$

For ion velocities lower than v_e, the friction force should increase linearly, and for higher velocities, it should decrease as $1/v^2$, which can be approximated as

$$F = F_{\max} \frac{v}{v_e} \quad \text{and} \quad F = F_{\max} \frac{v_e^2}{v^2} \tag{10.4}$$

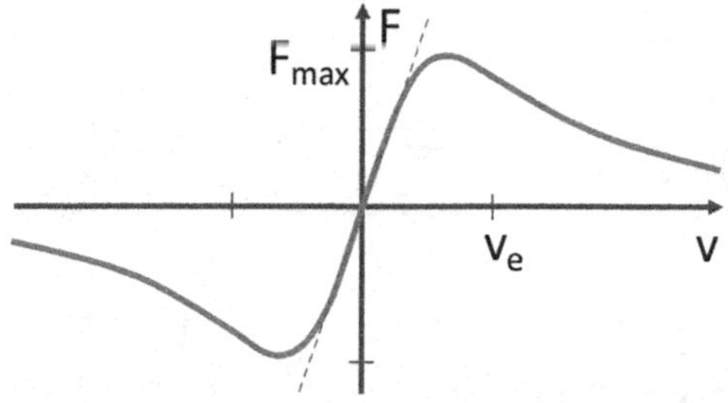

Figure 10.7 Approximate behavior of magnetized electron cooling friction force.

The ultimate performance of a magnetized electron cooling friction force is defined by the longitudinal temperature of electron gas and can reach the value of $F_{\max} \approx e^2 n_e^{2/3}$ for positive ions, and about a three times higher value for negative ions.

or, very approximately

$$F \approx F_{\max} \frac{\pi v_e^2 v}{(v^2 + v_e^2)^{3/2}} \tag{10.5}$$

which is illustrated in Fig. 10.7. See references[1] for accurate formulae that take into account longitudinal and transverse velocities of ions, as well as the intermediate regime of magnetization.

The maximum cooling rate for a particle with mass M, in cases where the cooling happens in the linear part of the friction force, can be correspondingly estimated as

$$\frac{1}{\tau} \approx \frac{F_{\max}}{M v_e \eta} \tag{10.6}$$

where η is the fraction of accelerator circumference taken by the electron cooler.

a) e^- → → H^+ b) e^- ⇄ H^-

Figure 10.8 Magnetized electron cooling interaction for positive and negative ions.

Magnetized cooling with low temperatures of electron beams and a low relative velocity of ions exhibits one more interesting feature — a difference between the friction forces for positive and negative ions. Recall that, for classic Rutherford scattering, the formulae for energy loss depend on the charge of particle squared; i.e., they do not depend on the sign of the charge of the particle. It is not the case for magnetized electron cooling, where an electron moving along the magnetic field line can get reflected by a negative ion, as illustrated in Fig. 10.8, resulting in a large momentum transfer, while when passing by a positive ion, the electron will first accelerate and then slow down, producing a much smaller momentum transfer. As a result, the maximum friction force for negative ions can be a factor of several higher

[1] See, e.g., "Electron cooling: physics and prospective applications," V.V. Parkhomchuk and A.N. Skrinsky, Rep. Prog. Phys., 54, 919 (1991)

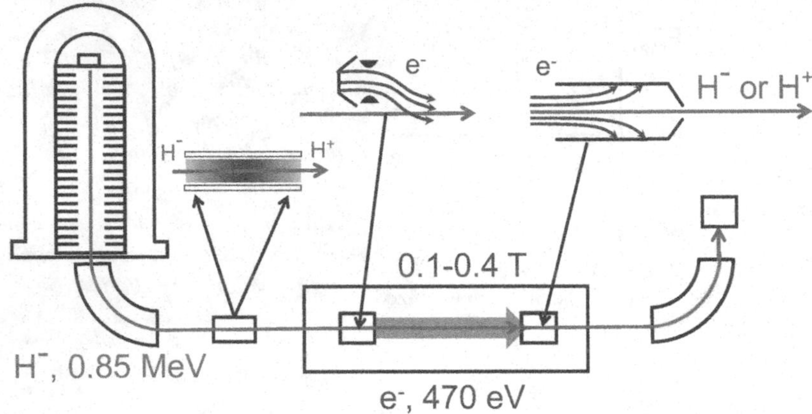

Figure 10.9 Experimental installation for studies of one-pass electron cooling and charge dependence.

than F_{max} defined above. One of the authors is grateful for colleagues with whom he had the opportunity to study the ultimate possibilities of electron cooling[2] at the experimental installation illustrated in Fig. 10.9.

10.1.5 ELECTRON COOLING, ELECTRON LENS AND GABOR LENS

In order to stimulate our TRIZ-inspired discussion, we will now draw a parallel between the three different techniques, which have completely different purposes but have, nevertheless, similarities.

Electron cooling, as we discussed, uses an electron beam co-propagating with the proton beam with equal velocities in order to enable their energy exchange. The electron beam is guided by a magnetic field. It is essential for the e and p velocities to be equal. The conceptual layout of an electron cooler is shown in Fig10.10.

An electron lens[3] is geometrically a very similar device wherein the proton (or antiproton) beam passes through the electron beam and receives additional focusing due to the fields of the electron beam; the lens aims to mitigate variation of betatron tunes between different bunches of proton beams. As the main effect of the lens comes from the electric field of the electron bunch, the direction of the e-beam travel is not essential (usually, counter-propagating beams are used). The e-lens concept is shown in the same Fig. 10.10.

We will also mention one more arrangement involving an electron beam and a magnetic field — the *Gabor lens*. Suggested[4] in 1947, this technique recently

[2]N.S. Dikansky et al., "Ultimate possibilities of electron cooling," Novosibirsk Preprint INP 88-61, 1988.

[3]V. Shiltsev et al., PR-ST-AB, 2, 071001 (1999).

[4]D. Gabor, Nature 160, 89–90 (July 19, 1947).

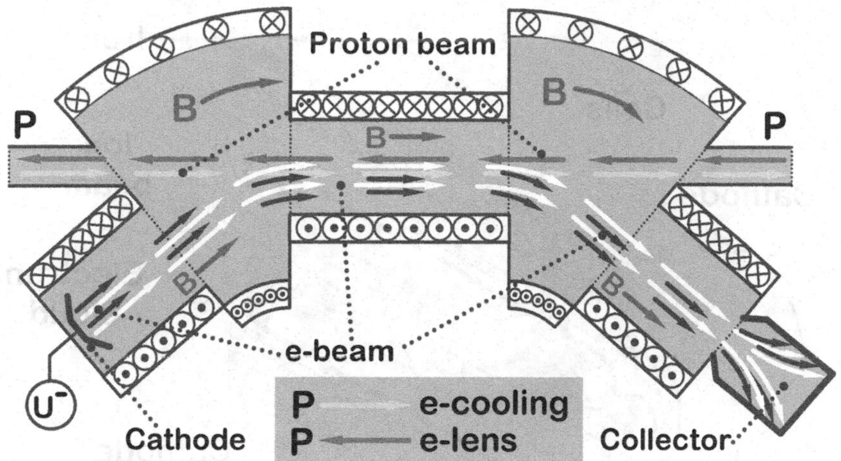

Figure 10.10 Electron cooling or electron lens.

attracted attention[5] in connection to its use for laser plasma ion acceleration, for energy selection of accelerated protons.

A conceptual schematic of a Gabor lens is shown in Fig. 10.11. In this lens, the electron beam is formed by a cathode with a hole in the center. Electrons trapped in the center of the lens are attracted to the anode and are contained by the magnetic field.

A Gabor lens can potentially accumulate a large electron charge. In a steady state, the electrons rotate around the axis and the electrostatic repulsion — together with the centrifugal force — will balance the radial Lorentz force produced by the magnetic field. Using this assumption, the maximum density of electrons in the Gabor lens can be estimated as

$$n = -\frac{B^2}{8\pi m_e c^2} \tag{10.7}$$

which can be high enough to be considered for use as a lens created by the radial field of the stored electrons.

The electrons stored in the Gabor lens move longitudinally in both directions. In Fig. 10.12, we summarize the relationships between the velocities of the proton (antiproton, ion) beams and the velocities of electrons in these three devices.

In the above, we have considered three systems that involve electron beams or stored electrons. These systems are aimed at different applications and details of their functions are very different. We assume, however, that it is useful to look at these systems simultaneously, as discussion of analogies between these systems is helpful for our TRIZ analysis (which runs in parallel with the main accelerator-laser-plasma story of this book).

---

[5]J. Pozimski et al., Laser and Particle Beams, v. 31, 04, pp. 723–733, (2013).

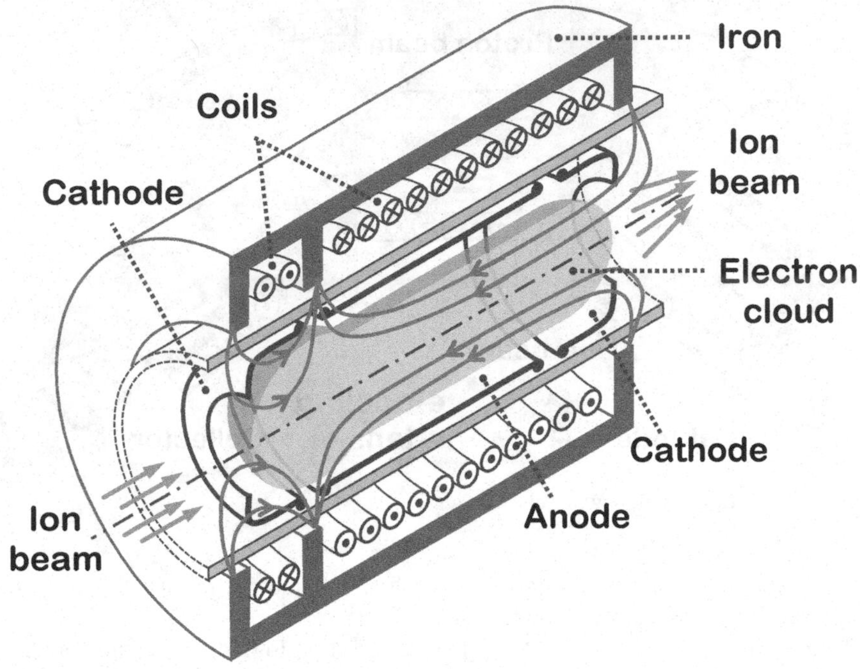

Figure 10.11 Conceptual schematic of a Gabor lens.

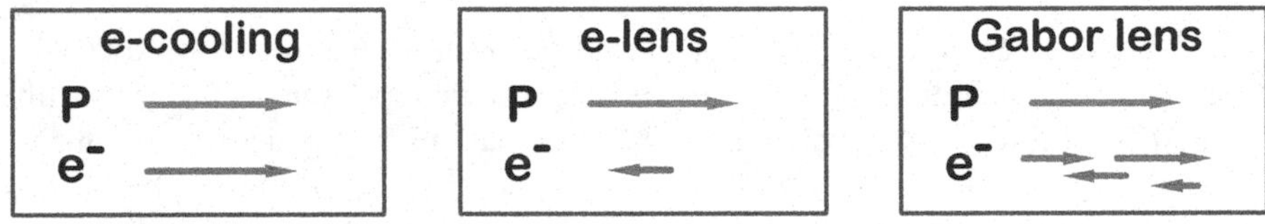

Figure 10.12 Relations of velocities of proton and electron beams in different configurations: electron cooling, electron lens, Gabor lens.

10.1.6 LASER COOLING

We conclude this section on cooling methods with laser cooling of ions. The method relies on the *Doppler frequency shift*, which is the key to laser cooling.

The cooling process involves the following steps (see Fig. 10.13). First, the laser photon coming from a certain direction hits the atom and is absorbed. The excited atom then re-emits a photon into a random direction.

The laser is in resonance with the atoms only when they are moving toward the laser, but not if they are moving sideways or away as shown in Fig. 10.14. This condition ensures the eventual cooling of the ions.

Figure 10.13 Laser cooling steps. Absorption of a photon by an atom (a); excited state of the atom (b); emission of a photon (c).

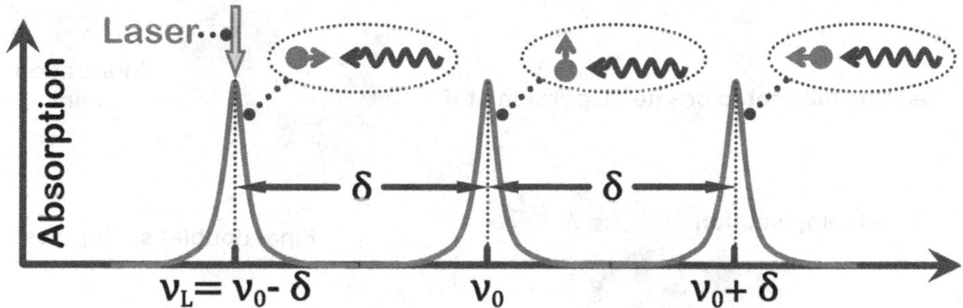

Figure 10.14 Relation between laser wavelength and Doppler shifted resonance absorption of an atom moving in different directions.

10.2 LOCAL CORRECTION

In accelerators — as well as in any other technical fields — any unwanted disturbances are corrected either locally or non-locally. The universality of the approaches can also be observed with help from TRIZ, which includes relevant inventive principles (principle of *preliminary anti-action*).

Non-local correction is often used when it is not possible to correct the disturbance at its origin. The general issue with this approach lies in its non-locality — a correction needs to propagate and be properly preserved from the point of preliminary correction to the point where it needs to act.

Local corrections, if they can be used, are often superior, as they correct the disturbance at the origin. In this section, we will discuss several examples of local corrections.

10.2.1 FINAL FOCUS LOCAL CORRECTIONS

A *final focusing (FF) system* of any collider aims to produce a small beam size at the *interaction point* (IP). The final lenses of the system (usually arranged in a *final doublet* — FD) are the strongest and produce the largest chromaticity. The values of chromaticity are usually very large and need to be compensated for (if left uncorrected, they increase the beam size at the IP by orders of magnitude).

For example, let's assume that the final focusing system illustrated by Fig. 2.20 has $L^* = 5$ m, beta function at the IP $\beta^* = 0.1$ mm and the energy spread $\sigma_E = 0.5\%$. In this case, the chromatic dilution (see Eq. 2.56) is $\Delta\sigma/\sigma \approx \sigma_E L^*/\beta^* \approx 250$, i.e., the beam size is largely dominated by the chromaticity of the final lenses and must be compensated for.

In a traditional final focus design, the chromaticity correction is performed in dedicated optical sections upstream from the IP. In these sections, sextupole magnets are installed in dispersion sections and create some chromaticity there, which will then propagate to the FD, where they will cancel the FD chromaticity.

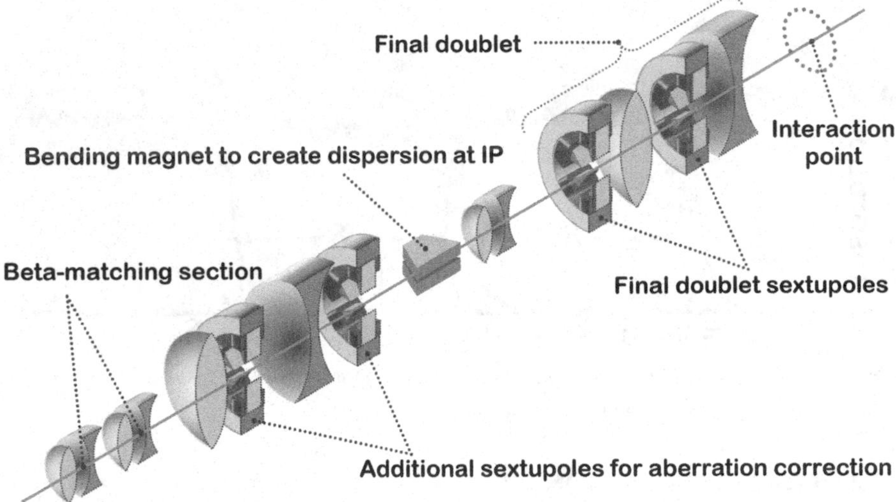

Figure 10.15 Final focus with local chromaticity correction.

The performance of such an FF with non-local chromaticity compensation is often limited, as various additional disturbances (e.g., the SR-generated energy spread), which occur between the chromatic correction section and the FD, can alter the cancellation conditions.

An alternative design[6] exists: a final focus with a local chromatic correction, shown conceptually in Fig. 10.15.

In this design, chromaticity is canceled locally by two sextupole magnets interleaved with FD, while an upstream bend generates the dispersion (which is necessary for the sextupoles to generate chromaticity) across the final doublet. The value of dispersion in the FD is usually chosen so that it does not increase the beam size in the FD by more than 20–30% for a typical energy spread of the beam.

Local chromatic compensation by FD sextupole magnets needs to compensate for the FD chromaticity without introducing other unwanted aberrations. In a vertical plane, the chromaticity compensation is straightforward, and therefore, the vertical FD sextupole magnet should be tuned to completely cancel the vertical FD chromaticity.

However, the FD horizontal sextupoles also introduce horizontal second-order dispersion. Second-order dispersion is produced by FD quadrupoles as well as FD sextupoles, but the latter produce just half as much of second-order dispersion as the

[6]P. Raimondi and A. Seryi, Phys. Rev. Lett., 86 (17), 3779, (2001).

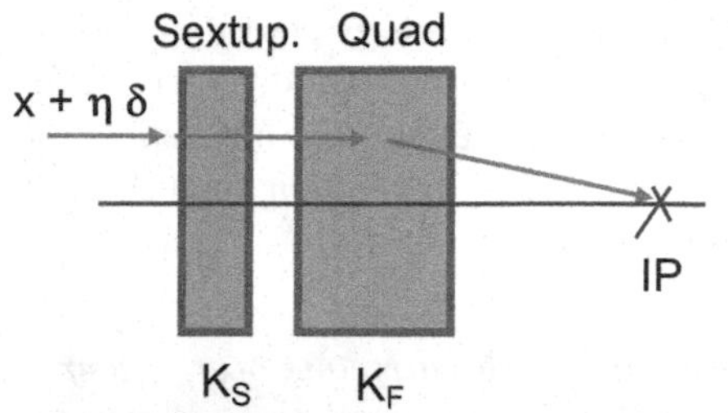

Figure 10.16 Final quadrupole with local chromaticity correction sextupole.

Local chromaticity correction is an illustration of the inventive principle of cancelling one harmful effect with another harmful effect and, since part of the second-order dispersion needs to be brought from upstream, of the principle of beforehand cushioning.

FD quadrupole. Therefore, if we tune the FD horizontal sextupole to cancel the FD chromaticity, Half of the uncompensated-for second-order dispersion will remain.

To illustrate this, let's write the horizontal angular kick for a particle with betatron coordinate x and with an additional offset due to dispersion η and energy offset δ. The kick due to the quadrupole with strength K_F:

$$\Delta x' = \frac{K_F}{(1+\delta)}(x+\eta\delta) \quad \text{which is} \quad \approx K_F(-\delta x - \eta\delta^2) \tag{10.8}$$

and the kick due to the sextupole with strength K_S:

$$\Delta x' = \frac{K_S}{2}(x+\eta\delta)^2 \quad \text{which is} \quad \approx K_S(\delta x + \frac{1}{2}\eta\delta^2)$$

We can see from the above expressions that if we require $K_S\eta = K_F$, then the horizontal chromaticity will cancel out, but half of the second-order dispersion (term with δ^2) will remain uncanceled.

A solution to this issue consists of allowing the beta-matching section (shown schematically in the beginning of the beamline in Fig. 10.15) to produce as much horizontal chromaticity as the final doublet, so that the horizontal FD sextupoles will run twice as strong and simultaneously cancel the second-order dispersion and horizontal chromaticity.

This can be expressed via the effective strength of β-matching quadrupole strength $K_{\beta-\text{match}}$ added to the equation Eq. 10.8 as follows:

$$\Delta x' = \frac{K_F}{(1+\delta)}(x+\eta\delta) + \frac{K_{\beta-\text{match}}}{(1+\delta)}x \quad \text{which will become} \quad \approx K_F(-\delta x - \frac{1}{2}\eta\delta^2)$$

if we demand that $K_{\beta-\text{match}} = K_F$. And, we can also see, that if we request $K_S\eta = 2K_F$, then both the horizontal chromaticity and second-order dispersion will cancel out, while the horizontal chromaticity and horizontal sextupole strength are twice higher. However, since for a typical final focus with a flat beam the horizontal chromaticity is much smaller than the vertical one, such an increase is acceptable.

Sextupoles added to the focusing structure will also introduce higher-order aberrations. Geometric aberrations of the final doubled sextupoles can be canceled in this design by two more sextupoles placed in phase with the FD sextupoles and upstream from the bend, which generates dispersion.

The final focus system can be reversed and the IP considered as an entrance point that captures a strongly diverging beam, such as the one coming out of the laser plasma accelerating bubble. This beam will also have an energy spread and associated chromaticity effects in the capture optics. A very small but divergent beam corresponds to a small beta function at the origin; therefore, the chromatic effect for such a beam should be significant. The local chromaticity correction approach can thus also be applicable to laser-plasma capture optics.

Inverted final focus system with local correction — the principle of the other way around — can be used to capture high-divergency beam coming out of a plasma acceleration bubble.

10.2.2 INTERACTION REGION CORRECTIONS

The interaction region of linear colliders exhibits many design inventions that can illustrate several TRIZ inventive principles.

A typical layout of a detector and beamlines is shown in Fig. 10.17. One can see here that an anti-solenoid is inserted into the beamline. This anti-solenoid is needed to cancel the coupling effects that the main solenoid produces on the beam.

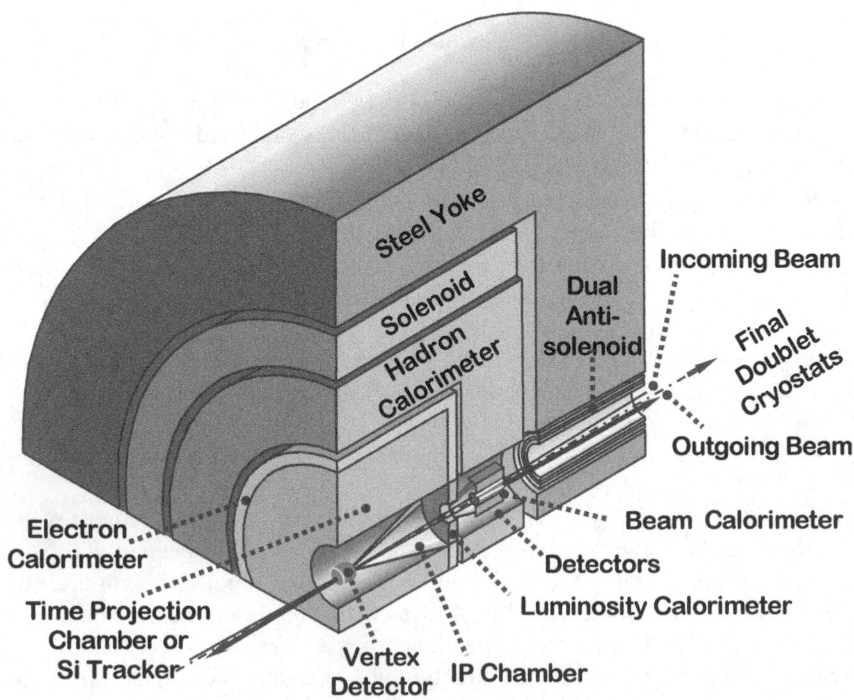

Figure 10.17 Conceptual layout of experimental detector and beamlines in the interaction region of a linear collider.

The problem with the anti-solenoid is that a huge force will be exerted on it because of the main solenoid. A solution implemented[7] in IR design is to use a dual anti-solenoid, where the outer coil with current $I_2 = -I_1 (R_1/R_2)^2$ cancels out the external field; this makes it force-neutral (see Fig. 10.18).

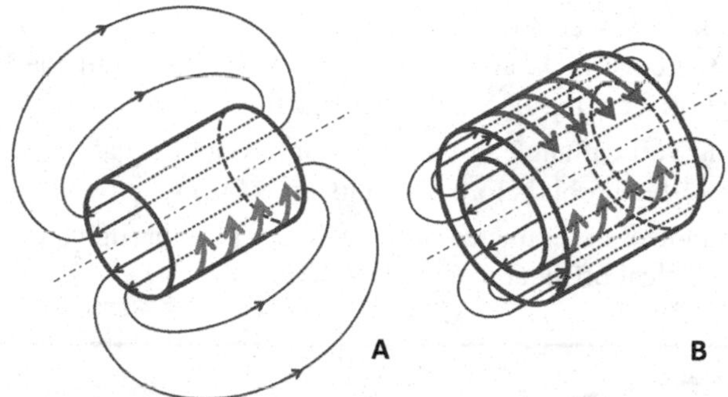

Figure 10.18 Standard solenoid (A) and interaction region dual solenoids (B).

From the point of view of TRIZ, the dual solenoid, where the external solenoid cancels all fields outside of the system, is an example of application of the combination of two inventive principles: the "nested doll" and the system/anti-system inventive principles.

A solenoid and its fields produce a lot of interesting effects on the beam that need to be compensated for locally. One of the examples is the compensation for the anomalous beam-size growth induced by the overlap of the solenoid field with the fields of the final doublet quadrupoles, as illustrated in Chapter 15 — see inventive principle #16 "partial or excessive action." This anomalous beam-size growth can be compensated for locally[8] almost entirely by shaping the solenoid field with the help of a weak anti-solenoid (exemplifying the "partial action" of the inventive principle).

Let's consider one more type of local correction in the interaction region: in this case, the correction of the trajectories of the beam which collide inside of the detector solenoid with a crossing angle.

With a crossing angle, when beams cross the solenoid field, a vertical orbit will arise. For an e+e- case, the orbit is anti-symmetrical and the beams still collide head-on, along the S-like vertical orbit bump around the IP, as shown in Fig. 10.19.

Let us consider a detector solenoid with sharp edges. In the case without compensation, the vertical deflection is caused by the edge kick

$$\Theta = \frac{\hat{\theta}_c B_0 L}{2B\rho} \tag{10.9}$$

[7]B. Parker, ca. 2002. A similar solution is used in nuclear magnetic resonance (NMR) scanners.

[8]Y. Nosochkov, A. Seryi, Phys. Rev. ST Accel. Beams, 8, 021001, (2005).

which occurs when the beam enters the solenoid off-axis at $\hat{\theta}_c L$, and also by the kick linearly distributed in the body of the solenoid. Here, $\hat{\theta}_c$ is half of the crossing angle (note that, elsewhere in the book, the full crossing angle is defined as θ_c, i.e. $\hat{\theta}_c = \theta_c/2$), L is the half-length of the detector solenoid, B_0 is the solenoid field, and $B\rho = pc/e$ is the magnetic rigidity of the beam.

The body kick integrated from the solenoid entrance to the IP is equal to -2Θ, which is twice the edge kick, and since the body kick has half the lever arm, the resulting vertical offset at the IP cancels out exactly.

The remaining vertical angle at the IP is nonzero and equals $-\Theta$. The maximal deviation of the vertical orbit before the collision is $\Theta L/4$.

The vertical angle of the extracted beam, which passes through the entire solenoid, is -2Θ and the vertical offset at the exit is $-3\Theta L$.

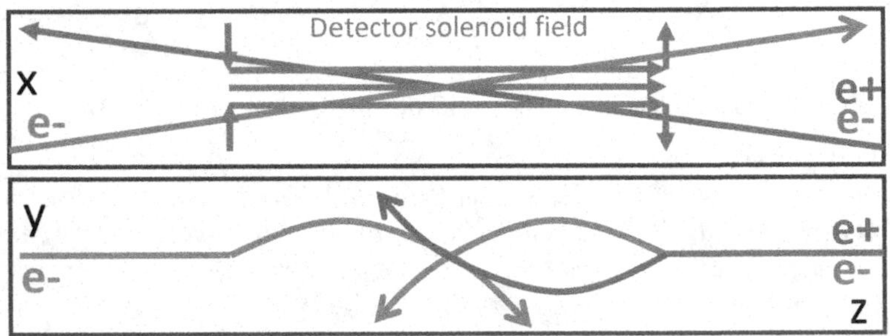

Figure 10.19 Collision of e+e- or e-e- beams in detector solenoid field with crossing angle.

In certain situations, the vertical angle of the orbit may be undesirable (e.g., to preserve the polarized beam spin orientation, or to preserve the e-e- or corresponding γ-γ luminosity), this angle can be compensated for locally[9] with the so-called Detector Integrated Dipole (DID), which introduces a sine-like transverse field inside of the detector (the transverse field is produced by additional coils integrated into the main detector solenoid).

Acting together with external orbit kicks, the DID can compensate for the vertical orbit angle at the IP for either e+e- or e-e- collisions, as illustrated in Fig. 10.20.

In cases where the vertical orbit angle at the IP is of no concern, an alternative exists; i.e., the negative polarity of DID may be used to reduce the angular spread of the low-energy e+e- pairs produced during collisions. This is the so-called anti-DID case. In this case, the vertical angle at the IP is somewhat increased, but the background conditions due to low-energy pairs can be improved, since the pairs will be directed, by the combined solenoid and anti-DID field, into the exit apertures of the focusing elements.

[9]B. Parker, A. Seryi, Phys. Rev. ST Accel. Beams 8, 041001, (2005)

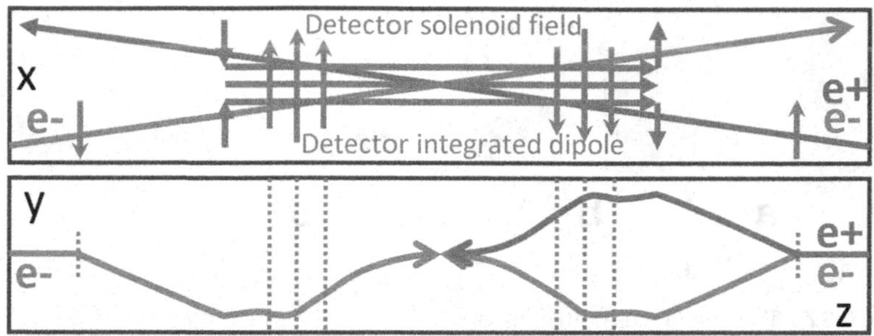

Figure 10.20 Collision of e+e- or e-e- beams in detector solenoid field with crossing angle when the orbit angle is locally compensated for by the Detector Integrated Dipole.

The compensation of crossing-angle collisions beam orbit with Detector Integrated Dipole is yet another example of the application of the combination of invention principles system and anti-system and "nested dolls."

10.2.3 TRAVELING FOCUS

The traveling focus regime[10] of beam collisions employs beam–beam focusing strengths in order to overcome the hourglass effect.

The *hourglass effect* (see Fig. 10.21) prevents reduction of the beta function at the IP to values lower than the length of the bunch, as further increases in focusing distort the bunch (into an hourglass shape) without increases to the luminosity.

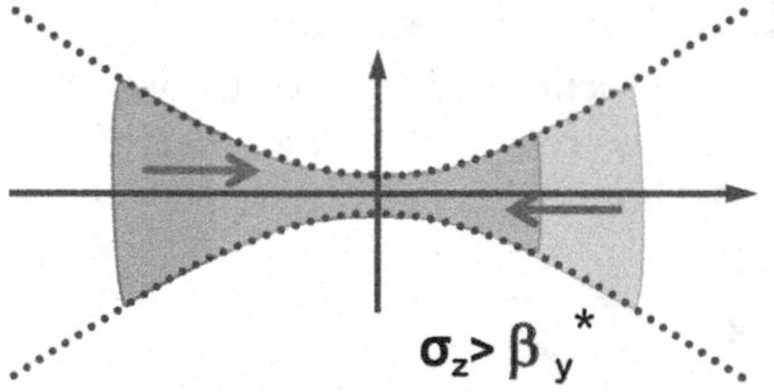

Figure 10.21 Hourglass effect.

Focusing force of e+ and e- bunches can provide additional local focusing at the IP for avoiding hourglass reduction of the luminosity — application of the inventive principle of using resources that already exist in the system.

In a traveling focus regime, the beam is focused to $\beta_y \ll \sigma_z$ and the location of focus dynamically changes during the collision. The focal point of the colliding bunches is made to coincide with the location of the head of the opposite bunch. This helps to optimally use additional focusing due to beam–beam forces and keeps the beams properly focused on each other during the entire collision.

Fig. 10.22 shows a simulation of traveling focus. The arrows show the position of the focus point during collision. This method hasn't yet been experimentally tested.

[10] V. Balakin, ca. 1991.

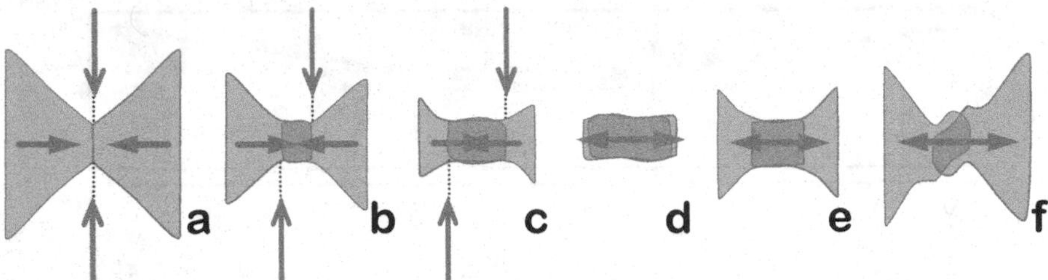

Figure 10.22 Traveling focus collisions.

One of the particular difficulties to this method consist of an increased sensitivity to imperfections, i.e., to the initial beam offset. This example gives us, however, additional information for a TRIZ-like analysis of the local correction approaches.

10.2.4 CRABBED COLLISIONS

The interaction region shown in Fig. 10.17 involves a collision of beams with a certain nonzero crossing angle θ_c. With crossing angle θ_c, the projected x-size is

$$\sqrt{\sigma_x^2 + \theta_c^2 \sigma_z^2} \approx \theta_c \sigma_z$$

Taking the IP beam parameters shown on Fig. 11.2 and taking $\theta_c \approx 15$ mrad, we can conclude that the projected horizontal size is equal to several micrometers — which is several times larger than the nominal size. This is illustrated in the upper part of Fig. 10.23, where incomplete overlaps of the beams are apparent.

The crossing angle collision will result, therefore, in a substantial (by several times) reduction in luminosity, unless this effect is locally compensated.

Compensation of the crossing angle effect can be achieved by giving the bunch a z-correlated kick in such a way that the beam starts to rotate in the horizontal plane and arrives at the IP properly overlapping with the opposite beam — as shown at the bottom part of Fig. 10.23.

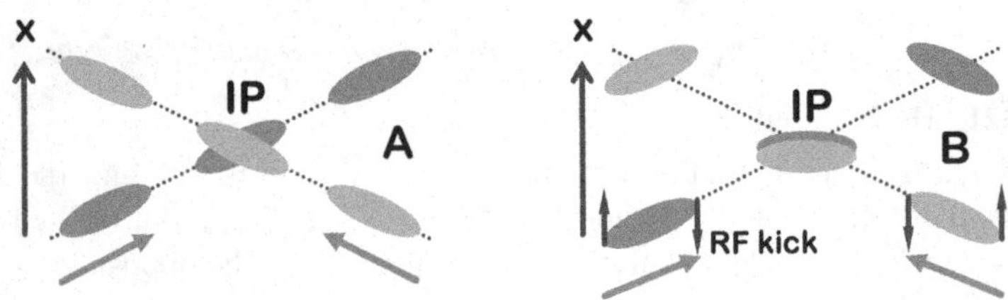

Figure 10.23 Normal (A) and crabbed (B) IP collisions of the beams with crossing angle.

The z-dependent kick on the bunch can be produced by a special transverse mode cavity called a *crab cavity*, which does not disturb the central particle of the beam, but kicks the head and the tail particles in opposite directions in the x-plane. The design of a crab cavity and its fields are shown in Fig. 10.24.

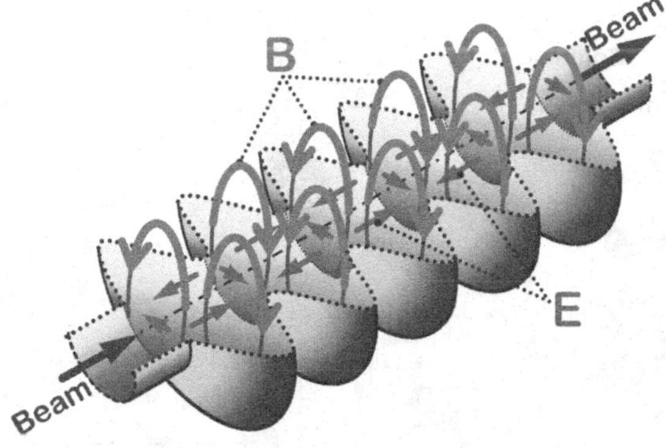

Figure 10.24 Crab cavity and its fields.

Crab cavity kicks can be applied either locally, when two cavities are placed around the IP and the crabbed motion will exist only between them, and can also be applied globally — when the crabbed motion will propagate around the entire ring in the form of a closed crabbed orbit.

10.2.5 ROUND-TO-FLAT BEAM TRANSFER

In the final section of this chapter we will discuss round-to-flat or flat-to-round beam transformation, suggested by Y. Derbenev.[11]

This method is primarily intended for the beams that have significant asymmetries between phase planes. In particular, flat beams are often generated in accelerators. For example, SR rings naturally have y emittance much smaller than the x emittance.

It has been shown by Derbenev that a triplet of skew quadrupoles can transform such a flat beam to a vortex, as illustrated in Fig. 10.25.

The beam with vortex-like phase space distribution can be transformed further, noting that an edge of the solenoid also creates a vortex-like phase space portrait for an initially parallel beam.

Therefore, sending the vortex beam created by the skew triplet into the solenoid with appropriate strength will result in creating a round beam with zero angles — as shown in the final step in Fig. 10.25.

Beam which is absolutely flat (zero size, nonzero angular spread) has zero emittance. The perfect vortex beam, such as after the skew-quad triplet of Derbenev's transformation, is also formally zero emittance beam. The edge of the solenoid transforms this vortex beam into exactly parallel beam inside of the solenoid, which is also zero emittance. From inventive principles viewpoint, the skew-quad triplet and the edge of the solenoid represent the system and its anti-system.

The transformation described above can be used in many accelerator systems. A particular example[12] uses this transformation to reduce space charge effects of flat beams in damping rings with long straight sections by employing flat-to-round and

[11] Ya. Derbenev, Adapting Optics for High Energy Electron Cooling, University of Michigan Report No. UM-HE-98–04, 1998.

[12] R. Brinkmann, Ya. Derbenev and K. Floettmann, Phys. Rev. ST Accel. Beams 4, 053501 (2001).

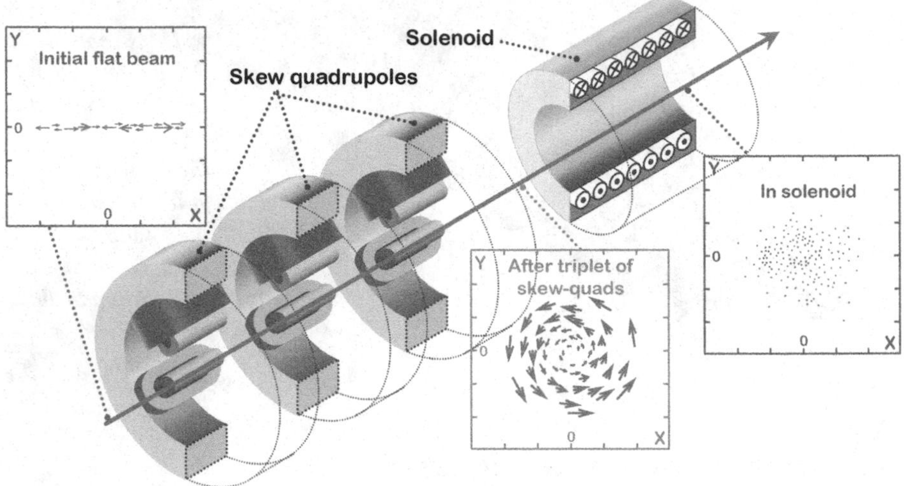

Figure 10.25 Beamline magnetic elements and phase-space portraits of the beam subjected to flat-to-round beam transformation. Initial flat beam, vortex, parallel beam in the solenoid.

round-to-flat transformations at the beginnings and exits of each section, respectively. The round beam in straight sections will have lower space charge effects, improving the beam stability.

10.3 LOCAL CHROMATIC CORRECTION – INVENTION CASE STUDY

Final focus local chromatic correction is a relatively recent invention. The focusing systems designed and implemented before said invention were based on non-local chromatic corrections. In this case, chromatic correction sextupoles were placed far upstream of the IP. Sextupole magnets were placed in pairs with a $M = -1$ matrix between them. In cases where dispersion functions are symmetrical in each sextupole of the pair (as shown in Fig. 10.26), the chromatic correction from the pair doubles, while most of the higher-order geometrical aberrations introduced by sextupoles would cancel out.

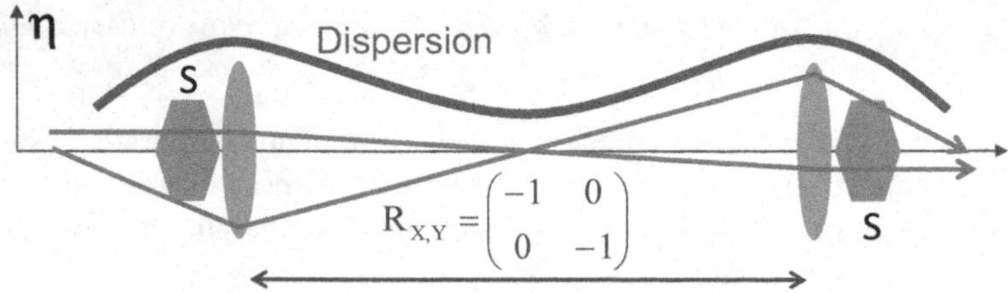

Figure 10.26 Sextupole pair with matrix $M = -1$ between them and symmetrical arrangement of dispersion was typically used for non-local chromatic correction.

In classic final focus systems with non-local chromatic correction, sextupole pairs were placed far upstream of the final doublet (FD), and the chromaticity W_y created in the pair would propagate for the entire length of the final focus L_{FF}, before canceling out in the FD; see Fig. 10.27. While this reminds us the inventive principle of preliminary anti-action, the FF length can be quite long (up to 1–2 km for high-energy e+e- colliders), and the preservation of such pre-compensation is a challenge.

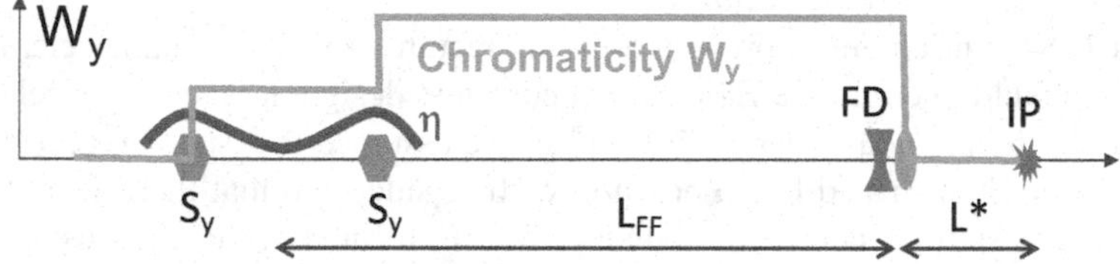

Figure 10.27 Classic final focus system with non-local chromatic correction.

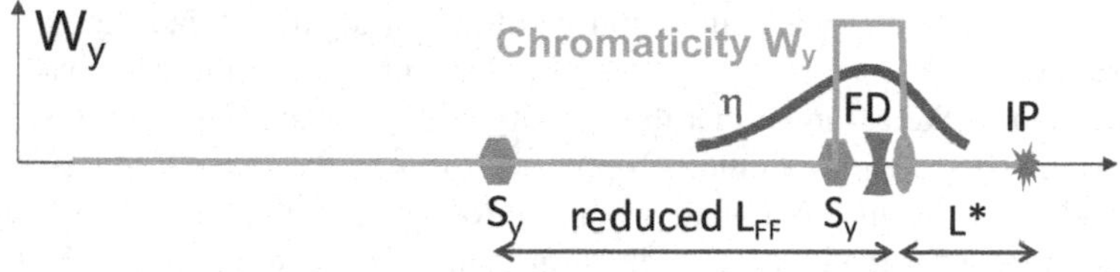

Figure 10.28 Compact final focus system with local chromatic correction.

On the other hand, the final focus system with local chromatic correction[13] employs the sextupole placed right into the final doublet; see Fig. 10.28. Another sextupole, placed slightly upstream, cancels geometric aberrations. The compensating chromatic function is local, it does not need to propagate, and its preservation is simpler. As a result, the length of such a final focus system can be significantly reduced. Experimental validation[14] of the compact final focus design opened the path for more efficient designs of high-energy linear colliders.

Chromatic compensation resembles the application of the preliminary anti-action inventive principle. Propagation of the pre-compensation chromatic function is prone to disturbances; i.e., it is a harmful process, which one would want to do fast, employing the skipping inventive principle. With propagation velocity limited to the speed of light, "skipping" would mean doing the pre-compensation as close as possible to the source of chromaticity (which is the Final Doublet) — as is implemented in the design of the final focus system with local chromatic correction.

[13] P. Raimondi and A. Seryi, Phys. Rev. Lett., 86 (17), 3779, (2001).

[14] G. R. White et al. (ATF2 Collaboration), Phys. Rev. Lett. 112, 034802 (2014).

EXERCISES

10.1. *Chapter materials review.*
Consider that corrections of chromatic and other aberrations are implemented perfectly, and the beams of an electron-positron 1 TeV CM linear collider are ready to collide with a crossing angle. Suggest other effects that may impact the collider's luminosity.

10.2. *Mini-project — continuation.*
In 1954, Enrico Fermi presented, in his lecture, a vision of an accelerator that would encircle the Earth. Let's continue designing such an accelerator. In the previous Chapter 2 exercise, we assumed that such an accelerator would be shaped like a polygon with N sides and that there would be N space stations launched around the Earth, located in the vertices of the polygon, and that each space station would carry one bending dipole magnet and one quadrupole magnet. We made assumptions about the sizes of the bending magnets and quadrupoles, and we assumed that between the space stations the accelerator orbit would be straight and that the particle beam would propagate in the open–space vacuum (i.e., without any vacuum chambers). We assumed that the polarity of the quadrupoles in the neighboring space stations would be opposite, forming the FODO optics of the accelerator. In this continuation, take into account synchrotron radiation, particularly for the case of an electron beam. Estimate the energy lost per turn, estimate the voltage of the RF cavities that need to be installed in every space station, and estimate the equilibrium emittance and the beam sizes in the quadrupoles and bends.

10.3. *Developing AS-TRIZ parameters and inventive principles.* Based on what you already know about accelerator science, discuss and suggest the possible additional parameters for the AS-TRIZ contradiction matrix, as well as the possible additional AS-TRIZ inventive principles.

10.4. *Practice in reinventing technical systems.* In the traveling focus regime, the focal points of the colliding bunches are moving during the collision in order to coincide with the location of the head of the opposite bunch, therefore optimally benefitting from the additional focusing from the opposite bunch. Suggest a way to dynamically move the focus point of each colliding bunch in this way. Remember the TRIZ principle *to use resources and energy that you already have in the system.*

10.5. *Practice in the art of back-of-the-envelope estimations.* Imagine that chromaticity compensation is arranged perfectly, and the size and angular spread of the 1 TeV energy electron beam in front of the final focusing quadrupole of $L_q = 2$ m is $\sigma_y = 100$ μm and $\sigma_{y'} = 10^{-9}$ rad. The distance between the final lens and the interaction point is $L^* = 4$ m (therefore, the ideal beam size at the IP is $\sigma_y^* \approx 5$ nm). Estimate the vertical size of the beam at the IP, taking into account synchrotron radiation in the final quadrupole. Suggest how the uncovered limitation can be mitigated. *(It is assumed that you can identify the most important effects playing roles in this task, can define the necessary parameters and set their values, and can get a numerical answer.)*

11 Beam Stability and Energy Recovery

In this chapter, we will discuss several topics related to the stability of beams, and we will also discuss the topic of energy recovery — the technique which is becoming more and more necessary for achieving the ever-increasing performance goals of accelerator-based facilities.

11.1 STABILITY OF BEAMS

Stability of particle beams or laser pulses is usually one of the most important requirements of any design. Any practical realization of a beamline or laser system has imperfections: static (caused by misalignments) or dynamic (caused, e.g., by vibrations) positioning errors. In this section, we will look at just a few selected examples from this wide topic.

11.1.1 STABILITY OF RELATIVISTIC BEAMS

Head–tail effects are often the cause of instabilities and deterioration of bunch properties, and occur when an initial offset of the bunch head creates much larger oscillations in the bunch tail.

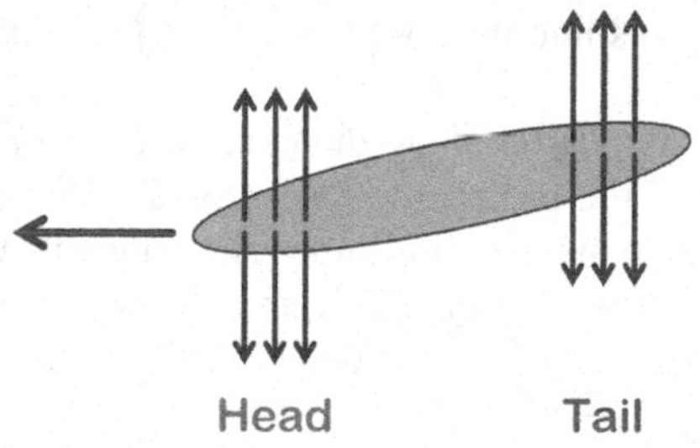

Figure 11.1 Fields of the bunch and head–tail effects.

The fields of a relativistic bunch are mostly transverse; therefore, in free space, the motion of the tail of the bunch would be independent of the motion of the head (e.g., whether the head has an offset and/or oscillations relative to the tail).

However, we need to take into account that the fields of a relativistic bunch are mostly transverse, as illustrated in Fig. 11.1, and the tail of the bunch "would not know" about the motion of the bunch head.

So, for the head–tail instability to develop, it is necessary to have an "agent" to carry the information about the offset from the bunch head to the bunch tail.

This so-called agent can be, for example, the fields induced by the bunch in the surrounding accelerating structures. The field of the opposite colliding bunch can also play the role of this agent. We will study our examples starting with the latter.

DOI: 10.1201/9781003326076-11

11.1.2 BEAM–BEAM EFFECTS

Modern electron–positron colliders aiming at the highest luminosity[1] require tiny beam sizes at their *interaction region* (IR). An example of the beam sizes for a 500 GeV CM e^+e^- ILC collider project is shown in Fig. 11.2.

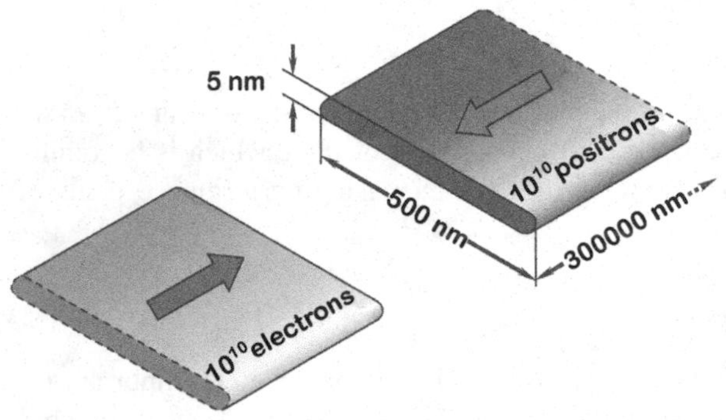

Figure 11.2 Flat beam collision in an interaction region of a typical linear collider.

Linear collider beams are extremely flat in order to ensure that the energy losses of the beams caused by synchrotron radiation during beam collision are significantly suppressed.

For a beam with uniform density distribution and transverse sizes $\sigma_{x,y}$, the maximum field at the boundary of the beam can be estimated using the Gauss theorem $\int E ds = 4\pi Q$, which thus gives for this maximum field:

$$E \approx \frac{eN}{(\sigma_x \sigma_z)} \tag{11.1}$$

where we take into account that, for the beam with uniform distribution in the longitudinal direction, the surface integral $\int ds$ turns into the contour integral $\int d\ell$ with charge Q taken per unit of length.

Taking the contour integral in the above derivation along the contour inside of the beam, we conclude that the field inside of the beam grows linearly. The fields outside of the beam can be estimated in the same manner. It is important to note that, if the beam is very flat, the integral would be almost independent of the offset from the beam in the y direction until the point when the vertical offset reaches values comparable with σ_x.

Therefore, for vertical offsets ranging between σ_y and σ_y, the field will be almost constant and after that will start to decrease as $1/r$, as illustrated in Fig. 11.3.

Fields of the beam cause synchrotron radiation in the particles of the opposite beam. As previously mentioned, this radiation is called *beamstrahlung* (Fig. 11.4). It results in an increase of the energy spread of the beams and smears out the luminosity spectrum. The produced SR photons create large numbers of e^+e^- pairs that create *background* in the experimental detector. *Beamstrahlung* energy losses can be estimated using the formulas from Chapter 3. We will leave these estimations for your exercises while we return to the transverse dynamic effect.

---

[1] International Linear Collider Technical Design Report, 2013.

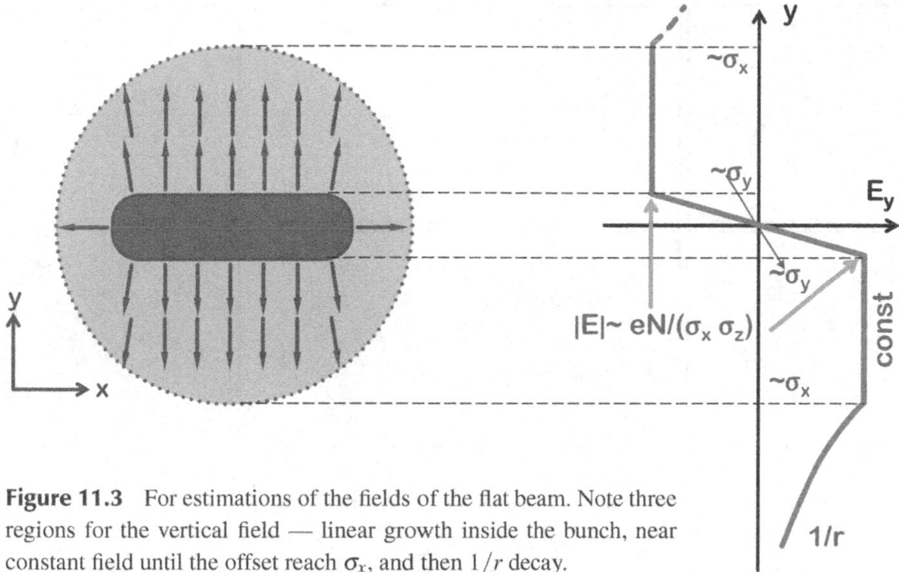

Figure 11.3 For estimations of the fields of the flat beam. Note three regions for the vertical field — linear growth inside the bunch, near constant field until the offset reach σ_x, and then $1/r$ decay.

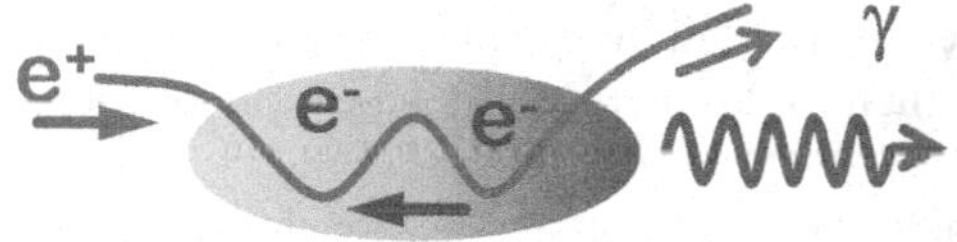

Figure 11.4 *Beamstrahlung.*

In an allusion to *bremsstrahlung* — electron radiation that passes through matter — radiation during beam-beam collisions is called *beamstrahlung*.

The transverse fields of the flat beam estimated above produce an angular kick for a particle traveling through the opposite colliding bunch. For Gaussian transverse beam distribution, and for a particle near the axis, the total angular kick of the particle after collision is estimated as

$$\Delta x' = \frac{dx}{dz} = -\frac{2Nr_e}{\gamma\sigma_x(\sigma_x+\sigma_y)} \cdot x$$

$$\Delta y' = \frac{dy}{dz} = -\frac{2Nr_e}{\gamma\sigma_y(\sigma_x+\sigma_y)} \cdot y \tag{11.2}$$

We can now introduce the notion of the *Disruption parameter*.

$$D_x = \frac{2Nr_e\sigma_z}{\gamma\sigma_x(\sigma_x+\sigma_y)} \quad \text{and} \quad D_y = \frac{2Nr_e\sigma_z}{\gamma\sigma_y(\sigma_x+\sigma_y)} \tag{11.3}$$

As follows from Eq. 11.2, the disruption parameter is inversely proportional to the focusing distance f_{beam} corresponding to the focusing field of the bunch: $D_y \sim \sigma_z/f_{beam}$.

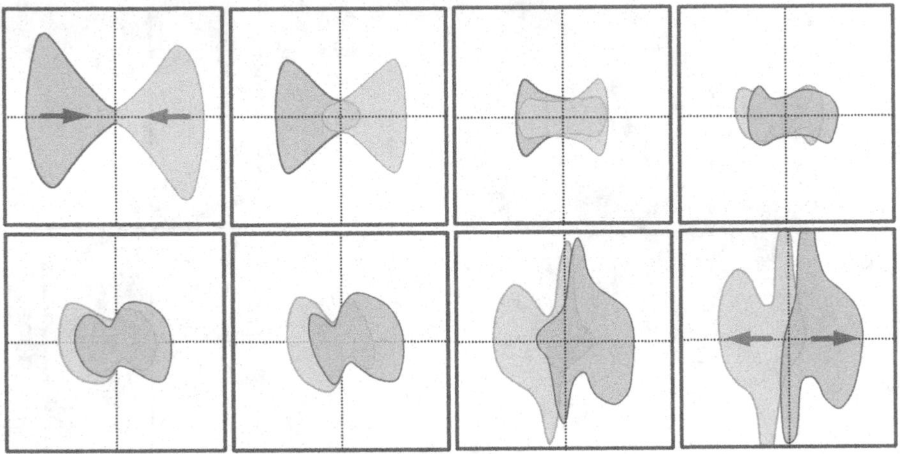

Figure 11.5 Consequent moments of high-disruption beam collision.

A low-disruption parameter $D \ll 1$ means that the bunch acts as a thin lens. Its high values ($D \gg 1$) link to the instance when particles oscillate in the field of the opposite colliding bunch. The number of oscillations is approximately equal to $\sqrt{D/(2\pi)}$, and if D is around or bigger than ~ 20, there will be a couple or more oscillations during collision.

Imagine now that the beams had small transverse offsets before the collision. The beams would attract each other and start to oscillate. As the collision developed further, the new portions of the beam would already have larger initial displacements. The oscillations would grow in an unstable manner, limited only by the finite length of the bunch.

An example of beam–beam collisions with $D_y \sim 24$ and with a $0.1\sigma_y$ initial offset between the beams is shown in Fig. 11.5. We can see that the second half of the collision is indeed noticeably disrupted.

High beam-beam focusing forces and a high disruption parameter regime can be both harmful and useful. They can be harmful by creating instability. They can be useful for additionally squeezing the beam and overcoming the hourglass luminosity reduction, illustrating the application of the inventive principle "turn lemons into lemonade."

Collisions in high disruption regimes have both their challenges and their advantages. As we can note in Fig. 11.5, the middle part of the collision shows that the beams are nicely focused on each other. The beam–beam focusing forces the beams to squeeze tighter and thus give an additional enhancement to the luminosity, which is an advantage. However, beam–beam instability may develop, and therefore the luminosity enhancement would be compromised by a higher sensitivity to the initial offsets.

We will later discuss one more collision scenario where the disruption regime may become useful — in the method aimed to overcome the *hourglass effect* (see Section 10.2).

11.1.3 BEAM BREAK-UP AND BNS DAMPING

The pursuit of high-charge and high-current beams leads to challenges caused by various imperfections. One particular issue is caused by *wakefields*.

The interaction of the charged beam with the accelerating structure, or the vacuum chamber in general, can generate electromagnetic fields — wakefields — that can act back on the bunch itself.

In the RF accelerating structure, these fields can build up resonantly and disrupt the bunch itself. This is called the *beam break-up* (BBU) instability and it can happen either for a single bunch or in a multi-bunch case.

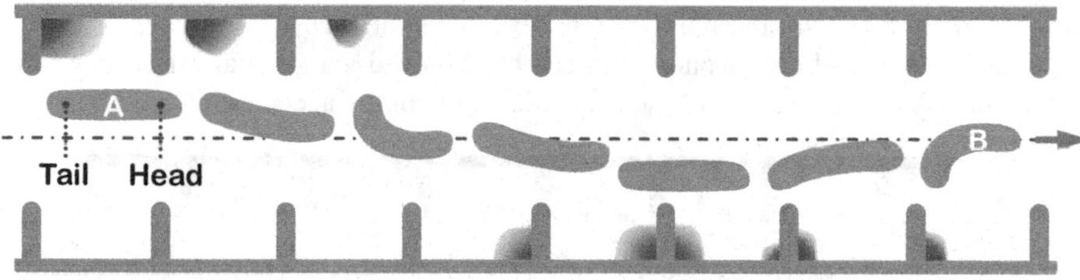

Figure 11.6 Beam break-up instability of a single beam. Fields left by the bunch are shown qualitatively. Beam evolution from the initial unperturbed shape (A) to the final BBU-distorted shape (B).

The single-beam break-up instability is illustrated in Fig. 11.6. The cure for the single-bunch BBU is called *BNS damping*, according to the names of its creators.[2]

The mechanism of how BNS damping mitigates single-bunch BBU is explained in the following: assume that the bunch has an offset with respect to the center of the accelerating cavity. The bunch head will excite a transverse dipole wakefield W (proportional to the offset) that will cause transverse deflection of the tail, which can consequently result in BBU.

We can note that the wake W acting on the tail is an additional *defocusing*. The BNS damping recipe is thus to introduce additional *focusing* to the bunch tail, which would then compensate the wakefield W acting on the tail.

In order to compensate the wake W acting on the tail, one needs to decrease the energy of the tail in such a way that the effectively increasing focusing via lenses in the accelerator channel exactly cancels out the wakefield's defocusing effect. The necessary energy difference between the head and the tail of the bunch is achieved for

[2] V. Balakin, A. Novokhatsky and V. Smirnov, in Proc. of the 12th Int. Conf. on High Energy Accelerators, Fermilab, 1983.

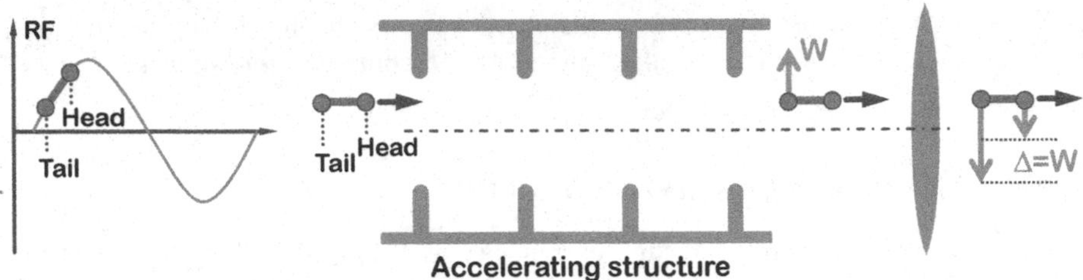

Figure 11.7 BNS damping method.

the BNS damping by placing a bunch off-crest in the RF pulse, which then creates corresponding optimal BNS energy spread over the bunch ($E - z$ correlation), as illustrated in Fig. 11.7.

BBU can also occur in trains of bunches, wherein accumulated wakefields act on the next bunch and the following bunches in the train, enhancing their oscillation. A possible method to cure the multi-bunch BBS is to enhance *decoherence* of transverse modes that the beam excites. This can be achieved via gradual variation of the parameters (such as iris radii) of the individual cells in the accelerating cavity.

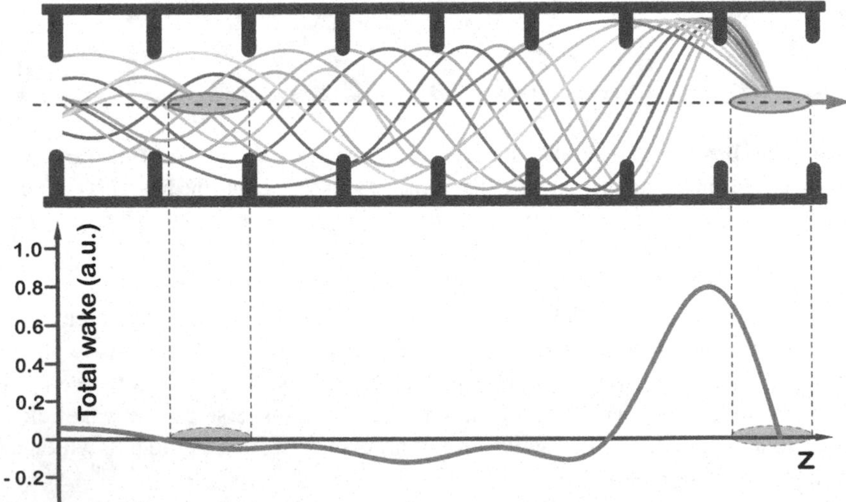

Figure 11.8 Detuned structure as a cure for multi-bunch BBU instability.

In this case, as qualitatively shown in Fig. 11.8, the time structure of excited modes will be slightly different, and the total sum wakefield will quickly lose coherence, minimizing the kick on the following bunch. Often, in addition to variation of accelerating cells, additional passive damping elements are inserted into the accelerating structures to absorb particularly the higher-order modes. Such structures are called *damped detuned structures* and can be suitable for accelerating trains of high-charge bunches with separation between bunches as minimal as just a few RF cycles.

Before moving on to the next section, we would like to make a few remarks. We have seen in this section that BNS damping prevents BBU instability by maintaining the conditions for coherent motion of the bunch head and tail (i.e., preventing relative oscillations of the head with respect to the tail). It is helpful to look at the word *damping*, in "BNS damping" (or the word *echo*, in "EEHG") broadly. The method to avoid BBU instability in multi-bunch cases involves making use of decoherence.

We will now look at one more mechanism — Landau damping, which is used both in plasma and accelerators, but has, in some cases, a different meaning within these two areas.

11.1.4 LANDAU DAMPING

Landau damping[3] is the mechanism first discovered in plasma. It describes a collisionless damping of collective plasma oscillations. This mechanism can be illustrated via the following approximate analogy.

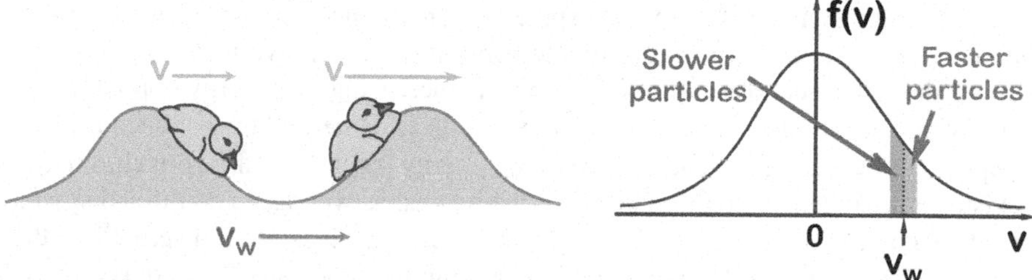

Figure 11.9 For illustration of Landau damping mechanism.

Let us consider an ocean wave that travels with phase velocity v_w. Assume that a duck floats along the wave with the velocity v, which is very close to v_w; see Fig. 11.9.

The duck would soon be "trapped" in the ocean wave, which means that if the duck was initially moving faster than the wave ($v > v_w$), it would slow down and thus the wave would gain energy from the duck. In the opposite case, if the duck was initially slower ($v < v_w$), the wave energy would be transmitted to the duck.

The overall damping of the wave can therefore occur if the distribution of velocities of ducks (or particles in plasma) decreases for larger values of velocity — which is indeed typically the case, as shown in Fig. 11.9 for Maxwellian distribution. As there are fewer faster particles than slower particles, the collective wave in plasma will be damped by the Landau mechanism.

It is important to note that, for the Landau damping to be possible, the distribution function should have a nonzero number of particles (ducks) at the wave velocity v_w. Therefore, if the initial spread of velocities of ducks or particles was not sufficient, increasing it as shown in Fig. 11.10 would help to enhance Landau damping.

[3]L. Landau, J. Phys. USSR 10 (1946).

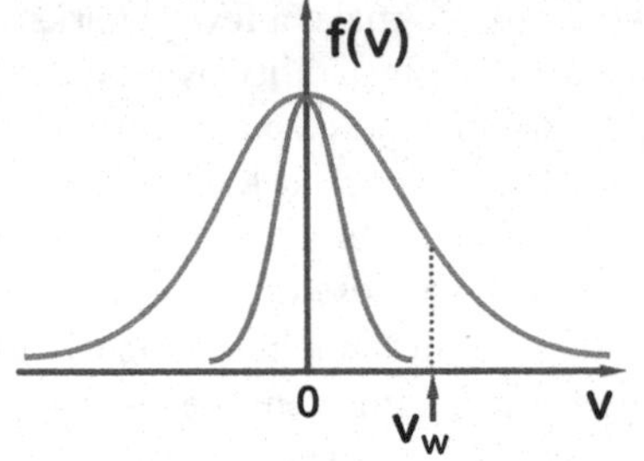

Figure 11.10 Velocity spread and Landau damping.

Landau damping in accelerators can be achieved by increasing the spread of certain beam parameters, such as the energy spread or betatron tune spread, resulting in improved beam stability.

Landau damping is a very important mechanism for accelerators too, as it helps to provide beam stability for either transverse or longitudinal motion. Increasing the spread of certain beam parameters, as was just mentioned above, is often also done in accelerators, as a possible way to provide improved stability.

In particular, increasing the energy spread in circular accelerators would result in, via the nonzero momentum compaction factor, the spread of revolution periods, thus helping the longitudinal stability of the beam. Increasing the energy spread of the beam with a *laser heater* may help in coping with CSR-caused microbunching.

Spread of betatron frequencies (caused by energy spread and nonzero chromaticity) is often also helpful to the stability of the beams, as well as an additional spread introduced by the so-called *Landau octupoles* — the magnets often inserted into the beamline specifically for the purpose of enhancing the tune spread and thus helping the beam stability.

We would like to mention here that increasing the spread of tunes or energy often helps accelerators simply because it enhances the decoherence, preventing particles from assembling into collective motion. The term *Landau damping* is sometimes used in cases where the term *decoherence* would be more appropriate.[4] Details of a particular physics setting define which of those two mechanisms is acting in each situation.

11.1.5 STABILITY AND SPECTRAL APPROACH

Beam or laser pulse stability issues are often analyzed based on a spectral approach. We will omit detailed and rigorous definitions and derivations here and will focus only on a few highlights.

The disturbance — e.g., the position of a particular magnetic lens or optical element of a laser (we will define it as $x(t)$) — often has the same characteristics as a *random process*. In this case, the notion of the *power spectral density* $p(f)$ (dubbed "power spectrum") should be used instead of the usual Fourier spectrum.

[4]Werner Herr, CERN Accelerator School, 2013.

The key property of the power spectrum is that its integral is equal to the variance σ of the signal x:

$$\sigma^2 = \langle x^2 \rangle = \int_{-\infty}^{+\infty} p(f)\, df \tag{11.4}$$

where we assume that the average of x is zero: $\langle x \rangle = 0$. The power spectral density is defined as

$$p(f) = \lim_{T \to \infty} \frac{1}{T} \left| \int_{-T/2}^{T/2} x(t)\, e^{-i\omega t}\, dt \right|^2 \tag{11.5}$$

Note that we have to use the power spectral density because the usual Fourier integral of a random process signal is unlimited, as can be observed from the above equation.

Disturbances to our beam or laser line often comes from certain frequencies. The power spectrum is useful here as it can identify the contribution from a particular frequency range according to the following (Fig. 11.11):

$$\sigma^2(f_1 < f < f_2) = \langle x^2 \rangle = \int_{f_1}^{f_2} p(f)\, df \tag{11.6}$$

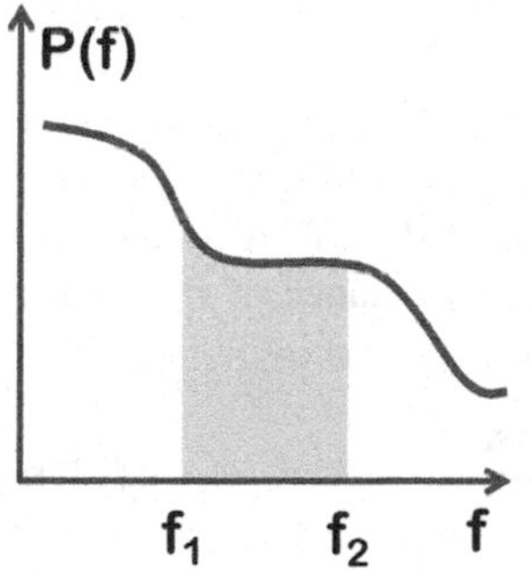

Figure 11.11 Power spectrum.

A spectral characteristic of a random signal $x(t)$ is its power spectral density $p(f)$ — power spectrum in short. It is rare to be interested in the displacement of just a single point, and we often need to consider the whole array or continuum of displaced points $x(t,s)$ and to analyze the RMS characteristics of their relative displacement. In this case, we need to use the inventive principle of *going to another dimension* and consider a *two-dimensional power spectral density* $P(\omega,k)$.

Quantitative analysis of the influence of disturbances often requires the knowledge of the expected relative displacement of a beamline element during a specific time duration τ. The associated variance (contributed to it from a frequency range $[f_1; f_2]$) can also be calculated from the power spectrum:

$$\langle [x(t+\tau) - x(t)]^2 \rangle_t = \int_{f_1}^{f_2} p(f)\, 2[1 - \cos(\omega t)] df \tag{11.7}$$

You may notice that the description given above is not sufficient for analyzing beamlines that have many elements distributed in space. In addition to time, we also need to take spatial information into account.

Therefore, our analysis should involve a spectrum that depends on frequency as well as on wavenumber — *a two-dimensional power spectrum*[5] $P(\omega,k)$.

[5] A. Seryi and O. Napoly, Phys. Rev. E. v.53, 5323, 1996.

The 2D spectrum (also called the PWK spectrum) can evaluate expected relative displacements of two beamline elements separated by a certain distance L and after a certain time interval T. Assuming that at $t = 0$ the beamline is perfectly straight, the variance of the relative misalignment after a time T of two points separated by the distance L is given by

$$\sigma^2(T,L) = \iint_{-\infty}^{+\infty} P(\omega,k)\, 4[1-\cos(\omega T)]\, [1-\cos(kL)] \frac{d\omega}{2\pi} \frac{dk}{2\pi} \tag{11.8}$$

Since the formula Eq. 11.8 can predict the stability of any two elements in our beamline, we can also evaluate the stability of the entire beamline by properly taking into account all of the elements.

Combining the coefficients that determine how the displacement x of an individual beam element or laser line contributes to, for example, the displacement of the beam/light at the focus x_{out}, we can construct a so-called *spectral response function* $G(k)$. The variance of the relative misalignment of the output beam/light after a time T depends on $G(k)$ as

$$\sigma^2(T) = \iint_{-\infty}^{+\infty} P(\omega,k)\, 2[1-\cos(\omega T)]\, G(k) \frac{d\omega}{2\pi} \frac{dk}{2\pi} \tag{11.9}$$

The variance of the misalignment defined above in Eq. 11.9 depends on the observation time T and typically grows with T. To stabilize the beam/light at the focus, a feedback would usually need to be applied, which can be described by the *characteristic function of the feedback* $F(\omega)$. In this case, the variance of the relative misalignment of the output beam/light is a constant, and it is given by

$$\sigma^2 = \iint_{-\infty}^{+\infty} P(\omega,k)\, F(\omega)\, G(k) \frac{d\omega}{2\pi} \frac{dk}{2\pi} \tag{11.10}$$

Examples of a 2D power spectrum, as well as a spectral response function and a characteristic of feedback for a particular beamline (final focus system beamline), are shown in Fig. 11.12.

A particularly useful and insightful approximation for a 2D spectrum is the one related to the *ATL law*, which was first suggested[6] to describe space-time motion of the ground (earth/soil), on which the focusing elements of a beamline are installed. A simple model depicting how ATL motion can happen is shown in Fig. 11.13.

For this "ATL law," the variance of the misalignment for elements separated by distance L and observed after time T (described above by Eq. 11.8) is simply

$$\sigma^2(T,L) = A \cdot T \cdot L \tag{11.11}$$

where A is the parameter of the model that could be estimated, for example, from ground motion measurements of our site (where the elements of beamline are installed) or perhaps of our optical table (where the laser elements are mounted).

[6] B. A. Baklakov et al., INP Report 91–15, 1991.

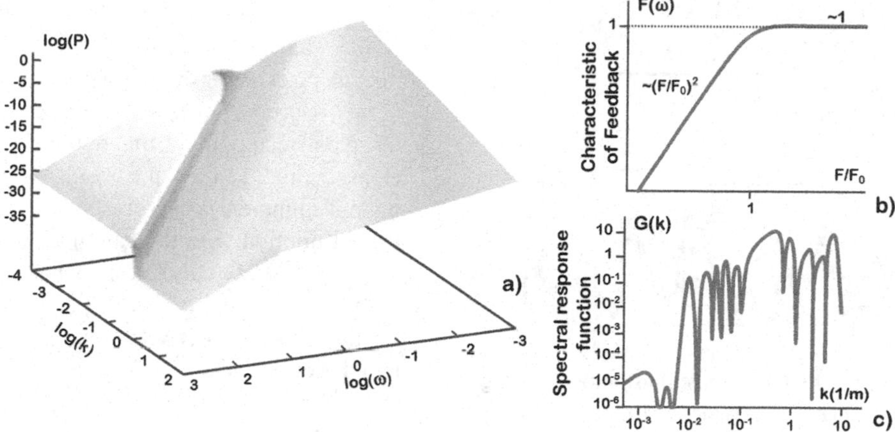

Figure 11.12 Examples of power spectrum $P(\omega,k)$ (left), spectral response function $G(k)$ and characteristic function of the feedback $F(\omega)$.

The 2D power spectrum corresponding to the "ATL law" is given by

$$P(\omega,k) = \frac{A}{\omega^2\, k^2} \tag{11.12}$$

The 2D spectrum shown in Fig. 11.12 includes the ATL component; however, it is included in a corrected way, at higher frequencies and wavenumbers, to ensure integrability of the equations.

The term "law" in "ATL law" should be taken cautiously here, as this is certainly a model, an estimate that is applicable in certain ranges of parameters and certain situations. It is, however, amazing how many different situations and wide parameter ranges[7] there are in which this approximation can be found valid.

11.2 ENERGY RECOVERY

A linear collider converts only a tiny fraction of energy of its beams into new particles. A free electron laser (FEL) converts a very small amount of its beam energy into useful radiation. An electron cooling beam may be only slightly disturbed after interacting with the proton or antiproton beam, which it is cooling. All these beams still carry almost all of their initial energies, even though they may not have the original qualities, like emittance or energy spread. Still, discarding these beams, at their full energy, would be a huge waste of resources. Taking the energy out from the used beams, before discarding them, is the method which boosts the efficiency of any accelerator facility, and is called *energy recovery*.

[7] V. Shiltsev, Phys. Rev. Lett. 104, 238501 (2010).

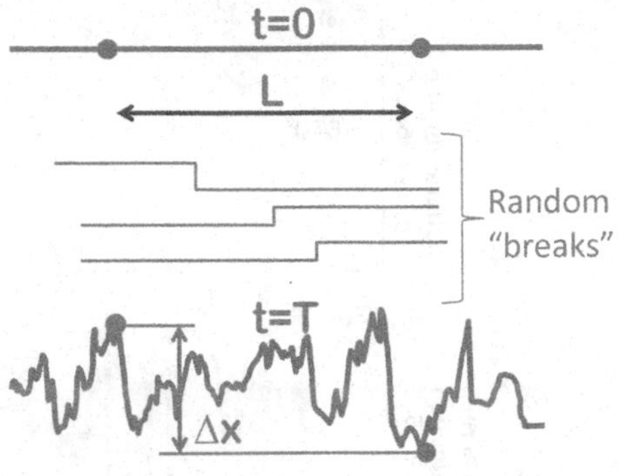

Figure 11.13 Simple illustration allowing to imagine how ATL motion happens.

If the number of random step-like displacements between two points is proportional to the distance between points L and to the elapsed time T, then the resulting misalignment will correspond to ATL motion with the variance $\langle \Delta X^2 \rangle = A \cdot T \cdot L$. This simple random breaks model can be implemented very easily in a computer simulation code.

11.2.1 ENERGY RECOVERY IN ELECTRON COOLING

Energy recovery was used in electron cooling installations from the very beginning of electron cooling invention. A typical scheme of a standard electron cooler for a low-energy range (~several tens MeV of proton energy) is shown in Fig. 11.14.

Imagine, as shown in this figure, that we need a 1 A electron beam current with 50 kV of electron beam kinetic energy to cool protons of about 100 MeV kinetic energy. Formally, the electron beam power is 50 kW, but it does not mean that we would need to have a 50 kW power supply to generate the electron beam.

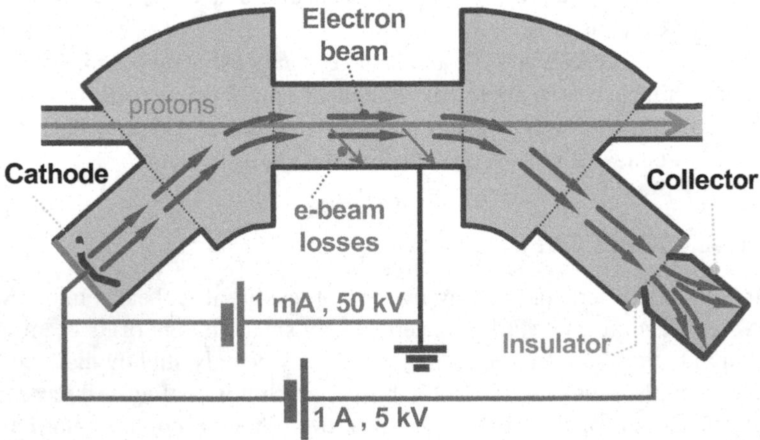

Figure 11.14 Electron cooling power supplies arrangement that enable energy recovery.

The typical arrangement of the electron cooler power supplies is as follows. The 1 A electron current is generated by the power supply with a much lower voltage, with the value sufficient to generate a 1 A current given the available perveance of the electron source — for example, a 5 kV power supply connected between the cathode and the anode, as shown in Fig. 11.14.

We still need to provide the 50 kV electron beam energy, and this is ensured by another power supply, connected between the ground and the cathode; however, the current that this power supply will need to provide would only need to match the possible losses of the current to the ground. In Fig. 11.14, the loss current is limited to 1 mA, and therefore the power of the 50 kV power supply can be just 50 W.

The energy recovery in this electron cooling case occurs, as we can see, due to the deceleration of the electron beam just before it comes to the collector to be dumped there. In this case, the energy recovery efficiency can be estimated as the ratio of the beam power to the maximum power of the involved power supplies, i.e., 50 kW–5 kW, or an order of magnitude. Therefore, 90% of the electron cooler beam energy is recovered.

11.2.2 ENERGY RECOVERY IN FREE ELECTRON LASERS

The fraction of energy taken out from the electron beam in an FEL, as we can see from Eq. 8.28, is roughly equal to 1.6 times the Pierce or FEL parameter ρ, which is typically much less than unity. Increasing the power of FEL radiation can be greatly enhanced by the energy recovery approach.

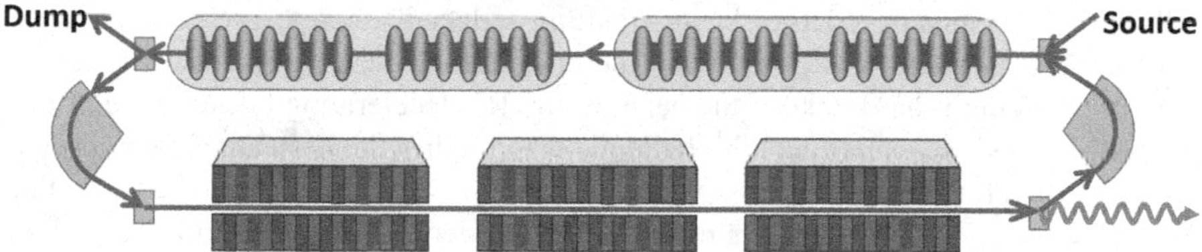

Figure 11.15 Free Electron Laser energy recovery.

The application of energy recovery to an FEL is illustrated in Fig. 11.15. The CW bunched beam is sent to a superconducting linac, accelerated, and sent, via the 180° bending optics, into an undulator, where useful FEL radiation is created. After exiting the undulator, the beam is turned back and sent once more to the superconducting linac–however, this time, the bunches are placed in the decelerating phase of the RF. After decelerating in the linac, the beam is dumped at the energy E_{dump}, as illustrated in Fig. 11.16.

With energy recovery, the overall efficiency η of electron power to FEL power conversion will be given by

$$\eta_{\text{ beam to FEL}} \approx 1.6\,\rho \frac{E_{\text{beam}}}{E_{\text{dump}}} \tag{11.13}$$

which can be greatly increased by reducing the energy at which point the beam is dumped. Such an energy recovery approach was very successfully used, in particular, in Jefferson Lab's FEL[8], which set a record of 14 kW radiation power in the infrared range.

[8]G.R. Neil et al., The JLab high power ERL light source, NIM A, V. 557, Issue 1 (2006).

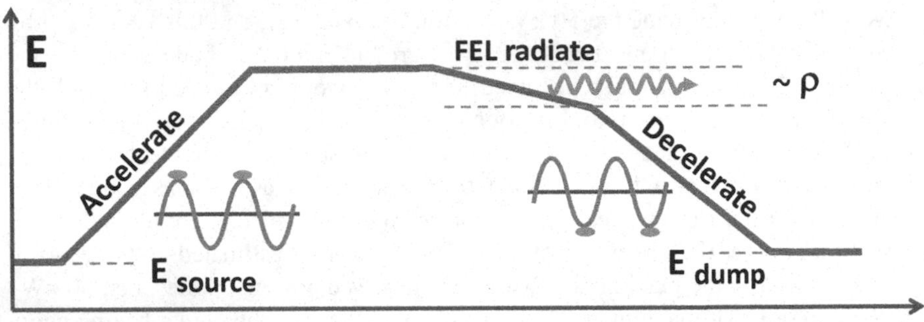

Figure 11.16 FEL energy recovery — evolution of the electron beam energy.

11.2.3 ENERGY RECOVERY IN COLLIDERS

The first proposal of using energy recovery in colliders[9] goes back to 1965. While energy recovery has been used by now in FELs, beam recirculating facilities, etc., its application to high-energy colliders is still under development and varieties of ideas have been suggested.

One of the challenges of using energy recovery in linear colliders is that, after colliding, the beams have large "beamstrahlung-induced" energy spread of a couple of percentages, which would relatively increase during deceleration toward energy recovery, making it hard to keep the beam in the RF decelerating bucket. One of the suggested ways[10] to overcome this challenge is to modify the IP beam parameters by using the traveling focus in such a way that beam disruption is increased while the "beamstrahlung" energy spread is reduced by an order of magnitude of about 0.2%, making it possible to decelerate the 250 GeV beam to a few GeV, recovering most of the beam energy.

Other suggestions[11] developed ideas of using dual-axis (twin) cavities (see an example of such a cavity in the illustration of the inventive principle asymmetry in Fig. 15.4 in Chapter 15), which separate accelerating and decelerating beams in a CW superconducting linac of a linear collider, enabling its energy recovery.

11.2.4 ENERGY RECOVERY IN PLASMA ACCELERATION

It has been proposed[12] that using a train of laser pulses can excite plasma waves for plasma acceleration — Multi Pulse LWFA or MP-LWFA. When laser pulses are separated by plasma wavelength λ_p, plasma amplitude is growing. However, it has also been shown in the paper cited above that an out-of-phase laser pulse can dampen the plasma wave, taking out its energy, therefore enabling the path to energy recovery in plasma acceleration — undoubtedly one of the most interesting and promising directions in the field of accelerators, lasers and plasma.

---

[9] M. Tigner, Il Nuovo Cimento, 37, 1228-1231 (1965).

[10] A. Seryi, NIM A, Volume 623, Issue 1, p. 23-28, (2010).

[11] V.I. Telnov, JINST 16, P12025 (2021); I.V. Konoplev et al., JINST 17, P01011 (2022).

[12] J. Cowley et al., Phys. Rev. Lett, vol. 119, no. 4, p. 044802, Jul 2017

11.3 HIGHER-ENERGY COOLING – INVENTION CASE STUDY

Energy recovery becomes very important for higher-energy electron cooling. Fermilab $E_k = 4.3$ MeV electron cooling[13] used $I_e \approx 0.5$ A and achieved $\tau_{\text{cooling}} \approx 0.5$ hours for 8 GeV antiprotons. The electron beam was accelerated and decelerated in a Pelletron, providing the energy recovery. The electron beam was guided by a 100 G solenoidal field inside of the cooling section, with quadrupole focusing outside. To compensate for the angular momentum that would be created by the edge field of the cooling section solenoid, the electron gun was immersed in a solenoid, providing an opposite sign of angular momentum — exemplifying the preliminary anti-action inventive principle.

Electrostatic accelerators encounter practical challenges to take electron cooling to higher energies. An alternative obvious approach is to use RF acceleration of bunched electron beams. Such electron cooling[14] has been successfully demonstrated and helped to increase RHIC collider luminosity for gold ions.

The beam energy in RF-accelerated bunched electron cooling can be increased, but the unfavorable scaling $\tau \sim \gamma^2$ prevents the use of such classic cooling at more than about 10 MeV. Coherent electron cooling has been suggested[15] where the initial disturbance (imprint) created by an ion in the electron beam is coherently amplified to provide much stronger, enhanced cooling; see Fig. 11.17.

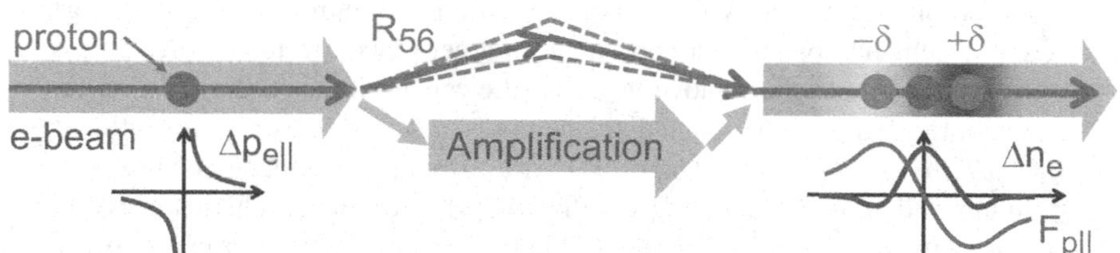

Figure 11.17 Coherent electron cooling. Proton creates energy modulation $\Delta p_{e||}$ in the electron beam, which after coherent amplification turns into electron density modulation Δn_e, creating longitudinal force $F_{p||}$ acting on R_{56}-chirped protons, reducing their energy spread.

Different amplification mechanisms have been considered for coherent cooling. In a recent study that focused on the FEL amplification mechanism, a plasma-cascade instability was revealed to deteriorate the initial quiet state of the electron beam. The inventive team, however, ingeniously turned this obstacle into an advantage, proposing the plasma-cascade-amplification-based coherent electron cooling[16].

> The *plasma-cascade amplification coherent electron cooling* concept emerged from the inventive principle *turn lemons into lemonade*, or *blessing in disguise*.

---

[13] S. Nagaitsev et al., JINST 10 T01001, (2015).

[14] A. Fedotov et al., Phys. Rev. Lett. 124, 004001, (2020).

[15] Derbenev ca. 1981, Litvinenko and Derbenev 2009, Ratner 2013.

[16] V. Litvinenko et al., Phys. Rev. Accel. Beams 24, 014402, (2021).

EXERCISES

11.1. *Chapter materials review.* Derive an estimate for SR energy losses (*beamstrahlung*) for colliding e^+e^- bunches. Justify the need to use flat beams in a high-energy e^+e^- collider. Estimate the energy losses for beam parameters from Fig. 11.2.

11.2. *Chapter materials review.* For the same beams from the previous exercise, estimate the number of emitted *beamstrahlung* photons per particle in the case when the oncoming positron bunch has either a 3- or 30-nm vertical offset with respect to the electron bunch.

11.3. *Mini-project.* Estimate the parameters of a single-pass FEL that can generate EUV 13.5 nm radiation for semiconductor lithography. Assume that your electron beam energy can be up to 3 GeV and that you have space for a 100-m undulator. Select suitable parameters of the FEL undulator and define requirements for the electron beam's quality. Estimate the average electron beam current needed to generate 10 kW of EUV radiation. Discuss how to modify this conceptual design into ERL-based FEL using a dual-axis cavity linac illustrated in Figs 15.4 and 15.25.

11.4. *Practice in reinventing technical systems.* Absolute ground motion (i.e., motion with respect to an inertial reference frame) is usually measured by a seismometer, which typically consists of a pendulum (mass on a spring) with a sensing coil, with a permanent magnet placed in its vicinity. Since the coil measures the relative velocity, the sensitivity (i.e., the ratio of the seismometer signal to ground displacement) of such a sensor is proportional to $\omega/(\omega^2 - \omega_0^2 - i\alpha\omega)$, where ω_0 is the resonance frequency of the pendulum and α is the damping coefficient. The resonance character of the sensitivity curve is not convenient (due to sharp resonance, large motion of mass inside of the sensor, low dynamic range and also a narrow frequency band where ground motion can be measured). Suggest a way to eliminate this resonance behavior without the loss of sensitivity. Also suggest a way to adjust ω_0 for a vertical seismometer. Remember the TRIZ principle *to use resources and energy that you already have in the system.*

11.5. *Practice in the art of back-of-the-envelope estimations.* Precise measurements of the beam energy in the 27-km perimeter LEP collider revealed[17] correlations of the beam energy with terrestrial tidal motion. The momentum compaction factor in LEP was $\alpha = 1.86 \cdot 10^{-4}$, which amplified small circumference changes and resulted in an observed beam energy variation of about $\pm 10^{-4}$, corresponding to the deformation of Earth as an elastic body with a maximal amplitude of about 0.5 m. The 2-mile-long Stanford Linear Collider (SLC), located on the Pacific Coast, was equipped with a "light pipe" — a pipe with a diameter of about a meter with insertable Fresnel lenses that could measure transverse deformation of the SLC linac (e.g., the relative transverse motion of the middle point of the linac relative

[17] L. Arnaudon , et al., CERN SL/94 07, (1994).

to its ends). Estimate the amount of tidal transverse deformation that was observed at SLC. *(It is assumed that you can identify the most important effects playing roles in this task, can define the necessary parameters and set their values, and can get a numerical answer.)*

12 Advanced Beam Manipulation

The scope of this chapter is a mixture of beam and pulse manipulation topics, selected based on the degree of the synergy they exhibit in four areas that we previously discussed. The first three areas are accelerators, lasers and plasma (see Chapter 4), and the fourth area is the methodology of inventiveness. We will start this chapter with the methods and techniques for creating short bunches of charged particles and also short laser pulses. We will follow with a discussion of various ways to enhance the intensity of both the particle beams and the laser pulses, aiming to connect these and other techniques to *Accelerating Science TRIZ.*

12.1 SHORT AND NARROW-BAND

The topic of bunch and pulse compression is an area that shows strong synergy between accelerators and lasers. This is a broad area that includes techniques of bunch or pulse compression as well as techniques for creating short bursts of radiation. We will also pay attention to some effects that may prevent the formation of ultra-short bunches — notably *coherent synchrotron radiation.* Connected to the topic of short pulse or bunch creation is the theme of narrow-band radiation, where techniques developed in optics and lasers merge with the accelerator-based techniques.

12.1.1 BUNCH COMPRESSION

Short electron bunches are typically produced by RF photo guns where the cathode is illuminated by a short laser pulse. Such photo injectors produce relatively long pulses (a few ps), while for many applications, such bunches would be too long and femtosecond bunches would be required instead.

Some compression of the beam may be performed already in the photo injector, via the technique of *velocity bunching*. In this case, by properly adjusting the moment when the laser illuminates the cathode with the RF field in the gun, one can create correlation between the energy of electrons and the longitudinal position within the bunch. The difference of energy for beams that are still weakly relativistic can create a difference in longitudinal velocities. If the faster particles are arranged to be at the tail of the bunch (as illustrated in Fig. 12.1), they can catch up, resulting in a shortening of the bunch.

Because velocity bunching is based on the velocity's dependence on energy, it can work only for weakly relativistic beams and this, in fact, is the method's main shortcoming. Weakly relativistic electron beams can, in particular, suffer strongly from space charge effects, limiting the degree of compression.

DOI: 10.1201/9781003326076-12

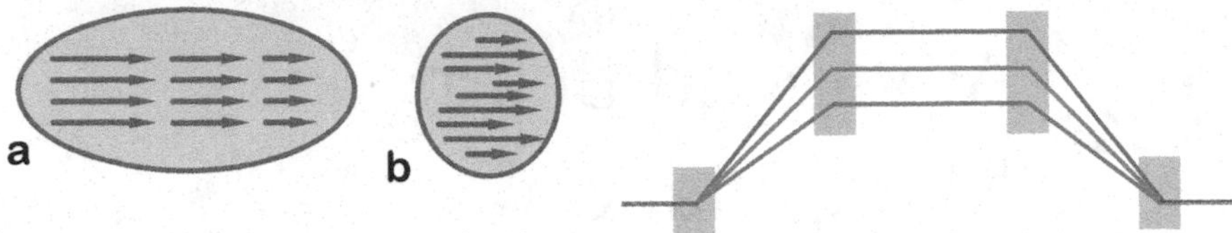

Figure 12.1 Velocity bunching. Initial beam (a) and compressed beam (b).

Figure 12.2 Four-magnet chicane.

Therefore, achieving ultra-short electron bunches is usually done at higher energies, when the beam is relativistic and space charge effects are less severe. Shortening the bunch length in this case is usually achieved with a magnetic compression system. When discussing bunch compressors, we will build on the foundation established in Chapter 4, further expanding it toward an analytical description.

A typical arrangement that can compress the bunch is a beamline made of four bending magnets of opposite polarity arranged as shown in Fig. 12.2 — called a *chicane*. In this chicane, the time of flight (or equivalently the path length) is different for different energies.

In order to exploit the dependence of the time of flight (or path length) on the particle's energy, we need to introduce an energy–time correlation within the bunch. This correlation can be created using the electric field of an RF cavity, properly phased with the beam.

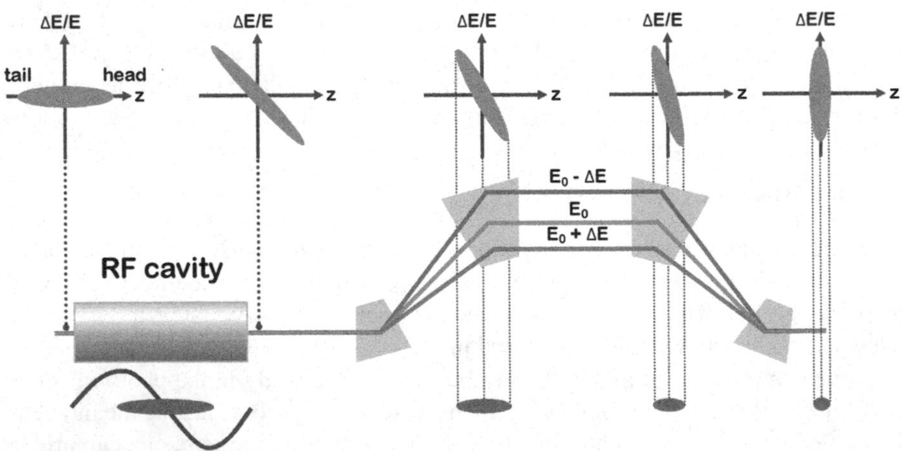

Figure 12.3 Energy–time correlation and bunch compression.

The RF cavity will create an energy chirp along the bunch — the necessary condition for bunch compression to work; see Fig. 12.3. The chirp is phased in such a way that the particles in the tail have a higher energy and will therefore travel on a straighter path through the chicane, catching up with the synchronous particle, and resulting in bunch compression.

The bunch compression process can also be described analytically. Let's use the linear transfer matrices to evaluate the evolution of the longitudinal position and relative energy offset (z, δ) as the particle propagates through the RF cavity and then the bunch compressor.

The first step is to evaluate how the coordinates (z, δ) change in the RF cavity. Passing through the RF cavity (assuming the cavity is thin), the longitudinal coordinate does not change, while the energy change depends on the initial position z_0 as follows:

$$\begin{aligned} z_1 &= z_0 \\ \delta_1 &= \delta_0 + \tfrac{eV_{RF}}{E_0} \cos\left(\tfrac{\pi}{2} - k_{RF} z_0\right) \end{aligned} \tag{12.1}$$

The above equations are equivalent to the following matrix transformation expressed in linear approximation in (z, δ) as follows (refer to Eqs. 2.59 and 2.60 for definitions of the indexes of the coordinate vector and of the transfer matrix):

$$\begin{pmatrix} z_1 \\ \delta_1 \end{pmatrix} \approx \begin{pmatrix} 1 & 0 \\ R_{65} & 1 \end{pmatrix} \cdot \begin{pmatrix} z_0 \\ \delta_0 \end{pmatrix} \tag{12.2}$$

where

$$R_{65} = \frac{eV_{RF}}{E_0} \sin(\varphi_{RF}) k_{RF} \tag{12.3}$$

The next step is to take into account the bunch compressor itself. In the chicane, the particle coordinate will change according to the following general expression (the higher-order terms are defined as in Eq. 2.61):

$$\begin{aligned} z_2 &= z_1 + R_{56}\delta_1 + T_{566}\delta_1^2 + U_{5666}\delta_1^3 \dots \\ \delta_2 &= \delta_1 \end{aligned} \tag{12.4}$$

which can be linearly approximated as

$$\begin{pmatrix} z_2 \\ \delta_2 \end{pmatrix} \approx \begin{pmatrix} 1 & R_{56} \\ 0 & 1 \end{pmatrix} \cdot \begin{pmatrix} z_1 \\ \delta_1 \end{pmatrix} \tag{12.5}$$

The full transformation is given by multiplying the matrices of each element, which can be computed to be given by the following:

$$\begin{pmatrix} z_2 \\ \delta_2 \end{pmatrix} \approx \mathbf{M} \cdot \begin{pmatrix} z_0 \\ \delta_0 \end{pmatrix} \qquad \mathbf{M} = \begin{pmatrix} 1 + R_{65}R_{56} & R_{56} \\ R_{65} & 1 \end{pmatrix} \tag{12.6}$$

Dependence of the final bunch length on the initial beam length is therefore given by the following equation

$$\sigma_{z_2} = |1 + R_{65}R_{56}|\sigma_{z_0} \tag{12.7}$$

We can see that, in order to achieve maximal compression, we should adjust the value of R_{65} (energy chirp induced by the RF cavity) in such a way that

$$R_{65}R_{56} \approx -1 \tag{12.8}$$

in which case the *compression factor* C will be the largest and the final bunch length

$$\sigma_{z_2} = \frac{\sigma_{z_0}}{C} \quad \text{where} \quad C = 1/|1 + R_{65}R_{56}| \tag{12.9}$$

will be minimal.

Let us now discuss the minimal achievable bunch length. While in the linear approximation, the condition $R_{65}R_{56} = -1$ suggests that C is infinite and the resulting final bunch length is zero, this cannot be achieved in practice.

First, taking second and higher-order terms into account will give us non-zero final bunch lengths for any initial parameters.

Second, there is a limitation caused by the longitudinal emittance of the beam. The linear transformation of the (z, δ) coordinates preserves the longitudinal beam emittance according to the *Liouville's theorem* (it's often said that the corresponding transformation is *symplectic*). The longitudinal emittance is given by the following:

$$\varepsilon = \sqrt{\sigma_z^2\sigma_\delta^2 - \sigma_{z\delta}^2} \tag{12.10}$$

which tells us that the minimum reachable bunch length is limited to the product of the beam energy spread σ_δ times the matrix element R_{56}.

Furthermore, additional limitations to the achievable compression come from the effects associated with the beam's high peak current that we have neglected in the linear approximation denoted above. These effects include longitudinal space charge, wakefields and *coherent synchrotron radiation* (CSR). When taken into account, these effects can produce serious degradation to the quality of the beams and limit the achievable minimum bunch length.

12.1.2 CSR – COHERENT SYNCHROTRON RADIATION

We will now evaluate the *coherent synchrotron radiation effect* in a back-of-the-envelope fashion, following the example of Chapter 3.

The characteristic frequency of SR

$$\omega_c = \frac{3}{2}\frac{c\gamma^3}{R}$$

defines the higher edge of the SR spectrum while radiation is emitted at any frequencies ω below ω_c.

Let us assume that we have a bunch with N electrons and its bunch length is σ. For certain frequencies ω of synchrotron radiation, its wavelength $\approx c/\omega$ can be longer than the bunch length σ; the beam at these frequencies will radiate coherently, as illustrated in Fig. 12.4.

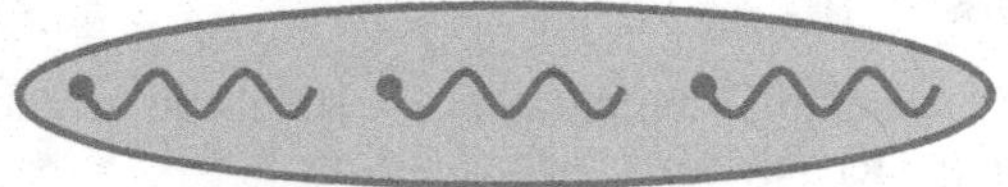
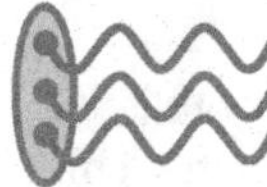

Figure 12.4 Incoherent radiation (left) and coherent radiation (right).

The two regimes can thus be defined as $c/\omega < \sigma$ — incoherent radiation — and $c/\omega > \sigma$ — coherent radiation.

For frequencies where c/ω is longer than σ, the beam radiates coherently, increasing radiated power proportionally to the square of the number of particles involved, i.e., as N^2. As the typical bunch population N is $\sim 10^9$ to $\sim 10^{10}$, this effect increases the radiated power tremendously.

Assuming that the spectrum of incoherent SR is given by $I(\omega)$, the CSR spectrum can be approximated as

$$I_b(\omega) = I(\omega) N \left[1 + N \exp\left(-\left(\frac{\omega\sigma}{c}\right)^2\right)\right] \tag{12.11}$$

A typical spectrum described by the above equation is shown in Fig. 12.5.

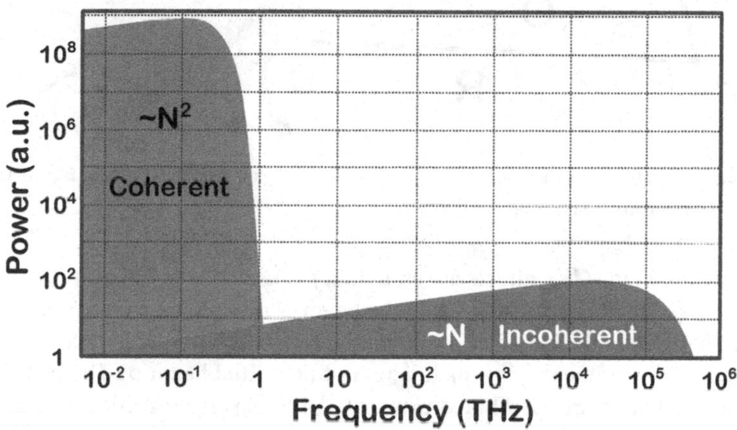

Figure 12.5 Qualitative comparison of the spectrum of coherent synchrotron radiation in comparison with the spectrum of incoherent SR.

12.1.3 CSR EFFECTS ON THE BEAM LONGITUDINAL PHASE SPACE

We will now make a back-of-the-envelope estimate[1] of the force that CSR exerts on the beam. We will use an approximation of the *two-particle model*; see Fig. 12.6.

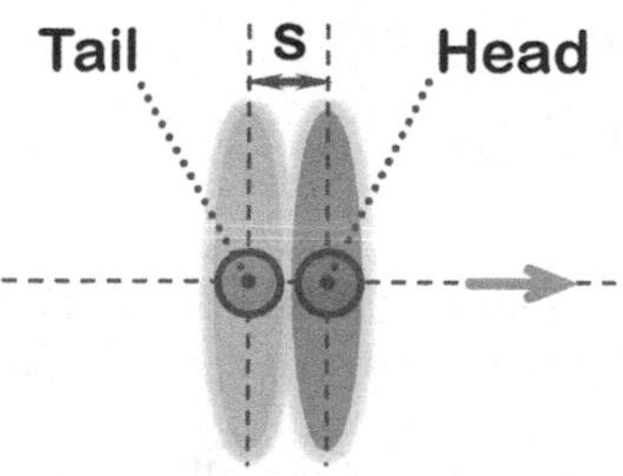

Figure 12.6 Two-particle model of the beam and its field. The beam fields are shown qualitatively.

An approximation of the *two-particle model* is a convenient way to estimate many effects related to beam dynamics. In this case, the beam is represented just by two particles — one at the head and one in the tail.

[1]Following derivations from Ya. Derbenev et al., Tesla-FEL Report 1995–05.

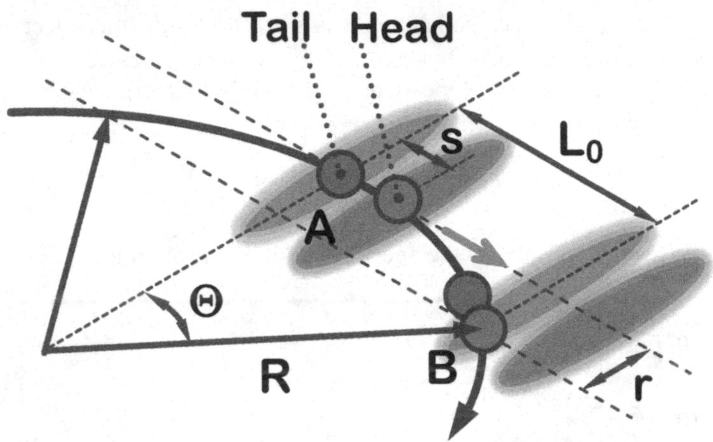

Figure 12.7 Illustration of the tail field overtaking the head of the bunch in the mechanism of coherent synchrotron radiation.

As we will see, the CSR effects are essentially caused by the possibility for the tail field to overtake the head particle while the beam is moving on a curved trajectory.

Let us thus define the bunch as a two-particle system with the length between the head and the tail s equal to *rms* bunch length of the initial beam σ. The EM fields of the relativistic bunch are distributed primarily in transverse directions, as indicated in Fig. 12.6.

Assuming that the beam is moving along the curved trajectory shown in Fig. 12.7, we can determine the moment at which the field of the tail, radiated at point A, and moving on the straight line, will catch up with the head of the beam. The condition for the catch-up (overtake) to happen in point B can be expressed as follows:

$$Arc\,(AB) - s = |\mathbf{AB}| \tag{12.12}$$

which can be rewritten as

$$R\,\theta - s = 2R\sin(\theta/2), \quad \text{which gives} \quad s \approx R\,\theta^3/24 \tag{12.13}$$

Thus the *overtaking distance* L_0 can be estimated as

$$L_0 = |\mathbf{AB}| = R\,\theta \approx 2\left(3sR^2\right)^{1/3} \tag{12.14}$$

Knowing the overtaking distance, we can determine the characteristic transverse distance r, which is equal to the distance between the head particle at the point of overtake and the axis of propagation of the SR fields that travel along the initial trajectory (Fig. 12.7).

The transverse characteristic distance can be estimated as follows:

$$r = L_0\,\theta/2 = 2\left(9s^2R\right)^{1/3} \tag{12.15}$$

We can now estimate the field of the bunch tail (radiated at point A), acting on the head at point B.

Assume that the beam is uniform and has the linear change density $eN\lambda$ where $\lambda = 1/\ell_b$ and ℓ_b is the length of the bunch.

The values of the transverse field at the characteristic distance for the linear charge density beam are as follows:

$$E_\perp = H_\perp \approx \frac{2N_e\lambda}{r} \tag{12.16}$$

In order to find the longitudinal field acting on the head, we need to multiply the transverse field by the angle: $E_\perp \cdot \theta$. The longitudinal force acting on the head can thus be estimated as

$$F_{||} = eE_\perp \cdot \theta = \frac{2Ne^2\lambda\theta}{r} = \frac{2Ne^2\lambda}{(3sR^2)^{1/3}} \tag{12.17}$$

Let us now assume that $s = \ell_b = 3^{1/2}\sigma$ (the latter assumes that the bunch distribution is Gaussian). The estimate for the longitudinal force thus becomes

$$F_{||} \approx \frac{2Ne^2}{3R^{2/3}\sigma^{4/3}} = \frac{2N\,r_e mc^2}{3R^{2/3}\sigma^{4/3}} \tag{12.18}$$

which is a rather accurate back-of-the-envelope estimate.

Accurate derivations of the CSR effects for a realistic Gaussian bunch can show that the longitudinal force acting on the beam is very close to the estimate Eq. 12.18 and contains an extra term: the shape function F_0, which depends on the position of the particle within the bunch.

$$F_{||} \approx \frac{2N\,r_e mc^2}{3R^{2/3}\sigma^{4/3}}\,F_0 \tag{12.19}$$

The shape of the CSR shape function is shown in Fig. 12.8. As we can see, its amplitude is close to one, confirming the back-of-the-envelope estimate. The shape function changes its sign. While the head of the bunch slightly accelerates due to the CSR effect, the major part of the bunch decelerates.

The effects of CSR are particularly important in bunch compressors and can prevent achieving ultra-short bunches. CSR can cause bunch instability and microbunching, and can therefore deteriorate the longitudinal phase space of the beam.

To conclude, one should note that the mechanism of CSR's creation — depicted in Fig. 12.7 — suggests that, in certain parameters, there may be a "cure" for CSR, because the vacuum chamber where the beam and fields propagate can partially shield the fields and reduce the CSR effects.

12.1.4 SHORT LASER PULSE AND Q-SWITCHING TECHNIQUES

The methods of generating short laser pulses often rely on so-called *Q-switching techniques.* In this case, the laser gain medium is placed in the optical cavity (the quality factor Q of this cavity can be controlled). In the initial moment, the Q factor

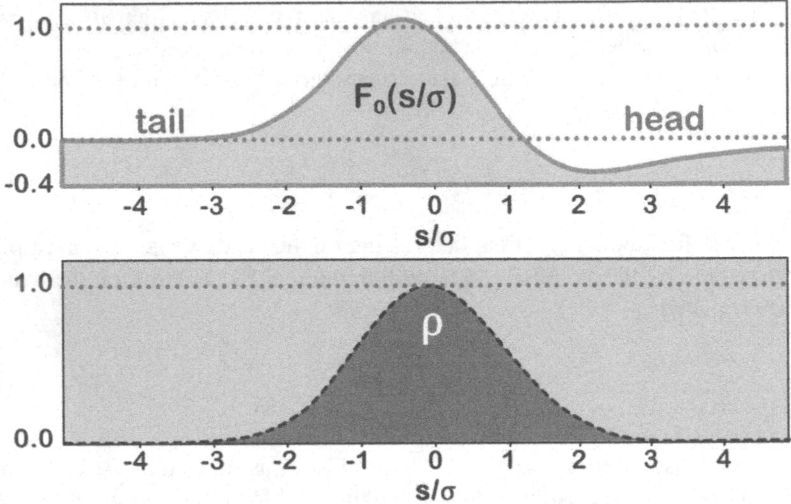

Figure 12.8 Shape function F_0 (top plot) of coherent synchrotron radiation for a bunch with Gaussian density profile (bottom plot).

is set to a low value, and pumping of the laser medium will then build up a large population inversion in the medium.

Once a sufficient inversion population is achieved, the Q factor of the laser cavity is suddenly increased, which results in quick buildup of the light in the cavity and an avalanche of stimulated emission. A giant and short pulse is thus emitted from the laser cavity. This technique is illustrated in Fig. 12.9.

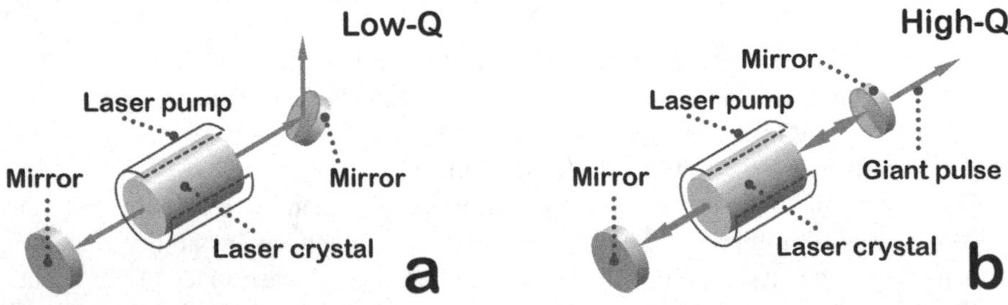

Figure 12.9 Q-switching technique. In step one (a) the pump builds up large inversion in the gain media. In step two (b) the laser cavity switches from low to high-Q.

12.1.5 Q-SWITCHING METHODS

The methods used to switch the Q factor in the laser cavity can be divided into two categories: passive and active. Active methods may involve mechanical effects, e.g., rotation of the laser cavity mirrors or applying an EM field to a substance inserted into the cavity (which changes its optical properties under the influence of the field). Passive methods involve special mirrors that can change the Q factor reflectivity when the power level of radiation reaches a certain level.

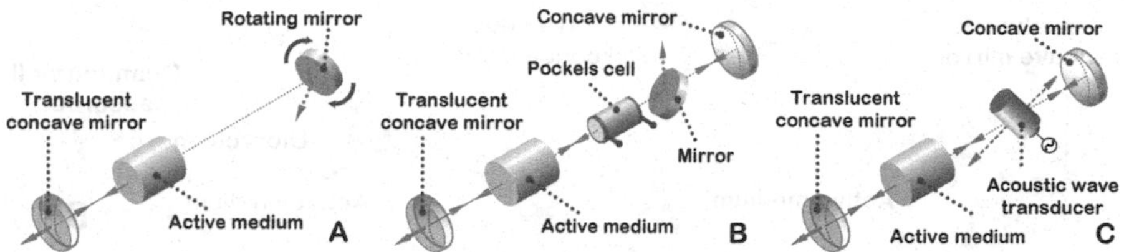

Figure 12.10 Examples of active Q-switching methods. Rotating mirror (A), Electro-optic (B) and Acousto-optic (C).

Some of the active methods are presented in Fig. 12.10. The first active Q-switching method is mechanical and involves the use of a rotating mirror. In this case, *lasing* will happen only at the moment when the mirror is parallel to the other mirror. A rotating cylinder — to which many mirrors are connected — is sometimes used because it generates many pulses for each rotation of the cylinder.

The speed of the mechanical active Q-switching methods is obviously limited by the mechanical strength of the movable objects, induced vibration and other factors. Active methods that minimize mechanical motion can avoid these limitations. The second active method is based on the use of the *acousto-optic* effect — the Bragg diffraction of light from the planes of a varied refractive index created in a crystal by the applied sound wave. When the acoustic wave transducer is switched on, a certain fraction of light diffracts away from the main light path in the laser cavity, which can then be used to change the cavity's Q value. The third active method is based on the *electro-optic* effect — the dependence of optical properties on objects such as absorption or refraction (called *Pockels effect*) on the applied electric field.

Similar to the active methods, various passive methods of Q-switching exist. Two examples shown in Fig. 12.11 include a saturable absorber and a SESAM. The saturable absorber uses the fact that certain absorbers become transparent when they reach saturation and cannot absorb radiation any further. At the moment when they become transparent, increasing the Q factor of the cavity can lead to lasing.

A saturable absorber can be built based on doped YAG crystals or GaAs, on media with immersed *quantum dots*, or semiconductors (called *semiconductor saturable absorber mirrors* — SESAM). Another possible design of a passive Q-switch is based on a *thin-film absorber* where a quarter-wave plate is combined with a *thin-film polarizer* (TFP), serving as a hold-off polarizer in the Q-switched laser cell.

12.1.6 REGENERATIVE AMPLIFIERS

The technology employed in Q-switching techniques is used in *regenerative laser amplifiers* (often called "regens"), which are designed to generate short, powerful laser pulses. In a "regen," a laser amplifier is placed inside of a Q-switched optical cavity (see Fig. 12.12) and it operates as follows.

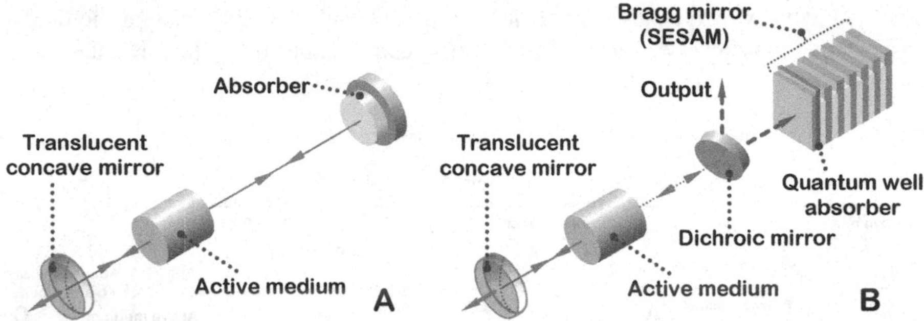

Figure 12.11 Passive Q-switching — saturable absorber (A) and SESAM (B).

First, the gain medium (amplifier crystal in Fig. 12.12) is pumped by the laser diode. A short initial pulse from the "master oscillator" is then injected into the input of the system through the thin-film polarizer and the Faraday rotator, both of which are switched on for this short moment in order to let the input pulse pass.

The initial pulse then undergoes many round trips through the system, its power level getting amplified each time it passes the gain medium until it reaches a high level. Finally, another switch is powered (Pockels cell in Fig. 12.12) and the amplified pulse is released from the system.

12.1.7 MODE LOCKING

Mode locking is the technique for achieving short laser pulses. In this method a rapid modulator is installed in the laser cavity, one that can open for short moments exactly in sync with the pulse's round-trip time around the cavity; see Fig. 12.13. Therefore, all the generated and amplified photons will be clustered only within those short moments when the modulator is open.

The amount of time of an average round trip in a laser resonator can be on the order of a nanosecond, while the modulator can open for periods of tens of femtoseconds. Correspondingly, the peak power in the mode-locked lasers can be of many orders of magnitude higher than the average power.

The name of the technique — mode locking — is a term derived from the analysis of this problem in the frequency domain. The laser cavity has a certain bandwidth that can support many longitudinal modes — all of which can coexist. In a normal laser, the phases between different modes are random. The mode locking created by the modulator would, in fact, create a certain fixed relation between the phases of all modes. In this case, there will be only one point in the laser cavity at each moment of time where the modes' electric fields will add together constructively — which is equivalent to the case of a single short pulse traveling inside the laser cavity.

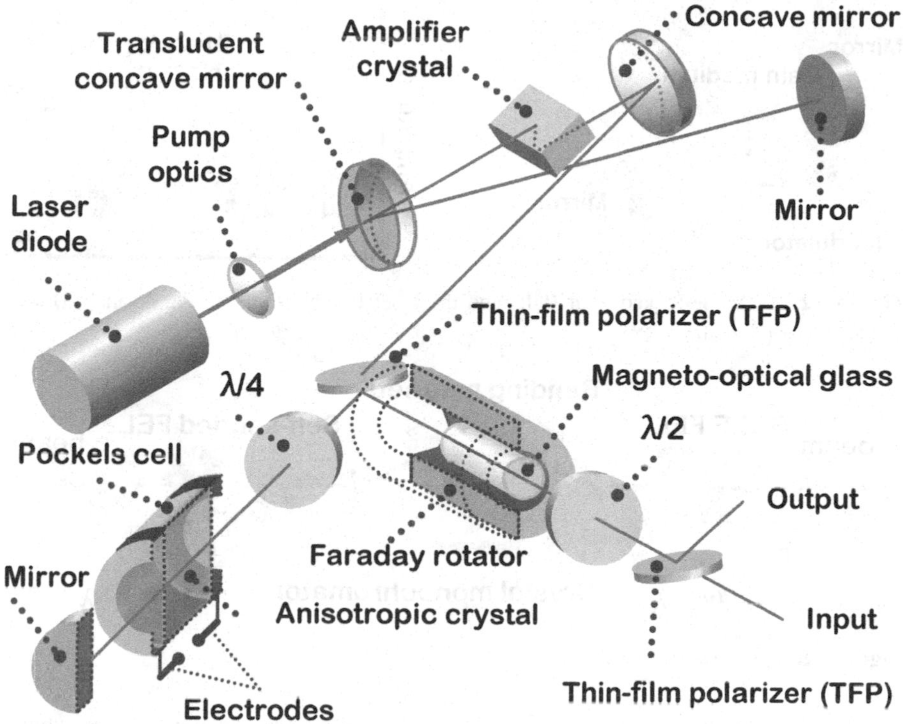

Figure 12.12 Schematics of a regenerative amplifier.

12.1.8 SELF-SEEDED FEL

Various techniques used in optics and particularly in laser systems are ideologically similar to techniques employed for creating short pulses of FEL radiation. The *self-seeding* technique described in this section has certain similarities with the mode-locking method described in the previous section.

Recall that, in the SASE regime, the FEL lasing starts from noise, and also remember that the length of the slice that contributes to radiation is much shorter than the length of the electron bunch.

This means that each individual slice can generate radiation at slightly different wavelengths near the resonant wavelength, and, moreover, the amplitude of the radiation coming from each slice can be slightly different, as illustrated in the $P(\lambda)$ spectrum on the left side of Fig. 12.14.

The output structure of the FEL can be considerably improved using the self-seeding approach. In this event, the FEL is split into two parts. Before entering the second part, radiation generated in the first part is passed through a crystal monochromator. In order to have the timing maintained, the beam is simultaneously passed through a four-bend chicane.

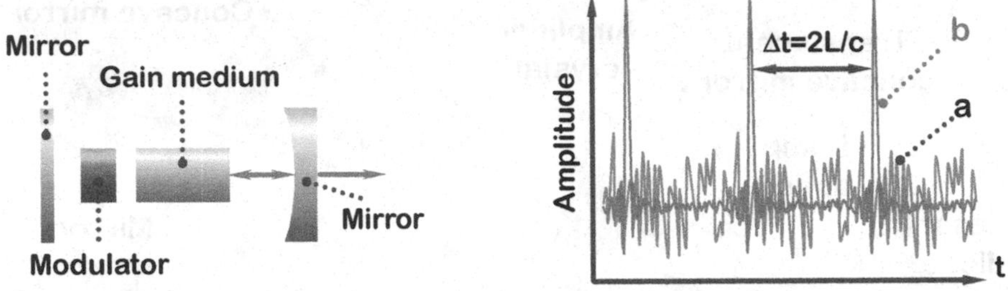

Figure 12.13 Mode-locked laser (left) and the laser output (right) in the normal (a) and mode-locked (b) regimes.

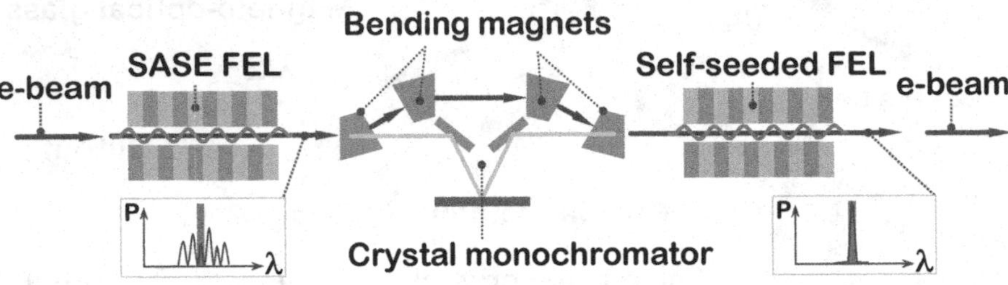

Figure 12.14 Self-seeded FEL.

The second part of the FEL is thus seeded with narrow spectrum radiation, which continues to be amplified along the way in the undulator; the resulting output spectrum of FEL then contains a narrow peak, as shown in Fig. 12.14 on the right side.

12.2 LASER–BEAM INTERACTION

Interaction of laser light with electron beams in wigglers offers a wide range of techniques that can be used to manipulate the electron beam or to improve the properties of FEL radiation.

We recall from Chapter 8 that the light of a resonant harmonic will be exponentially amplified in an undulator, taking the energy from the electron beam. The inverse can also be true — sending laser light at the resonant harmonic into the undulator will affect the electron beam (inverse FEL), creating variations of its energy.

The inverse FEL principle is the basis for many techniques used to manipulate electron beams, create short pulses of radiation, generate higher harmonics, create two-color FEL pulses and many others uses.

12.2.1 BEAM LASER HEATING

Short laser pulses can be used to improve the properties of short pulses of FEL radiation. The first example we will consider is the *laser heater*.

The generation of femtosecond-short X-ray pulses requires fs-short electron bunches. However, as we saw earlier in this chapter, the creation of short bunches in bunch compressors can be complicated due to CSR effects that can cause instability and microbunching.

Instabilities can often be suppressed (thanks to decoherence — see discussion of Landau damping in Section 11.1.4) if the beam has sufficient spread of its relevant phase space coordinates (the energy spread, in the case of CSR instability).

The issue that has been observed in FELs is that the beam energy spread coming from a photocathode gun is extremely small, and insufficient for suppressing CSR instability via Landau damping mechanism.

A method has been developed at SLAC to introduce additional uncorrelated energy spreads into the beam: laser heating (see Fig. 12.15).

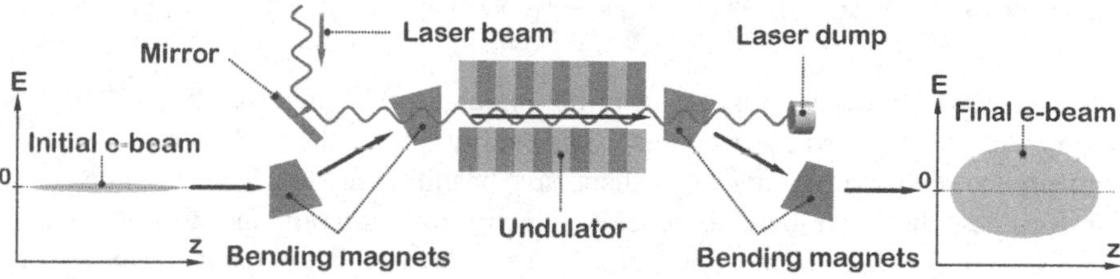

Figure 12.15 Laser heater.

In this case, the laser light of the resonant wavelength co-propagates with an electron beam through a wiggler. This system acts as an inverse FEL and introduces an additional energy spread to the beam, sufficient to cure the CSR instability.

A laser heater can increase the uncorrelated energy spread in the electron beam, thus enhancing the Landau damping, and suppress Coherent-Synchrotron-Radiation-driven instability, improving FEL generation.

12.2.2 BEAM LASER SLICING

Beam laser slicing is a technique that selects a femtosecond short portion of radiation from a much longer initial electron pulse.

In this method,[2] a very short laser pulse overlaps with the center of a longer bunch in the undulator or wiggler; see Fig. 12.16. The laser wavelength λ_L matches the undulator resonance condition

$$\lambda_L = \frac{\lambda_W}{2\gamma^2}\left(1 + \frac{K^2}{2}\right)$$

and the interaction of the light with the beam in the undulator will therefore produce modulation of energy in the short beam slice overlapped with the laser pulse.

[2]A. A. Zholents and M. S. Zolotorev, Phys. Rev. Lett. 76, 912 (1996).

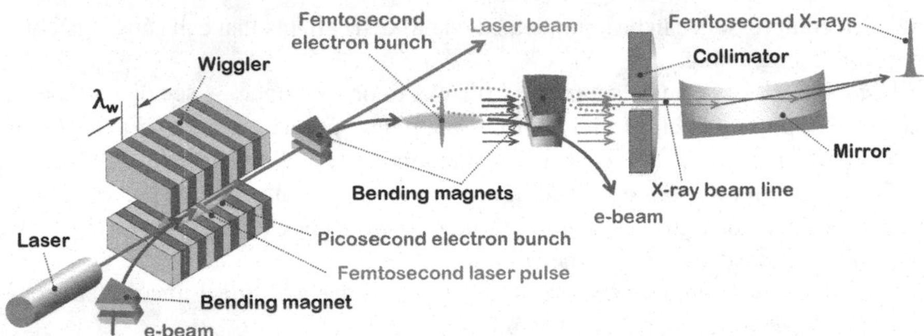

Figure 12.16 Beam laser slicing technique for producing femtosecond X-ray pulses.

The electron beam coming out from the undulator will therefore have a short fs region with an increased energy spread. The energy spread of this short region can be converted to spatial variation using a dispersive beamline section. The resulting beam can then pass through a bend to generate synchrotron radiation and, following that, the SR corresponding to the short portion of the electron beam can be transversely separated from the rest of the pulse using collimators.

The method described above can be especially suitable for ring-based SR sources, where the natural length of the electron bunch is a picosecond long. The application of the laser slicing technique can therefore help in generating femtosecond synchrotron radiation pulses, even for third-generation sources.

12.2.3 BEAM LASER HARMONIC GENERATION

Harmonic generation is the technique that can help to produce X-ray photons of much higher energies. We recall from Section 8.4.3 that odd higher harmonics can also resonate and can thus be generated in an FEL. An FEL would normally generate primarily the first harmonic. In order to primarily generate a higher-order harmonic, external seeding is required. There are, however, no conventional lasers of appropriately short wavelengths that can be used for such seeding.

A technique invented[3] by G. Stupakov is currently solving the problem: *echo-enabled harmonic generation* (EEHG). The term *echo* came from plasma physics and refers to the phenomenon of a spontaneous appearance of a wave with wavenumber k_3 in plasma, at a certain time after the initial excitation waves with wavenumbers k_1 and k_2 would decay via the collisionless *Landau damping* mechanism.

It is useful, however, to discuss the mechanism of EEHG without referring to the physics of plasma echo. The EEHG technique is based on laser-beam interaction in wigglers and distortion of the $z-E$ phase space in four-bend magnetic chicanes; see Fig. 12.17. Here, the first wiggler and laser with wavenumber k_1 create sine-like modulation of the $z-E$ phase space.

[3] G. Stupakov, PRL 102, 074801 (2009).

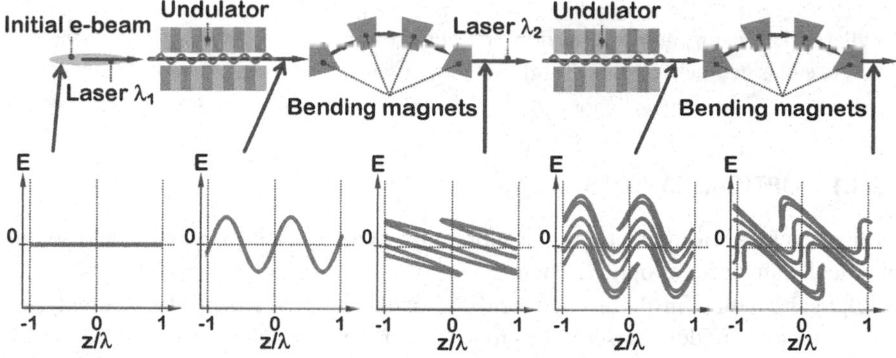

Figure 12.17 Echo-enabled harmonic generation scheme — EEHG.

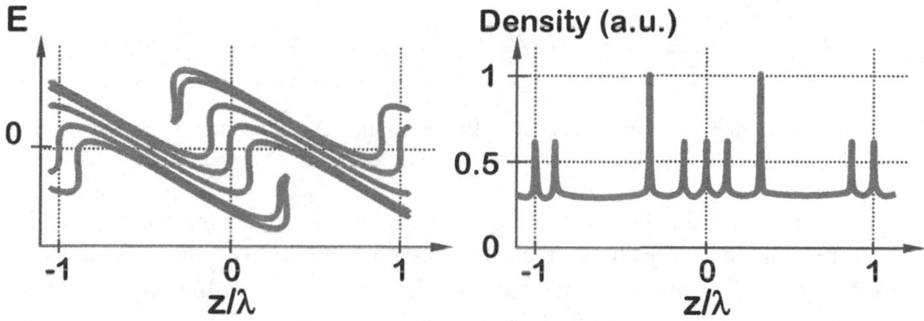

Figure 12.18 Phase space (left) and density profile (right) of an EEHG-modulated beam.

The first chicane applies $z = z + R_{56} \cdot \Delta E/E_0$ transformation that deforms the sine-like phase space into diagonally distorted lines. The second wiggler and laser k_2 modulate the beam again, and the second chicane creates another distortion, which, if properly adjusted, produces repeating vertical lines in the $z-E$ plane (see Fig. 12.18), which correspond to spatial density modulation of the beam at high harmonics of the lasers.

Spatial harmonics that can be created with this method are defined by $k = nk_1 + mk_2$, where the integers n and m can be large, granting the potential for seeding at very short wavelengths.

12.3 BEAM OR PULSE ADDITION

In this section, we will briefly discuss the methods of laser pulse addition in the optical cavities in comparison with injection methods used in accelerators.

Electrons and protons are *fermions* and obey the Pauli exclusion principle, while photons are *bosons*, and therefore the topic of beam or pulse addition is one where there is an important difference between the area of accelerators, where we use charged particles, and and the area of lasers, where we deal with light.

12.3.1 OPTICAL CAVITIES

Optical cavities consisting of two or more mirrors are a widespread element used in practically any laser or optics system.

Optical cavities, in the event when one of the mirrors is semi-transparent, are efficient tools for adding laser pulses together for their accumulation to higher intensities. Various arrangements of optical cavities are possible, and some examples are shown in Fig. 12.19.

Figure 12.19 Examples of optical cavities. Plane-parallel (A), concentric/spherical (B) and confocal (C) configurations.

An optical cavity suitable for hosting interaction between an electron beam and a laser (for Compton sources in particular) can be composed of two or more mirrors.

One of the important concerns is the cavity's stability and the light pattern inside of it, and also how it reacts to small deviations in its parameters.

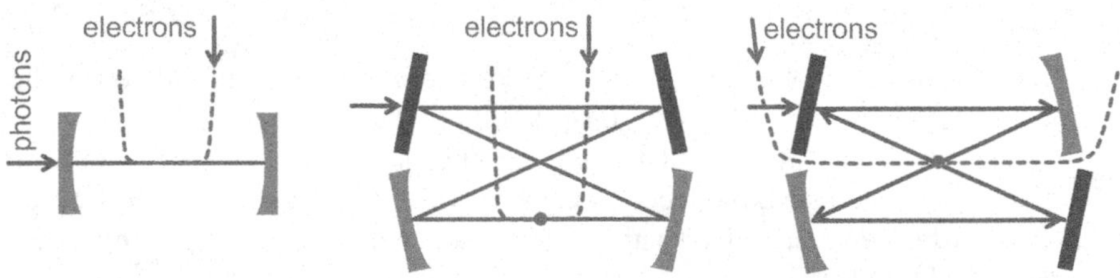

Figure 12.20 Four-mirror optical cavities suitable for electron beam–laser interaction.

It has been found that, while two-mirror systems are possible, the four-mirror optical cavities are more practical,[4] and are especially more impervious to disturbances.

Moreover, the four-mirror cavities can also be more suitably arranged when one needs to pass the electron beam around the cavities, as shown in the third case presented in Fig. 12.20.

There is an essential difference between the accumulation of pulses of light and the accumulation of bunches of particles, as we will discuss momentarily.

[4]A. Variola et al., ThomX — Conceptual Design Report, LAL, 2010.

12.3.2 ACCUMULATION OF CHARGED PARTICLE BUNCHES

The accumulation of bunches of charged particles (e.g., at injection) obeys *Liouville's theorem* (see its full definition in the grey box below).

The consequence of the theorem is the conservation of emittance ε; i.e., the area $\int p\,dq$ of the phase space beam portrait is constant (during accelerating, $\varepsilon_n = \gamma\varepsilon$ — called normalized emittance — is constant).

Beam cooling methods (synchrotron radiation cooling, electron, stochastic) do not contradict Liouville's theorem, as in these cases the cooling forces acting on the particle depend on its velocity.

One important consequence of the theorem is that we cannot stack bunches of charged particles on top of each other in the phase space. If we would like to accumulate them to increase the total intensity, we must inject the bunches into the neighboring areas of the phase space (Fig. 12.21).

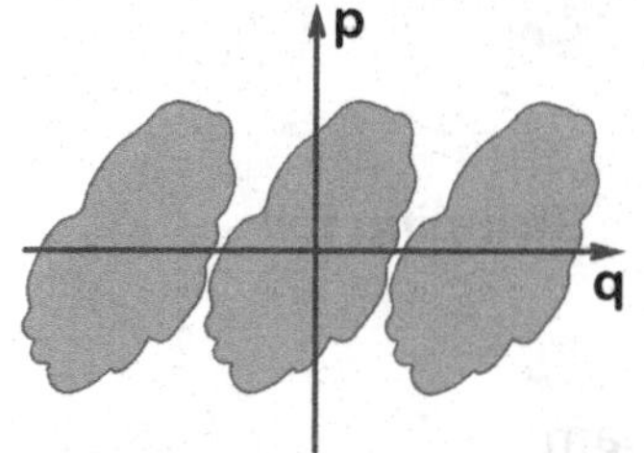

Figure 12.21 Phase-space stacking.

Liouville's theorem: in the vicinity of a particle, the particle density in phase space is constant if the particle moves in an external magnetic field or in a general field in which the forces do not depend upon velocity.

Phase-space stacking and accumulation of bunches during injection can be done in various ways. Fig. 12.22 shows an example of transverse stacking.

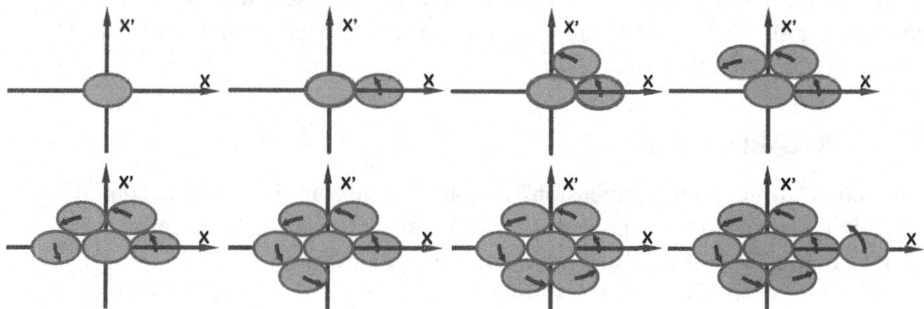

Figure 12.22 Transverse phase-space stacking. Consecutive moments.

Longitudinal stacking is also possible, as illustrated in Fig. 12.23. Fast-switching bending magnets or electrostatic deflectors — *kickers* — are used to perform bunch stacking during injection.

There is one particular injection mechanism that can overcome limitations imposed by Liouville's theorem — the *charge-exchange injection.* In this case, H^- ions are injected into the beamline through a thin foil. Ion interactions at the foil strip away two electrons and the ions are therefore converted (at the foil) to protons.

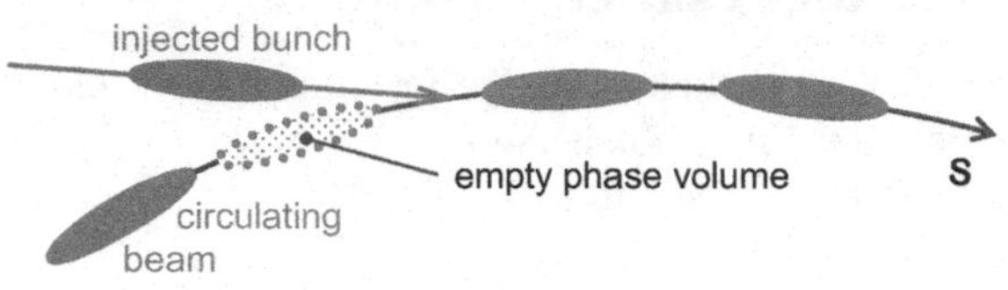

Figure 12.23 Longitudinal phase-space stacking.

Typical kickers have a uniform field. Nonlinear kickers with a zero field on their axes minimize any disturbance of the circulating beam during injection.

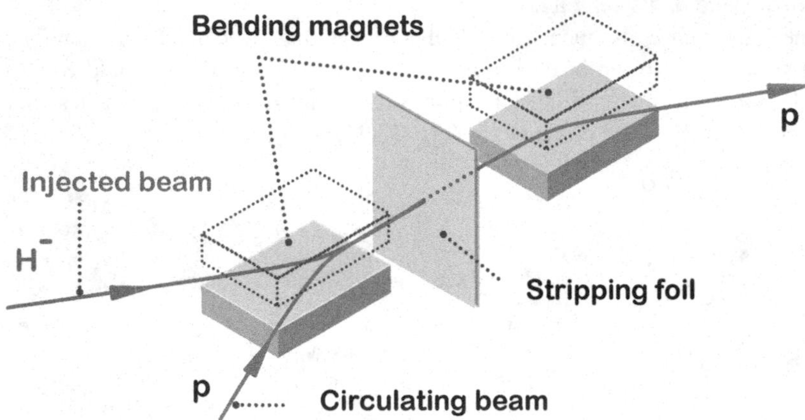

Figure 12.24 Charge-exchange injection.

This charge-exchange injection can be applied many times into the same phase space. Since the foil is inserted into the accelerator on the way to the main circulating beam (see Fig. 12.24), the number of injections is limited by the scattering of the circulating beam on this foil.

12.4 POLARIZATION

Let's now "go to another dimension" and discuss very briefly particle spin and polarization. Generation and preservation of beam polarization is essential for many designs of modern colliders and accelerator installations.

Spin motion in accelerators exhibits precession — in Fig. 12.25, a spin vector rotates around the guiding magnetic field, which is typically oriented in the vertical direction. The equation of motion for the spin $\vec{S}$ in an external magnetic field $\vec{B}$ in the particle rest frame is

$$\frac{d\vec{S}}{dt} = \vec{\Omega} \times \vec{S} \quad \text{where} \quad \vec{\Omega} = -\frac{g}{2}\frac{e}{mc}\vec{B} \tag{12.20}$$

where g is the Landé g-factor or gyromagnetic ratio. Note that Eq. 12.20 differs from the equation of motion of a particle in the magnetic field by the factor $g/2$. Correspondingly, one usually defines the difference as $G = (g-2)/2$, which depends

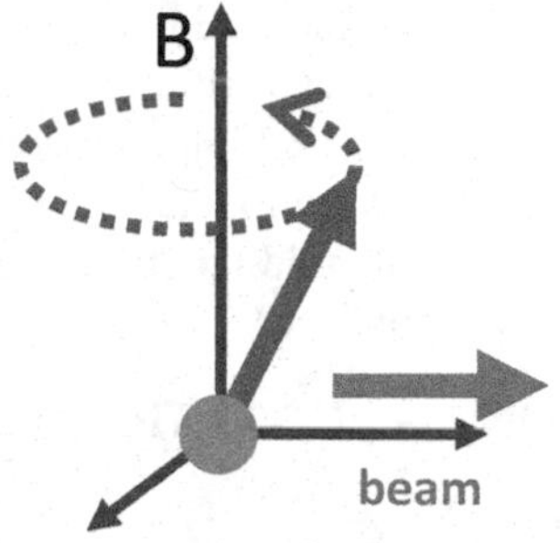

Figure 12.25 Spin vector precession.

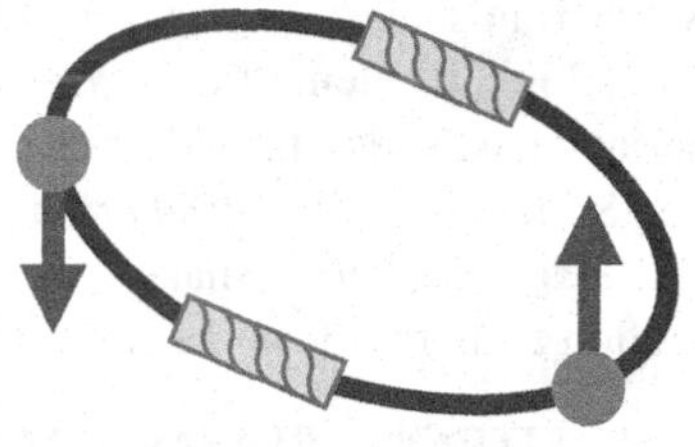

Figure 12.26 Two Siberian snakes for polarization preservation.

on the particle type:

$$G = \frac{g-2}{2} = 1.7928 \quad \text{for protons, and} \quad \frac{g-2}{2} = 0.00115966 \quad \text{for electrons}$$

The number of spin precessions in one orbital revolution is proportional to G and to the relativistic factor γ and will define the *spin tune* Q_s:

$$Q_s = \gamma G$$

Since the spin tune depends on energy, it can cross many resonances during acceleration, which can cause depolarization. The *imperfection resonances* are defined by the condition

$$Q_s = n$$

where n — integer, and said imperfection resonances are driven by various magnetic field imperfections spread around the orbit. On the other hand, the *intrinsic resonances* involve interaction with vertical betatron motion described by the vertical betatron tune Q_y:

$$Q_s = nP \pm Q_y$$

where P — the number of superperiods in the accelerator.

We can see that with $G = 1.793$ for protons, during their acceleration to any significant relativistic energies, many resonances will be crossed. Correspondingly, the initially polarized beam can lose its polarization during acceleration.

When the required acceleration is small (e.g., in intermediate booster accelerators) and one needs to cross just a few resonances, one can use the *spin tune jump* technique, which is conceptually similar to the gamma-jump technique used for crossing the transition energy discussed as an illustration of the inventive principle 21: *skipping* — see Fig. 15.21. A similar approach can help to preserve beam polarization.

A radical and ingenious method of avoiding crossing any resonances for arbitrary acceleration is to use *Siberian snakes*[5], which are special (e.g., helical) magnets that rotate the particle's spin (preserving their external orbit), as shown in Fig. 12.26. Due

[5] Ya.S. Derbenev, A.M. Kondratenko, Zh. Eksp.Teor. Fiz. 64: 1918-1929, (1973)

to how Siberian snakes spin and flip 180° at every half turn, their spin tune is equal to $Q_s = 1/2$ independent of energy — i.e., they avoid crossing any resonances.

Siberian snakes also provide full control of both intrinsic and imperfection resonances. Sometimes *partial Siberian snakes* are installed that flip and spin by less than 180°. Partial Siberian snakes can help to break coherent build-up of spin motion perturbation and provide some control of imperfection resonances.

12.5 POSITRON PLASMA ACCELERATION – INVENTION CASE STUDY

Creating a beam of positrons, particularly polarized positron beams for colliders or for nuclear physics or materials studies, involves a lot of advanced manipulations. We will, however, discuss only one topic: plasma acceleration of positrons.

There are two challenges for plasma acceleration of positrons in a plasma bubble created by normal matter (i.e., not anti-matter). The first is that plasma electrons, which create a density bump and a large accelerating field for electrons, will decelerate positrons if they are placed in the first main plasma bubble. The positrons can only be accelerated behind the density bump, i.e., in the beginning of the second plasma bubble. However, in the nonlinear blow-out regime, the second bubble is much smaller and there is not much space for positrons there. The second challenge is that the plasma ions, which give nice focusing in the case of electron beam acceleration, will give rise to defocusing in the case of a positron beam.

Thinking about these challenges for positron plasma acceleration, we can recall the inventive principle *separation*. Indeed, the electrons of plasma are useful, as they move fast and create a dense region which creates the accelerating field in the bubble. However, the ions of plasma are harmful — they defocus the positrons. If we could suggest how to *separate* these two factors — the useful and the harmful — we would solve the challenge.

Before continuing our discussion of this positron acceleration challenge, let's introduce the TRIZ inventive approach of *Smart Little People* — see Fig. 12.27.

Figure 12.27 Smart Little People inventive approach.

Instead of identifying yourself with an object like in synectics, in TRIZ, one can use the method of "Smart Little People," who can decide themselves how to move and what to do so that the desired inventive function will happen.

The method of Smart Little People (SLP) — who form the structure of the objects, fields or forces involved in a system, and who can themselves decide what to do and how to move for the desired result to happen — is a way to open and direct our imagination toward a possible inventive breakthrough. The well-known Maxwell's demon (illustrated in Fig. 12.28), where hot and cold molecules are collected in different parts of a box, can be considered an example of the Smart Little People method. While in the classical Maxwell's demon case, the demon would open the door at the right moment to sort the molecules, in the SLP method, the little people would know by themselves toward which side of the box to go. While this all sounds like pure fantasy, please hold your inner critic in the brainstorming mode, waiting for a more concrete example.

Figure 12.28 Maxwell's demon as an example of Smart Little People inventive approach.

One of the classic examples that TRIZ textbooks offer to illustrate the Smart Little People approach is the task of creating a more efficient ice-breaker — a ship intended to break thick ice in the Arctic Ocean.

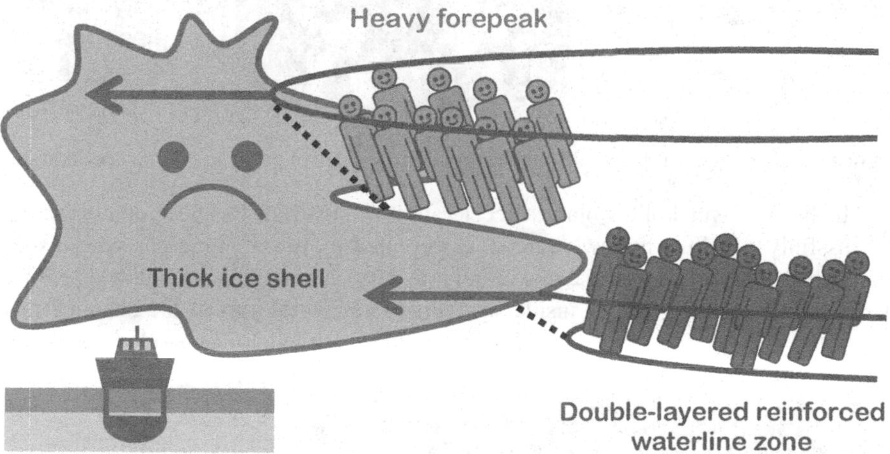

Figure 12.29 Smart Little People approach in application to a more efficient ice-breaker.

The ice-breaker contradiction is that it needs to be heavy and wide, but this would increase resistance from the ice. Ideally, the ice-breaker should exert pressure both below the ice and above the ice, where it is easy to move, yet without being bulky. So, the body of the ice-breaker, which we can assume consists of Smart Little People, should separate and rearrange themselves to be contained above and below the ice — see Fig. 12.29. Meanwhile, thin knifes connecting the upper and lower part of ice-breaker can keep it whole. This is an example from class TRIZ books. Maybe such an ice-breaker will not succeed in its other job — to create a path for other ships — but, by itself, it can pass through the ice easier. It also illustrates the SLP method.

Let's now return to the question of inventing positron plasma acceleration and imagine that we are considering a beam-driven (PWFA) approach, when the positron drive and witness bunches, separated by a short distance, are coming toward the plasma or to a gas (which can be pre-ionized by a laser).

We observed that ions are harmful for positron focusing. We also considered the need to separate the useful function of plasma electrons (acceleration) from the harmful factor of plasma ions (defocusing of positrons). Let's now imagine that ions of plasma are Smart Little People and just before the positron beam arrives, they step out of the way, or that they never appear there at the first place, performing the needed functions of separating the useful and harmful functions.

Once we stated the question in this way, it becomes clear that what we need is for the plasma ions — just like Smart Little People — to step out of the way before e+ beams arrive, or never appear. This will separate the useful function of plasma electrons (acceleration) from the harmful factor of plasma ions (defocusing of positrons). Outside of the metaphor of SLP, this solution is called hollow plasma channel positron acceleration — illustrated in Fig. 12.30.

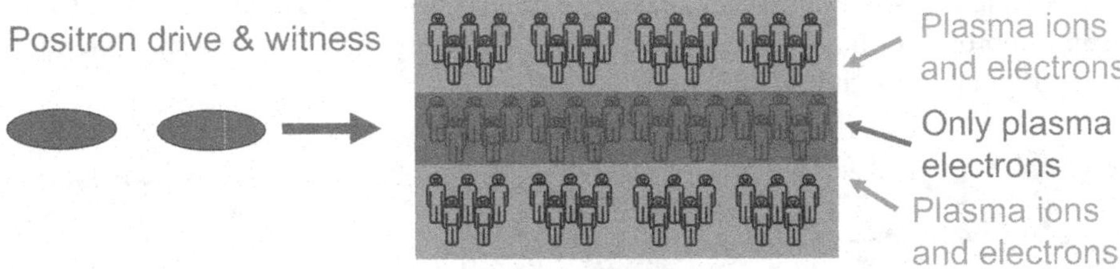

Figure 12.30 Smart Little People method and hollow channel positron plasma acceleration.

The beam-driven hollow plasma acceleration of positrons has been demonstrated successfully[6], and the hollow channel was created by pre-ionizing gas with a specially shaped laser pulse whose maximum intensity gathered in a circular ring, while its minimum intensity stayed inside said ring. The central part of the channel thus contained just gas and no plasma. The drive bunch attracted plasma electrons from the walls of the channel, creating an accelerating wakefield.

The invention of hollow channel beam-driven plasma acceleration of positrons demonstrates, retrospectively, the application of the inventive principle of separation and of the inventive approach "Smart Little People."

[6] S. Gessner et al., Nature Communications, v. 7, 11785 (2016)

EXERCISES

12.1. *Chapter materials review.*
Describe the typical approaches for preventing beam breakup instability in linacs. Discuss if these approaches can be used in other situations considered in this book.

12.2. *Mini-project.*
Define the approximate parameters of a 1 GeV compact SR source[7] ring aimed at 10 keV X-rays, based on top-off, on-orbit and on-energy injection by a laser plasma acceleration system. Discuss the research steps required for implementing such a concept.

12.3. *Analyze inventions or discoveries using TRIZ and AS-TRIZ.* Analyze and describe scientific or technical inventions described in this chapter (e.g., suppression of CRC radiation with help of vacuum chamber aperture) in terms of the TRIZ and AS-TRIZ approaches, identifying a contradiction and an inventive principle that were used (could have been used) for these inventions.

12.4. *Developing AS-TRIZ parameters and inventive principles.* Based on what you already know about accelerator science, discuss and suggest the possible additional parameters for the AS-TRIZ contradiction matrix, as well as the possible additional AS-TRIZ inventive principles.

12.5. *Practice in reinventing technical systems.* The *Wien filter* is a device with a combination of a transverse electric and a magnetic field that can act as a *velocity filter* for a particle beam. Suggest how this approach can be used for a) manipulating spin orientation of polarized beams; and for b) designing an interaction region of a linear collider with head-on collision of electron and positron bunches (i.e., zero crossing angle), while being able to separate the beams after the collision.

12.6. *Practice in the art of back-of-the-envelope estimations.* Imagine that an accelerator for polarized beams has a figure-8 shape; i.e., it looks like ∞ and is located in the horizontal plane. Estimate how the spin tune depends on the beam energy in this ring. *(It is assumed that you can identify the most important effects playing roles in this task, can define the necessary parameters and set their values, and can get a numerical answer.)*

12.7. *Practice in the art of back-of-the-envelope estimations.* Imagine that a very short relativistic electron bunch needs to be collimated to remove its transverse halo tails. The bunch is propagating on the axis of a cylindrical vacuum chamber. The collimation is performed by an iris-like insert placed on the axis. The inner radius of the insert is much smaller than the radius of the vacuum chamber, and the thickness of the insert is longer than the bunch length. Estimate the energy loss of the particles in the bunch after passing this collimator. *(It is assumed that you can identify the most important effects playing roles in this task, can define the necessary parameters and set their values, and can get a numerical answer.)*

---

[7] This task was the subject of USPAS-2016 mini-project. See the Solution guide for more details.

13 Advanced Technologies

In this chapter, we will briefly discuss a variety of topics related to technologies, including power sources like RF and lasers, superconducting technologies like magnets and RF cavities, and technologies for nonlinear injection, plasma-based CPA, and MRI-guided proton therapy, etc.

13.1 POWER SOURCES

Sinusoidal power ranging from a few kW to a few MW is needed to drive the accelerating structures. This is commonly achieved by using RF power amplifiers such as triodes and tetrodes (operating from a few MHz to a few hundred MHz), inductive output tubes (suitable for CW applications, providing tens of kW at a high efficiency) and klystrons (which typically operate above a few hundred MHz) and have proven to be the most effective power generators for accelerator applications at higher frequencies. We will also touch on the advanced idea to "power" plasma oscillations and another advanced concept to prepare powerful laser pulses for that.

13.1.1 IOT – INDUCTIVE OUTPUT TUBES

The *inductive output tube* (IOT) (invented by A. Haeff ca. 1939) is based on the principle that a toroidal cavity surrounding an electron beam of oscillating intensity could extract power from the beam without intercepting the beam itself; see Fig. 13.1.

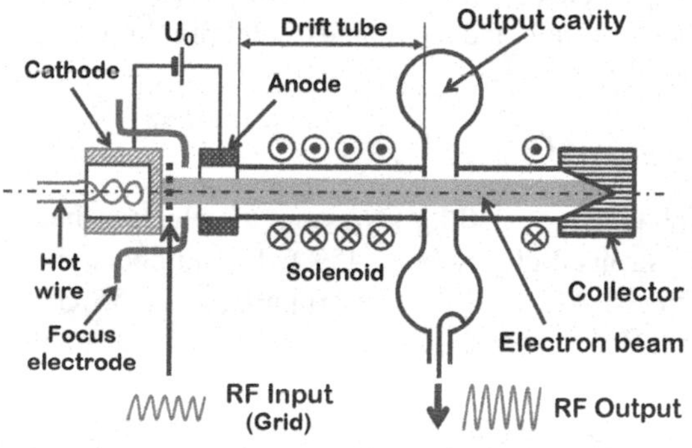

In an IOT, the oscillating EM fields carried with the beam excite the modes found inside the toroidal cavity, which allows RF energy to be transferred from the beam to a waveguide or coaxial cable connected to the resonator via a coupling loop.

Figure 13.1 Schematic of an inductive output tube.

13.1.2 KLYSTRON

Building on the success of the IOT, Sigurd and Russell Varian added a cavity resonator at the beginning of the beamline in order to provide a signal input for the inductive output tube. This input resonator acts as a pair of control grid electrodes

DOI: 10.1201/9781003326076-13

performing *velocity modulation* in the electron beam. This results in bunching developed in the beam by the time it arrives at the output resonator. In the latter the amplitude variation will be converted to energy extracted from the beam. The Varian brothers called their invention a *klystron*. See klystron's schematic design in Fig. 13.2.

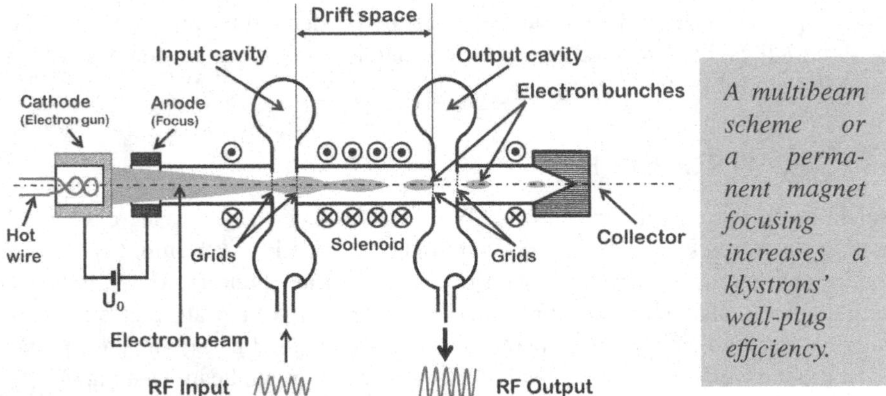

A multibeam scheme or a permanent magnet focusing increases a klystrons' wall-plug efficiency.

Figure 13.2 Schematic of a klystron.

Klystrons are similar to small linear accelerators, and operate as follows. First, electrons are emitted from a round cathode, which has a large surface area. Electrons are then accelerated by a voltage of a few tens of kV. A round beam is formed with a current from a few amperes to tens of amperes. The beam is focused by near-cathode electrodes and is further focused by the solenoid field (which is essential to ensuring effective beam transportation). The particles leaving the cathode pass through an input cavity operating at TM_{011} mode, which is fed from an external pre-amplifier.

The klystron output power is given by

$$P_{klystron} = \eta U_0 I_{beam} \tag{13.1}$$

where U_0 is the klystron supply voltage (e.g., 350 kV), I_{beam} is the electron beam current (e.g., 420 A) and η is the klystron efficiency (e.g., 45%). The numbers given above as examples correspond to the SLAC 5054 klystron running at 2.856 GHz. The klystron efficiency (which typically ranges from 45% to 65%) is one of the most important parameters and is the subject of continuous innovations — klystron designs of close to 90% efficiency have been suggested[1].

The trend of RF power-source technology changing from beam-based devices to solid-state semiconductor RF sources is a good illustration of TRIZ predictions about the evolution of technical systems.

[1] A. Baikov, C. Marrelli, I. Syratchev, in IEEE Trans. on Elect. Dev., v. 62, n. 10, 3406-3412, (2015).

IOTs, klystrons and similar RF power generators — while still extremely popular — are gradually being replaced by solid-state devices, for frequencies lower than 1 GHz in particular. The solid-state RF power systems have the advantage of compactness, higher efficiency and also reliability. The modular design of solid-state RF power sources, when each module contributes only a small fraction of the total power, makes it possible to significantly increase the reliability of the entire system.

13.1.3 MAGNETRON

Magnetron is an RF power source popular in particular for CW applications. In a magnetron the cylindrical cathode is located in the center and magnetic field is applied along the axis of the device. When the electrons are moving from the cathode to the anode, the magnetic field turns them and makes them move on spiral trajectories, creating azimuthal variations of the electron trajectories.

At certain parameters (of the voltage and magnetic field) the azimuthal variations of the electron trajectories will match the azimuthal spacing of the cavities arranged around the perimeter on the anode body, exciting fields in the cavities. The exited fields in their turn will enhance the velocity modulation and thus the spatial density modulation of the electron beam. Increased modulations will further increase the fields generated in the cavities. The magnetron amplitude will grow until saturation (due to effects related to the electron beam space charge) and remain constant.

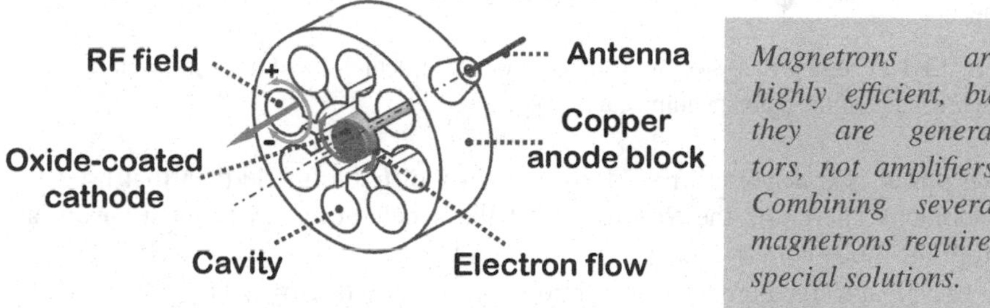

Magnetrons are highly efficient, but they are generators, not amplifiers. Combining several magnetrons requires special solutions.

Figure 13.3 Schematic of a magnetron.

A schematic of a magnetron is shown in Fig. 13.3. A magnetron, in contrast to klystrons and IOTs, is not an amplifier, but a generator, which starts from initial noise, i.e., the phase of the RF power generated by a magnetron is arbitrary in principle. Therefore, one of the difficulties of using magnetrons in accelerators when more than one magnetron is required is the difficulty of phase-locking those devices.

Efforts[2] of the community to phase lock one or more magnetrons — via RF power injection — and to combine their power, are bringing results — single phased locked magnetron driving a single superconducting cavity has been demonstrated, and phase locking two or more magnetrons is on a promising path.

---

[2]See, e.g., J. Benford et al., Phys. Rev. Lett. 62, 969, (1989); A. C. Dexter et al., Phys. Rev. ST Accel. Beams 14, 032001, (2011); H. Wang et al., IPAC2021, TUPAB348, 2021; G. Kazakevich et al., NIM A, v. 1039, 167086, (2022);

13.1.4 POWERING THE ACCELERATING STRUCTURE

An accelerating structure of a linac or other accelerator is typically fueled by pulsed or CW power RF tubes — klystrons. The inductive output tubes are also used, and most recently klystrons are being replaced more and more by solid state amplifiers.

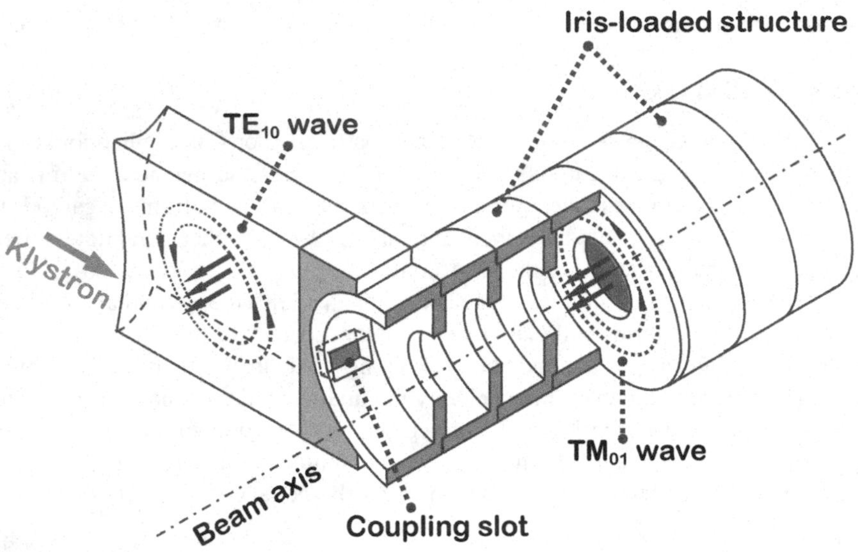

Figure 13.4 Feeding RF power into an accelerating structure. Field lines show electric and magnetic fields of the corresponding cavity modes.

RF power is transported from a klystron in a different mode than that used for acceleration. Therefore, the *RF coupler* of the acceleration cavity must ensure an appropriate mode conversion.

It is typical to feed the RF power into the linac structure by a TE_{10} mode of EM wave in a rectangular waveguide. The mode is then converted by a coupling slot into a cylindrical TM_{01} mode in the accelerating cavity, as illustrated in Fig. 13.4.

13.2 LASERS AND PLASMA

Let's discuss several topics related to laser and plasma technology — which exhibit connections to the inventive principle *merging*, as well as the principle of using *already damaged material* — and which can open the path toward plasma-based CPA and efficient plasma acceleration.

13.2.1 COHERENT ADDITION OF LASER PULSES

Unlike for electrons, which obey the Liouville's theorem and onto which we cannot add bunches on top of each other into the same phase space, such an action is perfectly possible for laser light. Returning to the laser theme, we would like to briefly mention a technique recently suggested[3] to coherently add laser pulses.

The technique is aimed in particular to fiber lasers that can produce pulses with high efficiency (tens of percent) and high repetition rates (tens of kHz and higher). The purpose of the technique, which is presently under development, is to coherently combine the output of many fiber lasers with the goal to produce laser systems with high repetition rates, high energy and high efficiency. The concept of coherent combination is illustrated in Fig. 13.5.

The technique of coherent fiber laser pulse combination strives to solve the deficiency of present high-peak power lasers (such as Ti:Sapphire or Nd:glass lasers), both of which have high-peak powers but very low repetition rates (ranging from one shot per hour to around a Hz) and low efficiency (less than 0.01%).

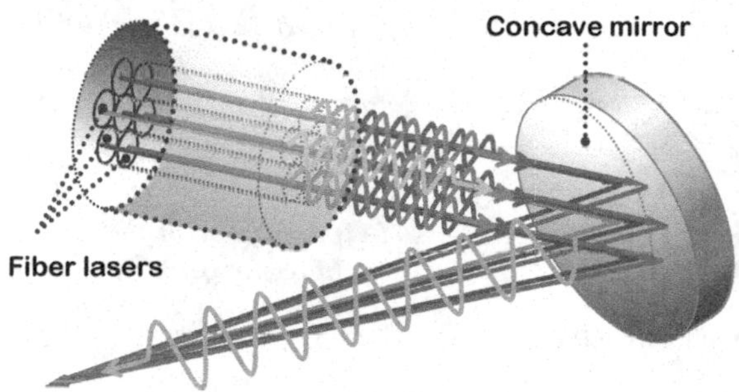

Figure 13.5 Concept of fiber laser coherent combination of pulses.

Laser fibers have very good efficiency, high rep rate, but low power. A coherent combination of many thousands of fibers may be a way to create an efficient high-rep-rate plasma accelerator.

This technique, once developed, can serve as an efficient laser driver for laser plasma acceleration-based light sources, which can be applied to the field of medicine, among others.

13.2.2 RESONANT PLASMA EXCITATION

We will mention one more accelerator technique that is currently being developed — *resonant plasma excitation.*[4]

In this case, the amplitude of plasma oscillation accumulates slowly, while plasma oscillations are excited by a train of low-energy laser pulses distanced at plasma wavelengths. Such a technique is particularly suitable for use in conjunction with fiber lasers because it can provide the driving laser trains with high efficiency and high repetition rates.

This technique, once finalized, can also offer the path to efficient laser-driven plasma acceleration light sources and to applications beyond accelerator physics.

---

[3]Gerard Mourou et al., Nature Photonics 7, 258–261 (2013).

[4]S.M. Hooker et al., J.Phys. B47, 234003 (2014).

13.2.3 TOWARD PLASMA-BASED CPA

Chirped pulse amplification (see Section 4.2.5) stretches the short laser pulse over time, amplifies the now-longer pulse and then compresses the amplified pulse. Gratings are used in CPA to spread the color-chirped laser pulses into different paths. Gratings of PW lasers can be a meter-long, which reduces power density and avoid getting damaged. Reducing the size of gratings, making more compact CPA systems, and increasing the CPA laser power may be possible[5] with the use of plasma transmission gratings, as illustrated in Fig. 13.6. In this case, the induced pattern of plasma density variation can act as the standard CPA grating, but in transmission mode. The plasma grating can survive very large power density, since it is made from *already damaged material*, and therefore, such CPA can be compact and can also withstand higher power.

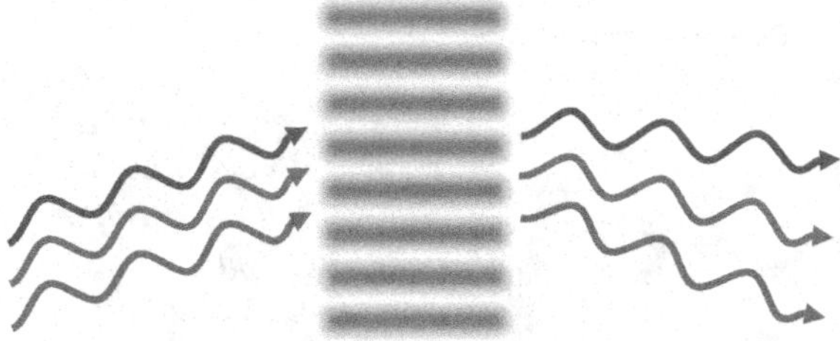

Figure 13.6 Concept of plasma grating for CPA.

Chirped pulsed amplification created a revolution in laser technology. Taking CPA to even higher power levels may be possible with the use of plasma gratings and plasma lenses.

The performance of high-power lasers can be limited by the damage threshold of their focusing elements. Overcoming this challenge may be possible with the use of lenses made from plasma. Holographic plasma lenses[6] can be created via the interference of two pulses from pump lasers that create a Fresnel lens-like pattern — see Fig. 13.7. The third laser pulse can be then sent to this Fresnel plasma lens and get focused. Building up on these two concepts may allow for compact PW lasers and efficient LWFA.

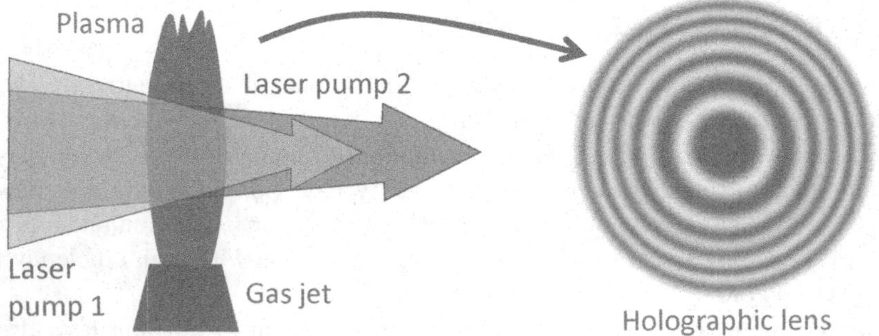

Figure 13.7 Concept of plasma holographic lens.

[5] M. R. Edwards and P. Michel, Phys. Rev. Applied 18, 024026, (2022).

[6] M. R. Edwards et al., Phys. Rev. Lett. 128, 065003 (2022).

13.3 TOP-UP AND NONLINEAR INJECTION

In the standard injection or extraction, a pulsed kicker or transverse cavity gives a small initial kick to the bunch, which then enters a septum magnet; see Fig. 13.8. The initial kick is increased in this special DC magnet, called a septum, which has two apertures — one with a deflecting magnetic field and another with an almost-zero field.

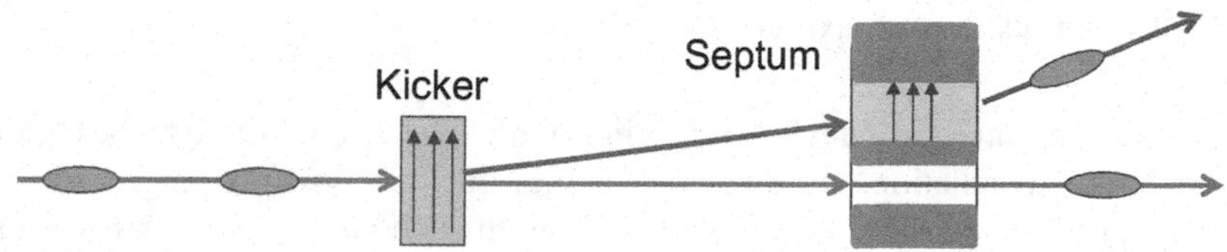

Figure 13.8 Standard septum-kicker combination for extraction of a bunch from the ring.

There are many types of septum magnets. The most common ones are shown in Fig. 13.9. In the figure, the first two are DC magnets, but the third one is pulsed and uses the Eddy currents excited in the right far conductor, preventing penetration of the field to outside of the septum.

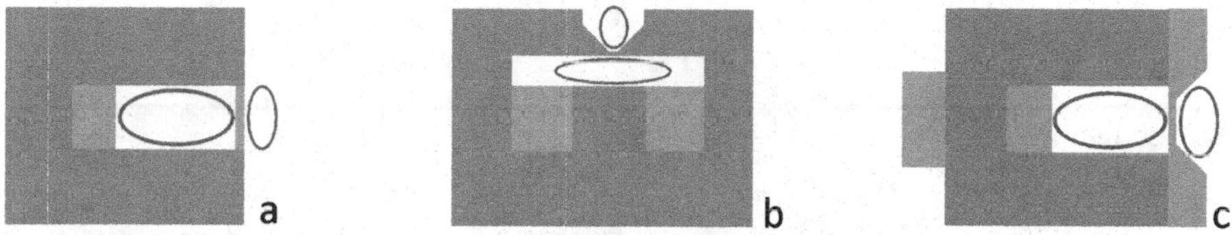

Figure 13.9 Septum types: a) DC drive, b) Lambertson, c) pulsed Eddy current septum.

Light sources with top-up injection impose specific demands on the injection system. The top-up injection gives several advantages to the light-source team: since the beam is replenished frequently, there is no issue with the beam's lifetime; stable temperature conditions are provided to the X-ray optics of users' beamlines. However, the top-up injection also introduces challenges: since the light source needs to operate for users continuously, the injection needs to be done with shutters (movable slits that protect the users' beamlines) open; furthermore, the injection needs to be done with the gaps of insertion devices (undulators, wigglers) closed.

Moreover, modern light sources, in pursuit of smallest emittance, use tight focusing intended to reduce the curly H function, creating the need for stronger sextupoles. The sextupoles and insertion devices impact the dynamic aperture (DA). A small DA will affect the injection's efficiency and lifetime, creating yet another challenge for top-up injections.

For a traditional off-axis injection (illustrated in Fig. 13.10), the injected bunch oscillates until it is damped by synchrotron radiation. If the DA is not sufficiently large, these oscillations will induce beam losses, reducing the injection's efficiency.

An improvement to the top-up injection can be achieved with the so-called on-axis swap-out injection (suggested originally by M. Borland) performed by a fast kicker.

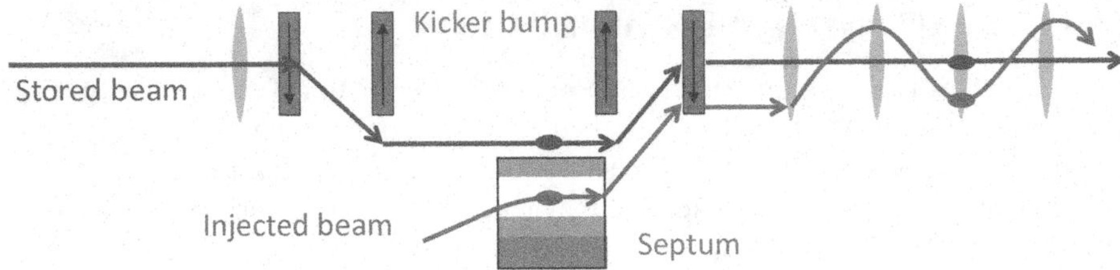

Figure 13.10 Traditional off-axis injection.

In this case, The injected bunch is replacing one of the circulating bunches almost without further oscillation, as illustrated in Fig. 13.11. Correspondingly, stronger focusing optics are possible for a light source with such an injection. A fast kicker needed for a swap-out injection is usually built in a strip-line configuration.

Top-up injections of light sources create many advantages, e.g., continuous use of X-rays and stable temperature conditions for beamline optics. Injection challenges can be solved with a swap-out or nonlinear kicker-based injection.

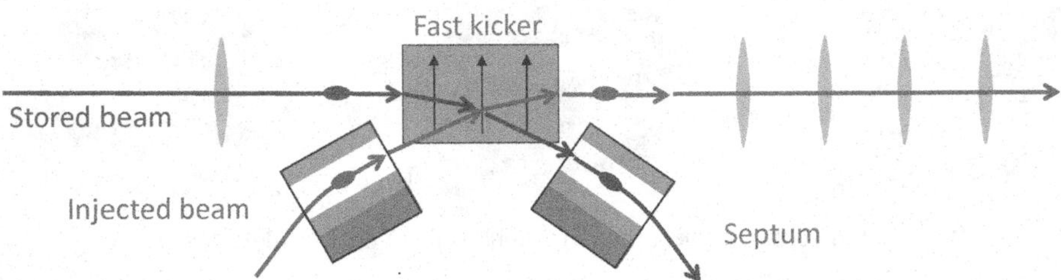

Figure 13.11 Swap-out injection.

Further improvements of top-up injections are possible with use of nonlinear kickers. While in a standard kicker there is some material (usually called "septum blade") between the aperture with the deflecting field and field-free aperture, in nonlinear kickers, there is one open continuous aperture. The distinction between "field" region and "field-free" region is soft and defined by the nonlinear dependence of the field on the transverse coordinate.

Fields in nonlinear kickers can be created by variety of ways[7]. In the case of Fig. 13.12, a) the nonlinear field is created by eight pulsed conductors, where the outer set of four conductors compensates for the field of the inner set of four conductors, creating a central near field-free region, where the stored beam will be located, followed by the steep increase of the field to the maximum, suitable for kicking the injected beam — see case c) in the same figure. Another version of the nonlinear kicker is based on magnetic octupole; see Fig. 13.12b.

[7]See, e.g., T. Atkinson, paper THPO024, IPAC2011, or S. White, paper MOPGW008, IPAC2019.

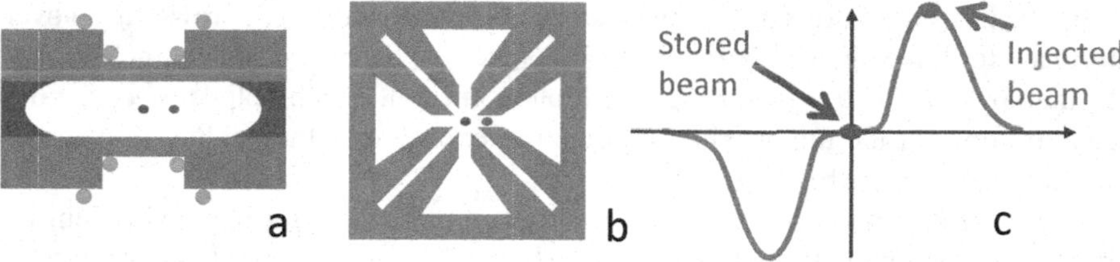

Figure 13.12 Nonlinear kickers: a) based on eight pulsed conductors, b) based on an octupole magnet. The deflecting field in the kicker aperture is illustrated in c).

Nonlinear injection reduces oscillation of the injected beam, reduces beam losses and improves injection efficiency, and also uses stronger low-emittance optics such as MBA or Hybrid MBA.

Non-linear injection kickers, where one set of conductors is compensating for another set of conductors, or where four poles of the octupole kicker magnet compensate for another four poles, is an illustration of the use of the inventive principle system and anti-system, combined with the inventive principle "nested dolls" — the powerful combination that we have already seen in many other examples.

13.4 MEDICAL SYSTEMS

Medical treatments that use X-rays rely on accurate imaging. The CT (*computed tomography*) X-ray is the key imaging technique for radiation treatment. The CT scan can be performed just before treatment. A CT scanner represents a system where the X-ray source moves around the patient in a full 360° scan, while the detectors on the opposite side of the circular system record the data feeding into the CT image.

CT has certain limitations, in particular: limited soft-tissue resolution and a limited ability to handle the patient's motion. Also, a CT scan cannot be done during radiation treatment. Moreover, CT gives an additional radiation dose to the healthy tissue.

Magnetic resonance imaging (MRI), on the other hand, has a superior performance. MRI has excellent soft-tissue contrast, and MR images can be taken in arbitrary planes and tissue volumes. However, MRI needs a solenoid with a large field.

It would then be natural to consider the inventive question of whether it would be possible to combine MRI and X-ray systems. The main contradiction in this task is that the large solenoidal field needed for MRI would be located just next to the electron linac needed for X-ray generation. This challenge was proven solvable, and the system which combines the 1.5 T MRI with X-ray therapy is now available[8].

An MRI-guided X-ray system is an example of the "nested doll" inventive principle.

[8] D. Winkel et al., Clin. Transl. Radiat. Oncol. 18:54-59, (2019).

As we just stated, the X-ray system can be combined with MRI, providing a novel approach for image-guided X-ray therapy. As we also discussed in Section 9.1, proton therapy benefits from the Bragg peak phenomenon, which helps to localize the radiation dose in the tumor while minimizing exposure to the healthy tissue, and therefore can provide better treatment.

A natural next question, then, is: can an even more superior therapy system be created, by combining proton therapy with MRI? The obvious challenge is that the MRI solenoidal field will affect the trajectory of the proton beam, which needs to be taken into account. Another challenge is that the proton beam needs to be angled in a different direction with respect to the patient.

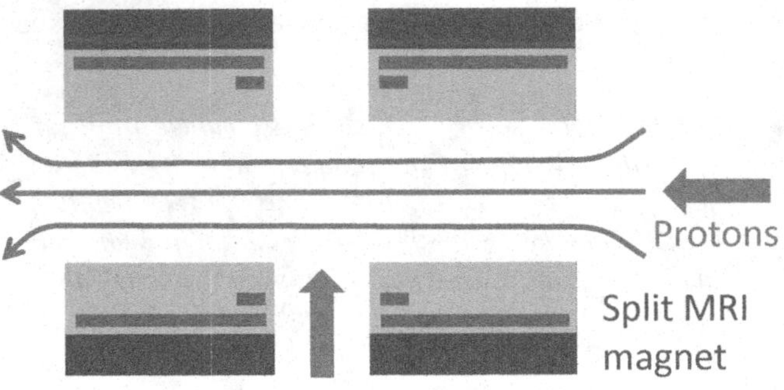

Figure 13.13 Concept of Magnetic Resonance Imaging combined with proton beam therapy system.

MRI-guided proton therapy is a promising new direction that is under development that combines the benefits of proton therapy with the superior MRI imaging.

The challenges of MRI-guided proton therapy are being actively studied and appear solvable[9]. The MRI solenoid magnet in this case will need to be split, as illustrated in Fig. 13.13, which will enable different orientations of the proton beam delivery, and additional coils will be added to control the homogeneity of the MRI solenoid field. The integrated design can also take into account and correct for the MRI solenoid impact on the proton trajectory and focusing.

In addition to providing a lot of interesting inventive tasks, MRI-guided proton therapy is a perfect concept for discussing the systems engineering.

13.5 SUPERCONDUCTING SYSTEMS

Superconducting technology is a very broad area. In this section, we will touch on just a couple of topics related to superconducting magnets and superconducting RF cavities. We will highlight some interesting inventions and recent developments, and we will lead the discussion toward the topic of systems engineering — the framework appropriate for discussing such complicated and interconnected technological areas.

[9] T. T. Pham, B. Whelan, B.M. Oborn et al., Radiotherapy and Oncology, v. 170, p.37-47, (2022).

13.5.1 SUPERCONDUCTING MAGNETS

Increasing the energy of accelerators, or making them compact, often requires the use of superconducting (SC) magnets. Let's briefly recall the meaning of low temperature superconductivity and review the main characteristics of superconductive materials.

According to the Bardeen-Cooper-Schrieffer theory, in a superconducting material, the electrons (which are fermions) become linked together into Cooper pairs via an attractive force which is produced from the interaction of the ion lattice with the material. The Cooper pairs act like bosons and can exhibit a Bose-Einstein condensation. This condensate acts as a single system, and the energy of a single electron cannot be changed without changing the energy of all of them. Moreover, due to an energy gap, there is no scattering and therefore no resistance, unless the energy transfer exceeds the threshold value. Superconductivity can be also described by the *two-fluid model* with electrons that are linked to Cooper pairs and those which are not. Above *critical temperature* T_c, all electrons are normal, and below T_c, there is a mix of normal electrons and those connected to Cooper pairs. In cases of direct current (DC), the Cooper pairs' fluid will carry all of the electric current, resulting in zero resistance. However, in cases of alternating current (AC), the oscillating field will carry both fluid components, resulting in a non-zero resistance for AC or SC RF applications.

There are a variety of SC materials, but only some of them are suitable for practical applications. Type I superconductors include various metals (e.g., aluminum, lead, mercury, tungsten, etc.). Such superconductors are perfect diamagnetic materials; i.e., they expel all magnetic fields. However, if the external magnetic field exceeds a certain critical value H_c, superconductivity will be lost. Type II superconductors have two critical fields, H_{c1} and H_{c2}. If the magnetic field is in between these two values, the magnetic flux will penetrate into the material in the form of *fluxons* — a quantized form of a magnetic field. Niobium (Nb) metal and many alloys belong to Type II.

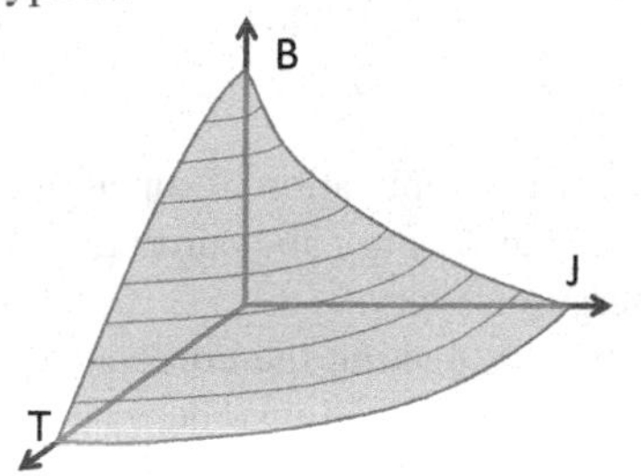

Figure 13.14 Critical surface concept of superconducting materials.

Typical range of the critical field is $H_c \approx 10^{-3}$ to 10^{-2} T in Type I superconductors, and several orders of magnitude higher for Type II superconductors: $H_c \approx 10$ to 10^2 T, making them much more useful for practical applications in accelerators and other technological areas.

One can also define the critical current density J_c above which the superconductor will *quench* into the normal state. The critical temperature, field and current define the edges of the 3D critical surface that characterize a superconductor; see Fig. 13.14.

The two most important materials for superconducting magnets are NbTi and Nb_3Sn, with their typical performance illustrated in Fig. 13.15.

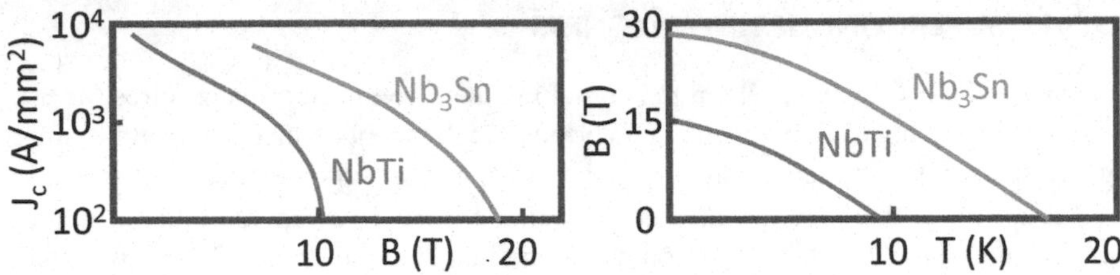

Figure 13.15 Typical comparison of NbTi and Nb_3Sn performance at $T \approx 2$ K.

Nb_3Sn materials help to achieve much higher fields in the magnets; however, higher fields impose much stronger forces on the magnets' components, creating the need to manage these stresses via their distribution to the materials surrounding the conductors[10] — see Fig. 13.16, where the superconducting conductor is split into smaller segments, with each of them passing the stresses on to the surrounding matrix, reducing stresses in Nb_3Sn to a tolerable level.

Figure 13.16 A standard NbTi dipole magnet a), and Nb_3Sn dipole magnet b) with stress management.

Stress management is an illustration of the *intermediary* inventive principle — passing stresses to structures around conductors.

In general, NbTi and especially Nb_3Sn magnets exhibit a number of inventive principles such as "nested doll," beforehand cushioning (applying pre-stress to magnet structures that will compensate for forces arising when the magnet is cooled and powered), and intermediary (distributing the stress to neighboring structure), among others.

13.5.2 SUPERCONDUCTING RF

Superconducting RF cavities are another very important application of superconductivity in accelerators; they allow for an increase in beam energy and current, making a more efficient machine, as well as enabling energy recovery.

Pure niobium is now by far the primary choice for making the conductive surface of the SC RF cavities. As described above, in RF applications, there is always nonzero resistance in superconductors. The behavior of this residual surface resistance vs. the temperature T can be given by

$$R_s \approx \frac{A}{T} f^2 \exp(-D(t)/T) + R_0 \tag{13.2}$$

where A, R_0 and $D(T)$ depend on the properties of a particular superconductor. As we can see, R_s increases with RF frequency f and decreases with temperature. This

[10] I. Novitski et al., in IEEE Trans. on Appl. Superconductivity, v. 32, no. 6, 4006005, (2022).

behavior explains why the pure Nb SC RF cavities of around 500 MHz frequency can be operated at $T = 4.2$ K, while the Nb cavities with 1.3 GHz and above require a 2 K operating temperature.

The main characteristic of a superconducting cavity is its quality factor Q_0 defined as the energy stored in the cavity divided by the energy loss in one RF period (see Eq. 5.17) as well as its accelerating field E. The Q_0 is obviously connected to the surface resistance R_s. For well-made cavities, Q_0 can exceed 10^{10}.

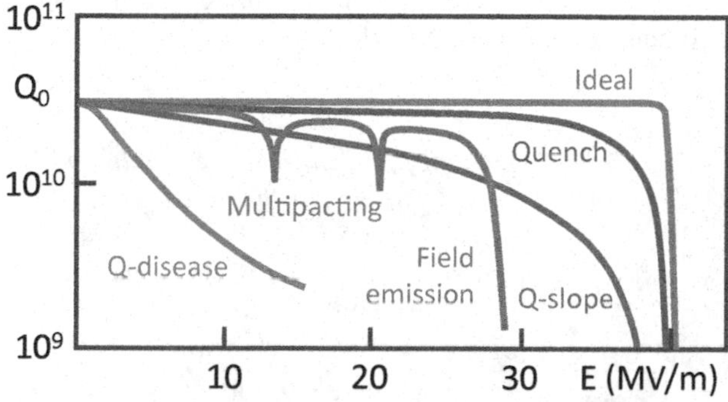

Figure 13.17 Qualitative behavior of superconducting cavity quality factor Q_0 vs. the accelerating gradient E.

The ultimate performance of SC cavities depends on material properties, cleanness of manufacturing and assembly, the shape of the cavities, magnetic hygiene, the rate of cavity cooldown, etc.

The behavior of Q_s vs. the accelerating gradient is an important characteristic of an SC cavity. An ideal curve (see Fig. 13.17), which would reach the critical magnetic field, is almost never achieved and instead the cavity would usually *quench* earlier. A number of effects can deteriorate cavity performance. *Field emission* can be caused by defects or contamination of the cavity's surface. The emitted current can reduce Q_0 and also heat the surface and create radiation. *Multipacting* can happen when there is field emission and also resonance conditions for certain trajectories, which can then persist in the cavity at certain energies; see Fig. 13.18. Both multipacting and field emission can be reduced by special processing of the cavity's surface (electropolishing, buffer chemical processing), as well as clean practices of cavity assembly.

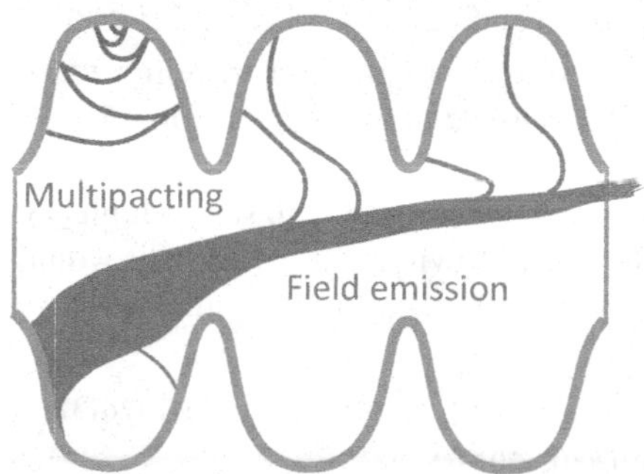

Figure 13.18 Field emission and multipacting in a superconducting cavity.

Multipacting is created by resonance conditions for electron trajectories and therefore happens at certain precise energies (multipacting barriers). It can be eliminated by changing the cavity shape, and sometimes can also be mitigated by continuous attempts of RF processing the cavity, which can help to remove or smooth the unwanted field emitters.

Other limiting effects illustrated in Fig. 13.17 include *Q-disease*, which is attributed to interstitial hydrogen in the niobium, which is responsible for forming hydrides when the cavity is cooling down through the region of 150 K and is not yet superconducting[11]. Mitigating the Q-disease can be achieved by heat treating the cavities before assembly and increasing the speed of the cavity cooldown. While the limitations illustrated as a *Q-slope* can be due to a number of reasons, including added impurities in the niobium, sometimes the "lemons can be turned into lemonade" — it was shown[12] that *Nitrogen doping* can significantly increase Q_0 of specially treated cavities in a large range of accelerating fields.

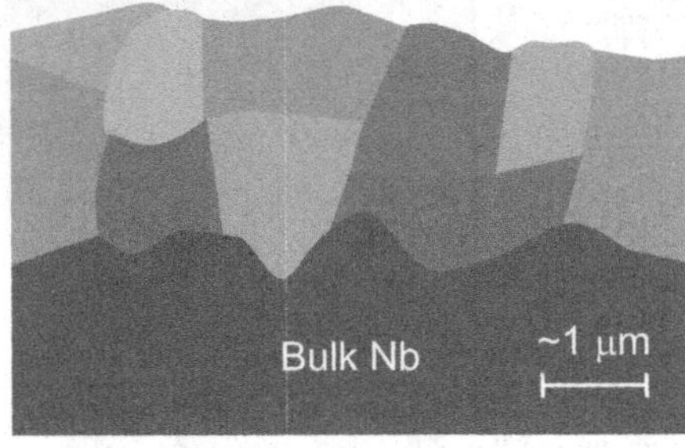

Creating an entire superconducting cavity out of Nb_3Sn is not feasible because Nb_3Sn is a brittle material; however, it is possible to deposit a thin film of Nb_3Sn onto the substrate of the bulk Nb — illustrating the inventive principle of "local quality."

Figure 13.19 Nb_3Sn thin film on bulk Nb.

Nb_3Sn offers a larger critical field and ability to operate at higher cryogenic temperatures. While pure Nb is ductile and easy to manufacture from, Nb_3Sn is a brittle material. The path forward[13] for using Nb_3Sn for superconducting cavities is to use it in thin film form deposited onto pure Nb substrate, as illustrated in Fig. 13.19. Such Nb-Nb_3Sn cavities are also illustrated in Fig. 15.3 in application with the "local quality" inventive principle. In that case, the cavity body has three layers: copper, niobium and the Nb_3Sn film, with Nb acting as a substrate for the Nb_3Sn film and the thick copper layer providing good thermal conductor connected to a cryo-cooler, enabling a compact superconducting linac design working at 4 K.

Impurities can decrease the quality factor of SC cavities, but can also increase it, illustrating the inventive principle of turning lemons into lemonade.

Finally, we would like to describe an interesting example of a traveling wave superconducting accelerating structure. As we discussed earlier (see Section 5.4.1), normal conducting structures are typically traveling waves, while superconducting structures are usually standing waves, for the simple reason that with such a high Q_0, the EM field can exist for a very long time in the SC cavity, and therefore it would just not be economical to let the RF field travel through the cavity too quickly.

[11] C. Reece and G. Ciovati, in Reviews of Accelerator Science and Technology, pp. 285-312 (2013).

[12] A. Grassellino et al., Supercond. Sci. Technol., 26 102001, (2013)

[13] Md Nizam Sayeed et al., Applied Surface Science, v. 541, 1, 148528 (2021); S. Posen et al., Supercond. Sci. Technol. 34 025007, (2021).

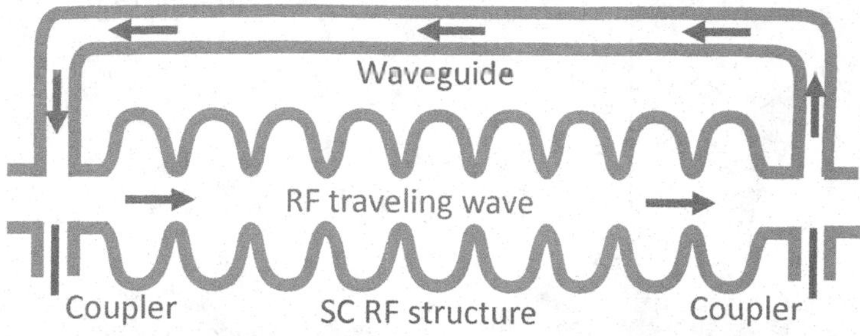

Figure 13.20 Superconducting traveling wave structure with recirculating waveguide.

However, this contradiction can be overcome if one would return the traveled RF wave back into the superconductive cavity[14] via an additional waveguide connecting the beginning and the end of the cavity[15], as illustrated in Fig. 13.20. This approach can increase the accelerating gradient (recall that a standing wave is a combination of two traveling waves going in opposite directions, and only one of them is synchronous with accelerating particle). However, the illustration in Fig. 13.20 also shows the obvious challenge: the system is completely symmetrical, and there is no preferred direction (clockwise or counter-clockwise) where the wave will travel. Breaking this symmetry and creating a preferred direction requires active feedback applied to tuners that act on the timescale of wave revolutions around the system.

Recirculating the traveling RF wave in a SC cavity illustrates the inventive principle of continuity of useful action. And, to make this idea work in practice, one would have to apply the inventive principle of anti-symmetry and feedback.

13.6 SYSTEMS ENGINEERING

Systems engineering (SE) is an approach that focuses on definition, design, and the realization of complex integrated systems over their entire lifecycles. The term SE originated from the Bell Telephone Laboratories back in the early 1940s.

Systems engineering can be applied to accelerator projects and in particular to the area of superconducting RF linacs, wherein different aspects of design (mechanical, thermal, electrodynamics, etc.) are extremely interconnected. As an example, let's consider the phenomenon of SRF cavity Lorentz detuning. The RF field, when it enters the SC cavity, exerts pressure on the walls of the cavity (electrodynamics design), resulting in its deformation (mechanical design), causing detuning; see Fig. 13.21. This effect is important for both pulsed and CW SC linacs. Increasing the rigidity of the cavities with stiffening rings, as shown in Fig. 13.21 on the right side, is a possible way to mitigate the effect; however, the rigidity increase should not be excessive, so that the tuners (Piezo or mechanical) can still be applied to deform the cavity and match its resonance to the nominal RF frequency.

---

[14]P. Avrakhov, A. Kanareykin and N. Solyak, in Proc. of PAC 2005, WPAT091, (2005).

[15]For normal conductive structures with traveling wave flowing outside of accelerating cavities see V. Dolgashev, LINAC2016, MOOP04, (2016); and V. Dolgashev, US Patent 9380695, (2015).

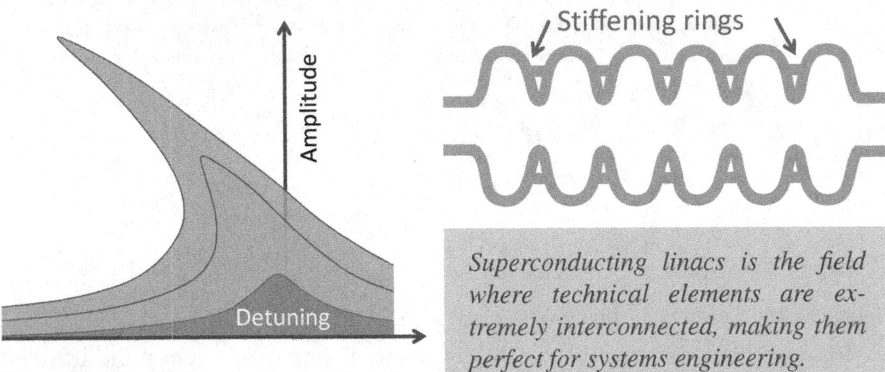

Figure 13.21 SRF cavity Lorentz detuning.

Superconducting linacs is the field where technical elements are extremely interconnected, making them perfect for systems engineering.

Systems engineering can be applied to the entire ecosystem of superconducting RF linacs, illustrated in Fig. 13.22, which includes sub-systems (such as SC RF cavities), cavities assembled into cryomodules, super-systems (such as entire linacs), cryogenics, controls, and low- and high-power RF, among others. The ecosystem also contains the entire cycle of research, development and production of SC RF cavities and other components.

Systems engineering can be applied to a much wider area too — to the entire scope of accelerators, laser and plasma technology, helping to build the compact plasma acceleration FEL, and to realize many other breakthrough ideas.

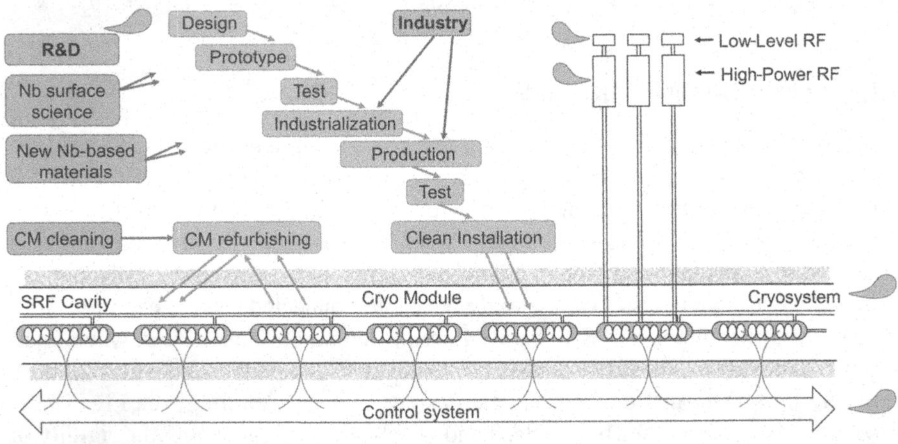

Figure 13.22 SC RF accelerator ecosystem — for discussion of Systems Engineering.

Systems engineering is the discipline that considers complex integrated systems over their entire lifecycles — conceptual design, preliminary and detailed design, production and construction, utilization and support, as well as their phase-out and disposal.

13.7 SUPERLATTICE PHOTOCATHODE – INVENTION CASE STUDY

Polarized electron sources are essential for a variety of scientific and technological applications. The evolution of polarized photocathode technology is a remarkable story of community efforts highlighted with many inventions. The story started from bulk GaAs, which provided polarization of around 30–40%. The next step involved creating a thin (about 100 nm) layer of strained GaAs[16] on top of a GaAsP buffer, which affected the energy level in the system in such a way that the extracted beam polarization jumped to around 70–85%. Further improvement came from sandwiching the strained layer into a superlattice[17], where a dozen GaAs/GaAsP layers were placed in the 100-nm surface layer, increasing polarization to around 85–90% together with an increase of quantum efficiency (defined as the ratio of emitted electrons per oncoming laser photons).

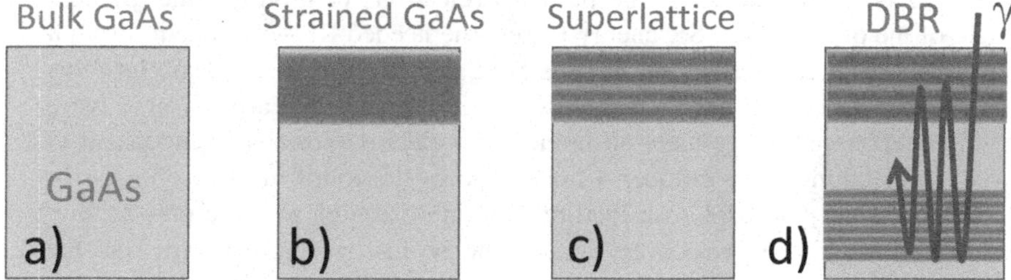

Figure 13.23 Evolution of polarized photocathode technology: a) bulk GaAs; b) strained GaAs; c) GaAs/GaAsP superlattice; and d) strained GaAs/GaAsP superlattice photocathode with Distributed Bragg Reflector.

Further improvements in terms of polarization and in particular in terms of quantum efficiency came from introducing the Distributed Bragg Reflector[18] (DBR) to the structure of the photocathode. The distance between the layers of the DBR correspond to the wavelength of the laser, so that the DBR acts like a mirror for the laser photons, which can then bounce several times from the superlattice, significantly increasing the efficiency of converting the photon energy to emission.

The development of polarized photocathode technology illustrates the application of inventive principles such as local quality and continuity of useful function. It also illustrates the TRIZ law of evolution of technical systems, when the two systems — strained superlattice and DBR — became parts of a supersystem — strained GaAs/GaAsP superlattice photocathode with a Distributed Bragg Reflector.

---

[16]T. Maruyama et al., Phys. Rev. B 46, 4261, (1992).

[17]M. Baylac et al., Phys. Rev. ST Accel. Beams 8, 123501, (2005).

[18]Wei Liu et al., Appl. Phys. Lett. 109, 252104 (2016).

EXERCISES

13.1. *Chapter materials review.* The resonant plasma excitation method may allow for the use of many low-energy pulses from fiber lasers to excite plasma oscillations for efficient laser-driven plasma acceleration. Discuss what effects can limit the maximum number of laser pulses that can be effectively used in this technique.

13.2. *Mini-project — conclusion**. In the exercises started in Chapter 2 and continued in Chapter 10, we estimated various parameters of the "Fermi accelerator" encircling the Earth, where N space stations were launched around the Earth, with each space station carrying one bending dipole magnet and one quadrupole magnet. As a result, we obtained the parameters for the Fermi accelerator, including the length of the bends and quadrupoles, bending field, the energy of the particles (electrons or protons), the gradient of the quadrupoles, SR energy losses, the needed RF voltage, equilibrium emittance, beam sizes, etc. In this final step of the project, please turn this Fermi accelerator into a triple-use facility around the Earth to serve these functions: a) a free-electron laser, b) an electron-positron collider and c) an electron-proton collider. Choose one of the following two options for designing your triple-use facility. Option A: design a single-pass facility working in energy recovery mode. The beams (e+ and e-) start from the injection space station, go around the Earth in different directions, getting accelerated to higher energies at each consecutive station, arrive at the undulator/detector space station where they produce FEL radiation or collide in the experimental detector, and go around the Earth for another half-turn, getting decelerated at each station and then dumped into the injector station. The RF cavity in each station gives energy to the beam it accelerates and takes energy from the beam it decelerates — this is energy recovery. Option B: design a storage ring accelerator facility where acceleration is slow and there is no energy recovery. The beams (e+ and e-) start from the injection space station and go around the Earth in different directions for many turns. Define as many parameters as possible for these space facilities.

13.3. *Practice in reinventing technical systems.* In PWFA, it is important to study plasma acceleration of positrons when the drive beam consist of electrons, and vice versa. The SLAC S-band linear accelerator has the capability to accelerate both e- and e+ bunches simultaneously, placing them 180° of RF phase apart, which is 5 cm, and not closer. However, for plasma acceleration, the drive and witness bunches need to be within 100 μm of each other. Suggest a beamline configuration that will allow us to study plasma acceleration with e- drive and e+ witness bunch.

13.4. *Practice in the art of back-of-the-envelope estimations.* A multi-layer insulation (MLI) is often used for the thermal insulation of components inside of cryomodules, where superconducting accelerating cavities are placed. The MLI is implemented by rolling N layers of aluminized mylar around the object that needs to be isolated. The temperature contacts between the neighboring layers of MLI are minimized by a special spacer net or just by slightly crumpling the mylar. Assuming that the dominant heat transfer mechanism can be described by the Stefan-Boltzmann law, estimate the heat transfer coefficient for 10 layers of mylar when the temperatures of the edge layers are 4 K and 70 K.

14 Inventions and Innovations in Science

We have seen, throughout this book, that the three technical fields we discussed — accelerators, lasers and plasma — sometimes have a lot in common, things such as scientific approaches and solutions, similar inventions and technological breakthroughs. We have seen several examples when the TRIZ, or AS-TRIZ, inventive principles can be retrospectively identified in various scientific inventions.

In this almost final chapter, we would like to return to the topic we began to discuss in the very first chapter of this book — the methodology of inventiveness.

We will start by reviewing the updated Accelerating Science TRIZ principles.

Following that, we will then look critically at some of the suggested AS-TRIZ inventive principles, and discuss whether these are indeed principles or rather trends of the evolution of scientific systems. In this discourse, we will recall the definition of the laws of the evolution of technical systems as described by the standard TRIZ principles. After observing the parallels between the evolution of radar and lasers, we will conclude that some of the suggested AS-TRIZ inventive principles should be redefined as the laws of the evolution of technical/scientific systems.

We will then continue to consider the inventive principles and the laws of the evolution of systems by taking a very general approach, and will look at various fields of science and even beyond those.

The inventiveness methodologies discussed and developed in this book will, we hope, help many researchers and engineers to be more inventive, as well as more successful in taking their inventions to practice.

The next topic we will discuss is the way to take these scientific inventions and apply them in practical settings. We will look at the linear and two-dimensional models of innovation and discuss the notion of the Pasteur quadrant in application to accelerator science in particular.

The well-known challenge of "crossing the Valley of Death" in technological innovation will be our next topic, and we will suggest a possible method for crossing this so-called valley in regards to the field of sci-tech, with help from a selection of compact light source projects.

In the next section we will offer a possible approach of using TRIZ to teach future generations of science students. In fact, the method we have used throughout this book has not been the ready-to-use standard TRIZ; instead, we immersed ourselves in the proactive process of developing an extension of TRIZ — Accelerating Science TRIZ.

Finally, we will aspire to apply these developed principles to new, challenging projects and their unsolved problems.

DOI: 10.1201/9781003326076-14

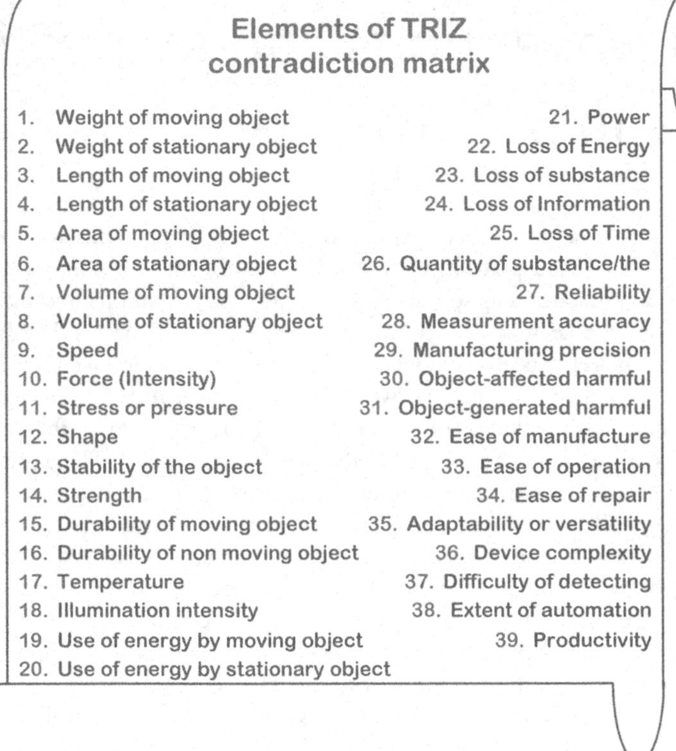

Figure 14.1 Elements of TRIZ contradiction matrix.

14.1 ACCELERATING SCIENCE TRIZ

First of all, we hope that the reader has already guessed the play of words in the term "Accelerating Science TRIZ." The word "accelerating" does not refer here to accelerators — the devices that help particles reach higher energies. Instead, the word "accelerating" in AS-TRIZ implies that science itself can be accelerated via the application of these inventive principles.

Throughout this book we have seen several recurring inventive principles that can be added into the tables of the AS-TRIZ extension. Shown below are the updated tables where a number of parameters for the AS-TRIZ matrix of contradiction (Table 14.1), as well as several inventive principles (Table 14.2) for the AS-TRIZ, are included.

In addition to the principles of "undamageable or already damaged" and "volume-to-surface ratio" mentioned already in Chapter 1, we have included the inventive principle related to local correction (based on various examples from the focusing systems in beams or lasers), transfer between phase planes (based, for example, on Derbenev's transformation), transition from microwave to optical (recall radar and CPA, or conventional and laser acceleration), and using time–energy correlation

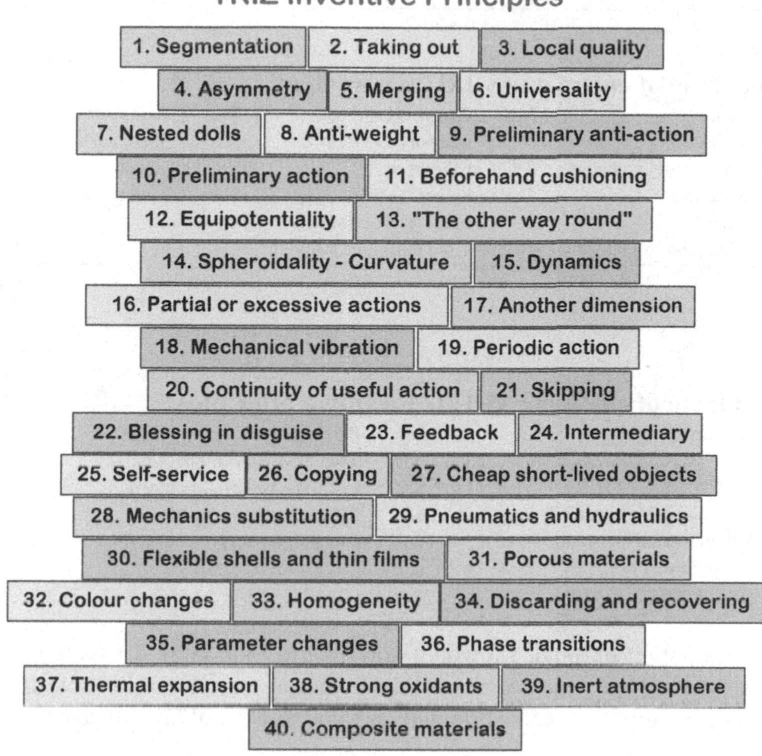

Figure 14.2 The list of forty inventive principles of TRIZ.

(laser pulse or bunch compression).

The reader can perhaps already add other inventive principles to this list — in fact, one of the purposes of keeping the unfilled spaces in the tables above is to fill them in, during the proactive process of learning.

14.2 TRENDS AND PRINCIPLES

Standard TRIZ describes not only the inventive principles, but also the laws (trends) of the evolution of technical systems. Knowledge of these evolution laws can help to create innovative inventions.

There are just a few laws/trends of the evolution of technical systems and they can be described in broad terms. The inventive principles, however, are more concrete, as they are unique approaches that help to solve contradictions — there are several dozens of these.

The lists of additional AS-TRIZ inventive principles described in Table 14.2 should not be interpreted as cast in stone. Compiling such tables should instead be considered as a process of proactive learning parallel to the application of the TRIZ method to science. A critical analysis is therefore appropriate.

Table 14.1
Updated table of emerging AS-TRIZ parameters

No.	Parameter	No.	Parameter	No.	Parameter
1.	Energy	6.	Intensity	11.	Spatial extent
2.	Rate of energy change	7.	Efficiency	12.	Sensitivity to imperfections
3.	Emittance	8.	Power	13.	Cooling rate
4.	Luminosity	9.	Integrity of materials	14.	...
5.	Brightness	10.	Time duration or length	...	...

Table 14.2
Updated table of emerging AS-TRIZ inventive principles

No.	Principle	No.	Principle
1.	...	6.	Transfer between phase planes
2.	...	7.	From microwave to optical
3.	Undamageable or already damaged	8.	Time–energy correlation
4.	Volume-to-surface ratio	9.	Energy recovery
5.	Local correction	...	...

With this notion in mind, we would now like to look more carefully at the inventive principle "from microwave to optical" that we introduced in Table 14.2, and discuss whether this is indeed a principle or rather a general law (trend) of the evolution of technical systems.

14.2.1 TRIZ LAWS OF TECHNICAL SYSTEM EVOLUTION

Standard TRIZ defines three types of laws of technical system evolution — *static*, *kinematic* and *dynamic*.

The three static laws are as follows. The TRIZ law of the *completeness of the parts of the system* states that a system should have the following four parts: an engine, a transmission, a working unit and a control element. The law of *energy conductivity of the system* says that every technical system is a transformer of energy and the energy should circulate freely and efficiently through these four main parts of the system. The law of *harmonizing the rhythms of the parts of the system* suggests that the frequencies and periodic motions of these parts should be in sync with each other.

The three kinematic laws are defined in standard TRIZ as follows. The law of *increasing the degree of ideality of the system* states that the "ideality," which is a qualitative ratio between all desirable benefits of the system and its cost or other harmful effects, should have the tendency to increase. The law of *uneven development of parts of a system* suggests that the different parts of a technical system will evolve differently, leading to new technical and physical contradictions. And finally, the *law of transition to a super-system* states that a system which has exhausted the possibilities of further significant improvement is included in a super-system as one of its parts.

Finally, the two dynamic laws are as follows. The law of *transition from macro to micro level* suggests that the development of working organs proceeds initially on a macro and then more and more on a micro level. And the law of *increasing involvement of fields in the system* states that the fields evolve from mechanical fields to electromagnetic fields.

We note that the phrase in the previous paragraph, "the fields evolve from mechanical fields to electromagnetic fields," can be extended and clarified as meaning that, within the electromagnetic spectrum, the evolution should follow from RF frequencies to optical frequencies.

We can conclude from this comparison and discussion that "from microwave to optical," which we initially identified as an inventive principle, is in fact better suited to be defined as a general trend of technical/scientific system evolution.

Let's highlight this conclusion by using an example of the invention of CPA in connection to radar.

14.2.2 FROM RADAR TO HIGH-POWER LASERS

In 1935, Robert Watson-Watt developed the RDF (Radio-Direction Finding), which later become known as RADAR (RAdio Detection And Ranging). Development of the radar device took place at Ditton Park in England, which later became known as the Appleton Laboratory and, after merging with the Rutherford Laboratory, became the Rutherford Appleton Laboratory — currently the home of many accelerators and lasers (and located just a half-hour drive from Oxford — the place where the first edition of this book was written).

The operation of radar involves chirped pulse amplification. The positive influence of radars on the span of the 20th century cannot be discounted. Indeed, we would not have been party to the tremendous and revolutionary advances in laser technology had it not been for the chirped pulse amplification — CPA, developed around 1985 by D. Strickland and G. Mourou.[1]

The development of CPA stimulated an exponential growth of available laser powers. From around 1990 to the early 2000s, the peak laser power increased by more than two orders of magnitude, reaching PW levels.

High-power lasers, in their turn, enabled progress in many technical and scientific areas. For instance, nuclear physics, the production of Giga–Gauss magnetic fields, fusion science, studies of material properties in extreme conditions, laboratory astrophysics, studies of turbulence, non-linear quantum electrodynamics and many other fields benefitted from the development of CPA lasers.

The example of radar and CPA once more asserts TRIZ's founding proposal that the same problems and solutions appear again and again, but in different disciplines. It also illustrates the general trend of technical/scientific system evolution (i.e., the transition from the microwave part of the EM spectrum to the optical range).

[1]D. Strickland and G. Mourou, Compression of amplified chirped optical pulses. Opt. Commun. 56, 219–221 (1985).

14.2.3 MODERN LAWS OF SYSTEM EVOLUTION

The laws/trends of the evolution of systems were defined by standard TRIZ several decades ago. The laws are very broad and will likely remain valid for a long time. Still, it is tempting to look at these laws from a present-day perspective and, initially, see whether the laws need to and/or can be updated, and second, attempt to apply these laws to other areas of science, and also to seemingly unrelated areas.

In order to illustrate the tendencies that can be redefined as the laws of the evolution of scientific systems, let us recall the example of tumor therapy wherein the effects of a proton beam can be combined and synergistically enhanced by using small compounds. The latter can be created specifically for a particular patient, based on a genetic analysis of their tumor.

This example shows, first of all, the trend of using more than one technique for achieving synergy and a superior overall effect. And second, this illustrates the universal trend of developing solutions and services tailored to specific persons.

The reader can certainly suggest other examples that illustrate these or other trends. The fascinating question of whether the noted trends are indeed modern general laws of scientific system evolution is better left for further discussion outside of the framework of this text.

14.3 ENGINEERING, TRIZ AND SCIENCE

There are many well-founded connections between science and engineering. We will take this time to discuss several particular aspects of these connections while highlighting their link to TRIZ methodology.

14.3.1 WEAK, STRONG AND COOL

Recall our discussion relating to the focusing in accelerators that is needed to keep particle trajectories near the center. The first accelerators had weak focusing with spatial periods greater than the perimeter of the accelerator. The trajectories of particles in a weak focusing accelerator had large transverse excursions, requiring extremely large vacuum chamber apertures.

Strong focusing was suggested as a better way to focus particles by involving the usage of a sequence of focusing–defocusing elements. This example in itself may serve as an illustration of the TRIZ inventive principle of system and antisystem.

When strong focusing was suggested, it was not totally clear if it would work in practice. The realization of the first strong focusing proton accelerator was led by Sir John Adams, who had the courage to cancel (in October, 1952) the already approved 10 GeV weak focusing accelerator for a totally innovative 25 GeV *Proton Synchrotron.* [2]

[2] *Strong FODO focusing may illustrate TRIZ inventive principle system–antisystem.*

The risk was worth it — in 1959, Sir John Adams announced that CERN's Proton Synchrotron had just reached 24 GeV and had surpassed Dubna's Synchrophasotron world record of 10 GeV.

Strong FODO focusing illustrates the TRIZ inventive principle system-antisystem.

Another example of scientific risk-taking relates to the development of electron beam cooling (which was especially necessary for antiparticles such as antiprotons). The method was suggested by G.I. Budker, the founder and first director of the Institute of Nuclear Physics in Novosibirsk. Budker was the mind behind many inventions in the field of physics, including the concept of electron cooling.

However, when electron cooling was first proposed by Budker, the general consensus from the scientific community was "brilliant idea, but unfortunately non-realistic." Luckily, Novosibirsk scientists did not listen to these predictions and successfully constructed the first e-cooler around 1974 and have built many more since then.

There is one more parallel between these two stories. Sir John Adams had a unique combination of scientific and engineering abilities, whereas Lev Landau once called Budker a "relativistic engineer." As it happens, the art of inventiveness (TRIZ) originated from engineering.

14.3.2 HIGGS, SUPERCONDUCTIVITY AND TRIZ

In the first chapter, we discussed the connection between the TRIZ inventive principle of the nested doll (*matreshka*) and Bryusov's poem describing the world of an electron and particle detectors, which are arranged like a *matreshka* doll — a system within another system.

The particle detectors are the devices that (in tandem with accelerators) helped to discover[3] the Higgs boson — the essential building block of the *Standard Model* of particle physics.

A particular connection we would like to highlight is the relationship between Higgs and superconductivity. In a recent article,[4] A. Pashkin and A. Leitenstorfer reminded us that "... the theoretical proposal of the Higgs mechanism was actually inspired by ideas from condensed matter physics ... In 1958, Anderson discussed the appearance of a coherent excited state in superconducting condensates with spontaneously broken symmetry... On page 1145 of this issue, Matsunaga et al. report direct observation of the Higgs mode in the conventional superconductor niobium nitride (NbN) excited by intense electric field transients."

---

[3]Francois Englert and Peter W. Higgs, Nobel Prize in Physics, 2013.

[4]A. Pashkin and A. Leitenstorfer, Science 345, 1121 (2014).

The above passage shows us once again that that the general conclusion of TRIZ that "the same problems and solutions appear again and again but in different disciplines" is applicable to science, too.

14.3.3 GARIN, *MATRESHKA* AND NOBEL

One of the first inspirations and predictions of a device similar to today's laser may have appeared in a 1926 novel by Aleksey Tolstoy, *The Hyperboloid of Engineer Garin*. In that story, a device was described that was capable of producing a ray of light of immense power. It is fascinating to note that technical drawings were included in this novel — a significant attraction for curious readers (even despite the fact that the drawings referred to non-existing materials).

The focusing mirror in Garin's hyperboloid was made from "shamonite" — an extremely durable material imagined by the writer. Fictitious and nonexistent in the early 20th century, such a material can perhaps be created in the 21st century thanks to the advent of new engineered materials.

The adventures described in Tolstoy's novel were extraordinary. This fictional and powerful ray[5] was responsible for many astounded remarks. One such was, "Can you imagine what opportunities are opening now? Nothing in nature can withstand the power of the beam of light — buildings, forts, battleships, airships, rocks, mountains, the earth's crust — everything could be penetrated, destroyed, cleaved with my beam." These are the words Tolstoy put into the mouth of the story's protagonist — the engineer Garin.

The lasers developed later in the century[6] did not adhere to the design in *The Hyperboloid of Engineer Garin*, and were luckily used for peaceful purposes.

In connection to lasers and TRIZ, we would like to mention here one particular recent invention, *stimulated emission depletion microscopy* (STED), which was developed by Stefan W. Hell and his colleagues.[7]

The stimulated emission depletion microscopy allows for an increase of resolution in optical microscopes by a factor of around five. It does this by applying two pulses of laser light to an object under study — an excitation pulse and, shortly afterwards, a de-excitation pulse.

A fluorescent dye introduced into the object is first excited and then depleted with the second pulse of an appropriately different wavelength. The key feature of this technique is the usage of the depleting laser pulse with a special spatial profile; i.e., its minimum intensity is located in the center, as illustrated in Fig. 14.3.

The minimum intensity region of the second laser can be several times smaller than the wavelength. Therefore, detecting the remaining fluorescence spot will

[5]Which also reminds us Jules Verne's "fulgurator" described in his novel "Facing the Flag."

[6]C. Townes, N. Basov and A. Prokhorov, Nobel Prize in Physics, 1964.

[7]E. Betzig, S. W. Hell and W. E. Moerner, Nobel Prize in Chemistry, 2014.

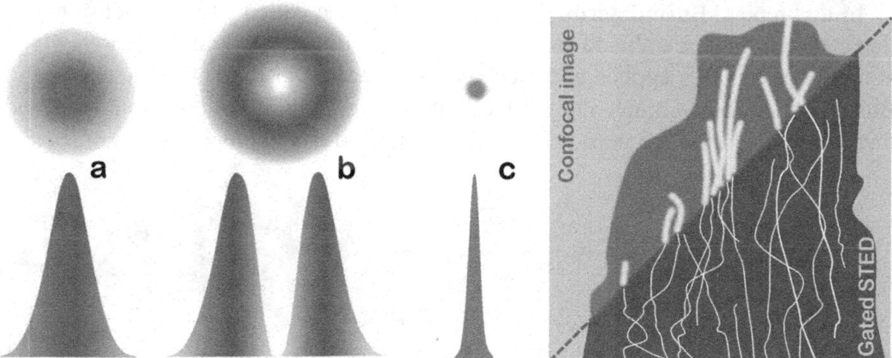

Figure 14.3 Stimulated emission depletion microscopy (STED) and TRIZ inventive principle of *matreshka* and system–antisystem. Excitation laser pulse (a), de-excitation pulse (b) and remaining fluorescence (c). Improvement of resolution of a protein imaging due to STED is shown qualitatively on the right.

result in improving the imaging resolution to values several times lower than the laser wavelength.

Considering this invention within the framework we have discussed in this book, we can also note a remarkable connection to TRIZ. From the perspective of the theory of inventive problem solving, the STED method is an illustration of the use of the principle of system and antisystem (excitation and de-excitation laser pulse), perhaps combined with the inventive principle of the *matreshka* (one laser pulse is located geometrically inside of the other pulse).

We have already observed examples where system-antisystem and *matreshka* inventive principles combined to create novel systems (e.g., dual-force neutral solenoids). This suggests that combinations of these or other inventive principles can be especially efficient in solving problems in various scientific fields.

14.4 AIMING FOR PASTEUR QUADRANT

Science has always been a driver for the economy. This is a commonly accepted statement — however, the mechanisms of its impact are complicated. Still, their analysis is necessary, and not only from a philosophical point of view, but also in order to optimize the research priorities and define the strategies for technological innovation.

One of the attempts to analyze the model for research and technology transfer was done by Vannevar Bush, who, during World War II, was instrumental in reorganizing the research and science community according to the needs of that difficult time. Vannevar Bush's 1945 report[8] has since defined post-war scientific policy in the United States and in many other countries (supposedly, for decades to come).

[8]*Science, the Endless Frontier: a Report to the President,* by Vannevar Bush, Director of the Office of Scientific Research and Development, July 1945.

In this report, Bush describes what would later be called a one-dimensional or linear model for research and technology transfer. Bush claims that research which is more basic is less applied and vice versa (illustrated in Fig. 14.4). According to Bush, applied research invariably drives out pure research if the two are mixed, and therefore, basic research must be completely isolated from *considerations of use*.

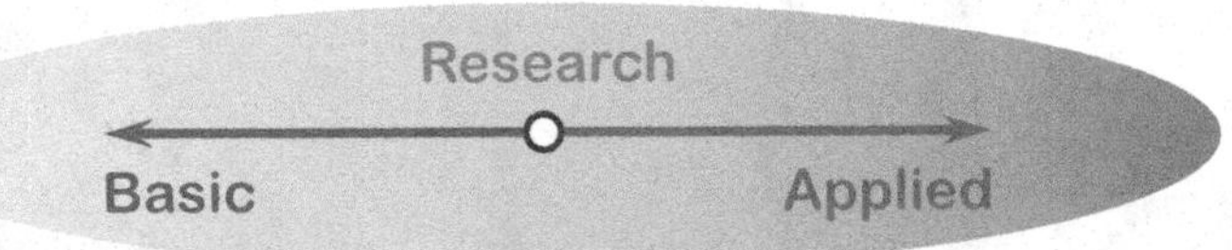

Figure 14.4 One-dimensional, linear model of research.

Correspondingly, the dynamic linear model of technology transfer resembles a pipeline wherein government funding stimulates basic research, which then in turn feeds applied research, and ultimately results in technology and product development, with society eventually benefiting from the process (see Fig. 14.5).

These views of the relationship between basic science and technological innovation have since then been analyzed, criticized, and a new model has been developed.

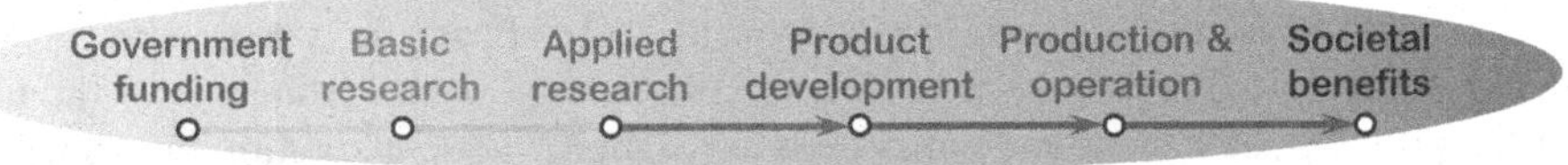

Figure 14.5 Dynamic linear model of technology transfer.

The contradiction between these linear models and practice can be illustrated via the example of accelerator science and technology. The invention of "strong focusing" in the 1950s was a revolutionary change in accelerator technology. It enabled numerous applications. This invention may have come about as a result of a pure fundamental interest — however, it was developed as a result of the pursuit of a certain concrete goal, and was made possible due to certain technologies available at that time.

A new model of research and technology transfer was suggested by Donald Stokes, who worked on the Advisory Committee on Research for the National Science Foundation.

In his report to the NSF, and in the book he subsequently published,[9]

"Research with consideration of use" is an effective way for science to provide an impact on our economy and society.

Donald Stokes argued against Bush's linear model and introduced the notion of use-inspired research — "research with consideration of use" — which redefined the

[9]Donald E. Stokes, *Pasteur's Quadrant — Basic Science and Technological Innovation*, Brookings Institution Press 1997.

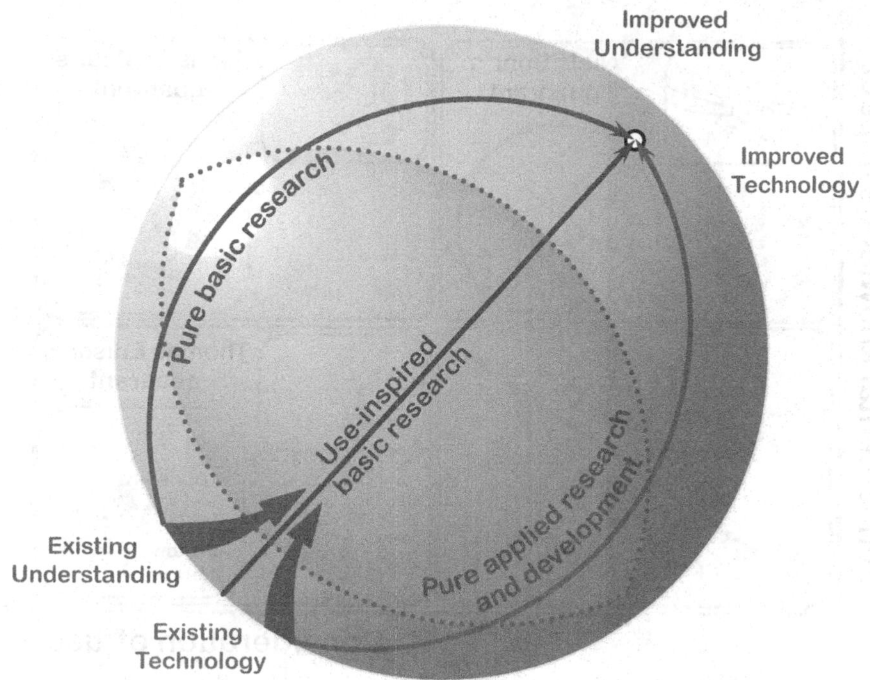

Figure 14.6 Revised dynamic model of research and technology transfer.

paradigm of the relationship between basic science and technological innovation. A revised dynamic model suggested by Stokes is illustrated in Fig. 14.6.

To portray his ideas, Donald Stokes suggested considering research on a two-dimensional plane, where the axes are fundamental knowledge impact and consideration of use.

Donald Stokes' plot is now called a *Pasteur quadrant* which we illustrated with examples from accelerator technology in Fig. 14.7.

A characteristic example of a purely fundamental scientific pursuit is the research works of Niels Bohr on the structure of nuclei, while another example of purely practical activity is Thomas Edison's developmental work on the filaments of light bulbs — as shown in Fig. 14.7.

The quantitative assessments of these examples — or of other research projects placed on this graph — can be done by evaluating the number of either the academic papers or patents, resulting from a particular research, as illustrated in Fig. 14.8.

Donald Stokes suggested, however, that an optimized approach should balance the fundamental pursuit of knowledge with consideration of use, which is illustrated by the works of Louis Pasteur — indicated in Fig. 14.7 as a preferred direction.

In the field of modern accelerator science and technology, characteristic examples for two sections of the quadrant are: colliders intended for the exploration of fundamental properties of elementary particles (on one axis), and accelerator-based devices for medical applications (on the other axis).

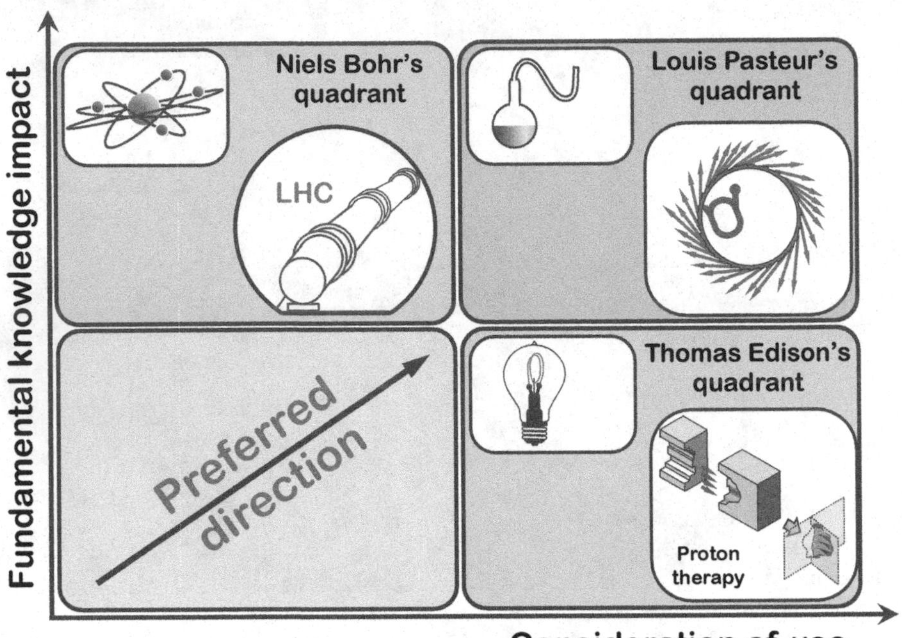

Figure 14.7 Pasteur quadrant and accelerator science.

The criteria suggested by Donald Stokes are universal and applicable to any scientific and technological area. By applying these criteria to accelerator science and technology, we can conclude that the preferred direction — which best balances the fundamental pursuit of knowledge with consideration of use — will be the direction to create novel light sources. Efforts made in this direction could potentially produce scientific instruments applicable to the investigation of protein structures or advanced materials, which may also be almost directly applicable to the creation of new medicines or metals with controllable properties.

This analysis also shows that — as in many other disciplines — accelerator science and technology, in synergy with lasers and plasma physics, can truly span the entire range of directions — from pure fundamental science to pure applied development.

Paster quadrant is the best framework for *the Year 2050 predictions*

The research and technology innovation model is indeed not just linear, but at least two-dimensional, as Stokes outlined in his revised dynamic model. In the 21st century, the driving forces of technological innovation — as well as the global pact between science and society — are different than they were in the middle of the last century. The revised criteria — as illustrated in this section via the example of accelerator science — are universal, and can be applied to any discipline, which can help us to optimize the impact of our research investments on our economy and society.

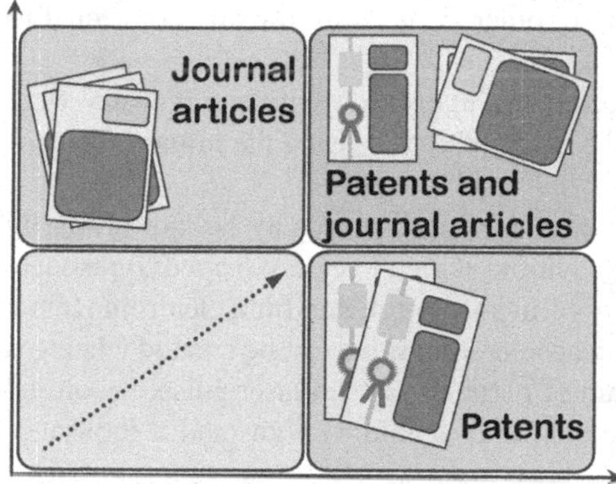

Figure 14.8 The units of quantitative assessments of research in the Pasteur quadrant.

Any research activity can be analyzed on a two-dimensional plane, where the axes are "fundamental impact of knowledge" and "consideration of use," typically measured by scientific articles or patents. An optimized approach should balance the fundamental pursuit of knowledge with consideration of use, which is illustrated by the works of Louis Pasteur.

14.5 HOW TO CROSS THE VALLEY OF DEATH

Powerful beams of light, heat rays — now called lasers — are features of H.G. Wells' and Alexey Tolstoy's science fiction stories that have excited many generations of future inventors and scientists. These stories thrilled crowds of kids (and adults) as they rambled along back streets dreaming of having lasers in their pockets.

The curiosity and imagination of the younger generation is the fuel that enables the development of our civilization. The challenge for governments, educational institutions and societies is to understand how to nurture and later harness these attributes.

The first visible light lasers (with a wavelength of light of about half a micrometer) were typically big when they were created half a century ago, even huge, and certainly not pocket size. Now, miniature lasers are in CD players, bar-scanners in shops, laser-pointers — practically everywhere. However, lasers with a much shorter wavelength, in the Angstrom range (light in the X-ray spectrum), have only just become available. Called free electron lasers, they are a kilometer long. Due to their short wavelength, X-rays are already indispensable for the analysis of protein structure, synthetic molecules, new materials and many other objects. The size and affordability of such X-ray lasers are major obstacles that hinder the widespread use of compact X-ray lasers. Creating compact X-ray lasers is the challenge that accelerator science now needs to confront.

Particle accelerators have already impacted many areas of our lives with their medical and industrial uses, as well as with their help in creating research instruments. Tens of millions of patients receive accelerator-based diagnoses and treatments each year, worldwide. The total annual market value[10] for all products that are treated or inspected by accelerators is more than $500B. Approximately 30% of the

[10] Accelerators for America's Future, Department of Energy, 2009.

Nobel prizes in physics, as well as many in other areas[11], are directly connected to the use of accelerators.

The ideas that enabled the use of accelerators in everyday life and industry were developed decades ago. New ideas will be essential for ensuring the future impact of this field.

Conventional accelerators, no matter how advanced they may be, are primarily based on the acceleration of particles in cavities — metal vessels shaped to resonate and create accelerating fields. The ability of metals to tolerate high electromagnetic fields is intrinsically limited. However, an accelerating wave can be created when gas is ionized and excited by an intense beam of particles or by a laser pulse, becoming plasma. Plasma is an indestructible medium and is able to withstand a thousand-times-higher accelerating gradient.

Accelerator science and technology is on the edge of a breakthrough brought on by synergy with laser and plasma physics. The most immediate outcome that this synergy will enable is the creation of novel, compact X-ray lasers and light sources. The direct collision of beam and laser light also opens up another opportunity for the creation of X-ray sources via the use of the Compton effect (when visible light photons are reflected from a relativistic electron beam and thereby decrease their wavelength down to Angstrom levels).

Science is indeed the driver of our civilization's progress. However, the journey from initial ideas and experimental demonstrations to widespread commercial applications is long and difficult. Various studies performed in different countries have all found a gap, a so-called "Valley of Death" in technology transfer. It is difficult to bridge the middle range of the technological readiness of ideas. On one end, the research institutions are usually not positioned to develop ideas into commercial applications, while on the other end, the risk is often too high for industry to pick up ideas that are too fresh and undeveloped.

The challenge originates from the different motivations, methods and timescales of three key players: academic institutions, industry and investors[12]. Their corresponding aims and motivations — the front-end fundamental scientific results, the development of commercial devices in the foreseeable future, and optimization of investments versus risk/return factors — are often incompatible (Fig. 14.9). Accelerators in synergy with lasers and plasma may, in fact, offer a solution for the academia–industry–investor puzzle through simultaneous, parallel work on a portfolio of three different types of compact X-ray light sources (Fig. 14.10).

A solution to the academia–industry–investor puzzle involving work on three designs of compact X-ray sources is illustrated in Fig. 14.11.

How to solve the academia-industry-investor puzzle? Working on a portfolio of compact light sources may provide an answer.

[11] E. Haussecker and A. Chao, Physics in Perspective, 2011.

[12] A government can be an investor.

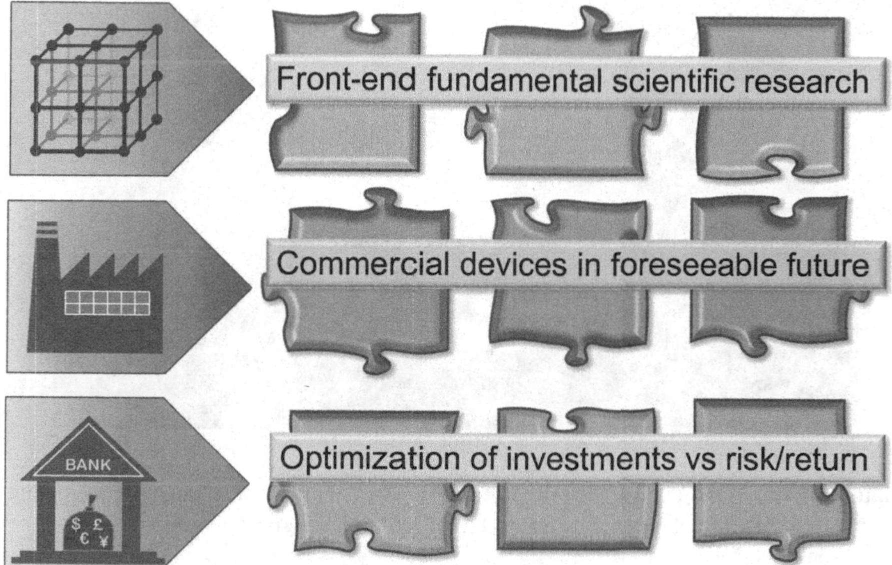

Figure 14.9 Academia–industry–investor puzzle caused by different motivations of the three participating groups.

Compton X-ray sources — the first type of source in the solved puzzle — are now actively being developed and are a lower-risk investment for industrial use. Yet a more challenging, but promising, Compton source requires superconducting acceleration to allow for a much higher electron beam current and X-ray brightness. This second option resides in the middle of the range for both the projected availability and risk/return.

The most challenging, but also the most promising source is an X-ray source based on laser plasma acceleration — initially a betatron source, and ultimately, perhaps in less than a decade* — a free electron laser. If we properly schedule the relative progression of the different stages of research and development for these three types of X-ray sources, we could balance the typical risks associated with the development of innovative products, and the opportunities they offer.

The phrase above was written in 2015 for the first edition. It is a pleasure to note that the first FELs based on plasma acceleration have already been demonstrated.*

Such work on compact X-ray sources fits into to the preferred direction of the Pasteur quadrant. Moreover, research and development in this direction may result in the creation of instruments that every university will aspire to have. The ability to obtain a compact X-ray laser would revolutionize contemporary science and technology yet again — not unlike how the spread of near-visible light lasers impacted science and industry during the 20th century.

Figure 14.10 Working on a portfolio of compact X-ray light sources can help in crossing the "Valley of Death" between accelerator science and technological innovation.

14.6 HOW TO LEARN TRIZ IN SCIENCE

In this section, we would like to convey our personal impressions of the best ways to learn (and also teach) the methodologies of inventiveness in the science departments of universities. These impressions are based on the development of training programs involving TRIZ at the John Adams Institute for Accelerator Science — JAI — at Oxford University, Imperial College, Royal Holloway University of London and, most recently, Old Dominion University, where these approaches were developed even further.

The JAI is a center of excellence in the UK for advanced and novel accelerator technology. JAI is focused on researching and developing accelerator science and techniques, as well as conducting graduate training. The cohort of PhD students usually number at around three dozen, and, on average, JAI educates about six PhD experts per year. Like any other research organization or graduate school, JAI is motivated to turn the results of research into having a positive societal impact, as well as endeavoring to find ways to be more inventive and innovative. For the latter, while at JAI, we began to consider TRIZ and developed a specific, scientifically oriented way to introduce this method to the PhD students.

Let us briefly review TRIZ and its connection to science graduate schools. TRIZ is the methodology of inventiveness that was specifically developed for engineering fields. It was created during the second half of the 20th century and gradually became one of the most powerful tools in the industrial world. TRIZ has become, according to Forbes, the bedrock of innovations of such companies as Samsung, as well as many others.

While the TRIZ method of industrial inventiveness was originally created for engineering, this methodology is universal and can also be applied to science. However, experience shows that knowledge of TRIZ is nearly nonexistent in the scientific

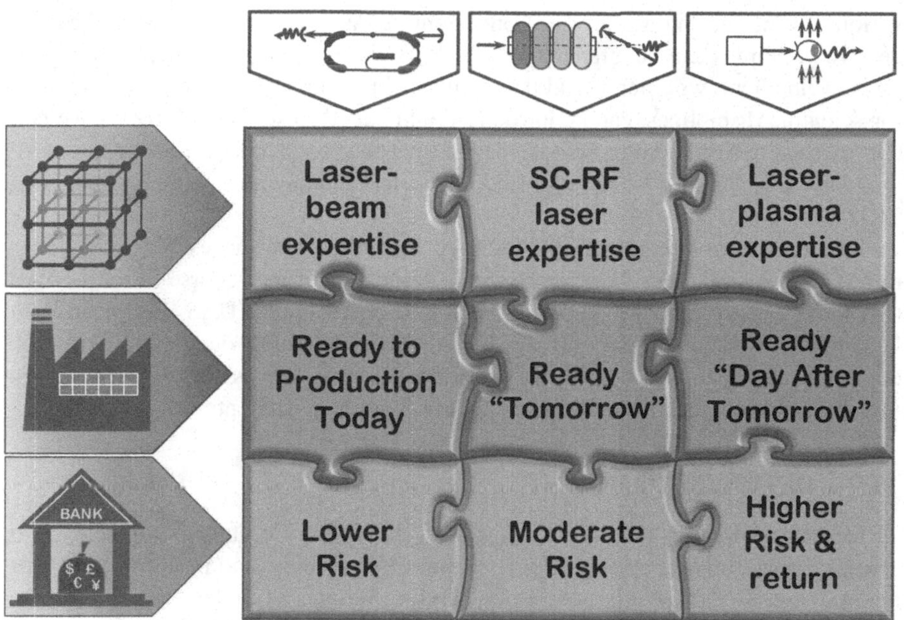

Figure 14.11 A solution to the academia–industry–investor puzzle — work on three designs of compact X-ray sources.

departments of Western universities.

Moreover, it is not rare to hear about unsuccessful attempts to introduce TRIZ into graduate courses of scientific departments of universities. Indeed, in many or most of these cases, the apparent reason for the failure was that the canonical version of TRIZ was introduced to science PhD students in the same way that TRIZ is taught to engineers in industrial companies. This appears to be a mistake, seeing as science students are rightfully more critically minded and justifiably skeptical about overly prescriptive step-by-step methods. (As we have seen above, TRIZ involves finding a pair of contradicting parameters in a problem, which then — using the TRIZ inventive tables created by TRIZ teams, based on the analysis of hundreds of thousands of past inventions — immediately leads to selecting just a few suitable inventive principles, narrowing down the choice and resulting in a much faster solution to a problem.)

The approach of introducing TRIZ to JAI and VITA[13] graduate students is different and takes into account lessons learned by its predecessors. Instead of teaching our graduate students the ready-to-use methodology, we are effectively taking them through the process of recreating parts of the TRIZ methodology by analyzing

[13]VITA is Virginia Innovative Traineeship in Accelerators, which now includes students from Old Dominion University, Hampton University and Norfolk State University

various inventions and discoveries from scientific disciplines, showing that these inventive principles can be efficiently applied to science. Moreover, in the process of this development, we often found that additional inventive principles, more suitable for scientific disciplines, can be introduced and added to standard TRIZ — we call this extension Accelerating Science TRIZ (the play on words is now apparent — the word "accelerating" is not referring to accelerators anymore, but highlights that TRIZ can help to boost science).

The approach to teaching TRIZ described above has been successfully introduced to JAI graduate students, was successfully implemented in a course in the USPAS (US Particle Accelerator School), has been introduced in the JUAS (European Joint Universities Accelerator School) and JAI APPEAL school (a one-day to one-week course for high-school teachers that we conduct every year as part of our outreach activity), and has also been introduced to VITA graduate students.

14.7 DESTINED TO INVENT

TRIZ methodology is another way to look at the world. Combined with science, it creates a powerful and eye-opening amalgam of knowledge and inventiveness. This methodology is particularly helpful for building bridges of understanding between completely different scientific disciplines, and so is also naturally useful to educational and research organizations that endeavor to break barriers between disciplines. Combined with the art of back-of-the-envelope estimations, this creates a powerful methodology that can help us in any areas of research or technology development.

The Breakthrough by Design or Innovation by Design approach is an amalgam of TRIZ inventive principles and laws of technical system evolution with the art of back-of-the-envelope estimations, applied to several neighboring areas of science and technology — by developing and applying this methodology, you will be destined to invent.

Ultimately, our recommendations on learning and teaching TRIZ in universities can be summarized as follows. Introducing TRIZ courses/lessons to university students is only the first step. While teaching these TRIZ courses/lessons, avoid the canonical, ready-to-use version of TRIZ. Instead, take the students through the process of proactively adapting TRIZ for science. Use back-of-the-envelope estimations as much as possible. Arrange a sequence of relatively short TRIZ lessons throughout the entire duration of education in the university so that students can gradually build up their ability to use TRIZ methods in order to build bridges of understanding between different disciplines that they will interact with during their education and research.

14.8 LET US BE CHALLENGED

Many of the examples of scientific inventions that we considered in this book have already been made, and we only analyzed the applicability of TRIZ post-factum.

It is natural to wonder whether TRIZ and AS-TRIZ can actually help to inspire and create new scientific inventions and innovations, especially in regard to projects that continue to manifest many unsolved obstacles.

One example of such a project is a circular collider currently being considered as a successor to the Large Hadron Collider at CERN — *the Future Circular Collider* (FCC). The FCC project is looking to create a 100-km tunnel infrastructure in order to house the e^+e^- collider and then the 100 TeV CM proton-proton collider.

This project has many scientific and technical tasks and challenges that need to be solved. Notably, one issue is that the synchrotron radiation at these high energies starts to influence the proton beams. The total energy in each circulating proton beam is expected to exceed 8 GJ, which is equivalent to a kinetic energy of an Airbus-380 flying at 720 km/h. Not only does such a beam need to be handled safely in the bending magnets, but it also needs to be focused in the interaction region to a micron spot size — which is practically the equivalent of having to pass through a needle's eye (Fig. 14.12).

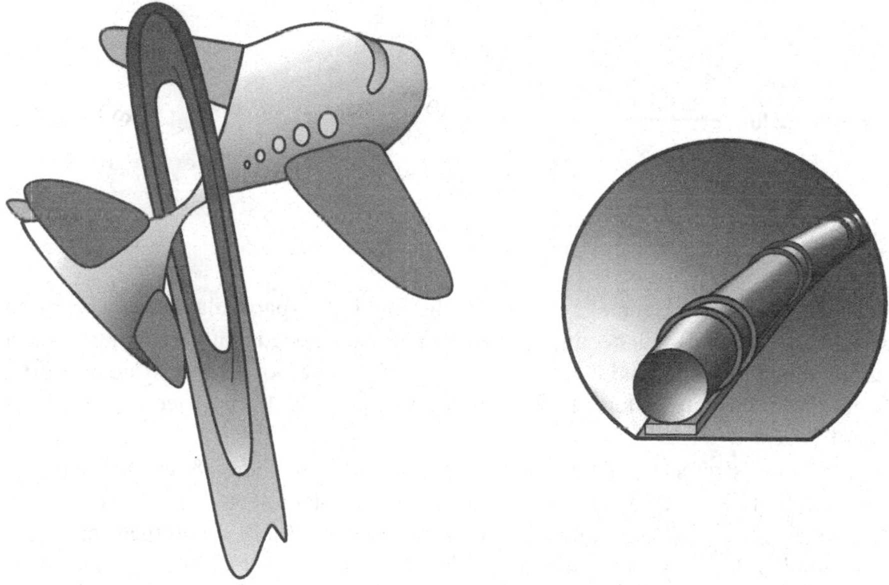

Figure 14.12 The FCC proton beam energy will be comparable to that of an airplane, while the beam will need to be focused at the interaction point to micron-scale size — akin to an analogy of a plane passing through a needle's eye.

It remains to be seen if TRIZ and AS-TRIZ methodology can be applied to such a large-scale project as the FCC, as it brings a whole array of new, difficult and exciting challenges to the table. Nonetheless, it is certainly a project that can only flourish with the application of our knowledge and inventiveness.

14.9 THE YEAR 2050 PREDICTIONS

Finally, in the spirit of *The Year 2000*[14], which was discussed in the first chapter and wherein predictions were made in 1968 for the year 2000, let's make a prediction for our area of science and technology for the year 2050.

Let's focus on the preferred direction of the Pasteur quadrant — i.e., light sources and FELs — and try to predict how they will evolve. Being inspired by the first demonstration of plasma acceleration-based FELs, let's assume that they will become widespread and eventually more compact, as illustrated in Fig. 14.13.

In order to predict the long-term perspectives and evolution of light sources and FELs, we can then apply the TRIZ general laws of evolution. The kinematic laws of standard TRIZ include the law of transition to a super-system that states, "a system exhausting possibilities of further significant improvement is included in a super-system as one of its parts."

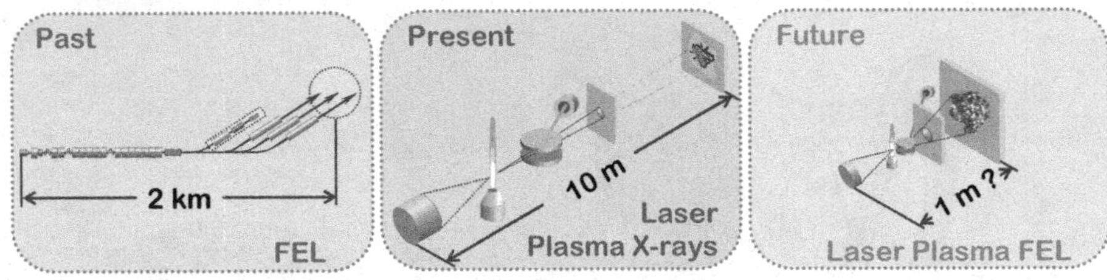

Figure 14.13 Vision for the evolution of FELs toward the year 2050.

Based on that, we can make the first prediction: FELs (particularly those based on plasma acceleration) will become so compact and developed that they will become a part of another system, and that system in turn will be part of a super-system. Before discussing of which system the FEL will become, let's discuss one of the recent Nobel Prizes.

The Chemistry Nobel Prize of 2016 was awarded to Pierre Sauvage, J. Fraser Stoddart, and Ben L. Feringa for the development of molecular machines. These "machines" have just some elements of what we would call a complete machine. First of all, TRIZ evolution laws help us predict what parts of molecular machines would be invented next. In particular, the law of the completeness of the parts of the system states that a machine needs to have four parts: an engine, a transmission, a working unit and a control element. Moreover, the law of energy conductivity of the system states that every technical system is a transformer of energy and should circulate freely and efficiently through its four main parts.

But, most importantly, TRIZ evolution laws state that these molecular machines can become a part of another super-system. We can therefore express our dreams and predictions for the year 2050. On the sub-system level, we have the compact

[14]*The Year 2000*, K. Herman, A. Wiener (editors), ISBN 978-0025604407, 1968.

and efficient laser-plasma FEL. On the system level, the FEL will become part of the system where it will analyze proteins synthesized by a molecular machine. Last, the entire super-system will produce patient-tailored molecular machines for DNA repair.

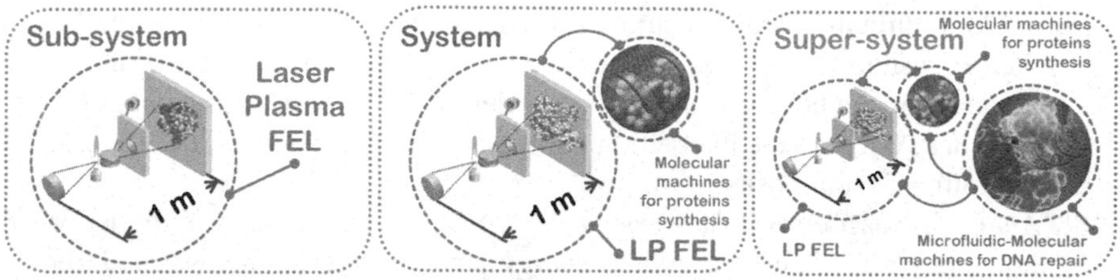

Figure 14.14 Prediction of the evolution of FELs toward the year 2050.

Among predictions for the year 2050 is a laser plasma FEL that is part of a super-system wherein it analyzes proteins synthesized by a molecular machine, while the entire super-system produces patient-tailored molecular machines for DNA repair.

Let's make this prediction and this dream — with the help of the Breakthrough By Design approach — a reality!

EXERCISES

14.1. *Mini-project.*
A proposal for a Higgs and top-quark factory is based on the design of an electron-positron collider (175 GeV for each beam) in a tunnel with a 50 km circumference. Assuming that the maximum power delivered by the RF system to each beam is limited to 50 MW, estimate the maximum current that can be stored in such a collider. Make an assumption about the focusing system at the interaction point, and evaluate beam emittances and the collider's luminosity.

14.2. *Analyze inventions or discoveries using TRIZ and AS-TRIZ.* A liquid jet or liquid target is often used in high-energy physics experiments, particularly for the production of antiprotons. A liquid jet can withstand the power of the incident proton beam. Analyze this technology in terms of the TRIZ and AS-TRIZ approach, identifying a contradiction and a general inventive principle that were used (could have been used) in this invention.

14.3. *Practice in reinventing technical systems.* Space-charge forces in the electron bunch emitted from a photo-injector will create a $\langle x \cdot x' \rangle$ correlation in the bunch (particles further away from the bunch center experience larger space-charge force). This effect can be large, since the beam is not yet relativistic, and can create large emittance growth of the beam. Optimally placing a solenoid on top of the photoinjector will cancel out (due to solenoid focusing) the aforementioned correlation, achieving *emittance compensation.* Suggest a way to create a strong solenoidal field in the plasma-accelerating bubble to achieve a similar emittance preservation for the plasma-accelerated bunch. Assume that any external solenoid (pulsed, superconducting, etc.) cannot create a field of sufficient strength and sufficiently sharp gradients. Try to use the inventive principle of *going to another dimension* and remember the TRIZ principle *to use resources and energy that you already have in the system.*

14.4. *Practice in the art of back-of-the-envelope estimations.* Imagine that you are building a *quantum computer* based on a storage ring of $^{24}Mg^{+}$ ions. Let's aim to store $N_i = 10^6$ ions, which will be cooled by electron or laser cooling to a low temperature, so that the ions will form a one-dimensional linear crystal in the storage ring. The qubits of this quantum computer will be formed by entanglement states of the ions, allowing us to get a very large number of qubits, aiming to achieve *quantum supremacy*. Estimate the longitudinal temperature of the ions to which they need to be cooled in order to form the one-dimensional crystal. *(It is assumed that you can identify the most important effects playing roles in this task, can define the necessary parameters and set their values, and can get a numerical answer.)*

14.5. *Practice in predicting evolution of technical systems*.* Based on what you have learned in this book about TRIZ laws of the evolution of technical systems, and based on your knowledge of the scientific or technical area you are an expert in or are interested in, make your own prediction in the style of *the year 2050.*

15 Forty Inventive Principles

Across many books, you can find countless illustrations of inventive principles based on engineering examples. However, examples related to science and advanced technologies, such as accelerators, lasers and plasma are rare. We will begin to fill in this gap with this chapter replete with inventive-principle illustrations[1] based on accelerator science, among other fields. You will notice that some of the standard definitions of TRIZ principles are redefined in this chapter; this is to better match the kinds of technological challenges we face in the 21st century.

The Theory of Inventive Problem Solving (TRIZ) is a powerful tool widely used in the engineering community. It involves identifying a physical contradiction in a problem and — based on the corresponding pair of contradicting parameters — selecting a few suitable inventive principles, ultimately narrowing down the choices and leading to a much faster solution of the problem.

It is remarkable that the TRIZ methodology can also be applied to scientific disciplines. Many TRIZ inventive principles can be identified *post factum* in various scientific inventions and discoveries. Additional inventive principles, more suitable for scientific disciplines, could be introduced and added to the standard TRIZ inventive principles, and some of these same principles could be reformulated to be better applicable to science — we call this extension Accelerating Science TRIZ, introduced in the first edition of this book.

In this chapter, we will describe and illustrate the AS-TRIZ inventive principles via scientific examples, identifying suitable discoveries and inventions originated from physics, biology, and other areas. This chapter, we believe, is yet one more step toward bringing the TRIZ methodology closer to the scientific community.

We have decided to keep the canonical number of 40 inventive principles, presenting them with illustrated examples and brief commentaries. However, we have renamed some of the principles or adjusted their descriptions in some cases. The process of adjusting the definitions and creating new examples should be continuous, and we encourage readers to proactively participate in this process, as this is the most efficient way to learn how to apply the TRIZ methodology to science.

(Part of the illustrations of inventive principles presented in the following pages were created for the USPAS-2016 course[2] "Unifying physics of accelerators, lasers and plasma." During this one-week course, students focused on a mini-project and, at the end of the course, were ready to submit NA-PAC abstract, which resulted in a scientific paper[3] authored by the team of students.)

---

[1]Some of these illustrations were first published in the essay "Accelerating Science TRIZ inventive methodology in illustrations," Elena Seraia, Andrei Seryi, July 2016, https://arxiv.org/abs/1608.00536

[2]USPAS-2016, https://uspas.fnal.gov/materials/16CSU/CSU-UnifyingPhysics.shtml

[3]Compact Ring-Based X-Ray Source with On-Orbit And On-Energy Laser-Plasma Injection, Marlene

DOI: 10.1201/9781003326076-15

15.1 SEGMENTATION

The inventive principle *segmentation* may involve dividing an object into independent parts, making the object easy to disassemble or increasing the degree of fragmentation or segmentation.

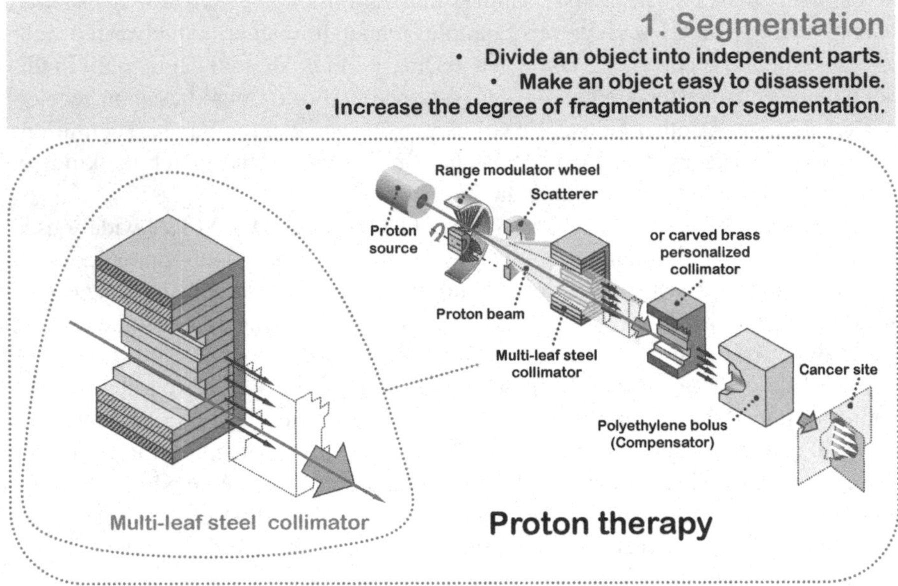

Figure 15.1 Inventive principle “Segmentation.”

An example illustrating this principle is a multi-leaf steel collimator used in a beamline for particle therapy. This collimator is needed in order to shape the proton beam in such a way that it will correspond to the shape of the target (e.g., cancer site). Sometimes, a solid, custom collimator can be used for this purpose; however, a new one would need to be machined every time. An adjustable collimator, made from segmented steel leaves, makes the treatment planning and delivery much more efficient.

15.2 TAKING OUT

The inventive principle *taking out* may involve separating and removing an interfering part or property from an object or singling out the only necessary part (or property) of an object.

An example that illustrates this principle is a collimator of the beam halo that localizes beam losses (which represent an interfering property in this case) in accelerators.

Turner, et al., NA-PAC, 2016, http://accelconf.web.cern.ch/napac2016/papers/tua3co03.pdf

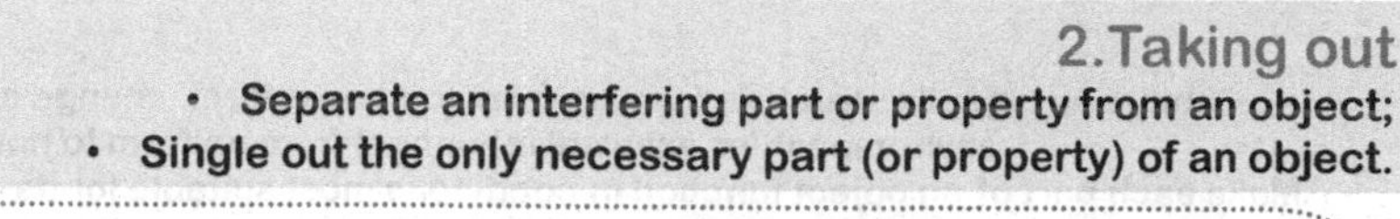

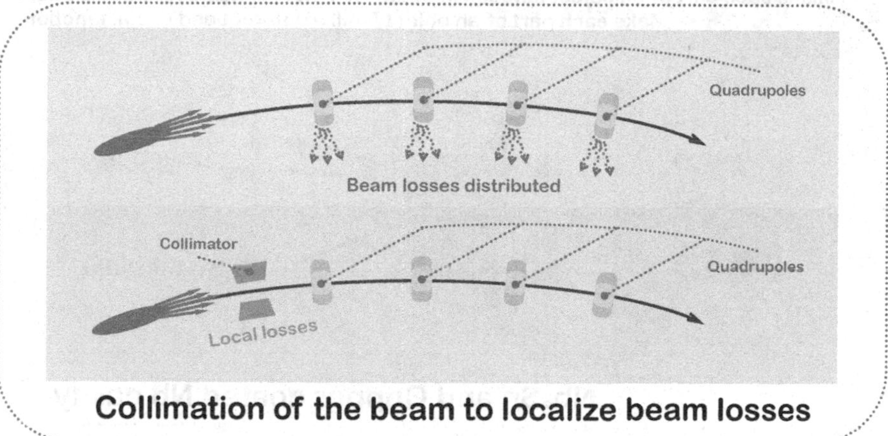

Figure 15.2 Inventive principle "Taking out."

The top half of Fig. 15.2 shows a beamline of an accelerator without a collimator. In this case, particles from the beam halo can be lost everywhere, creating problems associated with, for example, increased radiation in every location.

Inserting a collimator in one location (as shown in the bottom half of Fig. 15.2) would eliminate losses everywhere except in the collimator itself, which itself can be treated specially (e.g., additional radiation shielding can be installed around the collimator). Therefore, by inserting the collimator, we have taken out the interfering property (losses) from the system.

15.3 LOCAL QUALITY

The inventive principle *local quality* may involve changing an object's structure from uniform to nonuniform, changing an external environment (or external influence) from uniform to nonuniform, making each part of an object function in conditions most suitable for its operation, or making each part of an object fulfill a different and useful function.

An example illustrating this principle is a superconducting resonator cavity, the bulk of which is made from copper, save for the inner surface, which is covered by niobium.

While superconducting cavities are typically made entirely from niobium, an expensive material, a construction of copper and niobium, as shown in Fig. 15.3, would considerably save on material costs of such cavities, provided, of course, that results of certain ongoing studies[4] are successful.

---

[4] Enzo Palmieri, A.A. Rossi, R. Vaglio, *Experimental Results on Thermal Boundary Resistance for Nb and Nb/Cu*, Science, Oct 2014.

3. Local quality

- **Change an object's structure from uniform to non-uniform, change an external environment (or external influence) from uniform to non-uniform.**
- **Make each part of an object function in conditions most suitable for its operation.**
- **Make each part of an object fulfill a different and useful function.**

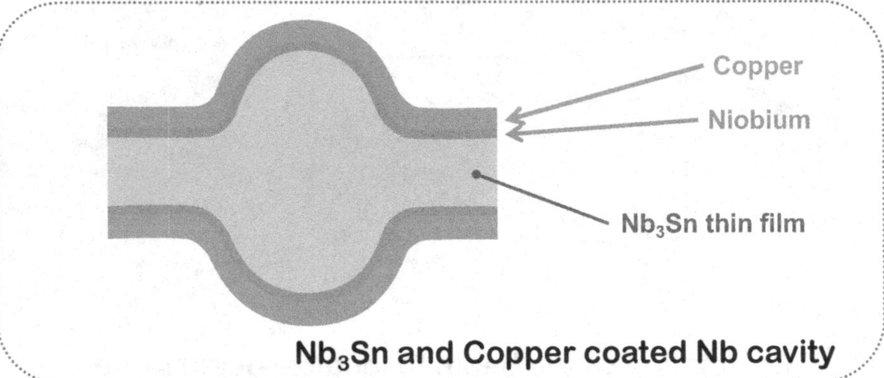

Figure 15.3 Inventive principle "Local quality."

15.4 ASYMMETRY

The inventive principle *asymmetry* may involve changing the shape of an object from symmetrical to asymmetrical, or, if an object is asymmetrical, increasing the degree of its asymmetry.

This principle can be found in the asymmetrical design of the dual-axis coupled cavities in the compact energy-recovery-based LINAC[5] as shown in Fig. 15.4.

In this LINAC, an accelerated electron beam, after radiation generation, comes back to the decelerating part of the cavity, where the beam returns its energy to the system. In order to avoid instabilities of the beam that can be created in this system, all high-order modes of the cavities need to be decoupled. This is achieved by introducing carefully designed asymmetry between every cell of the two cavities.

15.5 MERGING

The inventive principle *merging* may involve bringing closer together (or merging) identical or similar objects, assembling identical or similar parts to perform parallel operations or making operations contiguous or parallel, that is, bringing them together in time.

---

[5] R. Ainsworth, G. Burt, I. V. Konoplev, A. Seryi, *Asymmetric Dual Axis Energy Recovery LINAC for Ultra-High Flux sources of coherent X-ray/THz radiation: Investigations towards Its Ultimate Performance*, arXiv:1509.03675, physics.acc-ph, Sep 2015.

4. Asymmetry

- **Change the shape of an object from symmetrical to asymmetrical.**
- **If an object is asymmetrical, increase its degree of asymmetry.**

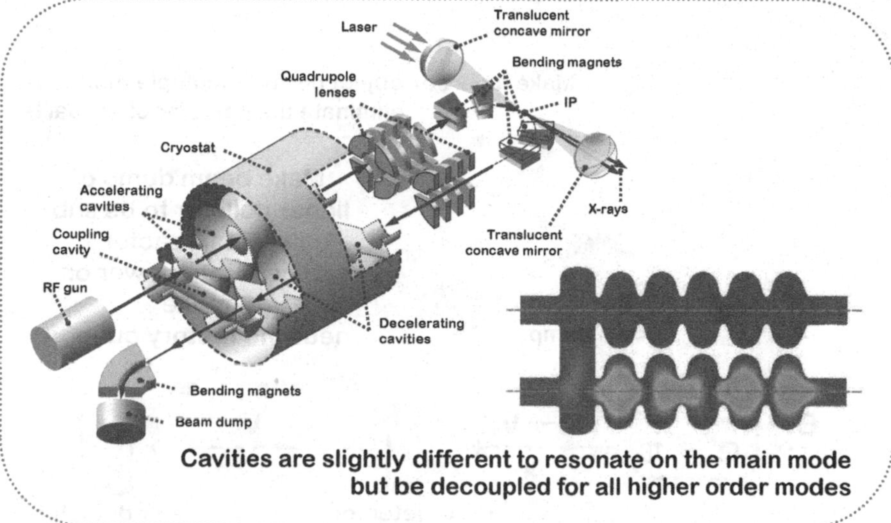

Figure 15.4 Inventive principle "Asymmetry."

5. Merging

- **Bring closer together (or merge) identical or similar objects, assemble identical or similar parts to perform parallel operations.**
- **Make operations contiguous or parallel; bring them together in time.**

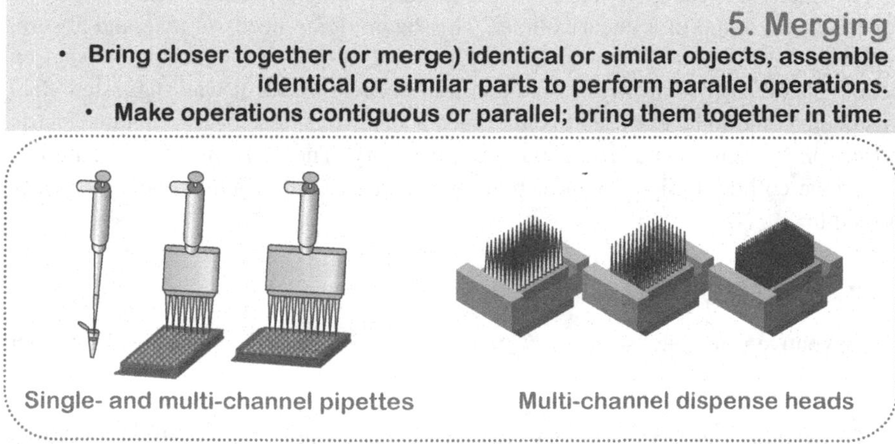

Figure 15.5 Inventive principle "Merging."

An example that illustrates this principle are multi-channel pipettes and modular dispensers that are now indispensable to biological studies wherein many samples, genes or variations of drugs need to be studied and analyzed in parallel.

15.6 UNIVERSALITY

The inventive principle *universality* may involve making a part or object perform multiple functions or eliminating the need for other parts.

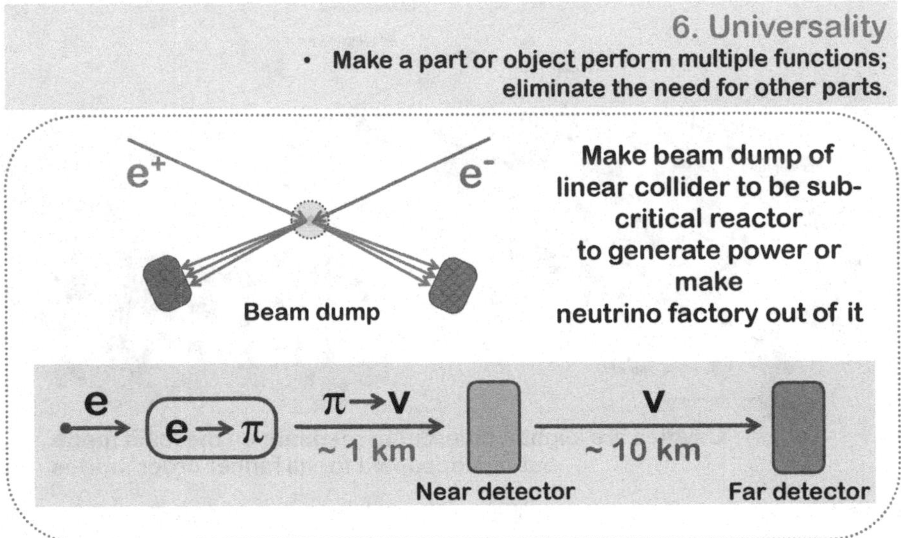

Figure 15.6 Inventive principle "Universality."

An example illustrating this principle is the following peculiar design proposal for the beam dump of a linear collider. This beam dump needs to take and absorb, typically, 10 MW of CW power in the form of 250–500 GeV electron or positron beam. This energy is mostly wasted and transforms to heat. It was suggested[6] that this beam could in fact be used to either feed a subcritical reactor to generate electric power or perhaps be made into a neutrino factory. Therefore, the beam dump of this linear collider design performs multiple functions and thus uses the universality principle.

15.7 "NESTED DOLL"

The inventive principle "*nested doll*" may involve placing one object inside another and then placing that combined object inside yet another object *ad infinitum*, making one part pass through a cavity in the other.

An example illustrating this inventive principle (which is also called the principle of *Russian dolls*) is the construction of a high-energy physics detector, where many different sub-detectors are inserted into one another to enhance the accuracy of elusive-particle detection.

[6]I.F. Ginzburg, *Beam Dump problem and Neutrino Factory Based on a e+e- Linear Collider*, arXiv: 1411.3295, physics.acc-ph, Oct 2014.

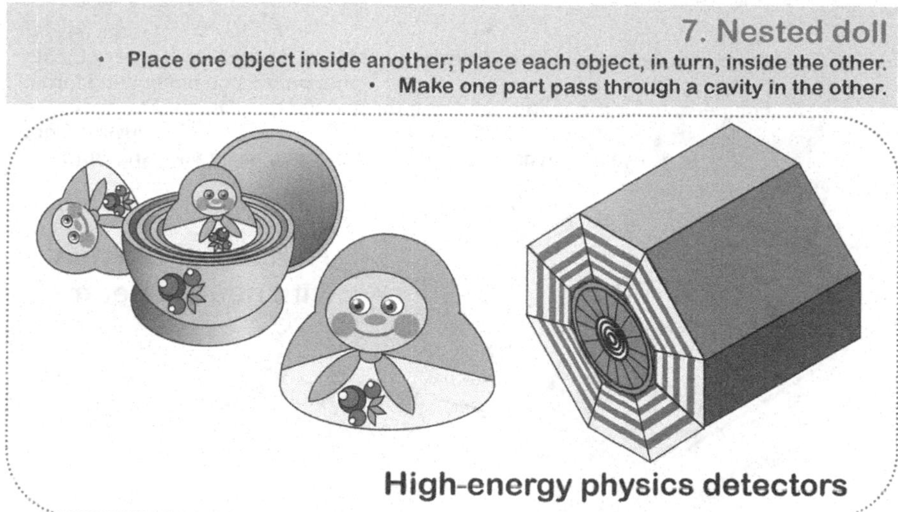

Figure 15.7 Inventive principle "Nested doll."

15.8 ANTI-FORCE

In standard TRIZ, this principle is called "anti-weight." However, for science applications, it needs to be redefined as "anti-force" because gravity usually plays a negligible role in situations involving particles or nano- and micro-objects, while electromagnetic forces can be much more important.

The inventive principle *anti-force* may involve compensating for the force on an object, merging it with other objects that provide compensating force, etc., as explained in Fig. 15.8.

We can illustrate this principle with the heating system for plasma in a tokamak, wherein an accelerated beam heats plasma. To avoid the beam sensing the field of solenoid or plasma, the beam is made of neutral atoms obtained by stripping electrons from the initial beam of negative hydrogen ions.

15.9 PRELIMINARY ANTI-ACTION

The inventive principle *preliminary anti-action* may involve replacing an action, which is known to produce both harmful and useful effects, with an anti-action to control those harmful effects.

An example that illustrates this principle is a final focus with local chromatic correction[7]. Any strong focusing optics suffer from chromatic aberrations, as shown on the left half of Fig. 15.9. Local chromatic correction involves dispersing the beam into energies *prior* to when it reaches the final lenses, and also involves inserting

[7] P. Raimondi, A. Seryi, *Novel Final Focus Design for Future Linear Colliders*, PRL **86**, 3779, 2001.

8. Anti-weight force

- **To compensate for the ~~weight of~~ force on an object, merge it with other objects that provide compensating force.**
- **To compensate for the ~~weight of~~ force on an object, make it interact with the environment (e.g., use aerodynamic, hydrodynamic, buoyancy and other forces).**

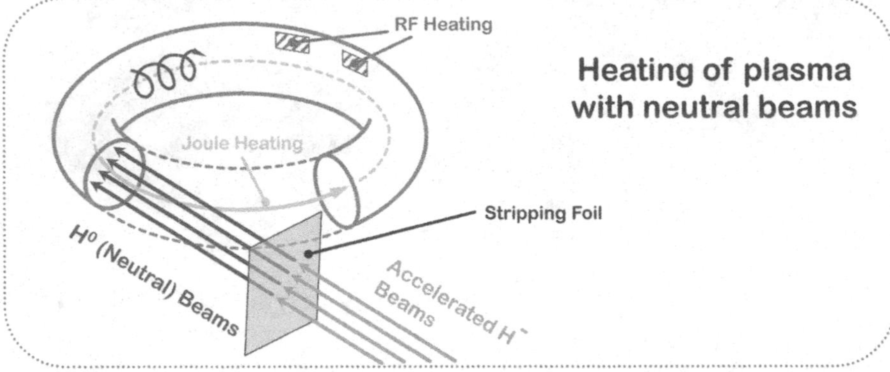

Figure 15.8 Inventive principle AS-TRIZ "Anti-force," named "Anti-weight" in standard TRIZ.

9. Preliminary anti-action

- **If it will be necessary to do an action with both harmful and useful effects, this action should be replaced with anti-actions to control harmful effects.**
- **Create beforehand stresses in an object that will oppose known undesirable working stresses later on.**

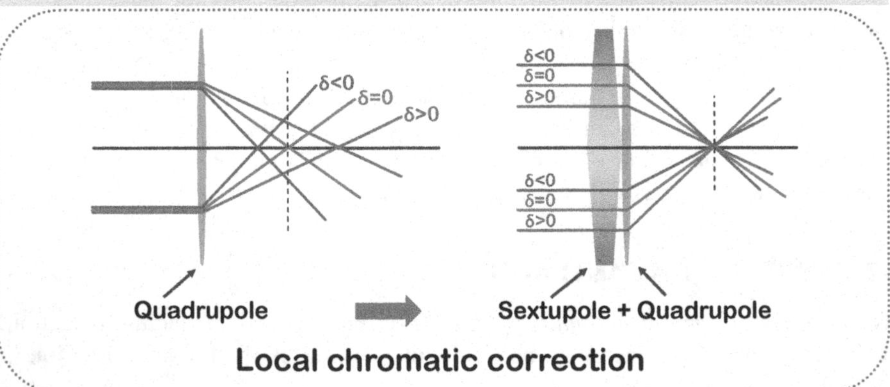

Figure 15.9 Inventive principle "Preliminary anti-action."

a nonlinear sextupole magnet next to the final lens, which cancels the chromatic aberrations, thus acting against is. As a result of this preliminary anti-action, the beam gets focused nicely into a tight spot as shown on the right half of Fig. 15.9.

15.10 PRELIMINARY ACTION

The inventive principle *preliminary action* may involve performing, before it is needed, the required change of an object (either fully or partially), or prearranging objects in such a way that they can come into action from the most convenient place and without losing time for their delivery.

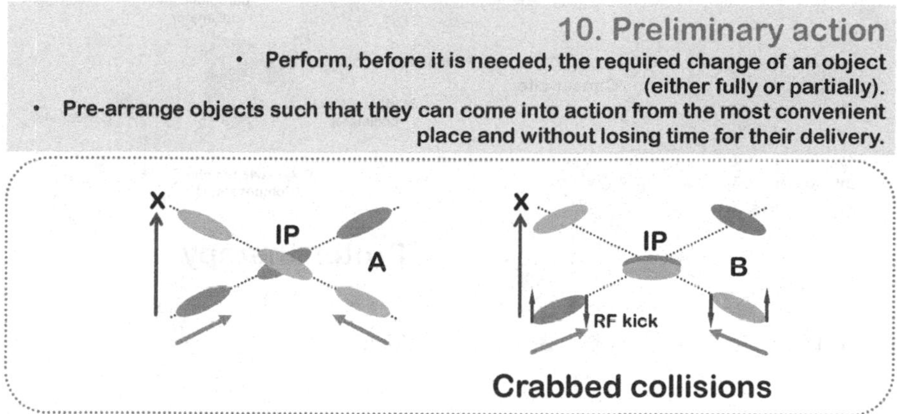

Figure 15.10 Inventive principle "Preliminary action."

An example that illustrates this principle is a crabbed collision. In a linear collider, the electron and positron beams need to collide with a small crossing angle, as shown on the left half of Fig. 15.10. However, their overlap during collision would then be incomplete and the luminosity would thus decrease. In order to prevent this loss, the beams, before colliding, can pass through a radio-frequency cavity, which would give the beams useful kicks in such a way that the heads and tails of the beams end up moving in different directions. The beams will therefore start to rotate and come to the collision point properly oriented, ensuring full overlap.

15.11 BEFOREHAND CUSHIONING

The inventive principle *beforehand cushioning* may involve preparing emergency means beforehand to compensate for the relatively low reliability of an object.

An example illustrating this principle is a bolus (compensator) for a proton therapy beamline. The thickness of the bolus varies depending on location and therefore modifies the energy of different parts of the proton beam, as well as correspondingly modifies the penetration depth of the protons, matching the shape of the cancer site.

The reader might argue that this example better suits the previous principle of preliminary action. A supplementary example, then, would be an emergency kicker that would dump the beam out of an accelerator in case of emergencies to prevent beam losses into precious superconducting magnets.

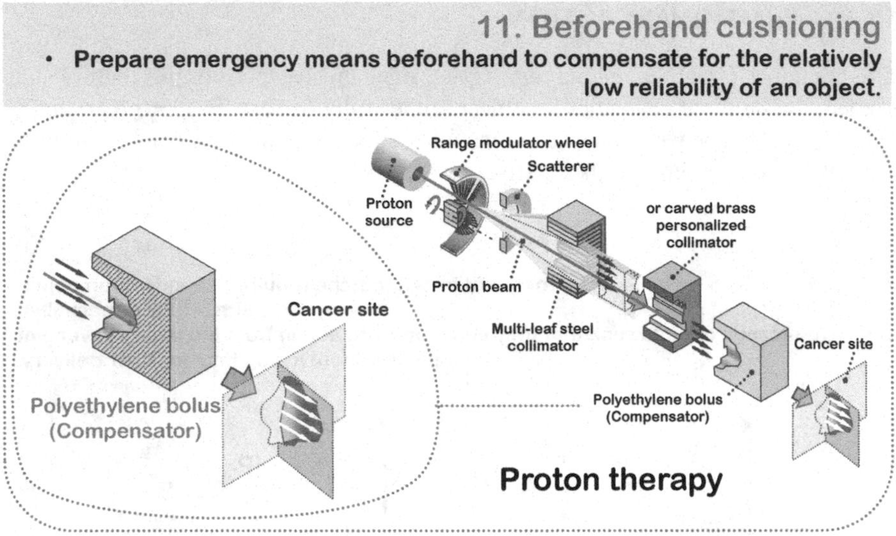

11. Beforehand cushioning
• Prepare emergency means beforehand to compensate for the relatively low reliability of an object.
Range modulator wheel
Scatterer
Proton source
or carved brass personalized collimator
Proton beam
Cancer site
Multi-leaf steel collimator
Cancer site
Polyethylene bolus (Compensator)
Polyethylene bolus (Compensator)
Proton therapy

Figure 15.11 Inventive principle "Beforehand cushioning."

15.12 EQUIPOTENTIALITY

The inventive principle *equipotentiality* may involve limiting the changes in position of an object in gravity or other potential field (e.g., changing operating conditions to eliminate the need to raise or lower the objects in a gravitational field).

12. Equipotentiality
• In a potential field, limit position changes (e.g., change operating conditions to eliminate the need to raise or lower objects in a gravity field).
Laser
Absorption
δ
δ
νL= ν0- δ
ν0
ν0+ δ
Laser cooling

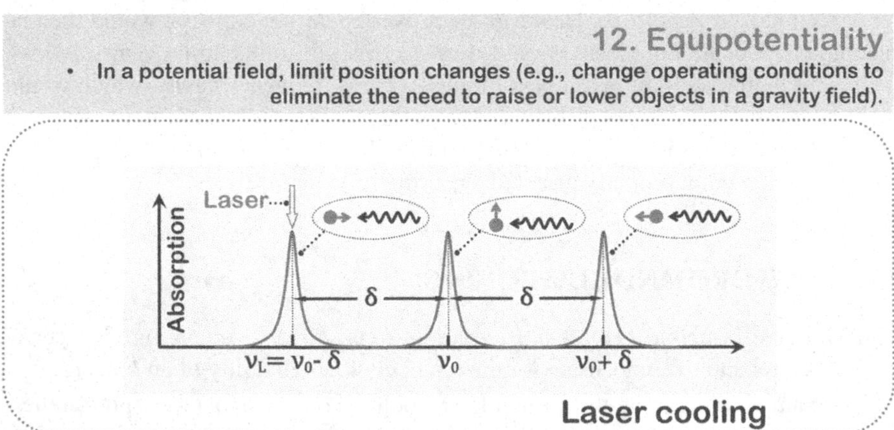

Figure 15.12 Inventive principle "Equipotentiality."

An example illustrating this principle is a laser cooling, where the interaction of a laser with an atom (prompting excitation of the atom) occurs only when they are "at the same potential," i.e., when the velocity of the atom is such that, due to the Doppler shift, the laser frequency corresponds to the energy of the atom excitation.

15.13 THE OTHER WAY AROUND

The inventive principle *the other way around* (which can be also be called the principle of the system and anti-system) may involve inverting the action(s) used to solve the problem or turning the object (or process) "upside down."

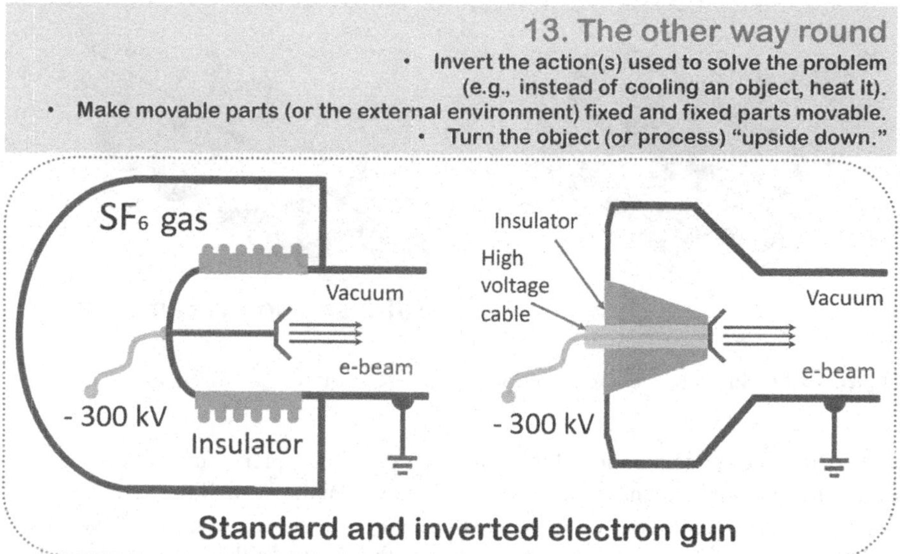

Figure 15.13 Inventive principle "The other way around" illustrated by the inverted electron gun.

TRIZ textbooks often cite Charles Wilson's cloud chamber (invented in 1911) and Donald Glaser's bubble chamber (invented in 1952) as examples of this principle of "system and anti-system." Indeed, the cloud chamber works on the basis of bubbles of liquid created in gas, whereas the bubble chamber uses bubbles of gas created in liquid. If the TRIZ inventive principle of the system and anti-system had been applied to the cloud chamber, the invention of the bubble chamber would have followed immediately and not almost half a century after the invention of the former.

However, we would like to offer another example: the inverted gun, shown in Fig. 15.13, which "pushes" the insulator inside, making the gun more compact and much more suitable for higher voltages.

15.14 SPHEROIDALITY – CURVATURE

The inventive principle *spheroidality — curvature* may involve using curvilinear parts, surfaces or as well as moving from flat surfaces to spherical ones, etc.

An example that illustrates this principle is the cavity resonator. A pill-box-style cavity resonator is shown on the left half of Fig. 15.14 in contrast with an elliptical cavity (where shapes are rounded and consist of various connected ellipses). The right half of Fig. 15.14 shows an example of a cavity that can produce a "crabbed

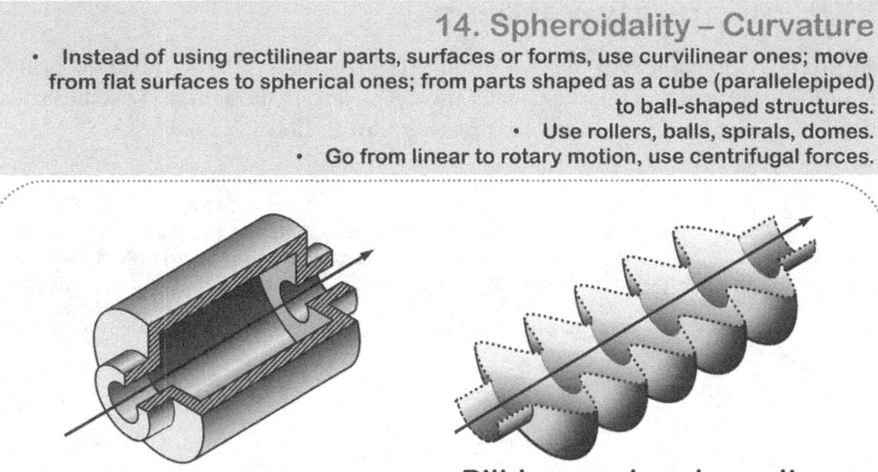

Figure 15.14 Inventive principle "Spheroidality — Curvature."

kick" (like in the one mentioned in section ten of this chapter), but it can also be an accelerating cavity. Rounding the shapes of the resonator in such a way results in a better and smoother distribution of fields and currents along the walls of the cavity and correspondingly higher fields generated by the cavity on the beam axis.

15.15 DYNAMICS

The inventive principle *dynamics* may involve allowing (by design) the change to the characteristics of an object, external environment or process to be optimal.

An example that illustrates this principle is a traveling focus proposal[8] which would increase the luminosity of linear colliders. The fields of the opposite beam during a collision of e+ and e- beams can create an additional focusing which can help to squeeze the beams even tighter. However, for this additional focusing to work optimally, the focal point for each beam needs to move during the collision in such a way that it would coincide with the location of the head of the opposite beam. The location of the focal point is shown by arrows of corresponding color in Fig. 15.15. Such a dynamic modification of the colliding beams would then result in some increased luminosity.

---

[8]Balakin, V., *Travelling Focus Regime for Linear Collider VLEPP*, Proc. of 1991 IEEE PAC, p.3273, 1991.

15. Dynamics

- Allow (or design) the characteristics of an object, external environment or process to change to be optimal or to find an optimal operating condition.
- Divide an object into parts capable of movement relative to each other.
- If an object (or process) is rigid or inflexible, make it movable or adaptive.

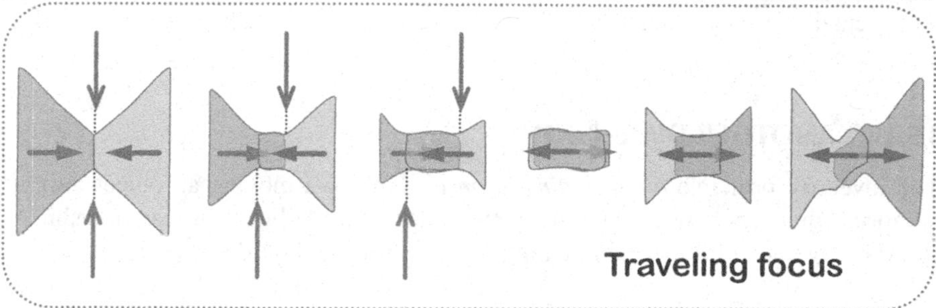

Figure 15.15 Inventive principle "Dynamics."

15.16 PARTIAL OR EXCESSIVE ACTIONS

The inventive principle *partial or excessive actions* may involve using "slightly less" or "slightly more" of a given method in cases where 100% of the effect is hard to achieve with said method, making the problem considerably easier to solve.

16. Partial or excessive actions

- If 100% of an object is hard to achieve using a given solution method then, by using "slightly less" or "slightly more" of the same method, the problem may be considerably easier to solve.

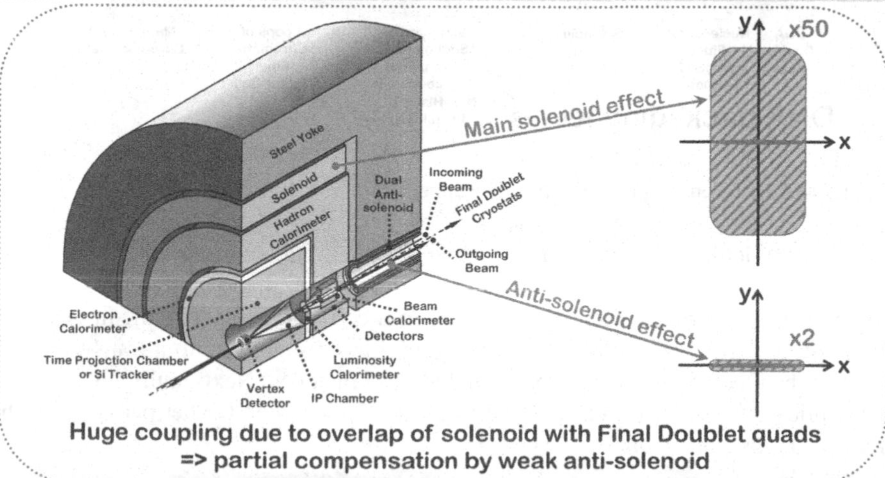

Figure 15.16 Inventive principle "Partial or excessive actions."

An example that illustrates this principle is the design concept of a weak anti-solenoid[9] which is intended to compensate for the beam X-Y coupling effects in the interaction region of a linear collider. The anomalously large coupling effects arise due to overlap of the field of the main solenoid with the final focusing lenses. A properly adjusted weak anti-solenoid can compensate for a large fraction of these detrimental effects, making the problem much easier to solve with upstream coupling correctors.

15.17 ANOTHER DIMENSION

The inventive principle *another dimension* may involve moving an object into an additional dimension, using a multi-story arrangement of objects instead of a single-story arrangement, tilting or reorienting the object; or laying it on its side, etc.

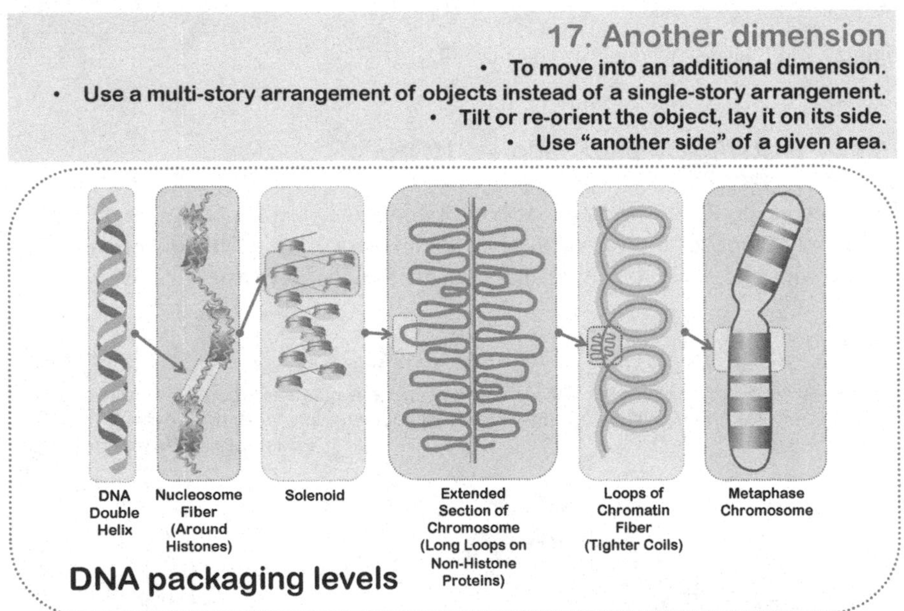

Figure 15.17 Inventive principle "Another dimension."

An example that illustrates this principle is something nature invented: the DNA packaging mechanism. DNA molecules, if straightened out, are quite long, reaching around 2 mm. However, they are packaged in a cell in a compact way. In general, there are five main different packaging levels. First, by going into a transverse direction, the DNA molecules are packaged around histones. Next, multiples of these assemblies are packaged in a solenoid-like way. This is then further packaged with multiple loops, and so on.

[9] Y. Nosochkov, A. Seryi, PRSTAB, 8, 021001, 2005.

15.18 OSCILLATIONS AND RESONANCES

In standard TRIZ, this principle is called "mechanical vibration." However, for science applications, it should be redefined as "oscillations and resonances." The inventive principle *oscillations and resonances* may involve causing an object to oscillate or vibrate, thereby increasing its frequency (e.g., from microwave to optical), or using an object's resonant frequency to one's advantage, etc.

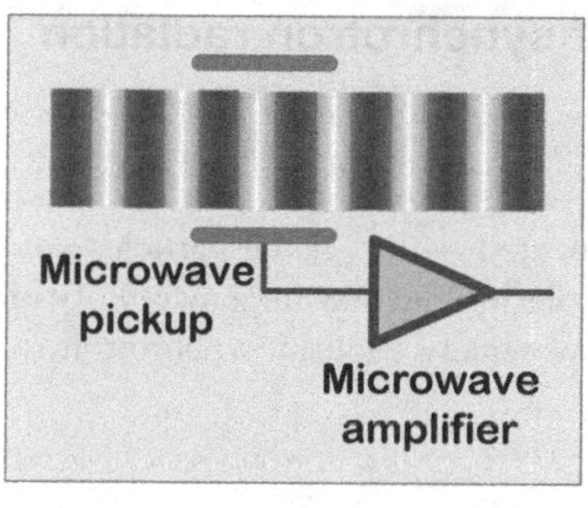

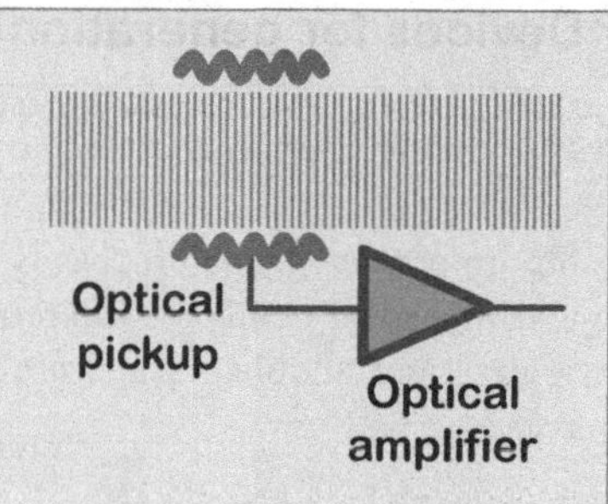

Figure 15.18 Inventive principle AS-TRIZ "Oscillations and resonances," named "Mechanical vibration" in TRIZ .

We can illustrate this principle with the conceptual design of optical stochastic cooling, an evolved version of stochastic cooling. Both of these methods are designed to decrease the phase-space volume of a charged particle's beam in accelerators. Stochastic cooling relies on the microwave range of frequencies to detect particles and act on them, while optical stochastic cooling relies on, correspondingly, optical frequencies.

15.19 PERIODIC ACTION

The inventive principle *periodic action* may involve, instead of continuous action, using periodic or pulsating actions.

We can illustrate this principle via devices that generate synchrotron radiation. This radiation is generated when relativistic charged particles move on a curved trajectory, losing parts of its electromagnetic field. The simplest way to generate such radiation is to pass particles via a bending magnet, as shown on the left side of Fig. 15.19. However, characteristics of radiation (such as brightness, etc.) can be

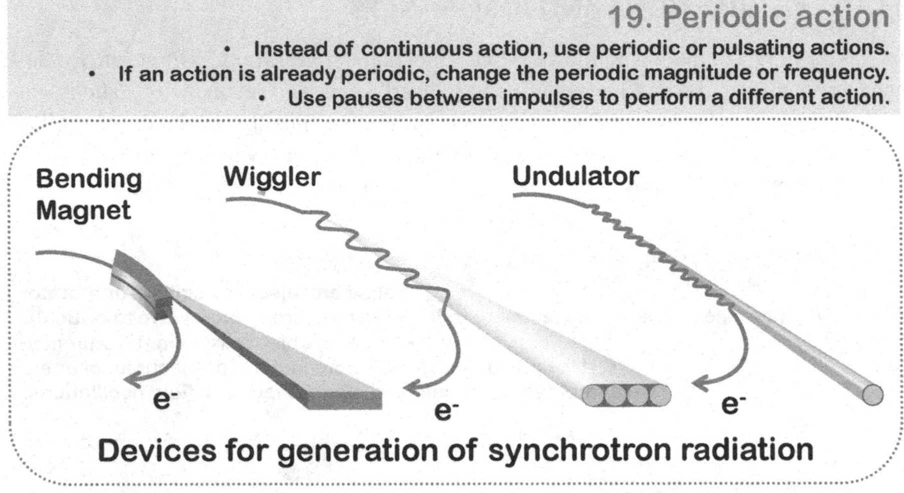

Figure 15.19 Inventive principle "Periodic action."

significantly improved if this process is repeated — i.e., if the particles are passed through a sequence of bends of different polarity. Such arrangements of bends are called wigglers and undulators and are now widely used in synchrotron radiation light sources.

15.20 CONTINUITY OF USEFUL ACTION

The inventive principle *continuity of useful action* may involve carrying on work continuously, making all parts of an object work nonstop at full capacity eliminating all idleness or intermittent actions.

We illustrated this principle via a concept of a top-off injection for synchrotron light sources in the earlier chapters of this book. In these light sources, useful synchrotron radiation is generated by circulating electron beams, which can decay due to losses, and thus the circulating current — as well as the intensity of the generated radiation — decreases with time, as shown on the left half of Fig. 15.19. The circulating beam is seldom renewed when the new beam is injected.

An alternate way to operate the light source is to arrange an almost continuous injection so that the fresh portion of the beam will be injected very often, considerably increasing the efficiency of the light source and also eliminating any thermal effects associated with variations of circulating current or intensity of the emitted radiation.

In this chapter, we have also selected to use the example of continuous it communication, which is important not only for keeping healthy relations with your team or organization (or even your family), but also for providing the most accurate beam corrections via any kind of accelerator feedback. As shown in Fig. 15.20, the ships that practiced constant communication avoided both the iceberg and whirlpool.

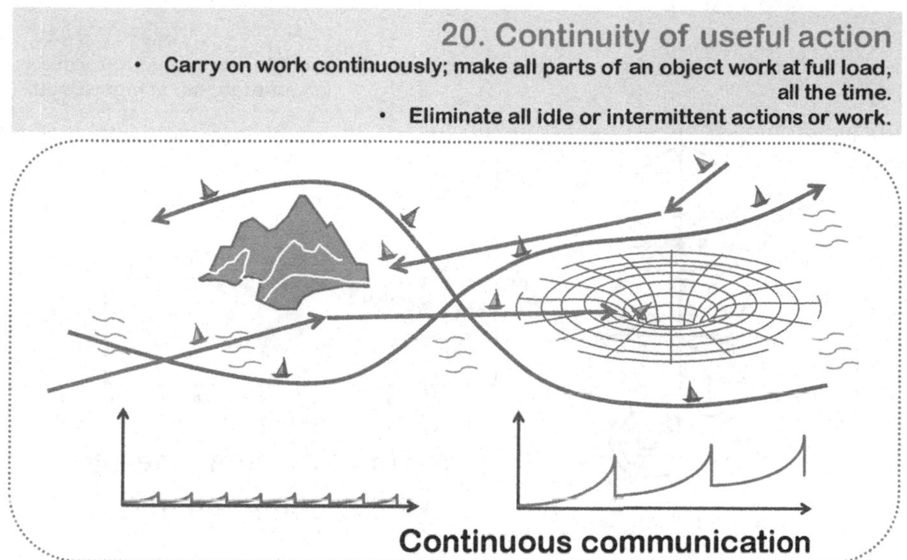

Figure 15.20 Inventive principle "Continuity of useful action" — continuous communication. The graphs show accumulation of problems vs. time.

15.21 SKIPPING

The inventive principle *skipping* may involve conducting a process, or certain stages of it (e.g., destructible, harmful or hazardous operations), at high speed. As you can imagine, it has a wide applicability.

An example that illustrates this principle is a gamma-jump technique used in accelerators. In proton circular accelerators in particular, there is a notion of transition energy — at this energy level (the relativistic factor corresponding to this energy is called γ_t), the stable phase (of the accelerating resonator) flips to the other side of the sine wave. Passing the critical energy during acceleration therefore requires jumping over the phase of the resonator to avoid losing the beam. Still, some beam disturbances unavoidably occur due to such a transition. This approach can be applied to the transition of critical energy, as well as to jumping over the de-polarization resonances.

The critical energy value depends on the properties of the focusing optics of the accelerator. It is possible, therefore, to program the optics change in such a way that the transition through the critical energy will happen much faster, significantly reducing the detrimental effects on the accelerated beam.

We cannot help but bring up one more illustration of the *skipping* inventive principle —running quickly over hot coals.

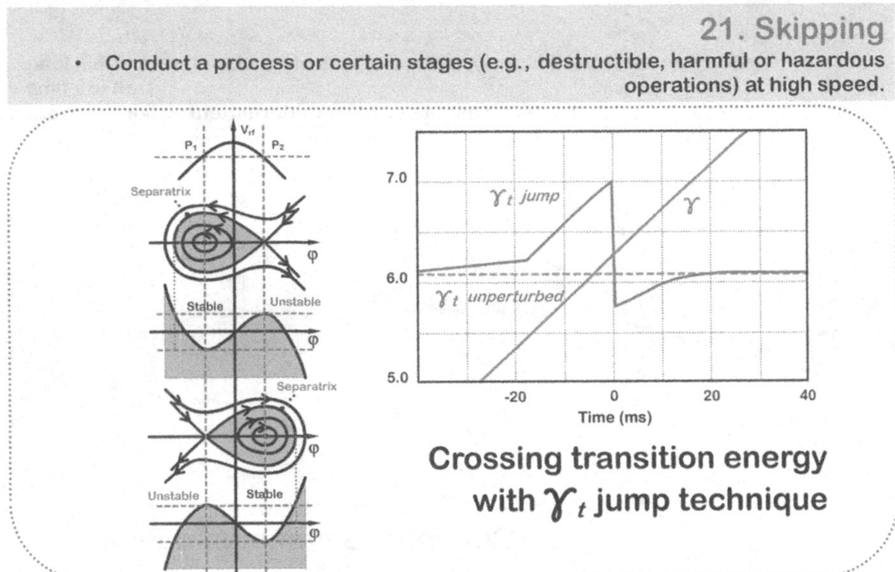

Figure 15.21 Inventive principle "Skipping."

For illustration of "skipping" inventive principle.

15.22 BLESSING IN DISGUISE

The inventive principle *blessing in disguise* (which is also called *turning lemons into lemonade*) may involve using harmful factors (particularly, harmful effects of the environment or surroundings) to achieve a positive effect, eliminating the primary harmful action by adding it to another harmful action to resolve the problem, or amplifying a harmful factor to such a degree that it is no longer harmful.

We can illustrate this principle via wake-fields in linacs, which are normally harmful, as they can deteriorate the quality of accelerated beam. However, in the case where this linac feeds a free electron laser, such wakes can turn into a useful

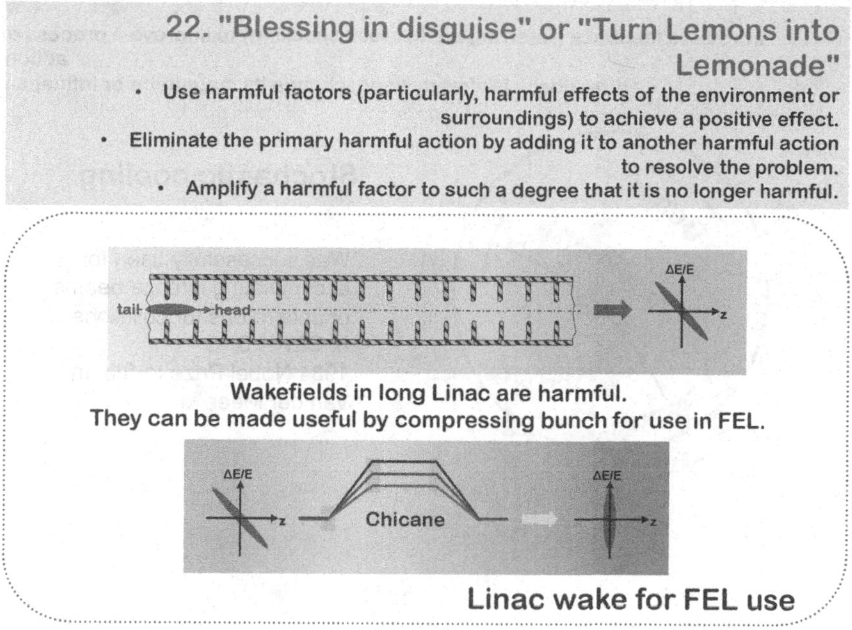

22. "Blessing in disguise" or "Turn Lemons into Lemonade"

- Use harmful factors (particularly, harmful effects of the environment or surroundings) to achieve a positive effect.
- Eliminate the primary harmful action by adding it to another harmful action to resolve the problem.
- Amplify a harmful factor to such a degree that it is no longer harmful.

tail head

ΔE/E z

Wakefields in long Linac are harmful.
They can be made useful by compressing bunch for use in FEL.

ΔE/E z Chicane ΔE/E z

Linac wake for FEL use

Figure 15.22 Inventive principle "Blessing in disguise" or "Turn Lemons into Lemonade."

factor, as they can produce an energy chirp along the beam that can serve to further compress the beam longitudinally, enhancing the generated radiation.

15.23 FEEDBACK

The inventive principle *feedback* may involve introducing feedback or feedforward connections into the process (referring back, cross-checking) to improve said process or action.

An example that illustrates this principle is the concept of stochastic cooling that reduces phase-space volume of the beam circulating in an accelerator. This is done by first detecting oscillations of particles in one location, then sending the amplified signal which carries information about these oscillations along a shorter path than the particle takes to travel, and then acting on the same particle with a kick that decreases its oscillation.

15.24 INTERMEDIARY

The inventive principle *intermediary* may involve using an intermediary carrier object or intermediary process, or temporarily merging one object with another (which can be easily removed).

An example that illustrates this principle is a three-level laser. In such a laser, the atoms of the active media are initially in a grounded state and then, due to excitation

23. Feedback

- **Introduce feedback (referring back, cross-checking) to improve a process or action.**
- **If feedback is already used, change its magnitude or influence.**

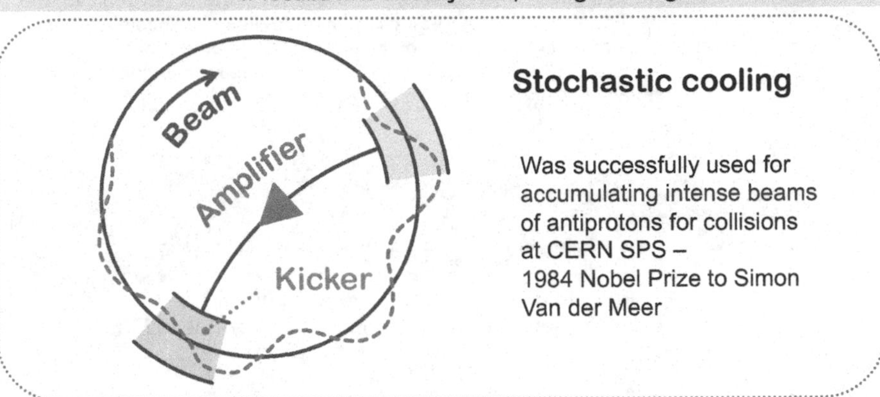

Figure 15.23 Inventive principle "Feedback."

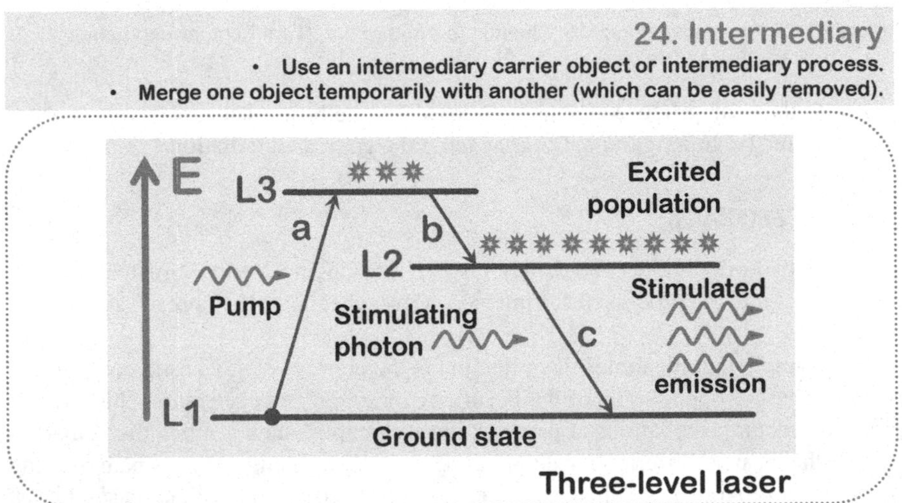

Figure 15.24 Inventive principle "Intermediary."

by a pump, rise to level L3, which has quite a short lifetime. So, the atoms quickly revert, via non-radiative decay, to level L2, which has, in contrast, a long lifetime. Any stimulating emission would then result in an avalanche of stimulated coherent radiation. Level L3, as we see in this case, plays the role of an intermediary.

15.25 SELF-SERVICE

The inventive principle *self-service* may involve making an object serve itself by performing auxiliary helpful functions or using waste resources, energy or substances.

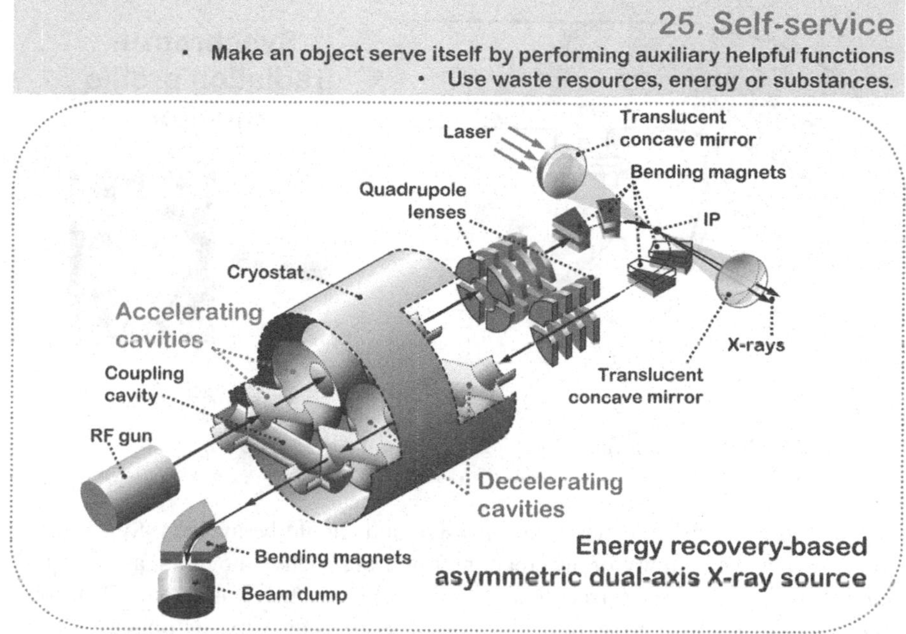

Figure 15.25 Inventive principle "Self service."

An example that illustrates this principle is a design of an energy-recovery-based asymmetric dual-axis X-ray source[10], where the cavities are arranged in such a way as to perform an auxiliary helpful function of decelerating the beam, which has already produced useful X-ray radiation, in order to recuperate its energy and thus allow for a more efficient system operation.

15.26 COPYING

The inventive principle *copying* may involve using — instead of an unavailable, expensive or fragile object — its simpler and inexpensive copies, or replacing an object or process with optical copies.

An example that illustrates this principle is the concept of a synchrotron light source beam-size monitor. The beam of charged particles, circulating in an accelerator, emits synchrotron radiation when it passes bending magnets. Beam-size monitors

---

[10] R. Ainsworth, G. Burt, I. V. Konoplev, A. Seryi, *Asymmetric Dual Axis Energy Recovery Linac for Ultra-High Flux Sources of Coherent X-ray/THz Radiation: Investigations towards Its Ultimate Performance*, arXiv:1509.03675, physics.acc-ph, Sep 2015.

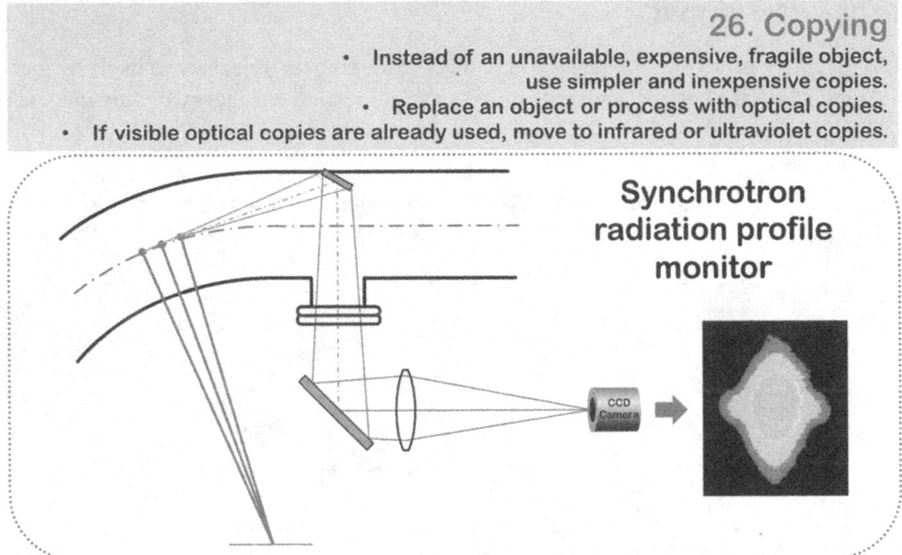

Figure 15.26 Inventive principle "Copying."

such as wires crossing the beams are invasive and should be avoided. Synchrotron radiation, however, allows for the creation of non-destructive monitors, as when this radiation is directed with mirrors to a detector and its profile is analyzed. Therefore, instead of a fragile beam being analyzed with a crossing wire, its optical copy is analyzed to determine its beam size.

15.27 CHEAP, SHORT-LIVED OBJECTS

The inventive principle *cheap, short-lived objects* may involve replacing an expensive object with a multiple of inexpensive objects comprising certain qualities (such as service life, for instance).

An example that illustrates this principle is the concept of laser plasma acceleration. In standard acceleration, the accelerating wave is excited inside of metal resonators. The accelerating gradient is therefore limited by the properties of materials and typically cannot exceed 100 MV/m. However, a short and powerful laser pulse can excite a wave in plasma, where the accelerating gradient can be a thousand times higher. This plasma wave can be used to accelerate particles. The plasma and its wave can therefore serve, in this case, as a cheap and short-lived object.

15.28 MECHANICS SUBSTITUTION

The inventive principle *mechanics substitution* may involve replacing mechanical means with electromagnetic or sensory (acoustic, taste or smell) ones.

27. Cheap short-living objects

- Replace an expensive object with a multiple of inexpensive objects, comprising certain qualities (such as service life, for instance).

1 m scale

E_z<100 MV/m

Accelerating structure, metal (normal conductive or super-conductive)

0.1 mm scale

$E_z = m_e c\ \omega_p / e \sim 100$ GV/m

"Accelerating structure" produced on-the-fly in plasma by laser pulse

Plasma acceleration

Figure 15.27 Inventive principle "Cheap short-living objects."

28. Mechanics substitution

- Replace a mechanical means with a sensory (optical, acoustic, taste or smell) means.
- Use electric, magnetic and electromagnetic fields to interact with the object.
- Change from static to movable fields, from unstructured fields to those having structure.
- Use fields in conjunction with field-activated (e.g., ferromagnetic) particles.

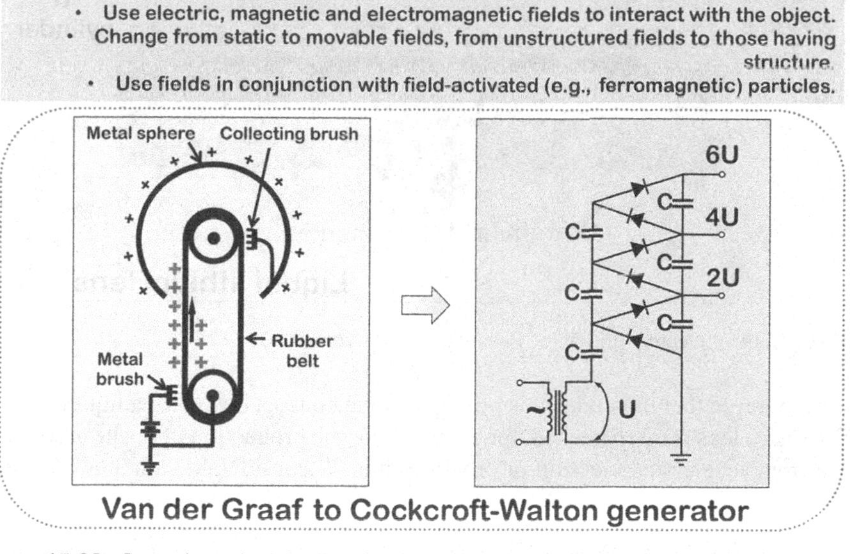

Figure 15.28 Inventive principle "Mechanics substitution."

We can illustrate this principle via electrostatic generators of two kinds. In the first type (shown on the left half of Fig. 15.28), the electric charges are delivered to the metal sphere by the moving rubber belt. The charges are deposited on the belt by a system of sharp needles. This is a Van der Graaf generator. In the second kind of generator, a sequence of diode-based rectifiers is used to amplify the voltage to a high level, sufficient for accelerating charged particles. This is called a Cockcroft-Walton generator.

15.29 PNEUMATICS AND HYDRAULICS

The inventive principle *pneumatics and hydraulics* may involve using gas and liquid parts of an object (e.g., inflatable, filled with liquids, air cushion, hydrostatic, hydro-reactive) instead of solid parts.

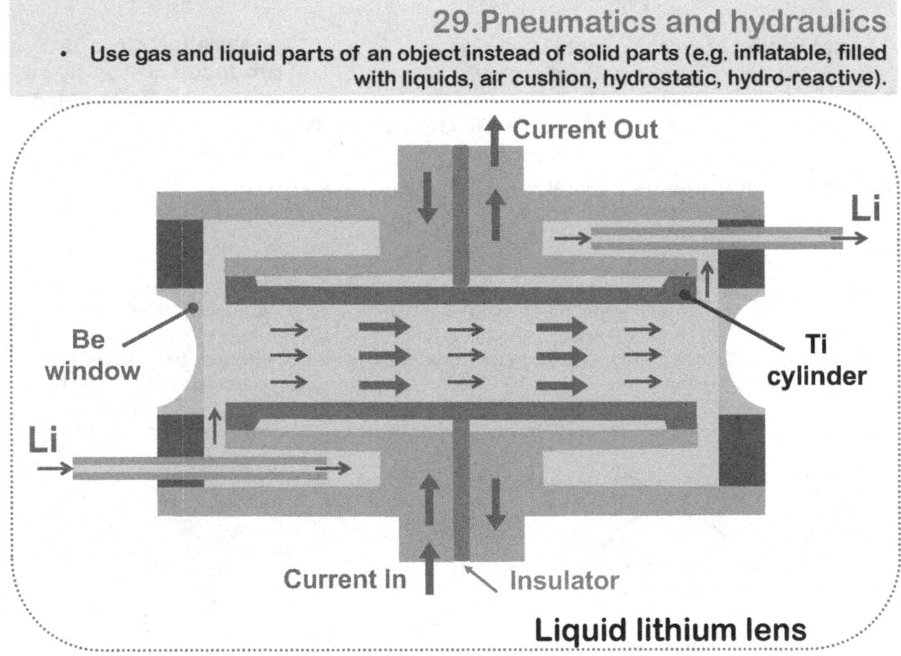

Figure 15.29 Inventive principle "Pneumatics and hydraulics."

An example that illustrates this principle is the concept design of a liquid lithium lens. Such a lens is often needed, for example, for the production of positrons or neutrons from accelerated electron or proton beams. A considerable amount of heat is deposited into the lens during its interaction with the beam. Preventing the destruction of the lens can be done by making it pulsed or, ultimately, already "destroyed," i.e., liquid.

15.30 FLEXIBLE SHELLS AND THIN FILMS

The inventive principle *flexible shells and thin films* may involve using flexible shells and thin films instead of three-dimensional structures or isolating the object from the external environment using flexible shells and thin films.

30. Flexible shells and thin films

- Use flexible shells and thin films instead of three dimensional structures
- Isolate the object from the external environment using flexible shells and thin films.

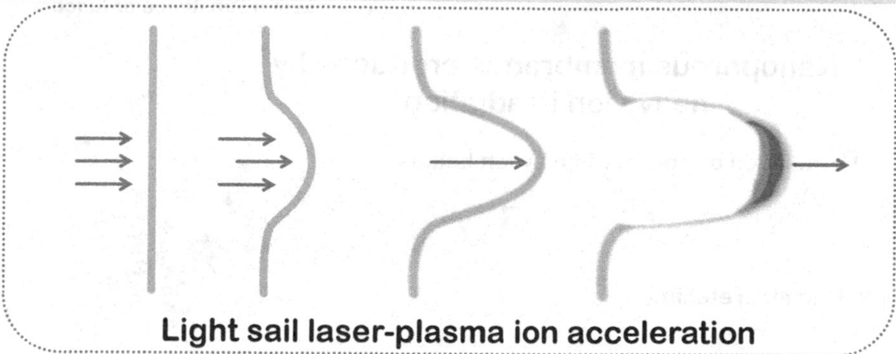

Figure 15.30 Inventive principle "Flexible shells and thin films."

An example that illustrates this principle is the concept of light-sail laser plasma ion acceleration. Laser plasma acceleration of ions is usually achieved via the interaction of powerful short laser pulses with a solid target. This method, however, does not produce a nice mono-energetic accelerated beam. Instead, shining the laser onto a very thin film creates photon flux pressure that propels the entire film as a whole, and correspondingly the ion beam accelerates in such a way that it will have much better properties.

15.31 POROUS MATERIALS

The inventive principle *porous materials* may involve making an object porous or adding porous elements (inserts, coatings, etc.) or, if an object is already porous, using the pores to introduce a useful substance or function.

An example that illustrates this principle is the concept of creating porous membranes using accelerated ion beams. In this case, ion beams accelerated typically in cyclotrons are directed into thin films, leaving ionized traces inside. Further chemical processing creates pores in such membranes, which can be then used for various filters or other applications.

15.32 COLOR CHANGES

The inventive principle *color changes* may involve changing the color of an object or its external environment, changing the transparency of an object or its external environment or changing the emissivity properties of an object subject to radiant heating, etc.

An example that illustrates this principle is the principle of OPCPA — Optical Parametric Chirped Pulse Amplification. In this method, a nonlinear crystal, which

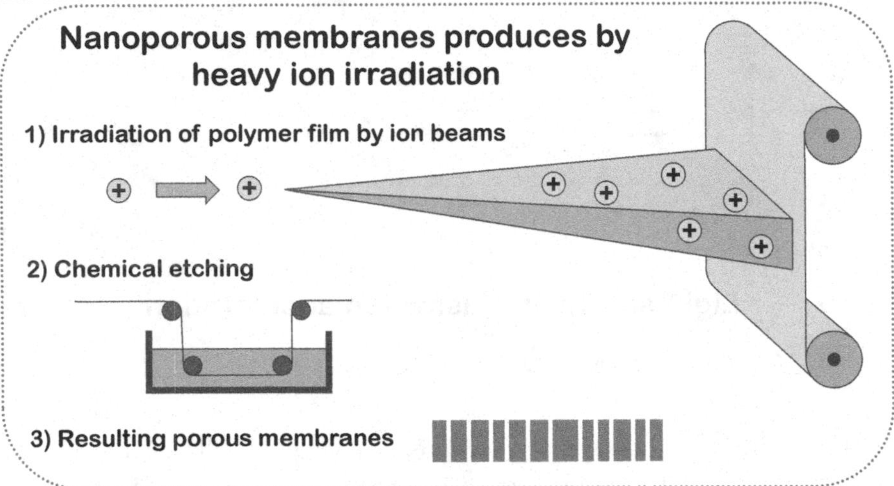

Figure 15.31 Inventive principle "Porous materials."

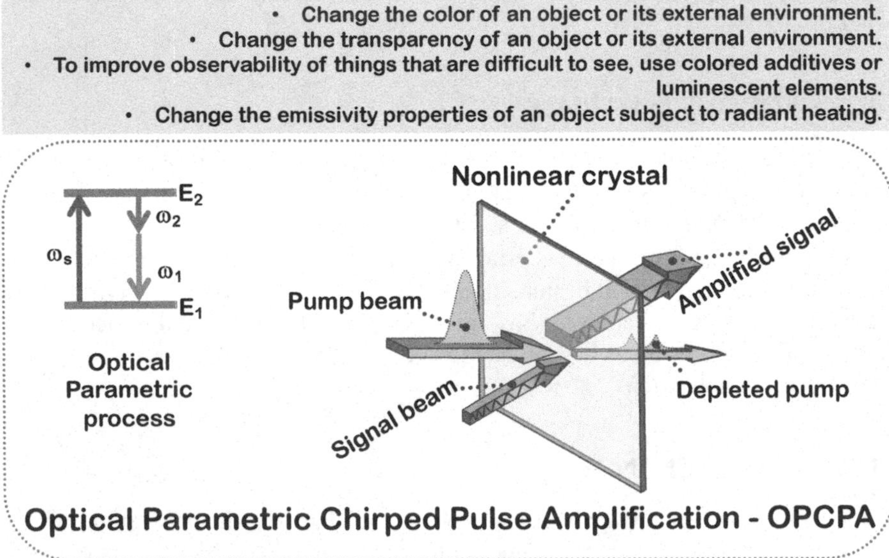

Figure 15.32 Inventive principle "Color changes."

emits two different wavelengths when pumped with a single wavelength, can be used in an optical amplifier. When a pump laser beam and a signal laser beam are sent onto the crystal, an amplified signal and depleted pump beams emerge from the crystal. If the signal beam is chirped (i.e., the signal frequency depends on time), the amplified output signal is also chirped. Via the optical parametric process, the nonlinear crystal changes the color of the oncoming light, thus illustrating the inventive principle.

15.33 HOMOGENEITY

The inventive principle *homogeneity* or, expressing it in Latin, *similia similibus curantur* may involve making objects interact with a given object of the same material (or a material with identical properties).

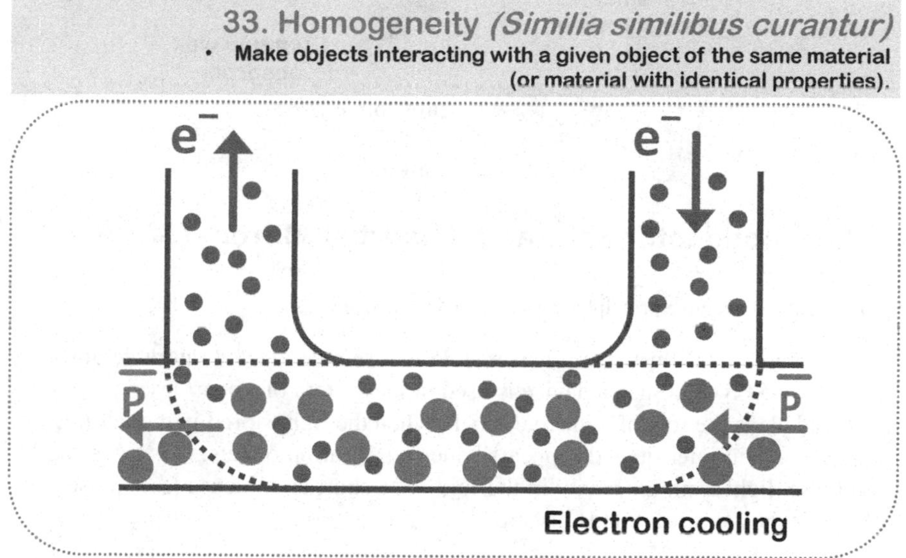

Figure 15.33 Inventive principle "Homogeneity."

An example that illustrates this principle is the concept of electron cooling. This cooling method is aimed at decreasing the phase-space volume of a charged particle beam; for example, a beam of anti-protons. The cooling is done by overlapping the anti-proton beam with the beam of electrons going in the same direction and with the same velocity. Hot anti-protons, colliding with colder electrons, will transfer their energy to the electrons and, after many passages through the electron beam, will cool down. So, here we cure like with like: we cool charged particles with other types of charged particles.

15.34 DISCARDING AND RECOVERING

The inventive principle *discarding and recovering* may involve removing portions of an object that have fulfilled their functions (discarding them through dissolution, evaporation, etc.), modifying these portions directly during operation or, conversely, restoring consumable parts of an object directly in operation.

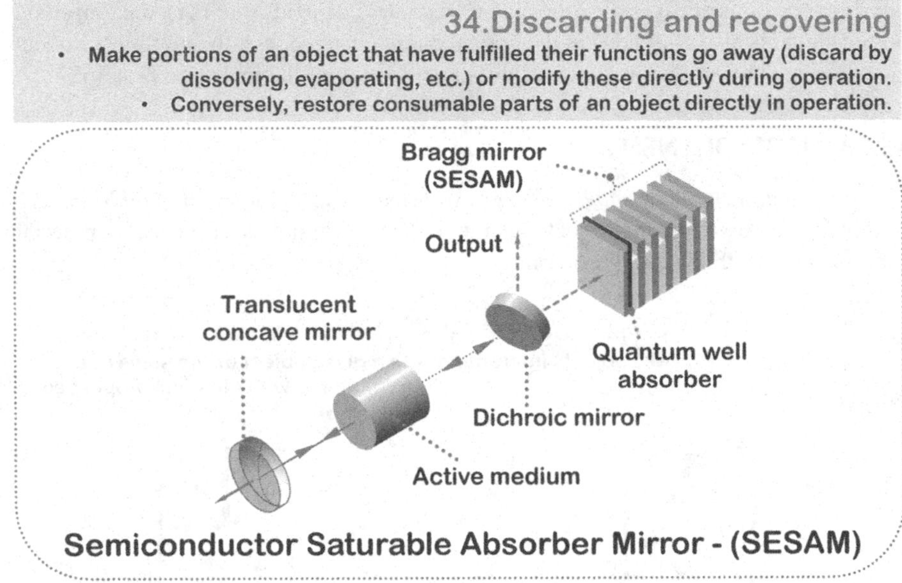

Figure 15.34 Inventive principle "Discarding and recovering."

An example that illustrates this principle is a semiconductor saturable-absorber mirror — or SESAM. Such a device is used in the system of a short laser-pulse generator and plays the role of a mirror, which, when the light stored in the laser cavity reaches a certain intensity, "disappears" due to saturation effects, thus releasing all stored laser light from the cavity in the form of a short and intense laser pulse. The SESAM is thus a mirror that is discarded at a proper moment.

15.35 PARAMETER CHANGES

The inventive principle *parameter changes* may involve changing an object's physical state (e.g., to a gas, liquid or solid), changing the concentration or consistency, changing the degree of flexibility or changing the temperature, etc.

An example that illustrates this principle is the variation of the ratio of the volume of an object to the surface area of the object. Maxwell or thermodynamic equations indicate that changing the volume-to-surface ratio can change an object's characteristics, such as its cooling rate or electromagnetic field. We can illustrate this via an example of a cat, who can change its surface area depending on the environmental temperature, to control its cooling rate (see Fig. 15.35), or via an example of fiber lasers, which, in comparison with lasers with standard geometric ratios of the active media sizes, have much higher surface areas and therefore better cooling, higher repetition rates, and higher efficiency.

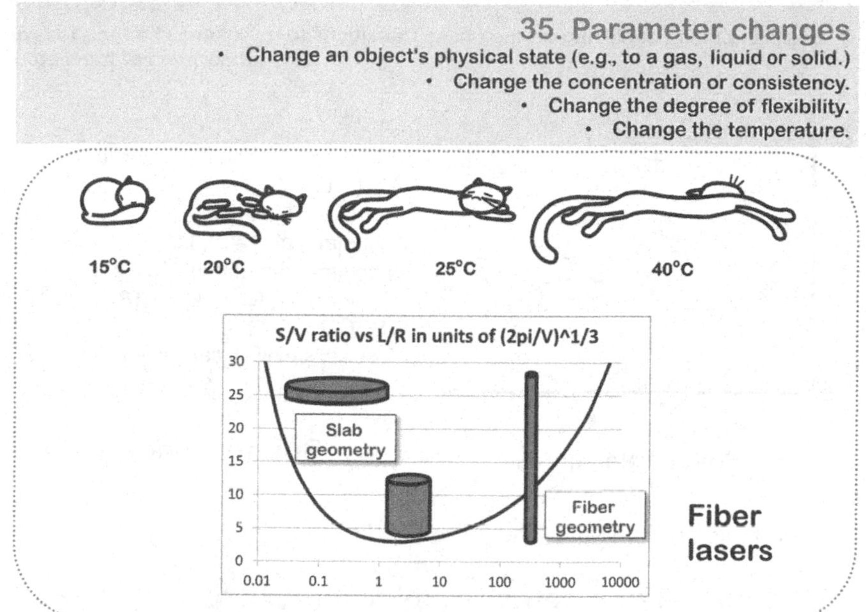

Figure 15.35 Inventive principle "Parameter changes."

15.36 PHASE TRANSITIONS

The inventive principle *phase transitions* may involve using phenomena occurring during phase transitions (e.g., volume changes, loss or absorption of heat, etc.).

An example that illustrate this principle is the phenomenon of superconductivity when the electrical resistance of certain materials can drop to zero when temperature decreases below the critical one.

15.37 THERMAL/ELECTRICAL EXPANSION OR PROPERTY CHANGE

The inventive principle *thermal/electrical expansion or property change* may involve using thermal or electrical expansion (or contraction) or other property changes of materials, or using multiple materials with different coefficients of thermal expansion (property changes) to one's advantage.

An example that illustrates this principle is an electro-optic effect — that is, the dependence of optical properties of objects such as absorption or refraction (Pockels effect) on the applied electric field. In the example shown in Fig. 15.37, this effect is used to create ultra-short laser pulses.

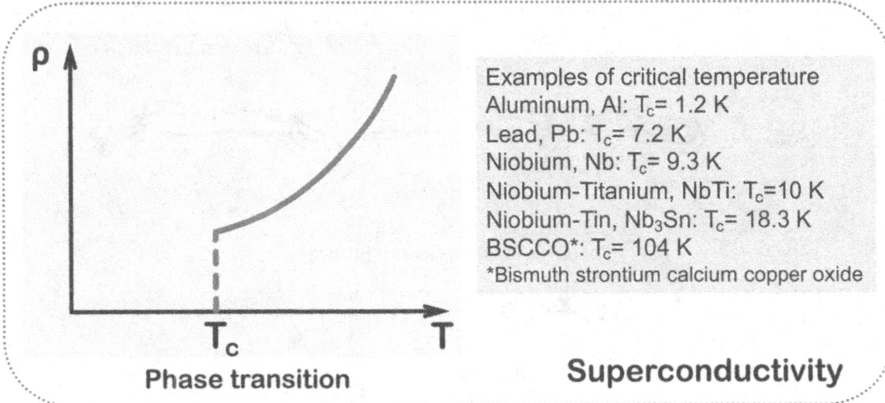

Figure 15.36 Inventive principle "Phase transitions."

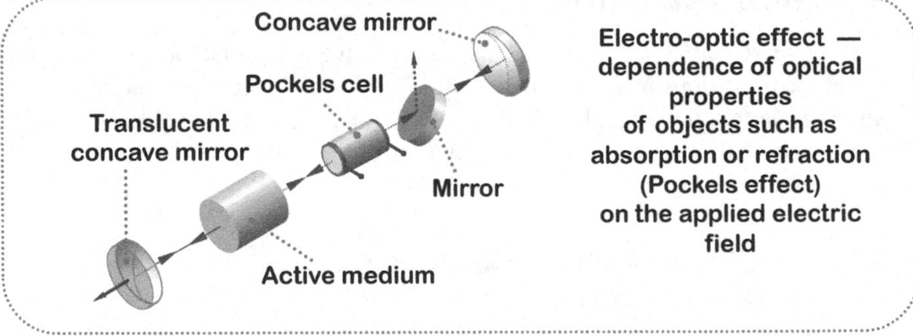

Figure 15.37 Inventive principle "Thermal or electrical expansion or property change."

15.38 STRONG OXIDANTS

The principle *strong oxidants* involves replacing common air with O_2-enriched air, replacing enriched air with pure oxygen, exposing air or oxygen to ionizing radiation, using ionized oxygen or replacing ozonized (or ionized) oxygen with ozone.

An example that illustrates this principle is a method for sterilizing food with a low-energy electron beam. Packaged food is deposited on a conveyor in a factory and passes through a raster scanning electron beam. Fig. 15.38 shows how this is done at a generic sterilization factory.

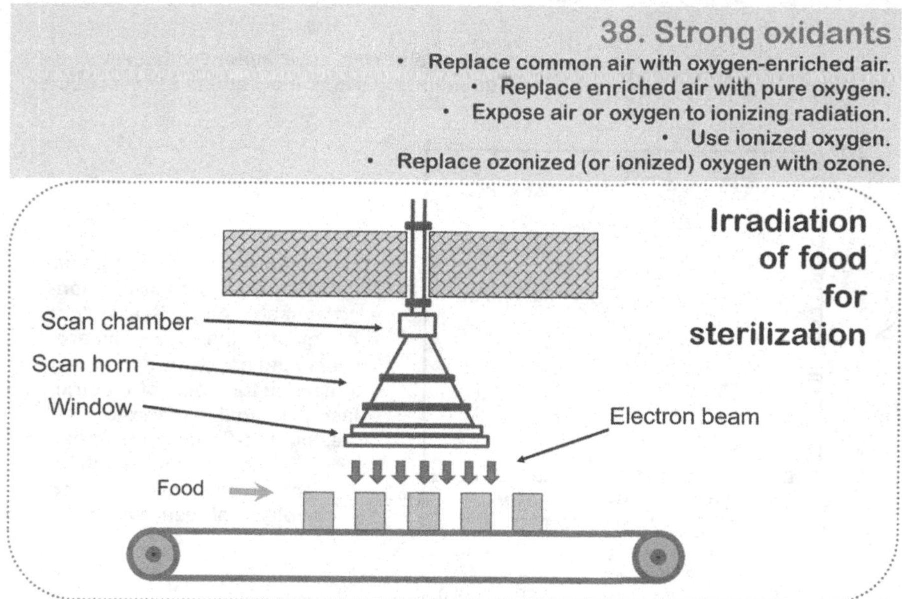

Figure 15.38 Inventive principle "Strong oxidants."

15.39 INERT ATMOSPHERE

The inventive principle *inert atmosphere* may involve replacing a normal environment with an inert one or adding neutral parts or inert additives to an object.

An example that illustrates this principle is a method involving sulfur hexafluoride (SF6 or Elegas), which is a colorless non-flammable gas with excellent electric insulating and arc-quenching capacity. This gas can be used to fill the interior of electrostatic generators (Van der Graaf or Cockcroft-Walton) in order to reach higher voltages and higher energies of accelerated particles. The curves on Fig. 15.39 show a comparison of electrical discharge voltage vs. the value of pressure multiplied by the gap between electrodes for SF6 in comparison with air.

15.40 COMPOSITE MATERIALS

The inventive principle *composite materials* may involve changing from a uniform material to a composite of multiple materials.

An example that illustrates this principle is the collimation system for the Large Hadron Collider[11], wherein the jaws of the collimators are made from composite materials in order to provide good thermal and electrical conductivity, strength, and radiation resistance.

[11]Novel Materials for Collimators at LHC and its Upgrades, A. Bertarelli, et al., CERN-ACC-2015-0173, in Proceedings of HB2014, THO4AB03.

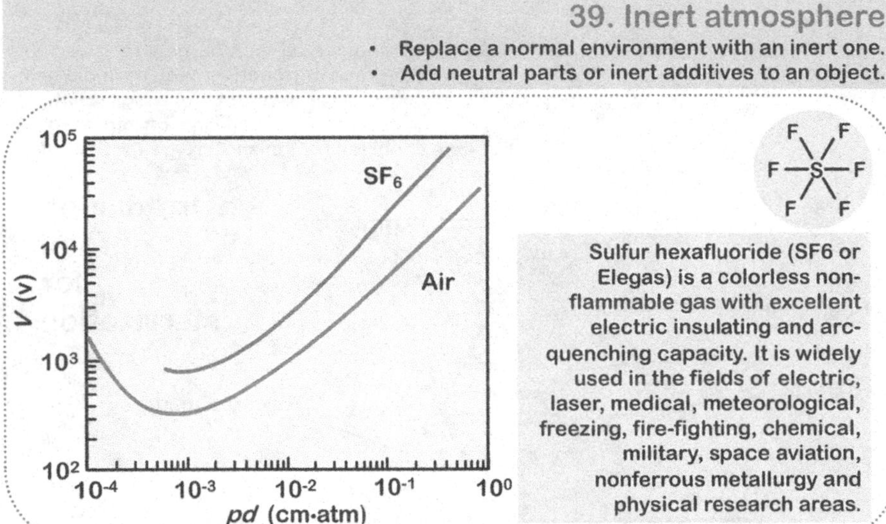

Figure 15.39 Inventive principle "Inert atmosphere."

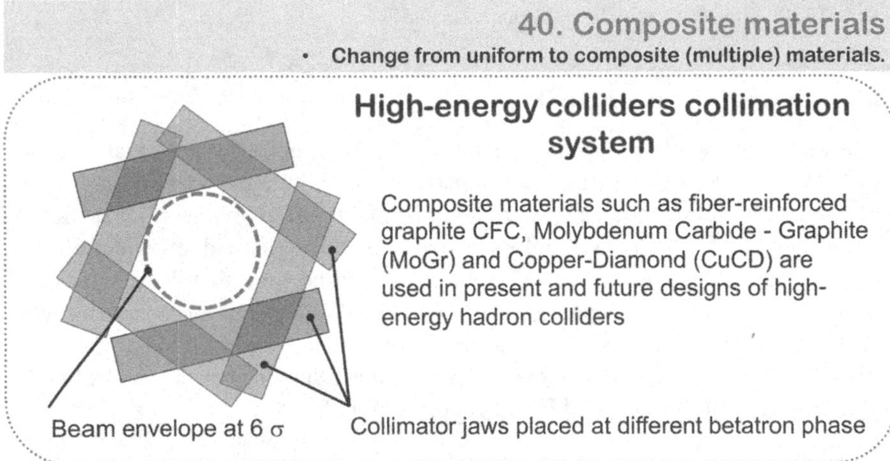

Figure 15.40 Inventive principle "Composite materials."

CHAPTER DISCUSSION AND CONCLUSION

We hope that our suggested examples of inventive principles will inspire readers to discover inventive ways to solve scientific problems they encounter in their research. While not all of our examples are ideal and could certainly be improved upon, we welcome our readers to participate in the creation of better case studies. Most importantly, we encourage readers to use the methodology of inventive problem solving in their research — and beyond.

Final Words

Writing this book was an absolutely wonderful and enjoyable experience, worth every hour of long nights and weekends spent creating it. Working on the second edition was an even greater pleasure, with our work fueled by the success of the first edition and by the community demand for the updated book.

The book is different from many (if not all) textbooks on accelerators, lasers and plasma because, first, it considers these areas altogether, and second, it mixes them with the methodology of inventiveness — the glue that pulls these areas to each other just as gluons connect the quarks in a Λ-Barion, as illustrated below.

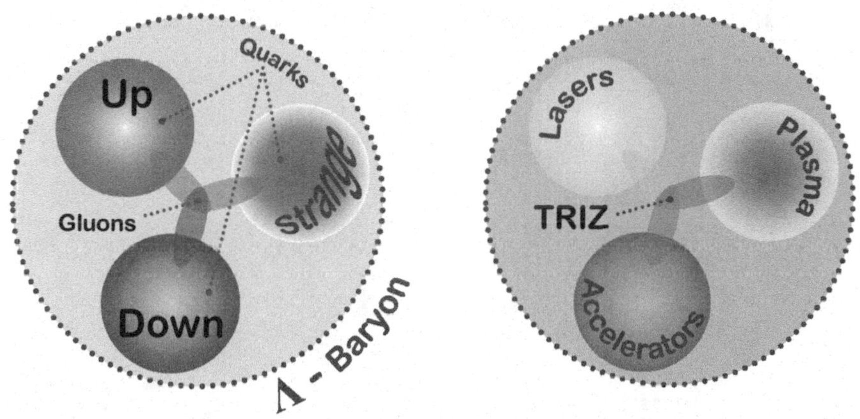

Even though this second edition includes four new chapters, there are still many scientific topics we would have wanted to discuss and include in this book. We will be looking forward for feedback from the readers, which will be invaluable for our work on possible future editions.

– Andrei Seryi and Elena Seraia

A Guide to Solutions of the Exercises

Many of the exercises included in this book, particularly the invention analysis tasks or the mini-projects, allow for a range of solutions. For these cases, rather than giving an exact solution, we have suggested below a step-by-step solution methodology. In various other tasks, an estimate is needed, and not an exact answer. Such cases, again, allow for a certain range of correct solutions. We will give just some possible examples of such solutions.

BASICS OF ACCELERATORS AND OF THE ART OF INVENTIVENESS

Exercise: 1.1 *Analyze the evolution of technical or scientific systems.* Discuss the evolution of any scientific or technical (or accelerator-science-related) fields, identifying successive technologies that arise, saturate and get replaced by new approaches and solutions.

Solution approach: This task allows for a variety of solutions. One particular example that can be considered in the analysis is the technology of particle acceleration, described on the Livingston plot — see Fig. 1.6.

Exercise: 1.2 *Analyze inventions or discoveries using TRIZ and AS-TRIZ.* A plasma mirror is often used when a standard metal mirror cannot withstand the power density of the laser. Analyze this technology in terms of the TRIZ and AS-TRIZ approach, identifying a contradiction and a general inventive principle that were used (or could have been used) for this invention.

Solution approach: The analysis of the metal mirror and plasma mirror is similar to the analysis of the carbon wire beam profile monitor that is replaced by the laser wire — the same contradictions and the same inventive principle can be applied.

Exercise: 1.3 *Analyze inventions or discoveries using TRIZ and AS-TRIZ.* Analyze and describe any scientific or technical invention/discovery (possibly related to accelerator science) in terms of the TRIZ and AS-TRIZ approaches, identifying a contradiction and an inventive principle that were used (or could have been used) for this invention or discovery.

Solution approach: This task allows for a variety of solutions. Taking into account the examples of inventions presented in this chapter, we recommend considering and analyzing the invention of strong focusing as well as the stochastic and electron cooling methods.

DOI: 10.1201/9781003326076-A

Exercise: 1.4 *Developing AS-TRIZ parameters and inventive principles.* Based on what you already know about accelerator science, discuss and suggest the possible additional parameters for the AS-TRIZ contradiction matrix, as well as the possible additional AS-TRIZ inventive principles.

Solution approach: This task allows for a variety of solutions. In order to come up with ideas of additional parameters for the matrix, we recommended considering those parameters that are specific for the beams and that are unlikely to be encountered in the engineering field. A particular example related to this chapter is the coherence or polarization of the beams of particles or photons. Additional inventive principles can be developed via analysis of Siberian snakes, polarizers, SR light sources, etc.

Exercise: 1.5 *Practice in reinventing technical systems.* Imagine that we need to provide 10 kW of electric power inside the top of a metal spherical electrode of the Van der Graaf accelerator, as shown in Fig. 1.8, to feed the electronics and the ion source. (It is not possible to bring this power via electrical wires from the low level). How do we provide this power?

Solution: The Van der Graaf accelerator charges the top electrode using the rubber belt, which is rotating around the axis – one at the bottom and one at the top. Once you remember to apply the TRIZ principle *use resources and energy that you already have in the system*, you just need to attach an electrical generator to the axis placed in the top electrode, and this will then become the required electrical power source.

Exercise: 1.6 *Practice in the art of back-of-the-envelope estimations.* In the book *The Three Body Problem* by Liu Cixin, the civilization of Alpha Centauri sent protons to Earth with embedded artificial intelligence, which penetrated high-energy physics experiments such as at the LHC, in order to spoil the results and stop scientific progress on Earth. Imagine that some of the AI-protons were captured, trapped, and Earth scientists decided to study their internal AI structure. They assumed that the AI was encoded into the interaction of gluons in the AI-protons and used fixed target experiments, similar to CEBAF, to study the AI-protons. Estimate the energy of the electron beam that Earth scientist needed to use to study the AI-protons.

Solution approach: Let's assume that an AI proton has the number of neurons N_n — the same as how many are in the brain of a fruit fly. The drosophila fruit fly's brain[1] has $N_n = 100000$ neurons.

Assume that, to encode one neuron, we need twice as many gluons. Therefore, the number of gluons is estimated as $N_g = 2 \cdot 10^5$.

With the proton's diameter $D = 0.85$ fm, or $0.85 \cdot 10^{-15}$ m, the volume per each gluon is $\pi * D^3/(6N_g)$ The characteristic size for each gluon cell is then equal to

[1] See e.g. https://elifesciences.org/digests/57443/reconstructing-the-brain-of-fruit-flies

$D*(\pi/(6N_g))^{1/3}$.

Let's equate the size of a gluon cell to the wavelength of the electron, which will study gluons and which is given by $\lambda = r_e/\alpha$. Thus, we will get, for the relativistic gamma factor of electrons:

$$\gamma = r_e/(\alpha D)*(6N_g/\pi)^{1/3}$$

For the above parameters, we get $\gamma = 32000$, which is equivalent to the energy of 16 GeV. This means that CEBAF can almost study the AI proton if its intelligence is akin to a fruit fly Drosophila's intelligence.

However, if the number of neurons in an AI proton matches the number in a human brain, i.e., 86 billion, then the required electron energy would increase to 1.5 TeV — one would need a linear collider to get to this energy range.

One would of course also need to remember to keep your sense of humor while reading this, and recall that you do not have to consider only serious scenarios for practicing the back-of-the-envelope estimations.

TRANSVERSE DYNAMICS

Exercise: 2.1 *Chapter materials review.*
Define the region of parameters where a pair of thin quadrupoles will focus the beam in both planes.

Solution: Let us start from Eq. 2.32 that describes the overall transfer matrix of the pair of thin quadrupoles. The effective focusing strength (the element m_{21}) of the overall transfer matrix for x- and y-planes will be given correspondingly by

$$\frac{1}{f_x} = \frac{1}{f_1} - \frac{1}{f_2} + \frac{L}{f_1 f_2}$$

and

$$\frac{1}{f_y} = -\frac{1}{f_1} + \frac{1}{f_2} + \frac{L}{f_1 f_2}$$

We would like to find the region of parameters when both f_x and f_y are positive, which would then correspond to focusing in both planes.

Let's assume that f_x and f_y are positive and also that $f_x > L$ and $f_y > L$ (i.e., we avoid over-focusing). Defining $x_1 = L/f_1$ and $x_2 = L/f_2$ allows us to rewrite the conditions for the positivity of f_x and f_y as follows:

$$x_1 - x_2 + x_1x_2 > 0 \quad \text{and} \quad x_2 - x_1 + x_1x_2 > 0$$

which we can rewrite as

$$x_2 < x_1/(1-x_1) \quad \text{and} \quad x_2 > x_1/(1+x_1)$$

which defines the area of parameters where the pair of thin quadrupoles will focus in both planes.

If the above two conditions are expressed in a 2D plot, the corresponding curves would remind us of a necktie diagram where the region of parameters we are interested in is in between the two lines defining the edges of the "tie" — such a plot is usually called *the tie diagram* for this reason.

Exercise: 2.2 *Chapter materials review.*

Prove Eq. 2.39, which defines the stability of a FODO beamline, geometrically, using the analogy of traditional geometrical optics.

Solution approach: Make a picture similar to Fig. 2.15 where you would first make a diagram of a repeating pattern of ray lines for the case of $f = L/4$. In this case, the rays will pass through zero at the defocusing lenses. After creating this picture, you can come to the conclusion that it is impossible to draw a non-growing finite ray envelope for the case of $f < L/4$.

Exercise: 2.3 *Chapter materials review.*

A parallel proton beam of E=200 MeV enters a beamline. It is necessary to focus this beam into a point at a 3 m distance from the entrance. Estimate the necessary parameters of a quadrupole system (gradients, lengths) that can perform this task.

Solution approach: The estimation can be done in a variety of ways. Let's assume that the two quadrupoles of the system are the same and that the distance between them (denoted as L) is smaller than 3 m, which we will denote here as L^*. Let's assume that $L = 1$ m. We can also assume that the lenses are thin and that their length is $L_q = 0.3$ m. We then can use Eq. 2.32 and have set $f^* = L^*$ find the requirement for the focusing distance of an individual quadrupole to be $f^2 \approx L \cdot L^*$ or $f \approx 1.7$ m. The proton beam of this kinetic energy has $B\rho \approx 2.1$ Tesla·m and therefore the required gradient of the quadrupoles is $G = B\rho/(L_q f^*) \approx 4.1$ Tesla/m. It is useful to double check that such a quadrupole is feasible. Suppose that the needed radius of the quadrupole aperture is 5 cm. The pole-tip field at this quadrupole is still around 2 kGs, which is just 10 % of the iron saturation field. Therefore, such a quadrupole is feasible.

Exercise: 2.4 *Mini-project.*

Consider the same proton beam as in the previous exercise, as well as the same focusing requirements. Assume that the focusing is performed by a continuous, cylindrical electron beam. Estimate the necessary electron density which can perform the focusing task. Select the electron beam energy, determine the electron current and discuss the optimization of the design of the electron beam system, as well as its feasibility.

Solution approach: The focusing distance of the electron beam with density n_e and length L, taking both electric and magnetic fields into account and assuming the direction of electron motion with respect to protons such that the effects of the field add together, can be derived to be given by the following:

$$f = \gamma_p \beta_p^2 / (1 + \beta_e \beta_p) \cdot 1/(2\pi n_e L r_p)$$

This focusing distance will be equal to 3 m for an electron density of $n_e = 1.3 \cdot 10^{16}$ m^{-3} where we assumed $\beta_e = 0.0625$ corresponds to 1 keV energy of electrons. If we assume that the radius of the electron beam is $r_{beam} = 1$ mm, then the corresponding beam current is $I_e \approx 0.12$ A and the beam power is about 120 W. The current density of this electron beam is around 4 A/cm^2. In determining the feasibly, the factors such as the overall power and current density should be considered at first. The typical perveance of an electron gun and the current density of typical CW cathodes need to be taken into account as well in the discussion of the feasibility of such an electron beam.

Exercise: 2.5 *Mini-project.*
In 1954, Enrico Fermi presented, in his lecture, a vision of an accelerator that would encircle the Earth. Design such an accelerator, assuming that it will be shaped like a polygon with *N* sides, that there will be *N* space stations launched around the Earth, located in the vertices of the polygon, and that each space station will carry one bending dipole magnet and one quadrupole magnet. Assume that, between the space stations, the accelerator's orbit will be straight, and the particle beam will propagate in the open-space vacuum (i.e., without any vacuum chambers). Determine, in particular, what will be the energy of the beam in this accelerator, for the case of an electron or proton beam.

Solution approach: Let's assume that the space stations will be similar in size to the *International Space Station* (ISS) and that the bending magnets and quadrupoles can be somewhat smaller in length than the ISS. We can also assume that the polarity of the quadrupoles in the neighboring space stations will be opposite, forming the FODO optics of the accelerator. As the first step, we can select a reasonably small number of space stations *N*, and select the orbit height for the space stations, making sure that the beam orbit at its closest to Earth will not touch the atmosphere (which starts at about 10 km from the Earth's surface). Next, we can select the length of the bending magnet which will be installed on every space station (e.g., making them equal to half of ISS's size), and select its magnetic field (e.g., using the same field as in Large Hardon Collider superconducting bending magnets, which have fields of about 8 Tesla). This will allow us to determine the energy of the beam that can be circulated in this accelerator. Next, we need to select the length of the quadrupole magnets which will be installed on every space station (e.g., half of ISS's length), and select their gradient (e.g., not exceeding the gradient of LHC superconducting quadrupoles that have gradient of about 200 Tesla/m). When selecting the quadrupole strength and length, we need to make sure that the designed FODO optics will be stable.

Exercise: 2.6 *Analyze inventions or discoveries using TRIZ and AS-TRIZ.* Analyze and describe scientific or technical inventions described in this chapter in terms of the

TRIZ and AS-TRIZ approaches, identifying a contradiction and an inventive principle that were used (could have been used) for these inventions.

Solution approach: This task allows for a variety of solutions. Taking into account the examples of inventions presented in this chapter, we recommend considering and analyzing the invention of the *Pierce electrode* or the *Faraday cup*.

Exercise: 2.7 *Developing AS-TRIZ parameters and inventive principles.* Based on what you already know about accelerator science, discuss and suggest the possible additional parameters for the AS-TRIZ contradiction matrix, as well as the possible additional AS-TRIZ inventive principles.

Solution approach: This task allows for a variety of solutions. In order to come up with ideas of additional parameters for the matrix, we recommend considering those parameters that are specific for the beams and that are unlikely to be encountered in the engineering field. A particular example related to this chapter is the technology of beam coupling. Additional inventive principles can be developed via discussion of skew quadrupole corrections, anti-solenoid corrections and so on.

Exercise: 2.8 *Practice in reinventing technical systems.* Suggest a way to make an achromatic beamline consisting only of focusing and defocusing quadrupoles and drift spaces, without any bends or nonlinear elements.

Solution approach: Creating an achromatic beamline only with quadrupoles is, in principle, possible. However, according to Eq. 2.62, the chromatic function W rotates with double betatron frequency. Therefore, two lenses with equal strength will compensate each other's chromaticity only if the phase advance between them is around $\pi/2$. Correspondingly, creating an achromatic beamline with many lenses, in case the lens-to-lens phase advance is around $\pi/2$ (i.e., focusing strength in the beamline is moderate), should be possible. However, for a typical case of strongly focused beamlines with large demagnification, betatron functions in the quadrupoles are large, and correspondingly the betatron phase advance between them is close to π. In this case, the W function makes a full 360° turn between the quadrupoles of the same strength, and their chromaticity will therefore add up, doubling the total chromaticity, and making it impossible to create an achromatic strongly focusing beamline only with quadrupoles.

Exercise: 2.9 *Practice in the art of back-of-the-envelope estimations.* A circular accelerator with a 600-m perimeter has a horizontal betatron tune equal to $Q_x = 5.173$. Estimate the average vertical betatron function in this accelerator.

Solution approach: The Eq. 2.53 allows us to estimate the average betatron function of a circular accelerator as $\langle\beta\rangle = \Pi/(2\pi Q)$, where $\Pi = 600$ m (the perimeter of the ring). We know $Q_x = 5.173$ and can therefore estimate that $\langle\beta_x\rangle \approx 18$ m. Without additional information, the only reasonable assumption is that the vertical tune is similar to the horizontal one, and therefore $\langle \beta_y \rangle \approx 18$ m as well.

SYNCHROTRON RADIATION

Exercise: 3.1 *Chapter materials review.*
A proton beam of E=50 TeV circulates in a 100-km perimeter ring. Estimate the synchrotron radiation energy loss per turn, the characteristic energy of the emitted photons and the cooling time.

Solution: The relativistic factor of the protons of this beam is $\gamma \approx 50$ TeV $/0.938$ GeV $\approx 5.33 \cdot 10^4$. Let's assume that the bending field fills the ring uniformly. In this case, the radius of the curvature is $R \approx 100\ \text{km}/(2\pi) \approx 1.6 \cdot 10^4$ m. The synchrotron radiation loss per turn can be estimated using Eq. 3.8, which will give us $U_0 \approx 3$ MeV (or, in relative terms, approximately a $dE/E \approx 6 \cdot 10^{-8}$ fraction of the proton energy).

The cooling time can be estimated as (see the derivation of Eq. 3.9) $\tau = 10^5\ \text{m}/c/dE/E \approx 5.5 \cdot 10^3$ s (or approximately 1.5 hours).

The characteristic energy of the photons can be estimated using Eq. 3.21, which gives $\varepsilon_c \approx 1.8$ keV.

One needs to note that a typical mistake that students make in these estimations is to use the proton mass in Eqs. 3.8 or 3.21. One can see from the derivations of these equations that an electron mass needs to be used together with the radius of electron r_e (or, equivalently, proton mass can be used together with the proton radius r_p).

Just for reference, we can note that the magnetic rigidity of these protons (see Eq. 2.14) is $B\rho \approx 1.67 \times 10^5$ Tesla·m. In our assumption of 100% filling factor, the required field in the bending magnets is $B \approx 10.5$ Tesla. However, if we make a more realistic assumption that the dipole filling factor of the ring is 75% (the rest is taken by drifts, quadrupoles, RF stations, etc.), then the needed dipole fields would increase to $B \approx 14$ Tesla. The SR estimations above will need to be correspondingly adjusted in this case. The loss per turn and the energy of the photons would increase by $\times 1/0.75$ and the cooling time will shorten by $\times 0.75$.

Exercise: 3.2 *Chapter materials review.*
Describe how one needs to change the optics of third-generation SR sources in order to approach the diffraction-limited SR source, particularly in the horizontal plane.

Solution: Imagine that the SR storage ring with radius R consists of a FODO focusing structure filled with bending magnets in the drift sections. Let's denote the length of the single FODO cell as L_{cell}. The overall number of cells in the ring thus $N_{cell} = 2\pi/L_{cell}$. Let's now look at Eq. 3.30, which describes the equilibrium horizontal emittance. The value of dispersion function in such a structure can be estimated as $\eta \sim L_{cell}^2/R$ while the beta-function will be around $\beta_x \sim L_{cell}$. Therefore, Eq. 3.30 can be rewritten (ignoring the numerical coefficient) as $\varepsilon_{x0} \sim \lambda_e \gamma^2 / N_{cell}^3$.

Correspondingly, in order to reduce the equilibrium emittance, we can reduce the energy or increase the number of cells. Decreasing the energy is often impractical, as it would undesirably affect the SR spectrum. Therefore, the only practical way to reach the ultimate diffraction-limited SR source is to increase the number of segmentations of its focusing structure, which in turn will result in the need to use much stronger focusing magnets, and so on.

Exercise: 3.3 *Chapter materials review.*
In a manner similar to how the equilibrium emittance was estimated in this chapter, derive the equilibrium energy spread of the beam.

Solution: Following derivations similar to those in Section 3.2.3, we can estimate the equilibrium energy spread to be around $(\Delta E/E)^2 \sim \lambda_e \gamma^2/R$.

Exercise: 3.4 *Mini-project.*
Define the approximate parameters (energy, sizes, fields in bending magnets) of a second-generation SR source aiming to achieve 10 keV of X-rays.

Solution approach: This task allows for a variety of solutions. Using the formulas from Section 3.1, we can for example find that a 3-GeV electron ring with a bending radius in the dipole magnets of 5.6 m would allow the SR spectrum to reach 10 keV. The required magnetic field in the bends is about 18 kGs, and if the fill-factor of the ring with bends is 30%, then the perimeter of the ring is about 120 m.

Exercise: 3.5 *Analyze inventions or discoveries using TRIZ and AS-TRIZ.* Analyze and describe scientific or technical inventions described in this chapter in terms of the TRIZ and AS-TRIZ approaches, identifying a contradiction and an inventive principle that were used (could have been used) for these inventions.

Solution approach: This task allows for a variety of solutions. Taking into account the examples of inventions presented in this chapter, we recommend considering and analyzing the invention of the wiggler and undulator.

Exercise: 3.6 *Developing AS-TRIZ parameters and inventive principles.* Based on what you already know about accelerator science, discuss and suggest the possible additional parameters for the AS-TRIZ contradiction matrix, as well as the possible additional AS-TRIZ inventive principles.

Solution approach: This task allows for a variety of solutions. In order to come up with ideas of additional parameters for the matrix, we recommended considering those parameters that are specific for the beams and that are unlikely to be encountered in the engineering field. A particular example related to this chapter is the spectral characteristics of radiation. Additional inventive principles can be developed via discussion of insertion devices such as wigglers and undulators.

Exercise: 3.7 *Practice in reinventing technical systems.* Synchrotron radiation emitted by the beam in bending magnets can hit the walls of the elliptical vacuum chamber, causing gas desorption, deterioration of vacuum and reduction of the beam's lifetime. Suggest a way to modify the design of the vacuum chamber to mitigate this issue. Try to use the inventive principle of separating out the negative factor.

Solution approach: Applying the principle of separating the function of propagating the beam in a vacuum chamber, and absorbing synchrotron radiation on a wall of the vacuum chamber, we can conclude that these should be two different vacuum chambers. The SR should still pass from the first chamber into the second; i.e., the two chambers should be connected by a narrow gap. So, we arrive at the standard "anti-chamber" vacuum system design of SR light sources.

Exercise: 3.8 *Practice in reinventing technical systems.* The FODO beamline, with focusing and defocusing quadrupoles, can focus beams in all directions. A long-wavelength adiabatic deviation of the direction of the FODO beamline will make the beam follow the deviated beamline trajectory. The LEP energy increase discussed in Section 3.4 shows that the quadrupoles with dipole corrections help to steer the beam in a circular accelerator. In both of these examples, the beamlines have certain limited energy acceptance — the beam with an energy spread will have larger sizes and may experience losses of particles. Taking these examples to an extreme, suggest a way to focus and transport the beam with nearly 100% energy spread on a circular trajectory.

Solution approach: Let's consider a FODO beamline which is not straight, but placed on a circle. If the beam has a very large energy spread, its low-energy particles will be over-focused, and high-energy particles will be under-focused. Long drift sections (the "O" elements of the FODO) create a problem — large beam size increase for over- and under-focused beams. Quadrupoles with shifted magnetic centers (like in the LEP energy increase example), can bend the trajectory. But the drift sections cannot bend the trajectory, as there is no field there. Therefore, as the first step, according to the inventive principle of *continuity of useful action*, let's eliminate the drifts and change the beamline to a FDFD structure — a sequence of quadrupoles shifted horizontally to provide a bending field for a central energy particle. The chain of short and strong quadrupoles of *fixed field alternating* polarity — called FFA — is in fact the system that can be optimized to transport the beam with a significant difference in energy[2] — e.g., with the ratio of minimum to maximum energy around a factor of two or more.

Exercise: 3.9 *Practice in the art of back-of-the-envelope estimations.* Synchrotron radiation was first observed around 1947 in a General Electric 70 MeV synchrotron. It is known from historical photos that the entire vacuum chamber of this synchrotron could fit on a small dining table. Estimate the wavelength of the light that was observed.

[2]S. Machida et. Al., Nature Physics 8, 243-247 (2012); S. L. Sheehy, arXiv:1604.05221 [physics.acc-ph], (2016).

Solution approach: Let's assume that the vacuum chamber size is around 0.6 m, and therefore the orbit radius is $R \approx 0.3$ m. With the electrons' energy of 70 MeV (i.e., $B\rho \approx 0.23$ T·m), this radius corresponds to the magnet field $B \approx 0.23/0.3 \approx 0.8$ T, which sounds reasonable (the iron of the magnet is not saturated). The wavelength of the radiation can be estimated as $\lambda = 4\pi R/(3\gamma^3) \approx 460$ nm, which corresponds to a visible blue color.

SYNERGIES BETWEEN ACCELERATORS, LASERS AND PLASMA

Exercise: 4.1 *Chapter materials review.*
A certain plasma's density is 10^{17} cm^{-3}. A laser light of which wavelength could still penetrate such a plasma? Also, estimate the corresponding plasma frequency.

Solution: Using equations from Section 4.2.7, we can find that the corresponding plasma frequency is $f_p \approx 2.8 \cdot 10^{12}$ Hz and that a laser with a wavelength equal to or shorter than $\lambda \approx 105$ μm can penetrate into this plasma.
Exercise: 4.2 *Chapter materials review.*
Taking into account a) Fermat's principle of the least amount of time for light propagation through an optical system and b) the observation that the diffraction angle should approach the law for reflection from a mirror as the wavelength becomes very short (and hence diffraction becomes less important), explain qualitatively why an optical telescope is needed inside the laser pulse stretcher, but not required in the laser pulse compressor.

Solution: Looking at Fig. 4.30, it is possible to see that if we send a short light pulse into the compressor, then at its exit the red light will emerge after the blue light, and hence the frequency of the output pulse will be negatively chirped (frequency decreasing in time). This indeed follows from the fact that the diffraction angle should approach the law for reflection from a surface as the frequency becomes very high (and hence diffraction becomes less important). According to Fermat's principle, light will take the shortest path through an optical system, so for blue light the path of the diffracted light is closer to the path which would be taken if the gratings were replaced with mirrors.

In order to get positive dispersion, we would need to have two gratings separated by *negative spacing*. The apparent impossibility of achieving this can be overcome by placing a 1:1 telescope in between the gratings, as shown in Fig. 4.29.
Exercise: 4.3 *Analyze inventions or discoveries using TRIZ and AS-TRIZ.* Analyze and describe scientific or technical inventions described in this chapter in terms of the TRIZ and AS-TRIZ approaches, identifying a contradiction and an inventive principle that were used (could have been used) for these inventions.

Solution approach: This task allows for a variety of solutions. Taking into account the examples of inventions presented in this chapter, we recommend considering and analyzing the invention of the laser pulse stretcher or compressor, CPA or optical stochastic cooling.

Exercise: 4.4 *Developing AS-TRIZ parameters and inventive principles.* Based on what you already know about accelerator science, discuss and suggest the possible additional parameters for the AS-TRIZ contradiction matrix, as well as the possible additional AS-TRIZ inventive principles.

Solution approach: This task allows for a variety of solutions. In order to come up with ideas of additional parameters for the matrix, we recommend considering those parameters that are specific for the beams of particles or light and that are unlikely to be be encountered in the engineering field. A particular example related to this chapter is quantum efficiency. Additional inventive principles can be developed via discussion of CPA or pulse compressors.

Exercise: 4.5 *Practice in reinventing technical systems.* Imagine that you need to measure the gas pressure inside of an electric light bulb very precisely. Can you suggest a method to do this? Breaking the bulb is not allowed. Hint: try to use one of the physical principles shown in illustrations to this Chapter 4.

Solution approach: Looking at the Paschen discharge curve shown in Fig. 4.11, where there is clear dependence of the discharge voltage on the gas pressure, we can come up with the idea to apply voltage between the electrodes of the light bulb and observe the discharge in order to measure the gas pressure inside of the light bulb very accurately.

Exercise: 4.6 *Practice in reinventing technical systems. Electron cloud* effects can arise in proton or positron storage rings, and can result in significant deterioration of the beam quality via a single bunch instability. The instability mechanism involves the circulating beam attracting electrons from the residual gas by positively charged beam bunches, with the residual gas electrons then bouncing in the vacuum chamber in resonance with the bunch frequency, leading to electron cloud density buildup and instability impacting the beam's quality. Suggest a way to mitigate electron cloud instability.

Solution approach: Electron cloud instability[3] buildup depends on the resonance between the bunch timing of the positively charged circulating beam, as well as the duration of the residual gas electrons bouncing transversely up and down the vacuum chamber. It seems sufficient, for breaking this ideal resonance condition, to *go to another dimension* by introducing a low-amplitude solenoidal field along the accelerator vacuum chamber. In this case, the trajectories of the low-energy residual gas electrons will change significantly, eliminating the electron cloud resonance buildup conditions.

Exercise: 4.7 *Practice in the art of back-of-the-envelope estimations.* An electron gun immersed in a 0.3 T solenoidal magnetic field is intended to produce a round 2 mm diameter beam for magnetized electron cooling of 1 MeV kinetic energy pro-

[3]See a review of the phenomena in F. Zimmermann, Phys. Rev. ST Accel. Beams 7, 124801, (2004).

tons. Estimate the electron current. Discuss how would you increase the electron current.

Solution approach: For 1 MeV kinetic energy of protons, the corresponding (equal velocity) kinetic energy of electrons is 544 V. Let's assume that the electron gun gap (see Fig. 4.3) is $d = 10$ mm. For the area A corresponding to the 2 mm beam diameter, we can estimate the perveance as about 0.07 μA/V$^{3/2}$, which, for this voltage, gives a current of about 0.8 mA. Reducing the gap will increase the beam current to $\sim 1/d^2$; however, it is unlikely that we can decrease the gap much further. The magnetic field value does not play a role in this estimation, except for the condition that both the electron gun and the proton beam cooling section need to be immersed in the magnetic field. Therefore, we can change the ratio of the magnetic field between these two areas and rely on magnetic field compression. In this case, the field in the electron gun area will be lower, and the size larger, while still providing 2 mm diameter beam in the cooling section. Correspondingly, the larger cathode diameter gun can generate a larger electron current.

CONVENTIONAL ACCELERATION

Exercise: 5.1 *Chapter materials review.*
Derive Eq. 5.34 for particle motion in the traveling wave.

Solution: Let us start from the second part of Eq. 5.30, i.e.,

$$\frac{d\mathscr{E}}{dt} = eE_0\dot{z}\cos(\omega t - kz)$$

and subtract Eq. 5.31 from it, taking into account the definition in Eq. 5.33, which will result in the following:

$$\frac{d(\mathscr{E} - E_s)}{dt} = eE_0\left(\dot{z}\cos\varphi - v_s\cos\varphi_s\right)$$

According to Eq. 5.32, the first term in the equation above needs to be substituted with $\dot{z} = v_s + \dot{u}$. However, we can note that the term $\dot{u}\cos\varphi$ will be either small or oscillating fast, and would not have much effect on changing the particle's energy. We will therefore keep only the leading terms and will also use $v_s d/ds$ instead of d/dt, arriving at

$$\frac{dW}{ds} = eE_0\left[\cos\varphi - \cos\varphi_s\right]$$

which is the first equation of Eq. 5.34 that needed to be proved. The second part of Eq. 5.34 can be proven by writing p_z as $\gamma m\dot{z}$ and substituting it into Eq. 5.30, using Eqs. 5.32 and 5.33, as well as the just proven equation.
Exercise: 5.2 *Chapter materials review.*
Discuss design approaches to beam optics that would result in achieving a negative value of the momentum compaction factor in a synchrotron.

Solution: According to the definition of the momentum compaction factor (see Eq. 2.52), the negative contributions to the overall integral value of α_c comes from the regions of the ring where $D(s)/\rho(s)$ is negative. Therefore, we can consider two options to make α_c negative. First of all, we can introduce reversed bends (negative $\rho(s)$) in the areas of large positive dispersion $D(s)$. The other, usually more effective method, is to modify the focusing structure in such a way that dispersion $D(s)$ through most of the bends will be negative. We can also combine these two methods to get more flexibility in the optics design.

Exercise: 5.3 *Mini-project.*

Define very approximate parameters (sizes, magnetic fields, parameters of RF system) of a 200 MeV rapid-cycling proton synchrotron capable of operating at a 10 Hz repetition rate. Assume injection at 1 MeV.

Solution approach: This task allows for a variety of solutions. We recommend splitting the task into several sub-systems and analyzing them separately, making sure that the interface parameters are compatible. The minimal set of sub-systems that can be considered in this mini-project is the following: 1) beam optics, 2) magnet system and 3) RF system. Correspondingly, three sub-teams should jointly work on this project. The first team will define the rough sizes of the ring, taking into account space needed for RF stations and injections, as well as the necessary length of the magnets. The second team will determine the size and length of the magnets, and also suggest the construction approach and the material choice for the magnets to allow their fast-cycling operation. The third team will suggest an optimal design of the RF system, taking into account the required acceleration rate, as well as the available length and the apertures of the chamber. Depending on the available time and the expertise of the team, further details of the design can be considered in addition to the minimal set of parameters outlined above.

Exercise: 5.4 *Analyze inventions or discoveries using TRIZ and AS-TRIZ.* Analyze and describe scientific or technical inventions described in this chapter (e.g., tandem, RFQ, bunch and pulse compressors) in terms of the TRIZ and AS-TRIZ approaches, identifying a contradiction and an inventive principle that were used (could have been used) for these inventions.

Solution approach: This task allows for a variety of solutions. Taking into account the examples of inventions presented in this chapter, we recommend considering and analyzing the invention of an RFQ accelerating structure or a klystron.

Exercise: 5.5 *Developing AS-TRIZ parameters and inventive principles.* Based on what you already know about accelerator science, discuss and suggest the possible additional parameters for the AS-TRIZ contradiction matrix, as well as the possible additional AS-TRIZ inventive principles.

Solution approach: This task allows for a variety of solutions. In order to come up with ideas of additional parameters for the matrix, we recommend considering those parameters that are specific for the beams of particles or light and that are unlikely to be encountered in the engineering field. A particular example related to this chapter is the phase-space area corresponding to stable motion. Additional inventive principles can be developed via discussion of an Alvarez drift tube linac.

Exercise: 5.6 *Practice in reinventing technical systems.* A special type of cavity, a so-called open cavity, is used in the systems of RF pulse compression. The cavity looks like an oak barrel without a lid and a bottom. An RF wave can circulate in this cavity in the plane of its largest diameter, for a long time and without decay, in the case where the RF wavelength is much smaller than the cavity's diameter. Suggest a way to excite such an RF wave in this open cavity.

Solution approach: Despite that such a cavity is open at the top and bottom, the field inside cannot be excited from those open ends, since these areas are not coupled to the field along the largest diameter by nature of the cavity's design (otherwise, the open cavity would quickly radiate out all of its energy). The required coupling can be provided via a coupling slot made in the middle of the cavity, at the largest diameter plane. We can also use the *periodic action* inventive principle and make a sequence of coupling slots, with the step equal to the RF wavelength, and excite the open cavity using a waveguide with many slots wrapped around the cavity and brazed to it along its largest diameter[4].

Exercise: 5.7 *Practice in reinventing technical systems.* Reinvent the way to deal with *Unidentified Flying Objects* in accelerators. The HERA electron-hadron collider in DESY suffered from difficult-to-identify events, attributed to UFOs — unidentified flying objects . The most popular hypothesis suggested that dust particles coming out of *ion pumps* were positively ionized and attracted to the circulating electron beams, creating an instability. Suggest a way to fix this problem.

Solution approach: The obvious way to solve this issue, if it can be afforded, is to apply *the other way around* inventive principle, i.e., to change the circulating electron beam to a positron beam, which would repel the UFO particles. Luckily, this was possible at the HERA collider and was indeed implemented[5]. The HERA collider working with the positron beams pushed this issue aside, helping to penetrate deeper into the science of the inner world of nuclei interaction.

Exercise: 5.8 *Practice in the art of back-of-the-envelope estimations.* LEP was a 27 km-long e+e- machine and, around the year 2000, was running with the energy of about 104 GeV per beam. The energy was limited by the energy losses due to synchrotron radiation and by the amount of voltage from the installed RF cavities. LEP optical structure included all typical magnets: bending dipoles, quadrupoles with

---

[4] I. Syrachev, EPAC-94, pp. 375-379, (1994).

[5] F. Zimmermann, DESY-HERA-93-08 (1993).

corrector coils, and sextupoles. Assume that you cannot install more RF to compensate for SR energy losses, but can power up the dipole correctors to smooth the orbit and reduce SR losses. Estimate how much you can increase the energy of the LEP collider using this approach.

Solution approach: Let's assume that in the LEP collider of perimeter $\Pi = 27$ km, the quadrupoles occupy $\xi = 10\%$ of the perimeter, their length is $\ell = 1$ m and their focusing length is $F = 20$ m. Let's assume that the bends occupy the rest of the perimeter (i.e., 90%), and let's thus estimate the radius of curvature in the bends as $R_B \approx 0.9 \cdot 27/(2\pi) \approx 3.87$ km.

We have to assume that the strength of the dipole correctors in the quadrupoles is limited. A reasonable estimate is to assume that the dipole correctors can shift the magnetic axis of the quadrupoles by about a millimeter — let's assume that $\Delta x \approx$ 1 mm. Therefore, the radius of the orbit curvature that the quadrupoles with powered-on dipole correctors can provide is $R_Q = F\ell/\Delta x \approx 20$ km.

Taking into account that the total angle provided by bends and quads with dipole correctors must be equal to 2π, we can estimate the new larger radius $\hat{R}_B$ of the orbit curvature in bending magnets as

$$\frac{1}{\hat{R}_B} \approx \frac{1}{R_B} - \frac{\xi}{R_Q}$$

The energy losses per unit length due to SR can be estimated as

$$\frac{dE}{ds} = \frac{2}{3}\frac{r_e \gamma^4}{R^2} mc^2$$

Assuming that the total energy loss (compensated by RF cavities) is the same in both cases, and taking only bends radiation into account, the ratio of the new gamma factor $\hat{\gamma}$ to the original one γ can thus be estimated as

$$\frac{\hat{\gamma}}{\gamma} \approx \left(\frac{\hat{R}_B}{R_B}\right)^{1/2} \approx 1 + \frac{0.1}{2}\frac{R_B}{R_Q}$$

For 104 GeV initial energy, this estimate then predicts an energy increase of about 1 GeV. While this estimate is close to the one implemented[6] at LEP, there are certainly a lot more factors to take into account for the practical case.

PLASMA ACCELERATION

Exercise: 6.1 *Chapter materials review.*
What laser intensity (in W/cm^2) would correspond to a normalized vector potential of $a_0 = 10$, and what are the maximum values of the electric and magnetic fields in the laser wave for a ruby laser or for a CO_2 laser?

[6] A. Beuret, S. Bidon, G. de Rijk, R. Genand, P. Raimondi, J. Wenninger, CERN-SL-2000-14 MS.

Solution: For a ruby laser with $\lambda \approx 0.694\ \mu$m, the equations 6.8–6.8 and 6.25 yield the following: $I \approx 2.8 \cdot 10^{20}$ W/cm^2, $E_{max} \approx 4.6 \cdot 10^{11}$ V/cm and $B_{max} \approx 1.6 \cdot 10^9$ Gauss. For a CO_2 laser with $\lambda \approx 10.6\ \mu$m, we get $I \approx 1.2 \cdot 10^{18}$ W/cm^2, $E_{max} \approx 3.0 \cdot 10^{10}$ V/cm and $B_{\max} \approx 1.0 \cdot 10^8$ Gauss.

Exercise: 6.2 *Chapter materials review.*
Assume that we get a 1-GeV electron bunch from a laser-plasma accelerator and would like to create an undulator from plasma using the focusing force of the plasma's ions. Suggest the plasma parameters and the amplitude of the beam oscillation that would correspond to the radiation at the boundary of the undulator regime, i.e., with the undulator parameter equaling $K = 1$.

Solution approach: This task allows for a variety of solutions. Let's recall the approximate definition of the undulator parameter given in Eq. 3.39 ($K \sim \gamma\lambda_u/R$) and take Eqs. 6.38 and 6.42 into account. This will give us the estimation for the undulator parameter

$$K \sim \frac{4\pi\sqrt{\gamma}r_b}{\sqrt{2}\lambda_p}$$

Let's consider the example with 1 GeV electron beam given at the end of Section 6.5.2. In this case, $\lambda_p = 0.03$ mm and $r_b = 0.001$ mm. According to the estimation given above, in this case, $K \sim 40$, while changing parameters, for example, to $\lambda_p = 0.1$ mm and $r_b = 0.0001$ mm, would bring K close to 1. According to Eq. 6.5, such a plasma wavelength would correspond to a plasma density of $n \sim 10^{17}$ cm^{-3}.
Exercise: 6.3 *Mini-project.*
Assume that the LHC proton beam of $E = 7$ TeV is going to be used as a driver for the plasma acceleration of electrons. Select the parameters for the plasma and estimate to what bunch length you would need to compress the LHC beam, so that it could be used for plasma acceleration. Define, roughly, the parameters of the corresponding bunch compressor.

Solution approach: This task allows for a variety of solutions and the amount of detail in the solution depends on the time available for the team. As a minimum, we recommend starting with identifying the longitudinal LHC beam parameters, the length (about a foot) and energy spread (around 0.01%). Assuming that the compressed beam should not have the energy spread more than 10%, we can conclude that the proton bunch, in principle, could be compressed to 0.3 mm bunch length. The plasma wavelength (see Eq. 6.5) thus could be around 1 mm and the plasma density around 10^{15} cm^{-3}, which corresponds to a possible accelerating gradient of 3 GeV/m (see Eq. 6.4).

Now let's consider the bunch compressor and assume that it has a triangular shape with three bends, where the edge bends give the angle θ and the central bend the angle -2θ and the distance between said bends is L. The difference of the straight path with respect to the path through the compressor is $L \cdot \theta^2$. Let's now assume that we compress the bunch by making an energy chirp along the bunch of the same amount mentioned above — 10%. That means that in order to compress the foot-long bunch, the energy-chirp-dependent path difference should be equal to the initial beam length, i.e., $0.1 \cdot L \cdot \theta^2 = 0.3$ m, which gives us the following requirement: $L \cdot \theta^2 = 3$ m.

Let's now assume that we use, for the bunch compressor, the same bends as used for the LHC ring. Those bends have a 8.33-Tesla field, they are 14.3 m long and, at this energy, they will bend the beam by around 0.005 rad. We can see that if we use just a single LHC dipole as the edge dipoles of the compressor, the required length will be $L \approx 100$ km — too long to be feasible. Using ten LHC dipoles as the bunch compressor outer dipole systems would reduce L to a kilometer, while the total length of the dipoles will be around $40 \cdot 14.3 \approx 570$ m — this option is already close to the optimum.

Exercise: 6.4 *Analyze inventions or discoveries using TRIZ and AS-TRIZ.* Analyze and describe scientific or technical inventions described in this chapter in terms of the TRIZ and AS-TRIZ approaches, identifying a contradiction and an inventive principle that were used (could have been used) for these inventions.

Solution approach: This task allows for a variety of solutions. Taking into account the examples of inventions presented in this chapter, we recommend considering and analyzing the invention of a particular method of self-injection of the beam into a plasma bubble, which involves the use of gases with different ionization potential (so-called *Trojan horse injection* method).

Exercise: 6.5 *Developing AS-TRIZ parameters and inventive principles.* Based on what you already know about accelerator science, discuss and suggest the possible additional parameters for the AS-TRIZ contradiction matrix, as well as the possible additional AS-TRIZ inventive principles.

Solution approach: This task allows for a variety of solutions. In order to come up with ideas of additional parameters for the matrix, we recommend considering those parameters that are specific for the beams of particles or light and that are unlikely to be encountered in the engineering field. A particular example related to this chapter is the energy spectrum of the beam accelerated in a plasma accelerator. Additional inventive principles can be developed via discussion of the capillary channel or downramp injection methods in plasma acceleration.

Exercise: 6.6 *Practice in reinventing technical systems.* One physics lab was studying plasma discharge. It was arranged in a cylindrical vessel that was placed horizontally. The team observed that the plasma of the discharge, being hotter and thus having a lower density than the surrounding gas, would move upward, disturbing the experiment. In order to keep the plasma in place, the team installed a solenoid around the plasma vessel. However, when the intensity of the plasma discharge was increased, it was observed that the solenoid could not keep the plasma in place anymore. One member of the team then suggested rebuilding the installation and using a much stronger solenoid. However, the head of the lab said that there is a much simpler way! And so, they made quick adjustments to the installation to fix the problem and keep the plasma centered. Try to recreate the solution to this challenge. Hint: one can use a device that can be found in any household.

Solution approach: The head of the lab[7] suggested using a household vacuum cleaner to create a circular motion of gas and plasma, which created centrifugal force, causing the plasma, which has a higher temperature and thus a lower density, to be pushed to the center of the vessel — illustrating the *anti-force* inventive principle.

Exercise: 6.7 *Practice in the art of back-of-the-envelope estimations.* In an electron-positron collider, the electron beam with an energy of 250 GeV is circulating in a storage ring with a 100 km perimeter. The average current of the circulating electron beam is 1 A. Estimate the total power needed to feed the RF cavities in such an accelerator.

Solution approach: For such a collider, in order to minimize the synchrotron radiation losses, we have to fill as much of the ring perimeter with bending magnets as possible. Let's assume that the bending magnets fill 95% of the ring's perimeter. Applying Eq. 3.8, where $R = 0.95 \cdot 100\ \text{km}/(2\pi) \approx 15.1$ km, we can estimate that the energy loss per turn for a single electron is about 22.8 GeV. With a 1 A beam current, we would need 22.8 GW of power to feed the RF cavities — this is so huge that we would have to consider a linear electron-positron collider for this energy instead of a circular one.

LIGHT SOURCES

Exercise: 7.1 *Chapter materials review.*
Describe a method to create a monochromatic X-ray beam in Compton sources.

Solution approach: To answering this question, we would need to refer to Section 7.3.2, where the use of a crystal monochromator is described, as well as to Section 7.4.3, where the dependence of the Compton photon energy on the angles is described, which makes it possible to create monochromatic beams via the use of collimation of the emitted protons.

Exercise: 7.2 *Chapter materials review.*
In laser-plasma acceleration, the final energy of an accelerated electron beam is 1 GeV. The wavelength of the laser used for laser-plasma acceleration is 800 nm. Part of the same laser pulse is redirected with mirrors to collide head-on with the accelerated electron beam. Estimate the energy of photons created in such a Compton source, as well as the angular spread of the photons.

---

[7] According to TRIZ stories, this head of the lab was P. Kapitsa.

Solution approach: This task allows for a variety of solutions. According to Eq. 7.4, the energy of produced gamma rays will be about 6 MeV. The angular spread can be estimated as $1/\gamma$, which is about $0.5 \cdot 10^{-3}$. Further and more detailed estimations may involve considerations, in particular, of the initial angular spread of the electron beam.

Exercise: 7.3 *Mini-project.*
Select a desired photon energy of a laser-plasma acceleration betatron X-ray source (e.g., from 1 to 100 keV) and devise a consistent set of basic parameters describing the source. Discuss the justifications for selecting particular values of certain parameters (for plasma or laser, etc.). Estimate the brightness of thc source.

Solution approach: This task allows for a variety of solutions. We can, for example, start from the parameters described at the end of Section 6.5.2, where 50 keV photons were created from a 1-GeV plasma-accelerated electron beam. Let's assume that the charge accelerated per one laser shot is around 16 pC (10^8 clcctrons) and the laser repetition rate is 1 Hz. According to estimations in Section 6.5.2, about 0.3 photons are emitted per period of oscillation λ, which is about 3.6 mm in this case, according to Eq. 6.38. The emitting area is about $r_b^2 \sim 10^{-6}$ mm^2 and the angular spread of the radiation due to betatron oscillations is about $r_b/\lambda \sim 0.25$ mrad, while the angular spread due to the natural opening of radiation is around $1/\gamma \sim 0.5$ mdar and therefore dominating. The maximal accelerating gradient in the bubble is around 1 GeV/cm in our case, which means that during the last oscillation period, the energy would increase by about 30%. Therefore, we will assume that the photons are emitted only during this last oscillation, which gives us $3 \cdot 10^7$ photons per pulse, emitted from a 10^{-6} mm^2 area and from a 0.25 mrad^2 solid angle per second, which gives us approximately 10^{14} photons/(mm^2 mrad^2), into full 100% bandwidth. Assuming, as an initial approximation, that the spectrum is uniform up to the maximum energy, we come to an estimation of 10^{13} photons/(mm^2 mrad^2 0.1% BW).

Exercise: 7.4 *Mini-project.*
Select a desired photon energy of a Compton X-ray source (e.g., from 1 keV to 10 MeV) and devise a consistent set of basic parameters describing the source. Discuss the reasons for selecting particular values of certain parameters (for electron beam, laser, etc.). Estimate the brightness of the source.

Solution approach: This task allows for a variety of solutions. We can, for example, start from the parameters used in Exercise 7.2, where 6 MeV photons were created from a 1-GeV electron beam and a 800-nm laser beam. We can assume that the electron beam is coming from a conventional accelerator, or we can assume that it is coming from a laser-plasma accelerator. In the latter case, we can take the electron beam parameters from the previous exercise and estimate the brightness following a very similar set of steps as in Exercise 7.3, taking into account that the number of Compton photons needs to be estimated using formulas from Section 7.4.3, and also making an assumption about laser powcr as well as the laser cavity amplification factor.

Exercise: 7.5 *Analyze inventions or discoveries using TRIZ and AS-TRIZ.* A liquid anode X-ray tube is a contemporary technology that increases the photon flux. Analyze this technology in terms of the TRIZ and AS-TRIZ approach, identifying a contradiction and a general inventive principle that was used (could have been used) in this invention.

Solution approach: The analysis of the liquid anode X-ray tube is similar to the analysis of the carbon wire beam profile monitor that is replaced by the laser wire or of metal mirrors and plasma mirrors — the same contradictions and the same inventive principle can be applied.

Exercise: 7.6 *Developing AS-TRIZ parameters and inventive principles.* Based on what you already know about accelerator science, discuss and suggest the possible additional parameters for the AS-TRIZ contradiction matrix, as well as the possible additional AS-TRIZ inventive principles.

Solution approach: This task allows for a variety of solutions. In order to come up with ideas of additional parameters for the matrix, we recommend considering those parameters that are specific for the beams of particles or light and that are unlikely to be encountered in the engineering field. A particular example related to this chapter is the biological effect of radiation, which in this case is illustrated by the example of the concept of the water window. Additional inventive principles can be developed via discussion of the pump-probe experiments or compact Compton sources, As well as the invention of the method of top-up-injection or phase contrast imaging.

Exercise: 7.7 *Practice in reinventing technical systems.* A hydrostatic level system (HLS) is often used for the alignment of synchrotron light sources. An HLS consists of many vessels of 10-15 cm height filled with water and connected by pipes. The pipes running from the bottoms of the vessels provide a connection to water, and the pipes running from the tops of the vessels provide a connection to the air. In an ideal case, the water levels in all the sensors are the same (with respect to the gravitational equipotential plane) and thus measuring these levels relative to the vessel (e.g., with capacitor sensors) would allow one to measure the misalignments of the SR source magnets to which the vessels are attached. However, there is a challenge: water has a very large thermal expansion coefficient (0.0002 at 20°C). If the height of the water column in the sensors is just 0.1 m, then variations of the temperature along the accelerator will give a measurement error of 20 μm/degree. Now imagine that you need to align the accelerator to better than 1 μm but you cannot stabilize the temperature to the required level. Suggest a way to change the design of the water level alignment system to solve this problem.

Solution approach: The harmful factor that causes the large temperature sensitivity of an HLS is the height h of the water column in the sensor. The ideal solution would be to eliminate this harmful factor entirely, i.e., make h as close to zero as possible. This can be achieved if the HLS uses a single, continuous water surface not interrupted by any pipes or sensor walls — i.e., *continuity of useful action* inventive principle. Simultaneously, the water and air pipes can be *merged* into a single pipe. The resulting HLS system will use the vessels connected by a single pipe half-filled with water, which will provide the uninterrupted water surface , and simultaneously the path for air. The temperature sensitivity of such a half-filled HLS will be significantly reduced[8] and a more accurate alignment of accelerators can be achieved.

FREE ELECTRON LASERS

Exercise: 8.1 *Chapter materials review.*
Evaluate the slice emittance requirements for an FEL based on a 1-GeV electron beam. Discuss the factors affecting the requirements for the slice energy spread of this FEL's beam.

Solution approach: This task allows for a variety of solutions. Let's aim for the FEL to shine in the "water window," i.e., with 0.4 keV of photon energy or $\lambda \approx 3$ nm. Then, according to Eq. 8.29, the requirement on the slice normalized emittance is $\varepsilon_N < 0.5$ mm·mrad.

Exercise: 8.2 *Chapter materials review.*
Discuss the phenomenon of filamentation in plasma acceleration in connection to the TRIZ inventive principle of preliminary action.

Solution approach: In the context of this inventive principle, we could note that the filamentation and corresponding large emittance growth can be avoided if we care about matching the beam phase-space (which can be done in variety of ways) *before* launching it into the propagation channel.

Exercise: 8.3 *Mini-project.*
Define, very approximately, the main parameters (energy, length, undulator field and step size) of a linac-based FEL aimed at 15 KeV X-ray energy.

Solution approach: This task allows for a variety of solutions. A possible set of parameters would include a normal conducting linac with an electron beam energy of 10 GeV ($\gamma \approx 2 \cdot 10^4$) which for the desired photon energy of 15 keV ($\lambda \approx 0.08$ nm) would require $\lambda_u \approx 4$ cm (see Eq. 8.15), if we assume that $K = 1$. This parameter, according to Eq. 8.6, would require undulators with a maximum field of approximately 0.26 Tesla. The undulator length can be estimated using Eq. 8.26, assuming that we need to allow for at least ten gain lengths.
Exercise: 8.4 *Analyze inventions or discoveries using TRIZ and AS-TRIZ.* Analyze and describe scientific or technical inventions described in this chapter in terms of the TRIZ and AS-TRIZ approaches, identifying a contradiction and an inventive

[8]S. Takeda et al., KEK Preprint 94-48, (1994); A. Chupyra, et al., in Proc. of IWAA2004, (2004).

principle that were used (could have been used) for these inventions.

Solution approach: This task allows for a variety of solutions. Taking into account the examples of inventions presented in this chapter, we recommend considering and analyzing the invention of an undulator or SASE process.

Exercise: 8.5 *Developing AS-TRIZ parameters and inventive principles.* Based on what you already know about accelerator science, discuss and suggest the possible additional parameters for the AS-TRIZ contradiction matrix, as well as the possible additional AS-TRIZ inventive principles.

Solution approach: This task allows for a variety of solutions. In order to come up with ideas of additional parameters for the matrix, we recommend considering those parameters that are specific for the beams of particles or light and that are unlikely to be encountered in the engineering field. A particular example related to this chapter is the coherence length. Additional inventive principles can be developed via discussion of the phenomena of micro-bunching or single-pass FEL.

Exercise: 8.6 *Practice in reinventing technical systems.* FEL radiation takes away only a small fraction of the beam energy. Still, at some point along the undulator, the beam energy offset may become noticeable and the resonant condition can be violated, resulting in reduced efficiency of radiation generation. Suggest a way to modify the system to prevent the loss of synchronism when the beam energy is decreasing along the undulator.

Solution approach: In order to keep the synchronism, we need to ensure that the resonance condition

$$\lambda = \frac{\lambda_u}{2\gamma^2}\left(1 + \frac{K^2}{2}\right)$$

will be maintained even if the beam energy, i.e., γ, will be decreasing along the undulator. We have a choice between changing the undulator wavelength λ_u or its parameter K (i.e., changing the field strength). Changing λ_u involves different mechanical designs and different manufacturing procedures of different parts of the undulator, and it is unlikely to be economical in comparison with adjusting K, which may involve easy shimming of permanent magnet cells or adjusting the current in the electromagnetic coils — i.e., *parameter change* inventive principle. Changing K along the undulator is called *tapering*. It can help keep the synchronism and increase efficiency of FEL radiation generation.

Exercise: 8.7 *Practice in reinventing technical systems.* An FEL requires a high-peak current, i.e., short bunches. Beams coming from a photoinjector are typically long, and bunch compression is required, which is typically performed with a sequence of magnetic chicanes together with an energy-position correlated chirp introduced in the bunch in an RF accelerating section. Suggest an alternative way to create an energy-position chirp in an electron bunch, using ideas and techniques discussed in this Chapter 8.

Solution approach: The energy-position chirp in an electron bunch can be created by sending the bunch to a wiggler together with a laser light at the resonance wavelength. The energy of the electron bunch will then be modulated via the *inverse FEL* process — illustrating *the other way around* inventive principle — and such a bunch can then be sent to a chicane to convert this energy modulation into a short density spike in the middle of the bunch, significantly enhancing the spontaneous generation of FEL radiation in the downstream undulator. This scheme is called enhanced self-amplified spontaneous emission — ESASE[9] — and can be used, in particular, to design an ultra-compact[10] X-ray FEL.

Exercise: 8.8 *Practice in the art of back-of-the-envelope estimations.* A free-electron laser with energy recovery is fed from an accelerator built with dual-axis cavities, as shown in Fig. 15.4. The FEL is designed to generate radiation with a wavelength 13.5 nm. Estimate the energy of the electron beam at the final energy of deceleration, just before the beam is directed to the beam dump.

Solution approach: Let's assume that the energy of the accelerated beam is 0.5 GeV. In this case, the undulator period needed to generate $\lambda = 13.5$ nm radiation is about $\lambda_u \approx 26$ mm, which sounds reasonable (we assumed that $K < 1$ and ignored the $0.5K^2$ term). Let's also assume that the undulator parameter $\rho = 0.001$. Since the fraction of the beam energy emitted to radiation (see Eq. 8.28) is about $1.6 \cdot \rho$, the beam energy spread at 0.5 GeV energy will be about $1.6 \cdot 10^{-3}$. During deceleration in the second path of the dual-axis cavities, which ensures the energy recovery, the beam energy spread will grow by $1/(\gamma\beta)$. Let's assume that the maximum energy spread that can still be stably kept in the RF separatrix and decelerated, is about 30%. The beam will reach this energy spread during deceleration to approximately 2.7 MeV, defining the energy at which the beam needs to be extracted from the dual-axis ERL accelerator and dumped.

PROTON AND ION LASER PLASMA ACCELERATION

Exercise: 9.1 *Chapter materials review.*
Discuss the advantages and challenges of particle beam therapy in comparison to X-ray therapy.

Solution approach: Discussing this question focuses in particular on the properties of the Bragg peak described in Section 9.1, which allows better localization of the irradiation area and minimization of the impact on healthy tissues.
Exercise: 9.2 *Chapter materials review.*

---

[9] A. A. Zholents, Phys. Rev. ST Accel. Beams 8, 040701, (2005).

[10] J. B. Rosenzweig et al., New J. Phys., 22, 093067, (2020).

Discuss the key requirements for the laser and the target that may result in 200 MeV mono-energetic beams of protons in laser-plasma acceleration.

Solution approach: This is an advanced question for which research is ongoing and where only a tentative answer can be given at this time. Following the discussion presented in Section 9.5.1, one can expect that the laser intensities needed for reaching 200 MeV are approximately 10^{21} W/cm^2. This task is closely related to the following exercise — 9.3.

Exercise: 9.3 *Mini-project.*
Discuss and develop a plan to create a 250-MeV proton source based on plasma acceleration, aiming for it to be applied in the medical field. Select approximate laser parameters and target parameters, and discuss their requirements. Discuss and select a method for energy monochromatization or energy collimation/selection. Describe why you selected these particular values of certain parameters (for target or laser, collimation or monochromatization system, etc.).

Solution approach: This task allows for a variety of solutions. This is also an advanced question for which research is ongoing and where only a tentative answer can be given at this time. We suggest building on the results from Exercise 9.2. A very important question is how to make the beam monochromatic. Experiments and simulations predict that using very thin targets can help create an accelerated beam with a large fraction of particles in the higher-energy-part of the energy spectrum. However, in order to use the advantage of the Bragg peak, one would need to reduce the beam energy spread and the uncertainty of the beam energy to numbers well below a percent (as 1% of energy error corresponds to 5 mm of range error for a beam of this energy). Therefore, various methods of collimation and energy selection will be necessary. One would also need to take into account that collimation and energy selection may produce additional widespread radiation originated from the near vicinity of a patient — this would need to be mitigated in the design.

Exercise: 9.4 *Analyze inventions or discoveries using TRIZ and AS-TRIZ.* Analyze and describe scientific or technical inventions described in this chapter (e.g., isochronous cyclotron) in terms of the TRIZ and AS-TRIZ approaches, identifying a contradiction and an inventive principle that were used (could have been used) for these inventions.

Solution approach: This task allows for a variety of solutions. Taking into account the examples of inventions presented in this chapter, we recommend considering and analyzing the invention of an isochronous cyclotron or of a range modulator used for particle therapy.

Exercise: 9.5 *Developing AS-TRIZ parameters and inventive principles.* Based on what you already know about accelerator science, discuss and suggest the possible additional parameters for the AS-TRIZ contradiction matrix, as well as the possible additional AS-TRIZ inventive principles.

Solution approach: This task allows for a variety of solutions. In order to come up with ideas of additional parameters for the matrix, we recommend considering those parameters that are specific for the beams of particles or light and that are unlikely to be encountered in the engineering field. A particular example related to this chapter is the probability of tumor recurrence after particle therapy. Additional inventive principles can be developed via discussion of the light-sail plasma acceleration technique.

Exercise: 9.6 *Practice in reinventing technical systems.* A FODO optical beamline system of focusing and defocusing magnetic quadrupoles can transport the beams of charged particles — for example, ions — stably and for a long distance. In an analogy with FODO, suggest a way to keep a charged ion, or a cloud of ions (that can be used as a target for laser-plasma acceleration or as a *qubit* for a *quantum computer*), stationary and stably in a fixed location. Hint: apply the inventive principle *the other way around.*

Solution approach: In the FODO case, the ion is moving relative to a static magnetic field of the focusing and defocusing quadrupoles; i.e., the fields vary in space, and in the reference frame of the ion, they vary in time. Applying *the other way around* principle will therefore give us a system where varying fields are applied to a stationary ion. The fields need to be transverse (like FODO focusing) and also longitudinal (like longitudinal RF focusing in an accelerator). This system, which we came up with via inventive principles, with the stationary ion and varying fields keeping it stable in a fixed position, is the well-known *Paul trap*[11]. Paul traps can be used, in particular, for holding the targets for laser-plasma acceleration, or for holding *entangled* ions forming *qubits* in *quantum computers*.

Exercise: 9.7 *Practice in reinventing technical systems.* A patient's motion during radiation treatment is one of the challenges that get in the way of providing accurate dose deliveries only to malignant tissues. Suggest a possible alternative way for providing the radiation dose to mitigate this challenge.

Solution approach: In order to eliminate the harmful factor, which is the patient's motion, we can apply the inventive principle *skipping*; i.e., we can perform the harmful action fast by delivering all radiation doses in a very short time interval. Possibly the most versatile way to deliver the entire dose at once is to use high-energy electrons (called VHEE therapy — very high-energy electrons). Combined with multi-directional beam delivery[12], this method of rapid delivery of the entire radiation dose is a potentially promising future direction of radiation therapy.

---

[11] W. Paul, Nobel Prize in Physics, 1989.

[12] B. W. Loo, P. G. Maxim, V. A. Dolgashev, US Patent 9018603 (2013).

Exercise: 9.8 *Practice in the art of back-of-the-envelope estimations.* Imagine that the beam size and angular spread of the ion beam produced by TNSA plasma acceleration are 10 μm and 0.1 rad and that the energy spread is 10%. Estimate where you would need to place the focusing lens with respect to the TNSA target to capture and focus such a beam.

Solution approach: For such a beam, the geometrical emittance is formally $\varepsilon \approx 10\,\mu\text{m} \times 0.1$ rad ≈ 1 mm·mrad, which is small. However, the value of the beta-function matched to such a beam is $\beta^* \approx 10\ \mu\text{m}/0.1$ rad ≈ 0.1 mm — a really tiny beta-function, close to the values at the IP of colliders. We can see from Eq. 2.56 that, if the energy spread is 10%, then in order to avoid chromatic dilution of the beam size, we need to have $L^* < 1$ mm. Therefore, the focusing of such a TNSA-produced beam should start almost immediately after coming from plasma acceleration — a millimeter from the target. Moreover, the focusing distance of the capture lens should also be around 1 mm. It is unlikely that such strong focusing can be produced by conventional elements. Plasma focusing[13], on the other hand, is a promising alternative approach.

BEAM COOLING AND FINAL FOCUSING

Exercise: 10.1 *Chapter materials review.*
Consider that corrections of chromatic and other aberrations are implemented perfectly, and the beams of an electron-positron 1-TeV CM linear collider are ready to collide with a crossing angle. Suggest other effects that may impact the collider's luminosity.

Solution approach: Among the effects that may impact luminosity are, in particular, synchrotron radiation in the final focusing quadrupoles, which may limit the IP beam size (*Oide effect* — considered in Exercise 10.5), and synchrotron radiation in the field of the detector solenoid. And there are certainly many others, which the reader can suggest.

Exercise: 10.2 *Mini-project — continuation.*
In 1954, Enrico Fermi presented, in his lecture, a vision of an accelerator that would encircle the Earth. Let's continue designing such an accelerator. In the previous Chapter 2 exercise, we assumed that such an accelerator would be shaped like a polygon with N sides and that there would be N space stations launched around the Earth, located in the vertices of the polygon, and that each space station would carry one bending dipole magnet and one quadrupole magnet. We made assumptions about the sizes of the bending magnets and quadrupoles, and we assumed that between the space stations the accelerator orbit would be straight and that the particle beam would

[13]See, e.g., R. Lehe, et al., Phys. Rev. ST. Acc and Beams, 17, 121301 (2014).

propagate in the open–space vacuum (i.e., without any vacuum chambers). We assumed that the polarity of the quadrupoles in the neighboring space stations would be opposite, forming the FODO optics of the accelerator. In this continuation, take into account synchrotron radiation, particularly for the case of an electron beam. Estimate the energy lost per turn, estimate the voltage of the RF cavities that need to be installed in every space station, and estimate the equilibrium emittance and the beam sizes in the quadrupoles and bends.

Solution approach: This task allows for a variety of solutions and we will give only brief guidance here. The analysis of the synchrotron radiation effects on this accelerator may show you that you need to go back to redefining the basic parameters of the magnets, beam energy or the number of space stations. Particular criteria to consider are the energy loss per turn, the voltage and the power required from the RF stations, as well as the beam sizes in the magnetic elements.

Exercise: 10.3 *Developing AS-TRIZ parameters and inventive principles.* Based on what you have learned in this book, discuss and suggest the possible additional parameters for the AS-TRIZ contradiction matrix, as well as the possible additional AS-TRIZ inventive principles.

Solution approach: This task allows for a variety of solutions. In order to come up with ideas of additional parameters for the matrix, we recommend considering those parameters that are specific for the beams of particles or light and that are unlikely to be encountered in the engineering field. A particular example related to this book in general are the relativistic factors or the connection between mass and energy. Additional inventive principles can be developed via discussion of the concept of muon or photon collider.

Exercise: 10.4 *Practice in reinventing technical systems.* In the traveling focus regime, the focal points of the colliding bunches are moving during the collision in order to coincide with the location of the head of the opposite bunch, therefore optimally benefitting from the additional focusing from the opposite bunch. Suggest a way to dynamically move the focus point of each colliding bunch in this way.

Solution approach: The traveling focus can be created in two ways. The first way is to combine a small uncompensated chromaticity with a coherent E-z energy-position correlated energy chirp along the bunch. The correlated energy chirp needs to be two-to-three times larger than the incoherent energy spread in the bunch. Using a small fraction of the uncompensated chromaticity is equivalent to the inventive principle of *using resources that already exist in the system.*

It is clear that additional energy variation along the bunch may impact the precision physics analysis in the linear collider. If this is a concern, then another method[14] can be considered. The second way to create a traveling focus is to use a transverse deflecting cavity, which produces a z-x correlation in one of the FF sextupoles, and thus results in z-correlated focusing. The cavity would be located upstream of the final doublet, at the $\pi/2$ betatron phase from the FD (which also includes sextupoles).
Exercise: 10.5 *Practice in the art of back-of-the-envelope estimations.* Imagine that chromaticity compensation is arranged perfectly, and the size and angular spread of the 1 TeV energy electron beam in front of the final focusing quadrupole of $L_q = 2$ m is $\sigma_y = 100\ \mu$m and $\sigma_{y'} = 10^{-9}$ rad. The distance between the final lens and the interaction point is $L^* = 4$ m (therefore, the ideal beam size at the IP is $\sigma_y^* \approx 5$ nm). Estimate the vertical size of the beam at the IP, taking into account synchrotron radiation in the final quadrupole. Suggest how the uncovered limitation can be mitigated.

Solution approach: The parameters of the ideal beam at the IP are: IP size is $\sigma_y^* \approx$ 5 nm, and IP divergence is $\theta^* = 100\ \mu$m / 5 m= 20 μrad. Note also that $\beta_y^* \approx \sigma_y^*/\theta^* \approx 0.25$ mm, and the estimate of the final doublet chromaticity is $\xi \approx L^*/\beta_y^* \approx 2\cdot 10^4$.

The radius of curvature of the trajectory in the final focusing quadrupole can be estimated as $R = L_q/\theta^* \approx 10^5$ m. The average energy loss due to synchrotron radiation is of no interest, as it can be compensated for. However, the beam energy spread obtained in the final quadrupole due to synchrotron radiation is important. It can be estimated as

$$\sigma_{\mathrm{E}}^2 \approx \frac{r_e \lambda_e \gamma^5 L_q}{R^3} \approx 1.3\cdot 10^{-4} \quad \text{for our parameters.}$$

Note that this energy spread times the chromaticity ξ is more than one. This additional energy spread generated in the final quadrupole will result in different particles focusing differently, due to different energy, in the final double. The chromatic compensation does not help here, since this energy spread is generated downstream of the chromatic correction. Correspondingly, the increase of the vertical beam size can be estimated as

$$\Delta\sigma^* \approx L^* \cdot \theta^* \cdot \sigma_{\mathrm{E}} \approx 13 \text{ nm} \quad \text{for our parameters.}$$

The actual beam size is therefore almost three times larger than the ideal one. The limit of the IP beam size due to synchrotron radiation in the final quadrupoles is called the *Oide limit*[15].

In order to discuss possible mitigation of this effect, let's rewrite our estimate for the beam size in terms of the beam emittance and β_y^*:

$$\sigma_y^* \approx \varepsilon\beta_y^* + r_e\lambda_e\gamma^5 \left(\frac{L^*}{L_q}\right)^2 \left(\frac{\varepsilon}{\beta_y^*}\right)^{5/2}$$

[14] R. Brinkmann and M. Dohlus, DESY-M-Report 95-11 (1995).

[15] Katsunobu Oide, Phys. Rev. Lett. 61, 1713, (1988).

The above expression is a function of β_y^* and its minimum value (omitting coefficients close to one) is

$$\sigma_{\min} \approx (r_e \lambda_e)^{1/7} (\gamma\varepsilon)^{5/7} \left(\frac{L^*}{L_q}\right)^{2/7}$$

We should note that $\gamma\varepsilon$ is the normalized emittance and, for a given injector, it does not depend on energy. The spectacular conclusion from this Oide limit consideration is that there is a fundamental minimum beam size that depends only on the normalized emittance and on the details of the focusing, i.e., on the ratio of L^* to L_q.

If we cannot change the normalized emittance, then our path to mitigation of the Oide limit is to drastically reduce L^*. Remembering the inventive principle of *continuity of useful action*, we can suggest filling the entire L^* distance with a focusing field — namely with a distributed plasma lens[16] that will provide adiabatic focusing starting immediately after the IP, and that can drastically eliminate the Oide limitation on the focused beam size.

BEAM STABILITY AND ENERGY RECOVERY

Exercise: 11.1 *Chapter materials review.*
Derive an estimate for SR energy losses (*beamstrahlung*) for colliding e^+e^- bunches. Justify the need to use flat beams in a high-energy e^+e^- collider. Estimate the energy losses for beam parameters from Fig. 11.2.

Solution: Using equation Eq. 11.2, as well as the definition of the disruption parameter, we can estimate that the radius of curvature of the particle trajectory (for a particle offset from the center on one σ either on the x- or y- axis) in the field of the opposite bunch is given by

$$R \sim \frac{\gamma \sigma_z (\sigma_x + \sigma_y)}{2 N r_e}$$

Using equation Eq. 3.7, we can find that the total relative energy loss for these edge particles can be estimated as

$$\frac{\Delta E}{E} \sim \frac{8 \gamma r_e^3 N^2}{3 \sigma_z (\sigma_x + \sigma_y)^2}$$

Recalling that the luminosity depends on the beam sizes as $L \propto 1/(\sigma_x \sigma_y)$, we can see that we can keep the luminosity constant, but decrease the energy loss, if we make the vertical size much smaller and make the horizontal one much larger.

Taking into account the beam sizes and beam population from Fig. 11.2, and assuming that the beam energy is 250 GeV, we can obtain, using the above defined formula, that the relative energy loss for such a beam is about 4%.

[16] P. Chen, K. Oide, A. M. Sessler, and S. S. Yu, Phys. Rev. Lett. 64, 1231, (1990).

Exercise: 11.2 *Chapter materials review.*
For the same beams from the previous exercise, estimate the number of emitted *beamstrahlung* photons per particle in the case when the oncoming positron bunch has either a 3- or 30-nm vertical offset with respect to the electron bunch.

Solution: First of all, looking at Fig. 11.3, we could note that either of these particles (with a 3- or 30-nm vertical offset) will be outside of the bunch and therefore the field acting on the particle will be the same and equal to the maximal field. The number of photons emitted per unit length is given by Eq. 3.22 and, in our case, the total number of emitted photons can be expressed as

$$N_\gamma \sim \frac{2N\alpha r_e}{\sigma_x}$$

This gives us approximately 0.8 photons emitted per particle for either of these cases (3 nm or 30 nm vertical offset).

Exercise: 11.3 *Mini-project.*
Estimate the parameters of a single-pass FEL that can generate EUV 13.5 nm radiation for semiconductor lithography. Assume that your electron beam energy can be up to 3 GeV and that you have space for a 100-m undulator. Select suitable parameters of the FEL undulator and define requirements for the electron beam's quality. Estimate the average electron beam current needed to generate 10 kW of EUV radiation. Discuss how to modify this conceptual design into ERL-based FEL using a dual-axis cavity linac illustrated in Figs. 15.4 and 15.25.

Solution approach: This task allows for a variety of solutions. Let's assume that the energy of the accelerated beam is $E = 0.5$ GeV. With an average accelerating gradient of 10 MeV/m, the length of the superconducting linac will be 50 m. With this beam energy, the undulator period required to generate $\lambda = 13.5$ nm radiation is about $\lambda_u \approx 23$ mm, which looks reasonable (we assumed that $K = 0.5$). Let's use 92 m for the undulator, which gives us the number of undulator periods to be $N_u = 4000$. The slice electron beam energy spread δ_E needs to be smaller than $1/(2N_u)$, i.e., $\delta_E < 1.3 \cdot 10^{-4}$. The electron slice normalized emittance ε_n needs to be smaller than $\gamma\lambda/(4\pi)$, i.e., $\varepsilon_n < 1$ mm·mrad. Let's assume that the undulator parameter $\rho = 0.001$. Since the fraction of the beam energy emitted to radiation (see Eq. 8.28) is about $1.6 \cdot \rho$, in order to generate $P = 10$ kW of radiation power at 13.5 nm, we need the average electron current $I = P/(1.6 \cdot \rho E) \approx 12.5$ mA. The average power of this beam is formally 6 MW, so we must use energy recovery to make this concept feasible. The energy recovery aspects of this task have already been discussed in the exercise after Chapter 8, where it was determined that the beam can be decelerated to 2.7 MeV and then dumped. In this case, the power of the dumped beam is only 34 kW, making this concept, due to the energy recovery, much more feasible.

Exercise: 11.4 *Practice in reinventing technical systems.* Absolute ground motion (i.e., motion with respect to an inertial reference frame) is usually measured by a seismometer, which typically consists of a pendulum (mass on a spring) with a sensing coil, with a permanent magnet placed in its vicinity. Since the coil measures the relative velocity, the sensitivity (i.e., the ratio of the seismometer signal to ground displacement) of such a sensor is proportional to $\omega/(\omega^2 - \omega_0^2 - i\alpha\omega)$, where ω_0 is the resonance frequency of the pendulum and α is the damping coefficient. The resonance character of the sensitivity curve is not convenient (due to sharp resonance, large motion of mass inside of the sensor, low dynamic range and also a narrow frequency band where ground motion can be measured). Suggest a way to eliminate this resonance behavior without the loss of sensitivity. Also suggest a way to adjust ω_0 for a vertical seismometer.

Solution approach: We can *use resources that we already have in the system* — specifically, the magnetic field with a gradient in which the pendulum with a sensor coil is moving. We can increase damping by adding one more coil — a damping coil shortened by a resistor. This will smooth the resonance, but will reduce the sensitivity. But, let's take it to the extreme and increase the damping even further, and use active *feedback* when the signal from the sensing coil is sent to an amplifier and then sent to the additional damping coil. The feedback signal will serve as the seismometer signal. Since the active damping coefficient α is now very large, the sensitivity, which is proportional to $\omega/(\omega^2 - \omega_0^2 - i\alpha\omega)$, will be constant in a large range (typically three orders of magnitude) of frequencies. Adjusting ω_0 can be done by *going to another dimension* — instead of placing the spring and the pendulum in the same vertical direction, make the pendulum a horizontal cantilever supported by a strong spring placed at a small angle with respect to the horizontal plane. Adjusting this small angle will allow us to tune the pendulum resonance frequency very accurately.

Exercise: 11.5 *Practice in the art of back-of-the-envelope estimations.* Precise measurements of the beam energy in the 27-km perimeter LEP collider revealed[17] correlations of the beam energy with terrestrial tidal motion. The momentum compaction factor in LEP was $\alpha = 1.86 \cdot 10^{-4}$, which amplified small circumference changes and resulted in an observed beam energy variation of about $\pm 10^{-4}$, corresponding to the deformation of Earth as an elastic body with a maximal amplitude of about 0.5 m. The 2-mile-long Stanford Linear Collider (SLC), located on the Pacific Coast, was equipped with a "light pipe" — a pipe with a diameter of about a meter with insertable Fresnel lenses that could measure transverse deformation of the SLC linac (e.g., the relative transverse motion of the middle point of the linac relative to its ends). Estimate the amount of tidal transverse deformation that was observed at SLC.

Solution approach: For the elastic Earth with radius $R_E \approx 6.3 \cdot 10^6$ m and tidal amplitude $a \approx 0.5$ m, the expected LEP relative energy variation is $\delta \approx 0.5a/(R_E\alpha)$, where α is the LEP momentum compaction. Numerically, this gives $\delta \approx 2 \cdot 10^{-4}$, corresponding to the measurements. For the SLC case with linac length $L \approx 3$ km, the transverse relative deformation can be estimated as $\Delta x \approx 0.5a(L/R_{\text{eff}})^2$, where R_{eff} cannot be equal to the radius of Earth, but needs to account for the proximity to the

[17] L. Arnaudon , et al., CERN SL/94 07, (1994).

Pacific Ocean, as well as the tidal motion of the ocean water that can amplify the local deformation of the Earth's surface. Let's assume that the area of a few hundred km is contributing to the effect and assume that $R_{\rm eff} \approx 500$ km, which gives $\Delta x \approx 10\ \mu$m, corresponding to measurements[18], where about 10 μm of tidal deformation of the SLC linac was observed — a thousand times larger than what can be expected for a uniform elastic Earth, due to the *ocean loading* effects.

ADVANCED BEAM MANIPULATION

Exercise: 12.1 *Chapter materials review.*
Describe the typical approaches for preventing beam breakup instability in linacs. Discuss if these approaches can be used in other situations considered in this book.

Solution approach: One of the approaches that prevent BBU instability by ensuring the coherent motion of the bunch is BNS damping, which involves creating an appropriate energy chirp along the bunch. Consider if this method can be applied to improve the stability of the beam accelerated in the plasma bubble.

Exercise: 12.2 *Mini-project.*
Define the approximate parameters of a 1-GeV compact SR source ring aimed at 10 keV X-rays, based on top-off, on-orbit and on-energy injection by a laser-plasma acceleration system. Discuss the research steps required for implementing such a concept.

Solution approach: This task allows for a variety of solutions[19]. This is also an advanced question for which research is ongoing and where only a tentative answer can be given at this time.

We recommend building on the results of Exercise 3.4. We can quickly realize that, if we rely on radiation from bends, we will have to make the bending radius about 0.2 m in order to reach X-ray energy of 10 keV. Therefore, insertion devices will need to be used. Let's set the ring size based on the field values achievable in superconducting magnets. A dipole field of 8 Tesla would correspond to a bending radius of about 0.4 m. This means that the ring can be about 1x2 m, where the longer sections are dedicated to injection and possibly RF and the other to an insertion device. The parameters of the insertion device can be evaluated using Eqs. 8.6 and 8.15. The on-axis laser-plasma injection into the ring can only be done using a gas jet (not capillary), and a low beta-function area needs to be arranged at the point of injection in order to match the divergent beam coming out of laser-plasma bubble. We would need to aim for using a gas jet with the lowest possible density and also

[18]A. Seryi, SLAC-PUB-8486, 2000; also in Proc. of EPAC 2000.

[19]This mini-project was the subject of USPAS-2016 class. Based on the efforts in the class, the participating students wrote a paper for the North American Particle Accelerator Conference. See here for details: M. Turner, et al., in Proc. of NAPAC-2016, TUA3CO03, (2016).

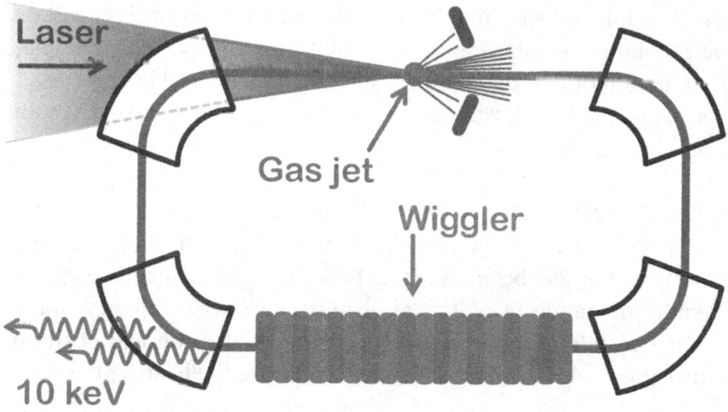

Figure A.1 For illustrations of 10 keV X-rays source based on on-orbit and on-energy injection by a laser-plasma acceleration system in Exercise 12.2.

well-defined in space. It is also very desirable for the jet to be pulsed, with the shortest pulse and minimal amount of gas density in pre- and after- pulses. The lifetime of the circulating electron beam in this compact ring will likely be limited by the interaction with the gas get. We suggest comparing the beam's lifetime defined by the gas jet with the synchrotron period and define the RF parameters based on that. The required R&D steps for this project would include design of the gas jet with minimal density, pulsed, with appropriate temporal profile of the plasma density. The other critical R&D items include a mini-beta insertion for capturing the injected beam, as well as fitting all these components into the overall design.

Exercise: 12.3 *Analyze inventions or discoveries using TRIZ and AS-TRIZ.* Analyze and describe scientific or technical inventions described in this chapter (e.g., coherent combination of laser pulses) in terms of the TRIZ and AS-TRIZ approaches, identifying a contradiction and an inventive principle that were used (could have been used) for these inventions.

Solution approach: This task allows for a variety of solutions. Taking into account the examples of inventions presented in this chapter, we recommend considering and analyzing the invention of a self-seeded FEL or an EEHG technique.

Exercise: 12.4 *Developing AS-TRIZ parameters and inventive principles.* Based on what you already know about accelerator science, discuss and suggest the possible additional parameters for the AS-TRIZ contradiction matrix, as well as the possible additional AS-TRIZ inventive principles.

Solution approach: This task allows for a variety of solutions. In order to come up with ideas of additional parameters for the matrix, we recommend to considering those parameters that are specific for the beams of particles or light and that are unlikely to be encountered in the engineering field. A particular example related to this chapter is the notion of two-dimensional spatial and temporal coherence. Additional inventive principles can be developed via discussion of the coherent combination of laser pulses.

Exercise: 12.5 *Practice in reinventing technical systems.* The *Wien filter* is a device with a combination of a transverse electric and a magnetic field that can act as a *velocity filter* for a particle beam. Suggest how this approach can be used for a) manipulating spin orientation of polarized beams; and for b) designing an interaction region of a linear collider with head-on collision of electron and positron bunches (i.e., zero crossing angle), while being able to separate the beams after the collision.

Solution approach: The precession of the particle spin depends on the magnetic field and is independent of the electric field. Therefore, setting the transverse $\vec{E}$ and $\vec{B}$ fields of the Wien filter such that $\vec{E}+\vec{v}\times\vec{B}=0$ will ensure that the straight trajectory of the polarized beam will not change, but the proportional increase of both fields will allow us to adjust the rotation of the beam spin at the exit of the filter. This answers the first part of the exercise. To answer the second part, we can recall the *system and anti-system* inventive principle and imagine that we place two Wien filters, one after the other, and flip the direction of the magnetic field in the second one:

$$\text{Wien}(E,B) \quad \text{then} \quad \text{Wien}(E,-B)$$

In this case, if the fields acting on the beam traveling from left to right cancel out (in terms of the beam trajectory) in the first filter, they will double in the second. In the same way, if the beam (either electron or positron) travels from right to left, since the velocity sign is flipped, then the effect of the right filter will cancel out, and the effect of the left one will double. Therefore, we can place the interaction point of the electron-positron collider in the middle between the two Wien filters. In this case, the trajectories of the beam before collision will not be affected, but after the collision, both beams will feel the deflecting field[20], allowing us to separate them and send the beams to the beam dump (or, even better, to the energy recovery channel).

Exercise: 12.6 *Practice in the art of back-of-the-envelope estimations.* Imagine that an accelerator for polarized beams has a figure-8 shape; i.e., it looks like ∞ and is located in the horizontal plane. Estimate how the spin tune depends on the beam energy in this ring.

Solution approach: In such a *figure-8 accelerator*[21], the beam will experience an exactly equal amount of clockwise and counterclockwise rotation for every turn in

---

[20] Such a configuration was proposed for the TESLA superconducting linear collider (ca 1995, DESY).

[21] Y.S. Derbenev, Univ. of Michigan Report UM HE 96-05 (1996); Y. S. Derbenev et al., Symmetry, 13, 398 (2021).

the accelerator. Correspondingly, the direction of the spin precession will change the sign and the net will be zero. Therefore, the spin tune will be exactly zero independently on the beam energy.

Exercise: 12.7 *Practice in the art of back-of-the-envelope estimations.* Imagine that a very short relativistic electron bunch needs to be collimated to remove its transverse halo tails. The bunch is propagating on the axis of a cylindrical vacuum chamber. The collimation is performed by an iris-like insert placed on the axis. The inner radius of the insert is much smaller than the radius of the vacuum chamber, and the thickness of the insert is longer than the bunch length. Estimate the energy loss of the particles in the bunch after passing this collimator.

Solution approach: We can make this estimation in the same way we would estimate energy loss for synchrotron radiation. Let's assume that the bunch length is σ_z, the bunch population is N, the radius of the vacuum chamber is R and the radius of the iris-like insert is a. For an ultra-relativistic bunch, the field is transverse and carries certain energy. When the bunch enters the vacuum chamber transition from radius R to a smaller radius a, part of its field located between these radii is stripped away and left in the larger-diameter chamber. After exiting the iris-like collimator, this field will grow again; however, the energy lost in the field left behind should result in an overall decrease of the energy of the particles in the bunch. The only way this energy decrease can happen is via the longitudinal decelerating field (wakefield) occurring in the vicinity of the iris collimator.

Let's now estimate the corresponding energy loss, which will allow us to also estimate the amplitude of the *longitudinal wakefield*. Let's, for the moment, assume that the bunch is uniform and has a linear charge density eNk, where k is the inverse of the bunch length (for Gaussian beam, we will later use $1/k = 3^{1/2}\sigma_z$). Using Gauss's theorem (which is $\int E ds = 4\pi Q$), we can estimate the transverse fields as $E_\perp = H_\perp \approx 2eNk/r$. Let's now estimate the amount of energy U in the field between radii a and R for the bunch length of $1/k$, since this will be the field cut out by the collimator:

$$U = \frac{1}{4\pi}\int_a^R E_\perp^2 2\pi r \frac{dr}{k} \approx 2e^2N^2\, k \ln\left(\frac{R}{a}\right) \approx \frac{2e^2N^2}{\sqrt{3}\sigma_z} \ln\left(\frac{R}{a}\right) \approx N^2 m_e c^2 \frac{r_e}{\sigma_z} \ln\left(\frac{R}{a}\right)$$

In the last expression, we also dropped the numerical factor about unity, which is likely beyond the accuracy of our estimation. The above corresponds to the change of the energy of the entire bunch with N particles. The amplitude of the corresponding decelerating longitudinal *wake potential* experienced by an average central particle in the bunch is then

$$W \approx 2N \frac{m_e c^2}{e} \frac{r_e}{\sigma_z} \ln\left(\frac{R}{a}\right)$$

where the factor of two came from considering that the head particles should experience no energy loss (due to causality in cases of ultra-relativistic particles), while the tail particle should see a larger field than the central particles. In other words, the wake potential is a function of longitudinal coordinate $W(s)$, which should rise from the head to the tail and then decay. Evaluation[22] of both longitudinal and transverse wake potentials is of particular importance for RF linear accelerators, since wakefields critically define the stability of beams and the achievable performance of the accelerators.

ADVANCED TECHNOLOGIES

Exercise: 13.1 *Chapter materials review.*
The resonant plasma excitation method may allow for the use of many low-energy pulses from fiber lasers to excite plasma oscillations for efficient laser-driven plasma acceleration. Discuss what effects can limit the maximum number of laser pulses that can be effectively used in this technique.

Solution approach: One of the limiting factors affecting the maximum number of laser pulses relates to relativistic effects. Plasma electrons become relativistically hot and the plasma frequency will change, deviating from the resonant spacing of the driving laser pulses, and thus limiting the number of useful pulses in the laser train. Understanding this limitation may also suggest the way to mitigate it via use of the *change of parameter* inventive principle — the time interval between the consecutive laser pulses can be slowly adjusted[23] along the train to keep in resonance with the slowly varying plasma frequency.

Exercise: 13.2 *Mini-project — conclusion*.*
In the exercises started in Chapter 2 and continued in Chapter 10, we estimated various parameters of the "Fermi accelerator" encircling the Earth, where N space stations were launched around the Earth, with each space station carrying one bending dipole magnet and one quadrupole magnet. As a result, we obtained the parameters for the Fermi accelerator, including the length of the bends and quadrupoles, bending field, the energy of the particles (electrons or protons), the gradient of the quadrupoles, SR energy losses, the needed RF voltage, equilibrium emittance, beam sizes, etc. In this final step of the project, please turn this Fermi accelerator into a triple-use facility around the Earth to serve these functions: a) a free-electron laser, b) an electron-positron collider and c) an electron-proton collider. Choose one of the following two options for designing your triple-use facility. Option A: design a single-pass facility working in energy recovery mode. The beams (e+ and e-) start from the injection space station, go around the Earth in different directions, getting accelerated to higher energies at each consecutive station, arrive at the undulator/detector space station where they produce FEL radiation or collide in the experimental detector, and go around the Earth for another half-turn, getting decelerated at each station and then dumped into the injector station. The RF cavity in each station

[22] See, e.g., A. Novokhatski, BINP preprint 88-39, Novosibirsk (1988); A. Novokhatski and A. Mosnier, NIM A, v. 763, p. 202-209 (2014).

[23] S.M. Hooker et al., J.Phys. B47, 234003, (2014).

gives energy to the beam it accelerates and takes energy from the beam it decelerates — this is energy recovery. Option B: design a storage ring accelerator facility where acceleration is slow and there is no energy recovery. The beams (e+ and e-) start from the injection space station and go around the Earth in different directions for many turns. Define as many parameters as possible for these space facilities.

Solution approach: This mini-project[24] allows for a variety of solutions, and we will give only cursory guidance here, leaving it for independent effort by the readers. In your work on this project, consider the following questions. What is the value of the wavelength that your FEL will generate? What is the center of mass of the energy of the collider? What are the parameters of the beamline elements (bends, quads, RF cavities)? What are the beam parameters (energy, sizes, emittances)? How large is the efficiency of the energy recovery? Do you need to rely on some beam cooling method to get to the desired beam quality? How can we create such cooling? For the latter, can you consider a stochastic cooling approach where the feedback signal will be sent through the Earth via the modulated neutrino beam?

Exercise: 13.3 *Practice in reinventing technical systems.* In PWFA, it is important to study plasma acceleration of positrons when the drive beam consist of electrons, and vice versa. The SLAC S-band linear accelerator has the capability to accelerate both e- and e+ bunches simultaneously, placing them 180° of RF phase apart, which is 5 cm, and not closer. However, for plasma acceleration, the drive and witness bunches need to be within 100 μm of each other. Suggest a beamline configuration that will allow us to study plasma acceleration with e- drive and e+ witness bunch.

Solution approach: The electron and positron bunches will naturally separate into different paths in a bending magnet and can then be sent to different beamlines, which can have different path lengths, matched to the initial distance between the electron and positron bunches. The two beamlines can be laid out as chicanes of different shape — the top chicane in a Λ-shape and the bottom chicane as a w-shape. The two beamlines will merge again at the end, bringing the bunches almost on top of each other (or with the needed 100 μm of distance between them). Such a configuration of dual chicanes was suggested[25] for the FACET project and was called a *sailboat chicane* due to its characteristic shape.

Exercise: 13.4 *Practice in the art of back-of-the-envelope estimations.* A multi-layer insulation (MLI) is often used for the thermal insulation of components inside of cryomodules, where superconducting accelerating cavities are placed. The MLI is implemented by rolling N layers of aluminized mylar around the object that needs

---

[24]This series of Fermi accelerator mini-projects was the subject of the 2022 VITA graduate course. As mentioned in Chapter 14, VITA is Virginia Innovative Traineeship in Accelerators, which now includes students from Old Dominion University, Hampton University and Norfolk State University

[25]C.I. Clarke, et al., IPAC-2011-WEOAB02, SLAC-PUB-14563, (2011).

to be isolated. The temperature contacts between the neighboring layers of MLI are minimized by a special spacer net or just by slightly crumpling the mylar. Assuming that the dominant heat transfer mechanism can be described by the Stefan-Boltzmann law, estimate the heat transfer coefficient for 10 layers of mylar when the temperatures of the edge layers are 4 K and 70 K.

Solution approach: Let's assume that the temperature is increasing from $T_{\rm min}$ at the layer index $i = 1$ to $T_{\rm max}$ at $i = N$. The heat flux J between layers $i-1$ and i, and layers i and $i+1$, must be equal. Therefore, $J/(\varepsilon\sigma_{\rm SB}) = (T_{i-1}^4 - T_i^4) = (T_i^4 - T_{i+1}^4)$ where $\sigma_{\rm SB} \approx 5.7 \cdot 10^{-8}$ W·m^{-2}·K^{-4} is the Stefan-Boltzmann constant and ε is the emissivity coefficient of the MLI surface. The above equation gives us the approximation $d(T^4)/dI = J/(\varepsilon\sigma_{\rm SB})$, which allows us to estimate the temperature behavior as $T^4 = (I-1)J/(\varepsilon\sigma_{\rm SB}) + T_{\rm min}^4$. At the final layer, i.e., at $i = N$, the temperature must be equal to $T_{\rm max}$, which gives us the estimate for the heat flux:

$$J \approx \varepsilon\sigma_{\rm SB}\frac{T_{\rm max}^4 - T_{\rm min}^4}{N-1}$$

Let's assume that $\varepsilon \approx 0.1$ (somewhat conservatively). Then, for the parameters given in the task, we can estimate that $J \approx 1.4 \cdot 10^{-2}$ W/m^2, and we can also observe that the MLI is an example of the *"nested doll"* inventive principle.

INVENTIONS AND INNOVATIONS IN SCIENCE

Exercise: 14.1 *Mini-project.*
A proposal for a Higgs and top-quark factory is based on the design of an electron-positron collider (175 GeV for each beam) in a tunnel with a 50-km circumference. Assuming that the maximum power delivered by the RF system to each beam is limited to 50 MW, estimate the maximum current that can be stored in such a collider. Make an assumption about the focusing system at the interaction point, and evaluate beam emittances and the collider's luminosity.

Solution approach: This task allows for a variety of solutions. This is also an advanced question for which research is ongoing and where only a tentative answer can be given at this time. Assuming that the ring is 100% filled with dipoles, we can estimate using Eq. 3.8 that each circulating electron or positron contributes approximately 10^{-5} W toward the SR losses. The total number of e^+ and e^- particles in the ring therefore cannot exceed $5 \cdot 10^{12}$ particles. Assuming that there are two beams (e^+ and e^-), the total population of each is half of that number. The beam emittances can be estimated to be equilibrium emittances in this ring (a certain value of x-y coupling needs to be assumed). The focusing system at the IP needs to be designed to take into account the limit on the number of particles in the beams and the desire to reach high luminosity.

Exercise: 14.2 *Analyze inventions or discoveries using TRIZ and AS-TRIZ.* A liquid jet or liquid target is often used in high-energy physics experiments, particularly for the production of antiprotons. A liquid jet can withstand the power of the incident proton beam. Analyze this technology in terms of the TRIZ and AS-TRIZ approach, identifying a contradiction and a general inventive principle that were used (could have been used) in this invention.

Solution approach: The analysis of this task is very similar to the analysis of carbon wire beam profile monitor replaced by the laser wire — the same contradictions and inventive principles are applicable.

Exercise: 14.3 *Practice in reinventing technical systems.* Space-charge forces in the electron bunch emitted from a photo-injector will create a $\langle x \cdot x' \rangle$ correlation in the bunch (particles further away from the bunch center experience larger space-charge force). This effect can be large, since the beam is not yet relativistic, and can create large emittance growth of the beam. Optimally placing a solenoid on top of the photoinjector will cancel out (due to solenoid focusing) the aforementioned correlation, achieving *emittance compensation.* Suggest a way to create a strong solenoidal field in the plasma-accelerating bubble to achieve a similar emittance preservation for the plasma-accelerated bunch. Assume that any external solenoid (pulsed, superconducting, etc.) cannot create a field of sufficient strength and sufficiently sharp gradients.

Solution approach: A solenoid field is produced by currents flowing on a spiral. In LWFA, when a laser pulse pushes the plasma electrons out due to ponderomotive force, the electrons flow around the newly formed plasma bubble, carrying large currents — these are *resources that we already have in the system.* These currents, however, are distributed symmetrically and do not create a solenoidal field. However, we can introduce an asymmetry by going to *another dimension* — instead of a symmetrically shaped laser pulse, we can use a spiral-shaped pulse. The laser pulse shape (i.e., the shape of its spatial intensity distribution) in this case can look like a slightly flat bean twisted 90° along its longer side. Such a spiral-shaped laser pulse, entering into plasma, will create spiral motion in the plasma electrons, which will now flow around the plasma bubble in a solenoid-like manner, creating a strong solenoidal field. It has been shown[26] that a million-Tesla solenoidal magnetic field can be created inside plasma bubbles. Optimization of the spiral shape of the driving laser pulse can create optimal conditions for emittance preservation of the plasma-accelerated electron bunch.

Exercise: 14.4 *Practice in the art of back-of-the-envelope estimations.* Imagine that you are building a *quantum computer* based on a storage ring of $^{24}Mg^{+}$ ions. Let's aim to store $N_i = 10^6$ ions, which will be cooled by electron or laser cooling to a low temperature, so that the ions will form a one-dimensional linear crystal in the storage ring. The qubits of this quantum computer will be formed by entanglement states of the ions, allowing us to get a very large number of qubits, aiming to achieve *quantum*

---

[26] Zs. Lécz, I. Konoplev, A. Seryi and A. Andreev, Scientific Reports, v. 6, 36139 (2016).

supremacy. Estimate the longitudinal temperature of the ions to which they need to be cooled in order to form the one-dimensional crystal.

Solution approach: First of all, we would need to consider what is the energy and size of this storage ring. It would be useful to be able to place this accelerator in a university lab, without special requirements to the room size and radiation shielding. Therefore, the energy of the ions in the storage ring needs to be low. Let's assume that the kinetic energy of ions is $E_k = 300$ kV. Such ions are non-relativistic, and their $\beta = v/c \approx (2E_k/(A \cdot m_p c^2))^{1/2} \approx 0.005$. Correspondingly, the momentum is $p \approx 0.12$ GeV/c and the magnetic rigidity is $B\rho \approx 0.38$ T·m. We can see that, with a reasonable magnetic field of about 0.25 Tesla, we can make a storage ring with a radius of about $R \approx 1.6$ m and a perimeter $\Pi = 10$ m (nicely fitting into a typical university lab). Therefore, the distance between ions in this ring is then $a \approx \Pi/N_i \approx 10\ \mu$m. The ions will be locked in their position in the linear crystal if their longitudinal temperature is lower than the Coulomb potential $U_{||}$ corresponding to the distance a, i.e., (for a single-charge ion) $U_{||} \approx e^2/a = m_e c^2 r_e/a \approx 1.4 \cdot 10^{-4}$ eV, which defines the requirements for the electron or laser cooling system for this quantum computer.

The encouraging experimental results[27] of electron and laser cooling of ions and the observation of one-dimensional ion crystals, as well as visionary design considerations[28], indicate that quantum computers based on accelerator technology may indeed be a promising future direction. We hope to see it realized by the year 2050.

Exercise: 14.5 *Practice in predicting evolution of technical systems**. Based on what you have learned in this book about TRIZ laws of the evolution of technical systems, and based on your knowledge of the scientific or technical area you are an expert in or are interested in, make your own prediction in the style of *the year 2050*

Solution approach: Obviously, this task is left to our readers for their independent consideration.

[27]M. Steck et al., Phys. Rev. Lett. 77, 3803, (1996); T. Schatz, et al., Nature 412, 717-720, (2001).

[28]T. Shaftan and B. Blinov, Phys. Rev. Accel. Beam 24, 094701, (2021).

Bibliography

Altshuller, G. (1999). *Innovation Algorithm: TRIZ, systematic innovation and technical creativity* (first ed.). Worcester, MA: Technical Innovation Center, Inc.

Amaldi, U. et al. (2013). *Elementary Particles*, Volume **21**: Subvolume C: Accelerators and Colliders , Group I: Elementary Particles, Nuclei and Atoms. Berlin, Heidelberg, New York: Springer-Verlag.

Badelek, B. et al. (2004). *The photon collider at TESLA*. International Journal of Modern Physics A **19**(30), pp.5097–5186. New Jersey, London, Singapore, Beijing, Shanghai, Hong Kong, Taipei, Chennai: World Scientific Publishing Co. Pte. Ltd.

Baklakov, B. and others (1991). *Investigation of seismic vibrations and relative displacements of linear collider VLEPP elements*. In Proc. of 1991 IEEE PAC. p.3273. San-Francisco, CA: Proc. of 1991 IEEE PAC.

Balakin, V. (1991). *Travelling Focus Regime for Linear Collider VLEPP*. In Proc. 7th ICFA Workshop on Beam Dynamics. Los Angeles, CA: Proc. 7th ICFA Workshop on Beam Dynamics.

Balakin, V., Nokhatsky, A. and Smirnov, V. (1983). *VLEPP: Transverse Beam Dynamics*. In Proc. of the 12th Int. Conf. on HEACC. Batavia, IL: FERMILAB Print.

Bellan, P. (2006). *Fundamentals of Plasma Physics*. New York: Cambridge University Press Inc.

Blumenfeld, I. et al. (2007). *Energy doubling of 42 GeV electrons in a metre-scale plasma wakefield accelerator*. Nature **445**, pp.741–744. London: Macmillan Publishers Limited.

Borghesi, M. et al. (2006). *Fast Ion Generation by High-Intensity Laser Irradiation of Solid Targets and Applications*. Fusion Science and Technology **49**(3), pp.412–439. La Grange Park, IL: American Nuclear Society.

Borghesi, M. et al. (2008). *Laser-driven proton acceleration: source optimization and radiographic applications*. Plasma Phys. Control. Fusion (**50**, 124040). Bristol, Philadelphia, Beijing, Tokyo: IOP Publishing Ltd.

Bosco, A. et al. (2008). *A Two-Dimensional Laser-Wire Scanner for Electron Accelerators*. Nuclear Instruments and Methods in Physics Research Section A: Accelerators, Spectrometers, Detectors and Associated Equipment **592**(3), pp.162–170. Elsevier.

Bourgeois, N., Cowley, J. and Hooker, S. M. (2013). *Two-Pulse Ionization Injection into Quasilinear Laser Wakefields*. Phys. Rev. Lett. **111**(15, 155004). Ridge, NY: American Physical Society.

Boyd, T. and Sanderson, J. (2003). *The Physics of Plasmas*. New York. Cambridge University Press Inc.

Brinkmann, R. et al. (2001). *A low emittance, flat-beam electron source for linear colliders*. Phys. Rev. ST Accel. Beams **4**(5, 053501). Ridge, NY: American Physical Society.

Budker, G.I. et al. (1976). *Experimental studies of electron cooling*. Particle Accelerators **7**, pp.197–211. United kingdom: Gordon and Breach, Science Publishers Ltd.

Burrows, M. and Sutton, G. (2013). *Interacting Gears Synchronize Propulsive Leg Movements in a Jumping*. Science **341**(6151), pp.1254–1256. Washington, DC: American Association for the Advancement of Science.

Bush, V. (1945). *Science, the Endless Frontier: a Report to the President*. Science Education **29**(4), p.218. Washington, D. C.: U.S. Government Printing Office.

Chao, A. et al. (2013). *Handbook of Accelerator Physics and Engineering* (second ed.). New Jersey, London, Singapore, Beijing, Shanghai, Hong Kong, Taipei, Chennai: World Scientific Publishing Co. Pte. Ltd.

Chen, F. (1974). *Introduction to Plasma Physics*. New York and London: Plenum Press.

Clark, E. et al. (2000). *Measurements of Energetic Proton Transport through Magnetized Plasma from Intense Laser Interactions with Solids*. Phys. Rev. Lett. **84**(18), p.670. Ridge, NY: American Physical Society.

Daido, H. et al. (2012). *Review of laser-driven ion sources and their applications*. Rep. Prog. Phys. (**75**, 056401). Bristol, Philadelphia, Beijing, Tokyo: IOP Publishing Ltd.

Danilov, V. and Nagaitsev, S. (2010). *Nonlinear accelerator lattices with one and two analytic invariants*. Phys. Rev. ST Accel. Beams **13**(084002). Ridge, NY: American Physical Society.

Davidson, R. (2001). *Physics of Nonneutral Plasmas*. New Jersey, London, Singapore, Beijing, Shanghai, Hong Kong, Taipei, Chennai: World Scientific Publishing Co. Pte. Ltd.

Delayen, J. R. (2012). *Superconducting spoke cavities for electrons and high-velocity proton linacs*. In Proceedings of LINAC2012, 03 Technology, 3A Superconducting RF, TH1A03, pp.758–762. Tel Aviv, Israel: Proceedings of LINAC2012.

Dendy, R. (1990). *Plasma Dynamics*. New York: Oxford University Press Inc.

Derbenev, Y. et al. (1995). *Microbunch Radiative Head-Tail Interaction*. In Tesla-FEL Report. Hamburg, Germany: DESY Print.

Derbenev, Y. (1998). *Adapting Optics for High Energy Electron Cooling*. In University of Michigan Report No. UM HE 98-04. Ann Arbor, MI: University of Michigan Report.

Diver, D. (2013). *Plasma Formulary for Physics, Astronomy, and Technology* (second ed.). Weinheim, Germany: WILEY-VCH Verlag.

Edwards, D. et al. (2004). *An Introduction to the Physics of High Energy Accelerators* (second ed.). Weinheim, Germany: WILEY-VCH Verlag.

Eliezer, S. (2002). *The Interaction of High-Power Lasers with Plasma*. Bristol, Philadelphia, Beijing, Tokyo: IOP Publishing Ltd.

Esirkepov, T. et al. (2004). *Highly Efficient Relativistic-Ion Generation in the Laser-Piston Regime*. Phys. Rev. Lett. **92**(7, 175003). Ridge, NY: American Physical Society.

Gabor, D. (1947). *A Space-Charge Lens for the Focusing of Ion Beams.* Nature **160**, pp.89–90. London: Macmillan Publishers Limited.

Gadd, K. (2011). *TRIZ For Ingeneers: Enabling Inventive Problem Solving.* Chichester, West Sussex: John Wiley & Sons, Ltd., Oxford Creativity.

Gibbon, P. (2005). *Short pulse laser interactions with matter. An introduction.* London: Imperial College Press.

Haberberger, D. et al. (2012). *Collisionless shocks in laser-produced plasma generate monoenergetic.* Nature Physics **8**(1), pp.95–99. London: Macmillan Publishers Limited.

Haseroth, H. et al. (1996). *Laser Ion Source Development for Heavy Ions.* In Proceedings of the XVIII International linear Accelerator Conference Linac96, v.2, WE202. pp.570–574. Geneva, Switzerland: CERN, European Organization for Nuclear Research.

Haussecker, E. and Chao, A.W. (2011). *The Influence of Accelerator Science on Physics Research.* Physics in Perspective (PIP). **13**(2), pp.146–160. Bristol, Philadelphia, Beijing, Tokyo: IOP Publishing Ltd.

Herr, W. (2013). *New Tools for Non Linear Dynamics I, II.* In The CERN Accelerator School 2013, Trondheim, Norway. Geneva, Switzerland: CERN, European Organization for Nuclear Research.

Hidding, B.B. et al. (2012). *Beyond injection : Trojan horse underdense photocathode plasma wakefield acceleration.* In Advanced Accelerator Concepts. AIP Conference Proceedings., pp.570–575. Melville, NY: AIP.

Hooker, S. and Webb, C. (2010). *Lasers physics (Oxford Master Series in Physics).* New York: Oxford University Press Inc.

Hooker, S. et al. (2014). *Multi-Pulse Laser Wakefield Acceleration: A New Route to Efficient, High-Repetition-Rate Plasma Accelerators and High Flux Radiation Sources.* J. Phys. B **47**, (234003). Bristol, Philadelphia, Beijing, Tokyo: IOP Publishing Ltd.

Huang, Zh. and Ruth, R.D. (1998). *Laser-electron storage ring.* Phys. Rev. Lett. **80**, pp.976–979. Ridge, NY: American Physical Society.

Jackson, J. (1999). *Classical Electrodynamics* (third ed.). Hoboken, NJ: John Wiley & Sons, Ltd.

Jaroszynski, D. et al. (2009). *Laser-Plasma Interactions.* Boca Raton, FL: CRC Press, Taylor & Francis Group.

Kneip, S. et al. (2011). *X-ray phase contrast imaging of biological specimens with femtosecond.* Applied Physics Letters **99**. Melville, NY: AIP Publishing LLC.

Landau, L. (1946). *On the vibration of the electronic plasma.* JETP (**16**), p.574. Russia: International Academic Publishing Company (IAPC) "Nauka/Interperiodica."

Lee, S. (2012). *Accelerator Physics* (third ed.). New Jersey, London, Singapore, Beijing, Shanghai, Hong Kong, Taipei, Chennai: World Scientific Publishing Co. Pte. Ltd.

Leemans, W. et al. (2006). *GeV electron beams from a centimetre-scale accelerator.* Nature Physics **2**, pp.696–699. London: Macmillan Publishers Limited.

Leemans, W. et al. (2014). *Multi-GeV Electron Beams from Capillary-Discharge-Guided Subpetawatt Laser Pulses in the Self-Trapping Regime.* Phys. Rev. Lett. **113**, (245002). Ridge, NY: American Physical Society.

Macchi, A. et al. (2013). *Ion acceleration by superintense laser-plasma interaction.* Rev. Mod. Phys. **85**, pp.751–793. Ridge, NY: American Physical Society.

Maksimchuk, A. et al. (2000). *Forward Ion Acceleration in Thin Films Driven by a High-Intensity Laser.* Phys. Rev. Lett. **84**(18), pp.4108–4111. Ridge, NY: American Physical Society.

Mangles, S. et al. (2004). *Monoenergetic beams of relativistic electrons from intense laser-plasma interactions.* Nature **431**, pp.535–538. London: Macmillan Publishers Limited.

Mendonca, J. (2001). *Theory of Photon Acceleration.* Bristol, Philadelphia, Beijing, Tokyo: IOP Publishing Ltd.

Mourou, G. et al. (2013). *The future is fibre accelerators.* Nature Photonics (**7**), pp.258–261. London: Macmillan Publishers Limited.

Najmudin, Z. (2014). *Laser Driven Systems.* In The CERN Accelerator School 2014, Geneva, Switzerland. Geneva, Switzerland: CERN, European Organization for Nuclear Research.

Padamsee, H., Knobloch, J., Hays, T. (2008) *RF Superconductivity for Accelerators 2nd Edition.* Wiley-VCH, ISBN: 978-3527408429.

Palmer, C.A. et al. (2011). *Monoenergetic Proton Beams Accelerated by a Radiation Pressure Driven .* Phys. Rev. Lett. **106**(1, 014801). Ridge, NY: American Physical Society.

Palmer, R.B. (1988). *Energy Scaling, Crab Crossing and the Pair Problem.* SLAC—PUB—4707. Menlo Park, CA: SLAC–PUB.

Parker, B. et al. (2007). *The Superconducting Magnets of the ILC Beam Delivery System.* SLAC—PUB—12832. Menlo Park, CA: SLAC–PUB.

Pashkin, A. and Leitenstorfer, A. (2014). *Particle physics in a superconductor.* Science **345**(6201), pp.1121–1122. Washington, DC: American Association for the Advancement of Science.

Pozimski, J. and Aslaninejad, M. (2013). *Gabor lenses for capture and energy selection of laser driven ion beams in cancer treatment.* Laser and Particle Beams **31**, pp.723–733. New York: Cambridge University Press Inc.

Qiao, B. et al. (2009). *Stable GeV Ion-Beam Acceleration from Thin Foils by Circularly Polarized Laser Pulses.* Phys. Rev. Lett. **102**(14, 145002). Ridge, NY: American Physical Society.

Qiao, B. et al. (2010). *Radiation-Pressure Acceleration of Ion Beams from Nanofoil Targets: The Leaky Light-Sail Regime.* Phys. Rev. Lett. **105**(15, 155002). Ridge, NY: American Physical Society.

Radovinsky, A. et al. (2013). *Variable Energy Acceleration in a Single Iron-Free Synchrocyclotron.* In Variable Energy Acceleration in a Single Iron-Free Synchrocyclotron. Cambridge, MA: Plasma Science and Fusion Center Massachusetts Institute of Technology PSFC/RR-13-9.

Raimondi, P. and Seryi, A. (2001). *Novel Final Focus Design for Future Linear Colliders.* Phys. Rev. Lett. **86**(17, 3779). Ridge, NY: American Physical Society.

Raimondi, P. et al. (2008). *Suppression of beam-beam resonances in crab waist collisions*. In Proceedings of EPAC08, Genoa, Italy, pp.2620–2622. Proceedings of EPAC08.

Saldin, E. et al. (2010). *The Physics of Free Electron Lasers (Advanced Texts in Physics)*. Berlin and Heidelberg, Germany: Springer-Verlag.

Schlenvoigt, H.-P. et al. (2008). *A compact synchrotron radiation source driven by a laser-plasma wakefield accelerator*. Nature Physics **4**, pp.130–133. London: Macmillan Publishers Limited.

Seryi, A. and Napoly, O. (1996). *Influence of ground motion on the time evolution of beams in linear colliders*. Phys. Rev. E **53**(5, 5323). Ridge, NY: American Physical Society.

Seryi, A., (2015). *Unifying Physics of Accelerators, Lasers and Plasma* CRC Press, Taylor & Francis Group.

Seryi, A.A., Seraia, E.I., (2016). *Izobretaia Instrumenti Nauki Bidushego* URSS Publishing company. In Russian. June 2016, ISBN 978-5-9710-3185-7

Sessler, A. and E. Wilson (2014). *Engines of Discovery. A Century of Particle Accelerators* (Revised and Expanded Edition ed.). New Jersey, London, Singapore, Beijing, Shanghai, Hong Kong, Taipei, Chennai: World Scientific Publishing Co. Pte. Ltd.

Shiltsev, V. et al. (1999). *Considerations on compensation of beam-beam effects in the Tevatron*. Phys. Rev. ST Accel. Beams **2**(071001). Ridge, NY: American Physical Society.

Shiltsev, V. (2010). *Observations of Random Walk of the Ground in Space and Time*. Phys. Rev. Lett. **104**(23, 238501). Ridge, NY: American Physical Society.

Schoerling, D. and Alexander V. Zlobin, A. (editors) (2019). *Nb3Sn Accelerator Magnets, Designs, Technologies and Performance*. Springer Cham, ISBN978-3-030-16120-0.

Silva, L. et al. (2004). *Proton Shock Acceleration in Laser-Plasma Interactions*. Phys. Rev. Lett. **92**(1, 015002). Ridge, NY: American Physical Society.

Snavely, R. et al. (2000). *Intense High-Energy Proton Beams from Petawatt-Laser Irradiation of Solids*. Phys. Rev. Lett. **85**(14), p.2945. Ridge, NY: American Physical Society.

Spence, D.J. and Hooker, S.M. (2001). *Investigation of a hydrogen plasma waveguide*. Phys. Rev. E **63**(1,015401(R)). Ridge, NY: American Physical Society.

Strickland, D. and Mourou, G. (1985). *Compression of amplified chirped optical pulses*. Optics Communications **56**(3), pp.219–221. Elsevier.

Stupakov, G. (2009). *Using the Beam-Echo Effect for Generation of Short-Wavelength Radiation*. Phys. Rev. Lett. **102**(7, 074801). Ridge, NY: American Physical Society.

Variola, A. et al. (2010). *ThomX - Conceptual Design Report*. In ThomX - Conceptual Design Report, pp.1–136. LAL RT 09/28, SOLEIL/SOU-RA-2678.

Wang, J.W. and Loew, G.A. (1997). *Field Emission and RF Breakdown in High-Gradient Room-Temperature Linac Structures*. SLAC—PUB—7684. Menlo Park, CA: SLAC–PUB.

Wiedemann, H. (2007). *Particle Accelerators Physics* (third ed.). Berlin and Heidelberg, Germany: Springer-Verlag.

Wilson, E.J.N. (2001). *An Introduction to Particle Accelerators*. New York: Oxford University Press Inc.

Wilson, E.J.N. (2011). *Sir John Adams: his legacy to the world of particle accelerators. John Adams Memorial Lecture, 2009*. Geneva, Switzerland: CERN, European Organization for Nuclear Research.

Yin, L. et al. (2011). *Three-Dimensional Dynamics of Breakout Afterburner Ion Acceleration Using High-Contrast Short-Pulse Laser and Nanoscale Targets*. Phys. Rev. Lett. **107**(4, 045003). Ridge, NY: American Physical Society.

Yu, T.-H. et al. (2010). *Stable Laser-Driven Proton Beam Acceleration from a Two-Ion-Species*. Phys. Rev. Lett. **105**(6, 065002). Ridge, NY: American Physical Society.

Zholents, A. and Zolotorev, M. (1996). *Femtosecond X-Ray Pulses of Synchrotron Radiation*. Phys. Rev. Lett. **76**(6), p.912. Ridge, NY: American Physical Society.

Index

Printed in the United States
by Baker & Taylor Publisher Services

Printed in the United States
by Baker & Taylor Publisher Services